ROUTLEDGE LIBRARY EDITIONS:
SOVIET POLITICS

Volume 6

HANDBOOK OF SOVIET SPACE-SCIENCE RESEARCH

HANDBOOK OF SOVIET SPACE-SCIENCE RESEARCH

Edited by
GEORGE E. WUKELIC

Routledge
Taylor & Francis Group
LONDON AND NEW YORK

First published in 1968 by Gordon and Breach Science Publishers

This edition first published in 2024
by Routledge
4 Park Square, Milton Park, Abingdon, Oxon OX14 4RN

and by Routledge
605 Third Avenue, New York, NY 10158

Routledge is an imprint of the Taylor & Francis Group, an informa business

British Library Cataloguing in Publication Data
A catalogue record for this book is available from the British Library

ISBN: 978-1-032-67165-9 (Set)
ISBN: 978-1-032-67405-6 (Volume 6) (hbk)
ISBN: 978-1-032-67425-4 (Volume 6) (pbk)
ISBN: 978-1-032-67424-7 (Volume 6) (ebk)

DOI: 10.4324/9781032674247

Publisher's Note
The publisher has gone to great lengths to ensure the quality of this reprint but points out that some imperfections in the original copies may be apparent.

Disclaimer
The publisher has made every effort to trace copyright holders and would welcome correspondence from those they have been unable to trace.

HANDBOOK OF SOVIET SPACE-SCIENCE RESEARCH

Written by the staff and consultants of
Battelle Memorial Institute, Columbus Laboratories

Edited by

GEORGE E. WUKELIC

GORDON AND BREACH SCIENCE PUBLISHERS

NEW YORK · LONDON · PARIS

150 Fifth Avenue, New York, N.Y. 10011

Library of Congress Catalog Card Number: 67–30479

Editorial Office for Great Britain:

Gordon and Breach, Science Publishers, Ltd.
8 Bloomsbury Way
London W.C.1

Editorial Office for France:

Gordon & Breach
7–9 rue Emile Dubois
Paris 14e

Distributed in Canada by;

The Ryerson Press
299 Queen Street West
Toronto 2B, Ontario

Printed in Great Britain by
Robert MacLehose & Company Limited
The University Press, Glasgow

Foreword

Science and technology by their very nature tend to transcend national boundaries. Few fields of science, however, have stimulated international interest as strongly as have space-science and associated research activities. Recognition of this situation in the United States has resulted in improved accessibility of foreign scientific literature in both original and translated versions. However, advances in this field have been so numerous and so rapid that few researchers have adequate time or resources to keep abreast of foreign developments. Accordingly this handbook has been prepared to acquaint the reader with Soviet space-science accomplishments over the past 10 years and to assist interested researchers in their use of the Soviet information which is available to us. Battelle, in keeping with its practice of serving science and society, is proud to have sponsored its preparation.

DAVID C. MINTON, JR., *Director*
Columbus Laboratories
Battelle Memorial Institute

Preface

The purpose of this handbook is to provide, in a single volume, a comprehensive and authoritative English-language summary of the first 10 years of Soviet space-science research. It is intended that this handbook serve primarily as a reference tool for students, engineers, and scientists in space research having limited access to Soviet literature. Accordingly, in the preparation of the handbook, the authors have attempted to identify, characterize, and document the more significant activities associated with Soviet space-science research since its formal inauguration in 1957. They have, however, in most cases, left final evaluation of the reported research activities to the handbook user. Also, with few exceptions, space science is considered in this compilation to be those areas of science directly associated with the scientific exploration of space, using rockets, unmanned and manned satellites, and space probes. Discussion of most of the related, ground-based activities is therefore not included.

The handbook is divided into five major parts. Part I is essentially a directory of all rocket, satellite, and space-probe payloads launched in the USSR between October 1957 and December 1966. Scientific equipment and results associated with these launchings are summarized in Parts II and III. Part II is concerned with physical studies, whereas Part III treats biomedical investigations. Soviet activities associated with meteorological, geodetic, and communications satellites are described in Part IV. Part V treats the related topics of space tracking, manned space-station concepts, and attitudes concerning the existence of extraterrestrial life.

The book is intended for use by both specialists and nonspecialists alike. For the nonspecialists, over 150 photographs and illustrations of Soviet spacecraft and associated instrumentation schemes are included to facilitate comprehension and stimulate interest. For the specialists, a narrative description of equipment and results associated with almost a decade of Soviet space-science research is provided along with numerous tables containing selected experimental data and over 800 references for further consideration. To eliminate the need for the user to read the entire manuscript, all 20 chapters are completely independent of one another. Also consistent use of the United States Board on Geographic Names (B.G.N.) transliteration system and the metric system of units was attempted throughout the handbook.

The preparation of this handbook has involved the collection and review of literally thousands of original and translated Russian items. Because of the quantity of information covered, inconsistencies contained in the original Russian publications, and errors introduced by translation, the reader is cautioned to be alert for errors that most assuredly have filtered into the final manuscript. Also, as with all books on the rapidly changing subject of space research, this handbook will suffer from the effect of instant obsolescence. Fortunately, much of the information consists of basic upper atmosphere and space environment data whose value may be regarded as somewhat permanent.

All referenced Russian-language material is available in the U.S., principally at the Library of Congress. Soviet items published in the English language or known to be available in translated form are so identified in parentheses following the original bibliographic references printed at the end of each chapter. Organizations from which referenced translations can be obtained are noted in Appendix A.

In addition to the contributing authors who donated much of their own time to the preparation of this manuscript, I wish to thank the management of Battelle-Columbus, in particular Mr. David C. Minton, Jr., and Mr. Frederick L. Bagby for having sponsored this project.

Likewise the interest, encouragement, and assistance of Dr. J. M. Pilcher, Mr. Howard C. Cross, Mr. G. S. Simpson, Jr., and Mr. R. O. Stith of the Battelle staff deserve acknowledgment as do the extensive information services provided by Miss I. Wheldon and Mrs. B. Rawles of the Battelle Slavic Library. Also appreciated were the numerous information items obtained from the USAF, NASA, the National Environmental Satellite Center, the Rockets and Satellites Data Center, and the Library of Congress.

I wish to express my indebtedness also to the reviewers of portions of the handbook. In particular the comments and suggestions offered by Dr. C. S. Sheldon II, of the National Aeronautics and Space Council; Dr. J. W. Chamberlain, Kitt Peak National Observatory; Dr. H. D. Edwards, Georgia Institute of Technology; Mr. W. J. Best, Air Force Office of Scientific Research; Lt. Col. B. M. Smith, Air Weather Service, USAF; Dr. F. J. Krieger, Rand Corporation; Mr. P. Kehler, Bell Aerosystems Company; and Mr. L. Beck, Aerospace Technology Division of the Library of Congress were most helpful.

Finally, I express deepest appreciation to two outstanding secretaries, Mrs. Nadine Doherty and Mrs. Margaret Swank, whose unwaning dedication to the project served as inspiration to us all.

George E. Wukelic

Columbus, Ohio
June 1967

Contributing Authors

Battelle Staff:

R. C. Behn—Senior Meteorologist
R. A. Duffee—Senior Biometeorologist
N. A. Frazier—Senior Geophysicist
H. T. Kemp—Research Microbiologist
A. G. Mourad—Senior Geodesist
J. G. Stephan—Senior Photogrammetrist
R. A. Wright—Senior Environmental Physiologist
G. E. Wukelic—Senior Physicist

Battelle Consultants:

N. T. Bobrovnikoff—Professor Emeritus (Astronomy)
The Ohio State University
H. H. Guendel—Research Department, Bell Aersyostems
Company

Contents

PART V

RELATED TOPICS

Part 1

Soviet Rockets, Satellites, and Space Probes

The purpose of Part I is to describe in general terms rockets, satellites, and space probes that have been launched through December 1966 in the Soviet Union for scientific exploration and exploitation of the upper atmosphere and interplanetary space. A graphic summary of the number, type, and dates of their launchings is presented on page 2. Since the discussion is based upon hundreds of technical and popular items published in the USSR, no attempt has been made to establish bibliographies for the three chapters comprising this section. Likewise, photographs and drawings of rockets, satellites, and space probes that have been reproduced and incorporated within these chapters appear in sources too numerous to reference. Several of the more significant technical publications utilized, however, are referenced in footnotes as they are mentioned in the text.

Chapter 1 contains separate summaries of Soviet vertical-rocket activities associated with upper atmosphere and biomedical studies. Emphasis in this chapter is on describing the types of vehicles, payloads, and scientific studies that have been involved in the Soviet program over the past 10 years. Most successful or partially successful unmanned and manned satellites orbited by the Soviets are briefly described by series in Chapter 2. The discussion notes first published payload characteristics of each event of the nine series, including payload photographs when

TEN-YEAR SUMMARY OF SOVIET ROCKETS, SATELLITES, AND SPACE PROBES

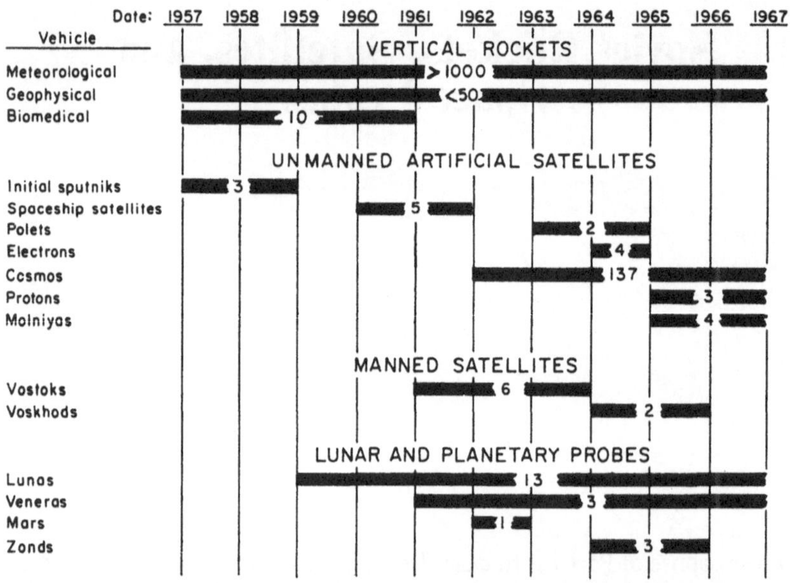

available, and secondly the principal mission features of the individual launchings of the series under discussion, e.g., orbital elements, active life, etc. Lastly, a brief summary noting the highlights of each of the series is presented. The third and final chapter in Part I contains a similar review of all announced Soviet lunar and planetary probes including the Luna, Venera, Mars, and Zond series.

The reader is reminded that all three chapters in Part I, like the remainder of the Handbook, are concerned almost exclusively with scientific (physical and biomedical) payloads, objectives, and results associated with Soviet efforts to research the upper atmosphere and space environment. Accordingly, readers interested in such related Soviet space exploration items as launch failures, booster characteristics and capabilities, military use of space, and comparisons of U.S. and USSR space programs are referred to existing publications by C. S.

Sheldon II, R. W. Porter, F. J. Krieger, W. Shelton, H. T. Simmons, and the Library of Congress.*

* 1. Richard W. Porter and Charles S. Sheldon II, "A Comparison of the United States and Soviet Space Programs", Program of Policy Studies in Science and Technology, George Washington University, June 1965, 85 pp.
2. Charles S. Sheldon II, "The Challenge of International Competition; U.S.-Soviet Space Programs", Third Manned Space Flight Meeting, Texas, 1964, 32 pp.
3. F. J. Krieger, "The Space Programs of the Soviet Union", Conference on "The Impact of Space Exploration on Society", August 18–20 1965.
4. F. J. Krieger, Various Rand reports on Soviet astronautics published since 1956.
5. Henry T. Simmons, "The Soviet Space Program", Space/Aeronautics, December 1965, pp 54–65, and "Russian Space Race", Astronautics and Aeronautics, June 1966, pp 3–9.
6. Selected Library of Congress publications prepared for the U.S. Senate Committee on Aeronautical and Space Sciences titled "Soviet Space Programs: Organization, Plans, Goals, and International Implications", 1962, and more recently "Soviet Space Programs 1962–1965's Goals and Purposes, Achievements, Plans, and International Implications", December 30 1966.
7. Also numerous items prepared by the Aerospace Technology Division of the Library of Congress containing reviews or surveys of selected foreign scientific and technical literature.
8. William Shelton, "The Russians Mean to Win the Space Race", Fortune, February 1966, pp 174–179.

Chapter 1

Vertical Research Rockets

G. E. Wukelic and R. A. Duffee

UPPER ATMOSPHERE STUDIES

Immediately following World War II, reports noting the use of vertically fired rockets in the Soviet Union for researching the upper atmosphere were nonexistent. Most Soviet publications contained frequent references to U.S. upper-air rocket experiments, which started in 1946, but made no mention of similar activities in progress in the USSR. The first indication of the existence of a similar Soviet program was the paper by S. M. Poloskov and B. A. Mirtov presented at the First International Congress of Rockets and Guided Missiles in 1956*. A later paper by A. A. Blagonravov in 1957† contained the first published

* S. M. Poloskov and B. A. Mirtov, "The Study of the Upper Atmosphere Through the Use of Rockets at the Academy of Sciences of the USSR", International Congress on Guided Missiles and Rockets, Paris, 1956.

† A. A. Blagonravov, "Investigation of the Upper Layers of the Atmosphere With the Aid of High Altitude Rockets", Vestnik Akademii Nauk SSSR, 1957, No. 6, pp 25–32.

description of the Soviet rocket program for researching the upper atmosphere. Scientific findings resulting from this effort, and descriptions of the research rockets employed since *circa* 1949 were not available until the International Geophysical Year (IGY) Rocket and Satellite Conference in Washington, D.C., September 30 through October 4 1957—almost concurrent with the successful Soviet orbiting of Sputnik 1.

Meteorological Rockets

Two papers by A. M. Kasatkin and P. P. Alekseyev, *et al.**, in 1957 are among the few items available which contain descriptions of Soviet meteorological rockets. The meteorological rocket, frequently referred to in Soviet publications as the "Meteo" or MR–1 series, was reportedly developed and flight tested in the 1948–1950 time period by the Central Aerological Observatory. This rocket was developed exclusively for researching the upper atmosphere. It is about 10 meters long, 50 cm in diameter, and weighs about one ton including a solid-fuel booster section. Its over 100-kg payload separates from the main body of the rocket at about 70 km with peak trajectory occurring between 80–100 km. Parachute recovery is made of both the nose cone and main-body sections of the rocket, sometimes within a kilometer of the launching site. Figures 1.1–1.6 show the meteorological rocket being launched, before and after booster separation, after parachute recovery, and the arrangement of the instrumentation in the nose cone. The simple steel launching pad arrangement makes it possible for MR–1 rockets to be launched from portable launching sites as well as from research ships. The launching stand is about 12 meters high and is tilted to compensate for wind effects. For determining the rocket's position during flight, the Soviets employ four synchronized cameras mounted aboard the rocket with their optical axes perpendicular to one another. A complete listing of the number of Soviet meteorological rockets launched per year from research ships (Shokal'skiy, Voyeikov, and Ob' operating primarily in the Pacific Ocean), Arctic observatories (Franz Josef Land, 80°N), and the central regions of the USSR (50°–60°N) between 1957 and 1966 is presented in Table 1.1. Thus for the 10-year period the Soviets have launched over 1000 meteorological rockets covering a geographic area extending from the Arctic to the Antarctic regions. These rockets have been used primarily for structural parameter profiling, i.e., for the routine

* A. M. Kasatkin, "USSR Meteorological Rocket", CSAGI Rocket-Satellite Conference, Washington, D. C., 1957; P. P. Alekseyev, *et al.*, "Rocket Investigations of the Upper Atmosphere", Meteorologiya i Gidrologiya, 1957, No. 8, pp 3–13.

determination of atmospheric density, temperature, pressure, and winds. They have upon occasion been associated with several geophysical studies of atmospheric optical properties, solar ultraviolet and X-ray radiation, cosmic radiation, and electron and positive-ion densities.

TABLE 1.1

Soviet Meteorological Rocket Launchings and Associated Launching Sites (1957–1966)

	Research Ships			Arctic Observatory		
Year	Black Sea	Pacific Ocean	Antarctic Expedition	Franz Joseph Land (Heiss)	Central Latitude European USSR	Yearly Totals
1957	—	—	1	3	10	14
1958	—	—	30	36	32	98
1959	—	12	—	19	15	46
1960	5	67	—	55	33	160
1961	—	43	—	53	17	113
1962	—	19	—	51	—	70
1963	—	22	—	10	78	110
1964	—	—	—	80	20	100
1965	—	68	—	51	31	150
1966	—	58	—	63	70	191

Geophysical Rockets

In addition to meteorological rockets the Soviets employ a so-called "geophysical rocket" which apparently has been in use for researching the higher layers of the upper atmosphere since 1949. Analogous to the meteorological rocket, geophysical rocket payloads are also recovered by using a combination payload-separation and parachute-recovery technique. To date, at least four different types of geophysical rockets exist. These have been designated A–1, A–2, A–3, and A–4 by the Soviets, the prefix A signifying the Academy of Sciences. Information on the smaller A–1 and A–4 rocket systems is extremely limited. These rockets are usually involved in launchings of under 750-kg payloads to altitudes around 100 km. The A–1 rocket, utilized in early (pre-IGY) Soviet upper atmospheric soundings (both geophysical and biomedical), is believed to resemble the German V–2 rocket in size and performance. The more recent geophysical rockets are shown in Figures 1.7–1.9. These rockets have also been associated with biomedical as well as geophysical payloads. The 20-meter long A–2 geophysical rocket, which was fired initially in 1957, is capable of carrying a 2200-kg payload to

Fig. 1.1. Launching of Meteorological Rocket

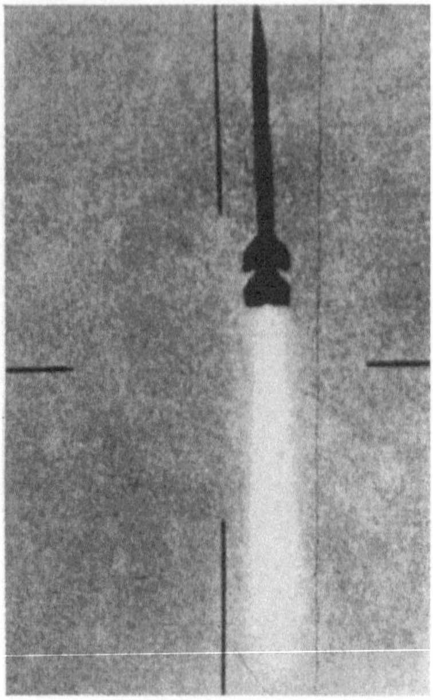

Fig. 1.2. Meteorological Rocket Before Booster Separation

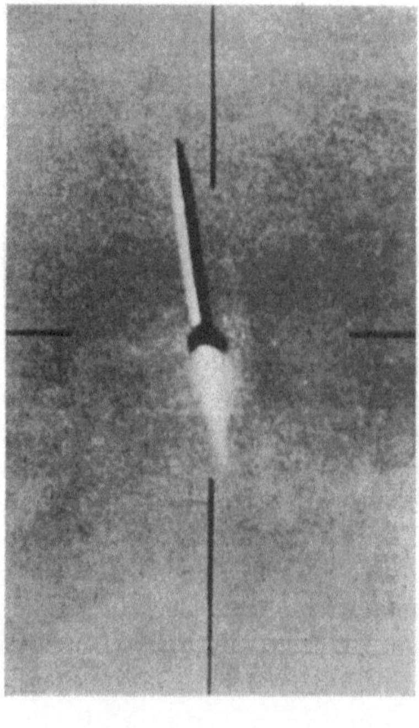

Fig. 1.3. Meteorological Rocket After Booster Separation

Fig. 1.4. Recovered Meteorological Rocket Nose Cone

Fig. 1.5. Recovered Meteorological Rocket Body

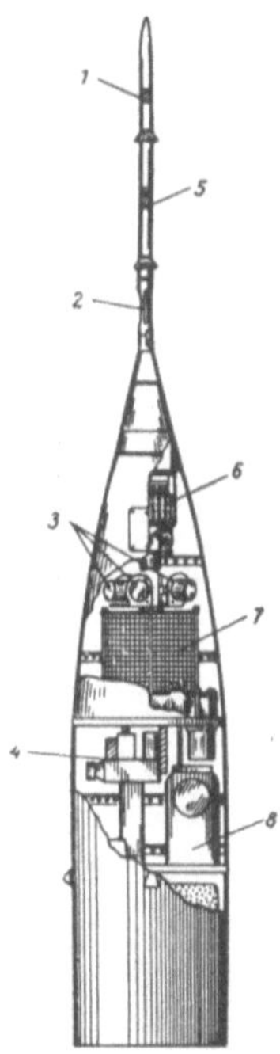

Fig. 1.6. Arrangement of Instrumentation in Meteorological Rocket Nose Cone

1. Hot-wire manometers for measuring pressure
2. Bolometer for measuring solar radiation
3. Diaphragm manometers
4. Transmitter
5. Resistance thermometer for measuring temperature
6. Commutator
7. Storage batteries
8. Photographic equipment

212 km. The A–3 geophysical rocket introduced in February 1958 has been used to lift 1600-kg payloads to peak altitudes between 400–500 km. Since May 1957 the Soviets have announced the launching of nearly 50 geophysical rockets. The launch dates, altitudes, and objectives of these flights are shown in Table 1.2. Discussions of the instrumentation schemes and results associated with these studies are contained in Chapters 4–12.

TABLE 1.2

Soviet Geophysical Rocket Launchings and Associated Investigations (Excluding Biomedical Studies), 1957–1966

Launch Data		Investigations													
Date	Altitude, km	Physical Properties	Optical Properties	Earth Surface Photography	Terrestrial Infrared Radiation	Solar UV, X-Ray, and Corpuscular Radiation	Chemical Composition	Electron and Positive Ion Densities	Micrometeorites	Geomagnetic Field	Cosmic Radiation	Electric Field (Electrostatic Charge)	Artificial Clouds	Studies During Eclipses	Aerodynamic/Reentry Studies
1957															
May 16	212	—	—	—	—	—	—	×	—	—	—	—	—	—	—
May 24	212	—	—	—	—	—	—	—	×	—	—	—	—	—	—
Aug. 25	212	×	—	—	—	×	×	×	×	—	—	—	—	—	—
Aug. 31	212	×	—	—	—	—	×	×	×	—	—	—	—	—	—
Sep. 9	212	×	—	—	—	—	×	×	×	—	—	×	—	—	—
1958															
Feb. 21	473	×	—	—	—	×	×	×	×	—	—	×	—	—	—
July 2	210	—	—	—	—	—	—	—	—	—	×	—	—	—	—
Aug. 2	212	×	—	—	—	—	×	×	×	—	×	×	—	—	—
Aug. 13	212	×	—	—	—	×	×	—	×	—	×	—	—	—	—
Aug. 27	451	×	—	—	×	×	×	×	×	—	—	×	—	—	—
Sep. 19	473	×	—	—	—	—	×	×	×	—	—	—	×	—	—
Oct. 4	110	×	×	—	—	×	—	×	×	—	—	—	—	—	—
Oct. 10	110	×	×	—	—	×	—	×	×	—	—	—	—	—	—
Oct. 31	473	×	—	—	—	—	×	×	×	—	—	—	—	—	—
Dec. 23	110	×	×	—	—	×	—	×	×	—	—	—	—	—	—
Dec. 25	110	×	×	—	—	×	—	×	×	—	—	—	—	—	—
1959															
June 22	212	—	—	—	—	—	×	—	—	—	—	—	—	—	—
July 2	212	×	—	—	—	×	×	×	×	—	—	—	—	—	—
July 10	212	×	—	×	×	×	×	×	×	—	—	—	—	—	—
July 14	212	—	—	—	—	—	—	×	—	—	—	×	—	—	—
July 21 A.M.	110	—	×	—	—	×	—	×	—	—	×	—	—	—	—
July 21 P.M.	110	—	×	—	—	×	—	×	—	—	×	—	—	—	—
July 22	195	—	—	—	—	—	×	×	—	—	—	—	—	—	—

TABLE 1.2 (Continued)

Soviet Geophysical Rocket Launchings and Associated Investigations (Excluding Biomedical Studies), 1957–1966

Launch Data		Investigations													
Date	Altitude, km	Physical Properties	Optical Properties	Earth Surface Photography	Terrestrial Infrared Radiation	Solar UV, X-Ray, and Corpuscular Radiation	Chemical Composition	Electron and Positive Ion Densities	Micrometeorites	Geomagnetic Field	Cosmic Radiation	Electric Field (Electrostatic Charge)	Artificial Clouds	Studies During Eclipses	Aerodynamic/Reentry Studies
1960															
June 15	212	×	—	×	×	×	×	×	—	—	—	×	×	—	—
June 24	212	×	×	×	×	—	×	×	—	—	—	×	—	—	—
Sep. 6	100	×	—	—	×	×	—	—	—	—	×	—	—	—	—
Sep. 16	210	×	—	—	—	×	×	—	—	—	×	—	—	—	—
Sep. 19	100	×	—	—	×	×	—	—	—	—	×	—	—	—	—
Sep. 21	100	×	—	—	×	×	—	—	—	—	×	—	—	—	—
Sep. 22	210	×	×	—	—	—	×	—	—	—	×	—	—	—	—
Sep. 23	212	—	× ?	—	—	—	×	—	—	—	—	—	—	—	—
Sep. 24	200	—	×	—	—	—	—	—	—	—	—	—	—	—	—
Oct. ?	212	—	—	—	—	—	—	—	—	—	—	—	—	—	—
1961															
Feb. 15	100	—	—	—	×	×	—	—	—	—	—	—	—	×	—
Sep. 23	100	—	—	—	—	—	—	—	—	—	—	—	—	—	—
Nov. 15	430	—	—	—	—	—	×	×	—	—	—	×	—	—	—
1962															
Oct. 18	500	×	—	—	×	×	—	×	—	—	×	—	—	—	—
1963															
June 6	515	×	×	—	×	×	—	×	—	—	×	×	—	—	—
June 18	510	×	×	—	×	×	—	×	—	—	×	×	—	—	—
Oct. 25	430	—	×	—	—	×	×	—	—	—	×	—	—	—	×
Dec. 24	422	—	×	—	—	×	×	—	—	—	×	—	—	—	×
1964		NO ANNOUNCED LAUNCHINGS													
1965															
Sep. 20	500	—	—	—	×	×	—	×	—	—	×	×	—	—	—
Oct. 1	500	—	—	—	×	×	—	×	—	—	×	×	—	—	—
1966															
Oct. 13	500	—	—	—	—	×	—	×	—	—	—	—	—	—	—

Geophysical Rocket Containers

In addition to instrumented nose cones the Soviets employ specialized geophysical containers which separate themselves from the geophysical rockets and are safely returned to earth by a parachute system. One such container is a nonoriented cylindrical container which is attached

Fig. 1.7. A–2 Geophysical Rocket

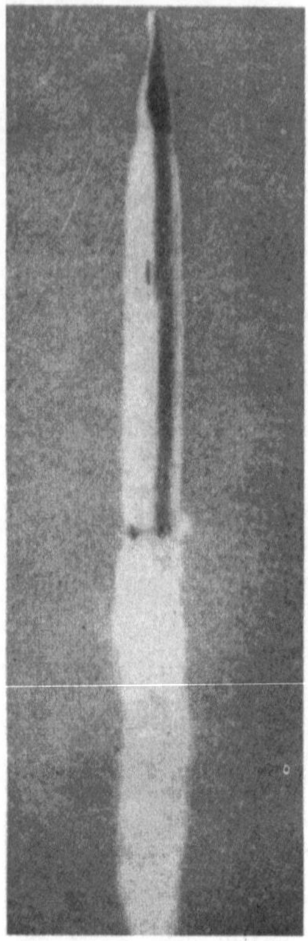

Fig. 1.8. A–3 Geophysical Rocket

Fig. 1.9. V–5 Geophysical Rocket

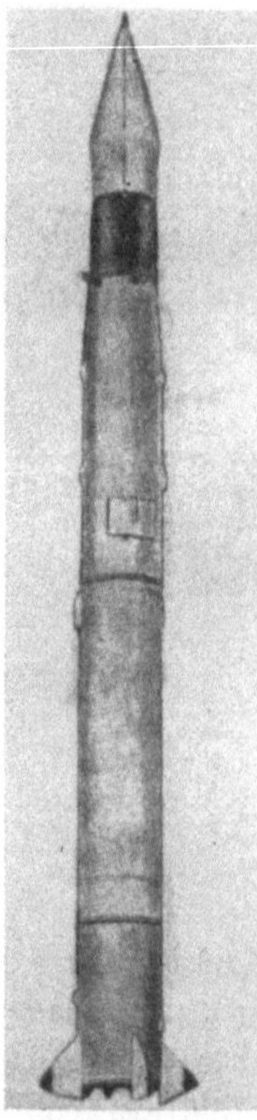

(usually in pairs) to the side of the A–2 type geophysical rocket (see Figure 1.7) and at a predetermined height is ejected at a very low speed. The container (Figure 1.10) is normally a metallic cylinder about 2 meters long and 0·4 meters in diameter and weighs about 250 kg when loaded. Before launch it consists of four sections, one each for the

Fig. 1.10. Nonoriented Geophysical Rocket Container (Before and After Launching)

parachute, sensing equipment, the hermetically sealed compartment for instruments and guidance apparatus, and the power supply. The earth, clouds, and surrounding space are photographed through windows to determine the container's orientation. A polished metallic plate is attached to the side of the power supply section for registration of meteorites. The sensing section is completely exposed and contains apparatus for making pressure measurements or obtaining atmospheric samples. The automatic opening of the parachute takes place at about 2 km above the surface of the earth. Spikes located in the nose portion of the container insure a vertical position upon landing. The force upon

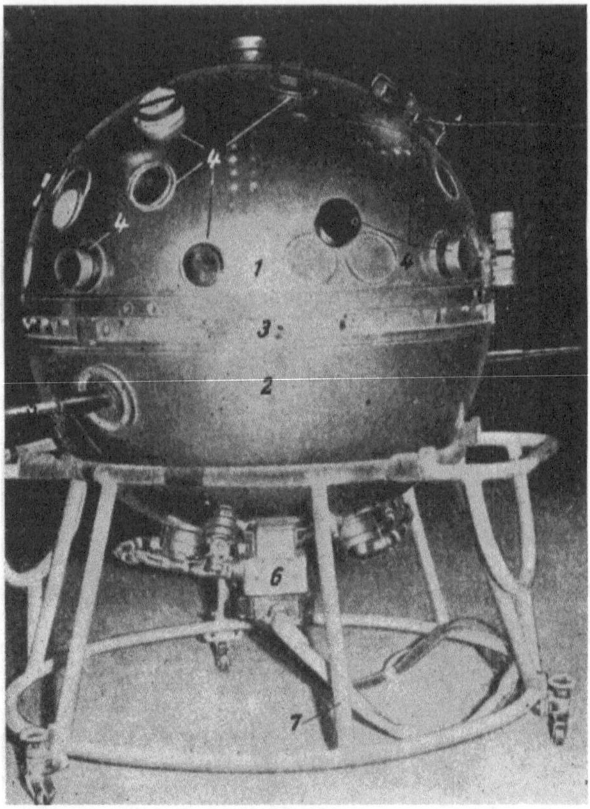

Fig. 1.11. High Altitude Geophysical Automatic Station (VGAS) or High Altitude Optical Station (VOS)

impact is reduced by a shock absorber system also located in the nose portion of the container.

For performing rocket measurements requiring a controlled orientation the Soviets employ the so-called "high altitude geophysical automatic stations" (VGAS). According to A. M. Kasatkin*, this container

* A. M. Kasatkin, "High-Altitude Optical Station for the Investigation of the Atmosphere", Iskusstvennye Sputniki Zemli, 1964, No. 15, pp 3–21.

(Figure 1.11) is a metal sphere one meter in diameter, weighs about 365 kg, and contains a variety of geophysical instruments, an automatic stabilization and orientation system, and a multichannel telemetry system. Apparently the station was initially installed in either the A–1 or A–4 geophysical rocket but more recently has been associated with geophysical rocket flights to over 500 km (Figure 1.9). The Soviets have modified several of these stations by adding windows for optical studies. In the latter case the Soviets frequently refer to the station as a high-altitude optical station (VOS). In flight the spherical container is normally separated from the rocket at a height of 65–67 km. After disposal of the protective cone, the velocity of the station is increased 1 m/sec. The auto-orientation system begins to operate at the moment of separation and operates until somewhere between the peak of the trajectory and its entry into the denser layers of the lower atmosphere at a height of approximately 40 km. Under normal conditions the station is initially oriented by coarse controls to a precision of $\pm 3°$ and maintains this orientation to within $\pm 0\overset{\circ}{.}5$. In special cases controls insuring more precise orientation have been utilized. The Soviets have noted that such geophysical containers have been used in measurements of solar, atmospheric, and terrestrial radiation, charged particle density, atmospheric structure and composition, electric and magnetic fields, micrometeorites, as well as medical and biological parameters.

BIOMEDICAL STUDIES

The first announcement of Soviet investigations of biomedical problems pertaining to manned space flight was a paper presented at an international meeting by A. V. Pokrovskiy in 1956*. This paper described in some detail Soviet methodology for investigating the effects on dogs of vertical flights in rockets to altitudes of 100–110 km, and presented results of physiological measurements (pulse and respiration frequency). In a 1957 publication by A. A. Blagonravov† the time period for these initial investigations was identified as 1951–1956. First detailed results of Soviet biomedical studies with dogs in vertical flights to altitudes of 212 km were published in 1958, after the successful launch-

* A. V. Pokrovskiy, "Study of the Vital Activity of Animals During Rocket Flights in the Upper Layers of the Atmosphere", International Congress on Guided Missiles and Rockets, Paris, December 1956.

† A. A. Blagonravov, "Investigation of the Upper Layers of the Atmosphere With the Aid of High-Altitude Rockets", Vestnik Akademii Nauk SSSR, 1957, No. 6, pp 25–32.

ing of Sputniks 1 and 2. These appeared in two articles* contained in the booklet "Preliminary Results of Scientific Researches on the First Soviet Artificial Earth Satellites and Rockets, No. 1", published by the Publishing House of the Academy of Sciences of the USSR. These articles established that the experimental program was initiated in 1949, with the first flight to 100 km occurring in 1951. The first flight with dogs to an altitude of 212 km occurred in 1957. Similar studies with dogs to altitudes of 450–473 km were conducted in 1958. A summary of Soviet biomedical investigations on vertical research rockets, a program that continued until at least 1960, is presented in Table 1.3. This table identifies the specific rocket flights involved and the associated physiological methods utilized.

Physiological Investigations

Soviet biomedical experiments aboard vertical research rockets have been conducted in three phases. The first phase, involving the launching of two dogs per vehicle in a hermetic capsule of 0·28 m^3 volume to an altitude of 100 km, was conducted in the 1949–1952 period with the first actual launching of the dogs "Tsygan" and "Dezik" occurring in 1951. As shown in Table 1.3, this initial series comprised six flights with nine dogs, three of the animals participating in two flights each. For this series two dogs, each weighing about 10 kg, were placed in the hermetic compartment and fastened on trays by a system of reins. Fluctuations of skin temperature and pulse were recorded on a four-channel recorder, along with cabin pressure and temperature. A movie camera was included in the instrumentation as a means of observing the subjects' behavior during the flight. Before and after the flights, an ECG, X-ray of the thorax, conditional food reflexes, and body weight were obtained.

The second phase of Soviet biomedical studies on vertical rockets extended from 1952 through 1956, with the first flight apparently occurring in 1955. This series comprised nine flights involving 12 dogs, with six of the animals participating in two flights each. The animals, each weighing four to five kg, were placed in a nonhermetic compartment in the nose section of small geophysical rockets and launched to altitudes of 100–110 km. Pressure suits with plexiglas helmets were used to encap-

* A. M. Galkin, *et al.*, "Investigations of the Vital Activity of Animals When Flying in Hermetically Sealed Cabins of Rockets Up to a Height of 212 km", "Preliminary Results of Scientific Researches on the First Soviet Artificial Earth Satellites and Rockets, No. 1", Moscow, 1958; B. G. Bugrov, *et al.*, "Investigations of the Vital Activity of Animals When Flying in Hermetically Sealed Cabins of Rockets Up to a Height of 110 km", ibid.

TABLE 1.3

Soviet Biomedical Investigations on Vertical Research Rockets (1951–1966)

Vehicular Data			Experimental Subjects		Physiological Measurements*												Cabin Environment Measurements					
Type of Rocket	Altitude, km	Total Number of Flights and Approximate Date	Type/ Average Weight	Number	Pulse	Respiration	Arterial Pressure	Body Temperature	Skin Temperature	Behavior (film)	ECG	X-Ray	Body Weight	Blood Analysis	Urinalysis	Conditioned Reflex	Wall Temperature	Air Temperature	Pressure	Acceleration	Humidity	Suit Temperature
A–1,, A–4	100	6 (1951–1952)	Dogs/ 10 kg	9	O	—	—	G	O	O	G	G	G	—	—	G	—	O	O	O	—	—
A–1, A–4	100–110	? 4 Feb. 1955 26 May 1956 1 June 1956 7 June 1956 14 June 1956 22 June 1956 20 July 1956 20 Dec. 1956	Dogs/ 4–5 kg	12	O T	O T	O	O T	—	O	G	G	G	G	—			T	T	O	—	O
A–2 A–3	212 450–470 450–470 450–470	16 May 1957 27 Aug. 1958 19 Sep. 1958 31 Oct. 1958	Dogs/ 5–7 kg Rabbits	14 2	O G T	O G T	O G T	G	—	O	O G	G	G	G	G	—	O T	O T	O T	O	—	—
A–2	212	2 July 1959 10 July 1959 15 June 1960 24 June 1960 16 Sep. 1960 22 Sep. 1960																				

* O = On-board measurement
G = Ground measurement, pre- and postflight only
T = Telemetered

sulate the dogs. Each animal was fastened to a movable tray which was installed within a welded steel carriage termed an ejection vehicle. As shown in Figure 1.12, a schematic flight diagram, one of the vehicles was ejected at an altitude of 75–90 km, with parachute descent from an altitude of 75–85 km. The other animal remained in the nose section of the rocket until ejection occurred at an altitude of 35–50 km. Physiological monitoring during this series consisted of continuous recording of pulse, respiration frequency, maximum and minimum arterial pressure, and body temperature. These data were telemetered to the ground, in addition to onboard recording. Behavior of the animals was again

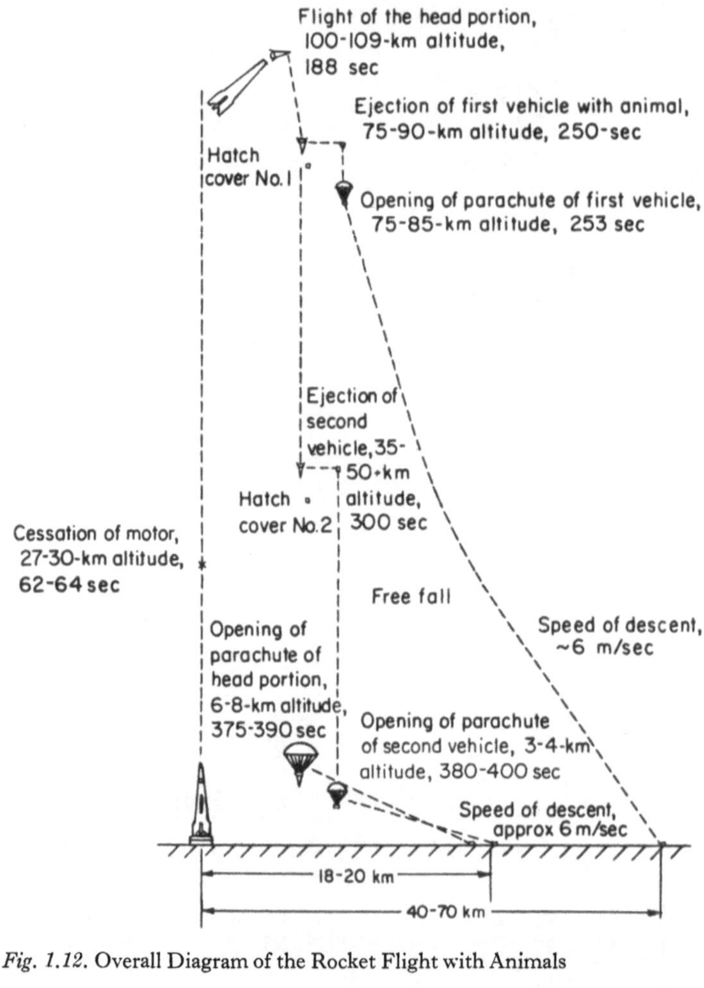

Fig. 1.12. Overall Diagram of the Rocket Flight with Animals

recorded by means of a movie camera with a five- to six-minute film supply. Preflight base lines of the physiological indices were obtained on the launch pad with the dogs in the vehicles, approximately 35 minutes before launch. In addition, one to two days before flight, ECG's, X-rays of the skeleto-muscular system, and blood samples were taken. Postflight analyses of ECG, X-ray, and the hemopoetic system were made several days after the flights. Blood samples, however, were not obtained for all subjects.

The third and final stage of the Soviet biomedical vertical rocket program began in 1955, with the first launching of two dogs in a geophysical rocket to an altitude of 212 km taking place on May 16 1957. Fourteen dogs, whose average weight was five to seven kg, were used as experimental subjects. A rabbit accompanied the two dogs on a flight to 208 km on July 2 1959, and again on a flight to 212 km on June 15 1960. At least three flights to altitudes of 450–473 km, again with two dogs per vehicle, were carried out in 1958. Several of the dogs made two or three flights each and one animal, named "Otvazhnaya", made six flights. During several experiments one of the two dogs was anesthetized by means of a 10 per cent solution of hexenal at a dose rate of 0·8–1·2 g/kg injected subcutaneously. The animals were placed in an hermetic compartment of 0·46 m^3 volume in the nose section of the rocket. During the flights the dogs' blood pressure, respiration frequency, and pulse were recorded continuously on board and telemetered to the ground. An ECG was also taken on board. Animal behavior was again monitored via movie film for a period of 11–12 minutes. Pre- and postflight examinations included clinical analysis of blood, X-ray of the thorax, ECG, blood pressure, recording of pulse and respiration, urinalysis for some of the animals, and measurement of body temperature and weight. Blood samples were examined both on the eve of the flight and immediately after the flight. Thorax X-rays were obtained on the eve of the experiment and the day following the flight. ECG's were taken two to three times per week prior to flight, on the launching pad and after landing.

Life-Support Techniques

The life-support method employed in the initial series of rocket flights to 100 km was a simple injection system. It consisted basically of a seven-liter steel globe filled with a mixture of 70% air and 30% oxygen. Gas pressure in the globe was not reported other than that it was adequate to maintain the dogs for three hours with a cabin pressure of 680–760 mm Hg and cabin volume of $0{\cdot}28m^3$. A soda lime cartridge was used to adsorb expired CO_2, and a silica gel cartridge was employed as a desiccant. Prior to launch, the hermetic compartment was cooled by expansion of liquid oxygen from 2·5-liter cylinders. The dogs were tied, separately, to trays by a system of belts.

For the second phase the animals were placed in pressure suits with plexiglas helmets. The suits were made of a three-ply rubberized fabric with two extensions for the dog's forefeet. The dogs were fastened within the suits by four belts on the bottom side of the suit interior. An additional four exterior belts were used to attach the suit to the catapult holder. The globular helmet had a 25-mm diameter porthole in the lower front section equipped with a valve which opened at an altitude of 4000 meters on the descending leg of the flight to admit air. A photograph of the suit and helmet is shown in Figure 1.13. Three two-liter

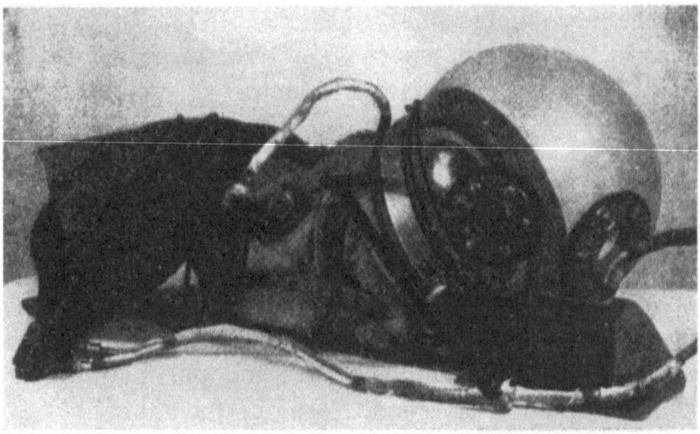

Fig. 1.13. Spacesuit for Dogs (110 km)

Fig. 1.14. Left Ejection Vehicle—Lateral View (110 km)

Fig. 1.15. Right Ejection Vehicle—Lateral View (110 km)

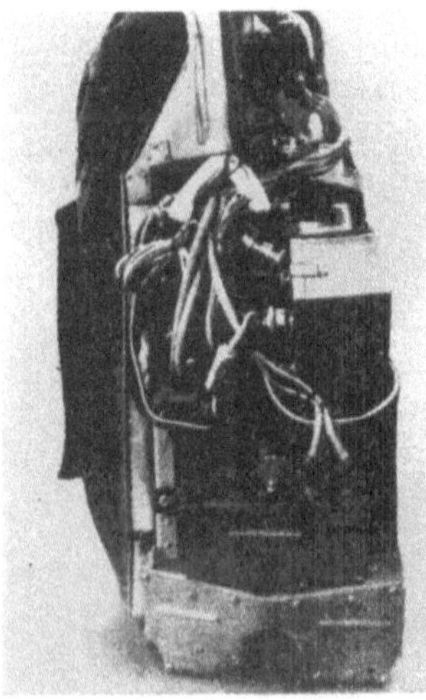

Fig. 1.16. Rear View of Ejection Vehicle

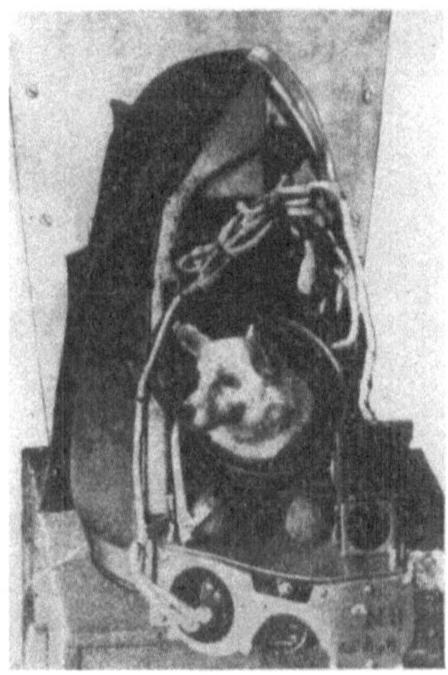

Fig. 1.17. Front View of Ejection Vehicle

Fig. 1.18. Nose Section of 110-km Rocket

cylinders filled with oxygen at a pressure of 150 atmospheres were used to ventilate the suit. Suit ventilation rate was six liters per minute with a suit pressure of 440 mm Hg. Accordingly a 900-liter oxygen supply could maintain the animal for a period of 2·0 to 2·5 hours. Suit pressure was controlled by pressure sensors and an excess-pressure valve installed in the suit lining.

Each animal was tied to a movable tray and placed within a welded steel carriage, termed an ejection vehicle. Each ejection vehicle contained separate oxygen supply, parachute system, and physiological sensors and recorders. Total weight of the ejection vehicle was about 70 kg. Photographs of the ejection vehicle showing the position of the dog are also presented in Figures 1.14 to 1.17. The nose section with the animal compartment is shown in Figure 1.18.

For the flights to altitudes of 212 km and higher the Soviets reverted to hermetically sealed cabins in the shape of a truncated cone. Three

Fig. 1.19. Recovered Nose Section of 212-km Rocket (A–2)

Fig. 1.20. Recovered Nose Section of 450-km Rocket (A–3)

seven-liter tanks, pressurized to 150 atmospheres with an air-oxygen mixture containing 40% O_2, were connected to a manifold which supplied the ventilation and breathing gases. Total gas supply was 2520 liters with a ventilation rate of 6·7 to 7·0 liters per minute from the tanks, sufficient for six-hour operation. The gas mixture was fed to the injector nozzle through a reducing valve with the output pressure regulated to between 2 and 20 atmospheres. The system was designed for a minimum cabin pressure of 460 mm Hg but reported cabin pressures were 780 mm Hg. Two kilograms of soda lime, in a vessel attached to the bottom of the injector, were used to adsorb expired CO_2. Moisture was adsorbed on 1·5 kg of silica gel.

Although the cabin is described as hermetically sealed, complete hermetization was not used, particularly because of the large door area in the compartment. The seven-liter-per-minute supply from the tanks was vented directly to the atmosphere through a critical flow orifice operating on a pressure differential of 760 mm Hg. Photographs of the recovery sections of the vehicles, launched to altitudes of 212 km and 450 km, are shown in Figures 1.19 and 1.20, respectively. A photograph illustrating the arrangement of the cabin for a flight to 212 km is presented in Figure 1.21.

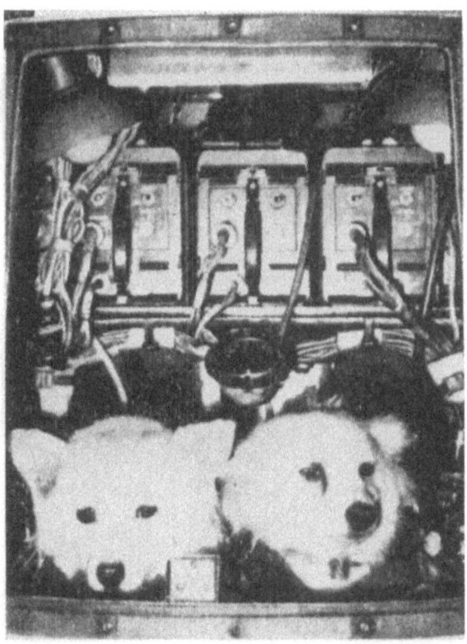

Fig. 1.21. Dogs in 212-km-Rocket Cabin

Chapter 2

Artificial Earth Satellites

G. E. Wukelic, R. A. Duffee, and R. C. Behn

The Soviets launched the first artificial earth satellite on October 4 1957. Because of its impressive size and early launching date, Sputnik 1 shocked scientists, politicians, and laymen the world over more than any other single space event to date. The reason that such a launching was not anticipated by the U.S. scientific community was two-fold. First, although the Soviets, like the U.S., had made an official commitment to launch an artificial satellite during the International Geophysical Year (July 1957–December 1958), they had not provided any preliminary data in either their technical or popular press relative to the planned type, size, weight, or schedule for such a significant event. Secondly, evidence of a sincere Soviet interest in the closely allied rocket technique for directly researching the upper layers of the atmosphere (as was increasing in popularity in the U.S.) was similarly absent from Soviet technical literature until just prior to Sputnik 1. This policy of not announcing in advance the types and dates of planned space events is still being practiced by the Soviets. Also, the amount of technically informative material contained in Soviet releases made concurrent with and immediately subsequent to most satellite launchings leaves much to be desired.

Since the successful launching of Sputnik 1, the Soviets have released varying amounts of information relative to some 125 successful or partially successful launches that have resulted in the orbiting of over 160 separate unmanned or manned satellites through 1966. The purpose of this chapter is to briefly summarize the technically significant information on each of these events contained in the large number of Soviet technical and popular releases. Discussion in this chapter has been arranged according to the following Soviet-designated satellite groups:

1. Initial IGY Series for geophysical and biomedical studies—Sputniks 1, 2, and 3.
2. Spaceship-Satellite Series for studies and tests relating to the development of a man-rated spacecraft—Spaceship-Satellites 1, 2, 3, 4, and 5.
3. Polet Series for testing spacecraft maneuverability—Polets 1 and 2.
4. Electron Series for extensive International Quiet Sun Year (IQSY) studies—Electrons 1/2 and 3/4.
5. Cosmos Series for geophysical, biomedical, meteorological, as well as various technological studies—Cosmos 1 through 137.
6. Proton Series for unique ultrahigh energy cosmic radiation studies—Protons 1, 2, and 3.
7. Molniya Series for developing a communication-satellite system—four Molniya-1 satellites.
8. Vostok Series for manned space flights of increasing duration—Vostoks 1 through 6.
9. Voskhod Series for multimanned space flights and extravehicular activities—Voskhods 1 and 2.

Reported payload characteristics (i.e., weight, size, objective(s), and photographic illustrations) and principal mission features (e.g., lifetime, orbital elements, and miscellaneous comments regarding degree of success) are given for the satellites representing each of the nine series. Also a general summary noting the highlights and significance of each of the series is provided. Details of the scientific results (physical and biomedical) associated with these events are reported in Parts II and III of this Handbook.

INITIAL SPUTNIK SERIES

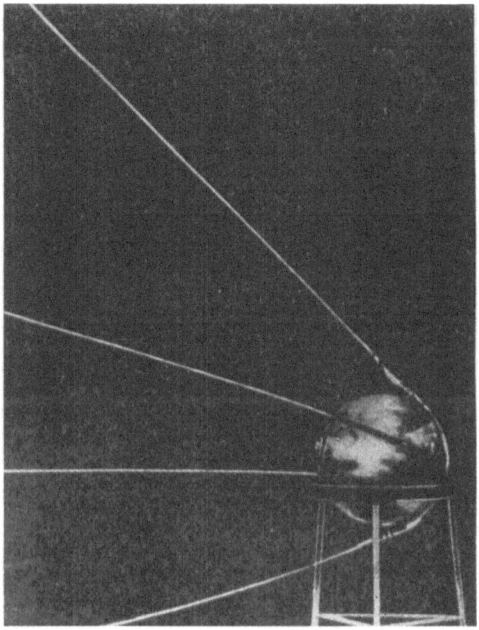

Fig. 2.1. Sputnik 1

Fig. 2.2. Sputnik 2

INITIAL SPUTNIK SERIES

Fig. 2.3. Sputnik 3

PAYLOAD CHARACTERISTICS

Sputnik 1

Launch Date: October 4 1957 Weight: 83·6 kg
Size: Sphere: 0·58 m, diameter
Scientific Objectives
Direct Studies: Temperature inside and outside satellite
Indirect Studies: Atmospheric density (from orbital data), ionospheric electron density (from radio tracking data)

Sputnik 2

Launch Date: November 3 1957 Weight: 508·3 kg
Size: Cone: 5·8 m, length; 1·2 m, diameter
Scientific Objectives
Direct Studies: Cosmic radiation, solar UV and X-ray, temperature and pressure, biomedical (Laika)
Indirect Studies: Atmospheric density (from orbital data)

Sputnik 3

Launch Date: May 15 1958 Weight: 1327 kg
Size: Cone: 3·57 m, length; 1·74 m, diameter
Scientific Objectives
Direct Studies: Solar and cosmic radiation, electric field, geomagnetic field, micrometeorites, ion density and composition, structural properties

TABLE 2.1

Principal Mission Features

Vehicle	Active Life: Launch Date	Active Life: Terminal Date	Orbital Elements: Apogee, km	Orbital Elements: Perigee, km	Orbital Elements: Inclina-tion,°	Transmitter Frequencies, mc
Sputnik 1	10/4/57	Decayed: 1/4/58	948	226	65·1	20·005 40·002
Sputnik 2	11/3/57	Decayed: 4/14/58	1679	226	65·3	20·005 40·002
Sputnik 3	5/15/58	Decayed: 4/6/60	1881	226	65·2	20·005 40·002

SUMMARY

The first three Soviet artificial earth satellites were launched during the International Geophysical Year. Sputnik 1 (Fig. 2.1), launched on October 4 1957, represented the world's first successful satellite launching. The impressive, 83-kg spherical satellite, with which only temperature was measured, remained in orbit exactly three months. Sputnik 2 (Figure 2.2) launched almost one month later, on November 3 1957, was over five times heavier and contained separate containers for some first-time attempts at satellite geophysical and biomedical (Laika) studies. It was during this event that the first traces of the then unknown radiation belts were recorded but not recognized. Although it became silent a few days after launch, on November 10 1957, Sputnik 2 remained in orbit 163 days, disintegrating on April 14 1958. The following month, May 15 1958, the Soviets launched the most sophisticated satellite of the entire IGY series, Sputnik 3 (Figure 2.3). It was launched into orbit inclined 65° to the earth's equator as were the two previous vehicles. The useful payload of this satellite totaled 968 kg and contained, for the first time, advanced telemetry apparatus which permitted data storage for subsequent transmittal during passage over USSR tracking stations and a variety of geophysical instruments for the first simultaneous satellite measurements of the space environment. The Soviets have indicated that radio contact with Sputnik 3 was maintained until its decay (691 days later) on April 6 1960.

In addition to the prestige associated with the early launching of these three impressive satellites, the Soviets profited technically by obtaining what was then unique data both in the geophysical and biomedical areas. The latter pertained mostly to the physiological response to long-term weightlessness experienced by Laika aboard Sputnik 2. The direct and indirect studies aboard Sputniks 1, 2, and 3 benefited geophysics by providing unique data on upper-atmospheric structure (principally density, pressure, and composition), ionospheric characteristics above the region of maximum electron density, micrometeorite flux, magnetic field intensity, and intensity and composition of primary cosmic rays. Descriptions of instruments and results associated with most of these studies have been published by the Soviets and are summarized in Parts II and III.

SPACESHIP-SATELLITE SERIES

Fig. 2.4(a). Ejectable Animal Capsule

PAYLOAD CHARACTERISTICS

Launch Dates: May 15 1960 through March 25 1961

Weight: All orbital payloads in this series weighed between 4·5 and 4·7 metric tons.

Size/Dimensions: Not reported specifically. Flights were claimed to be tests of Vostok vehicles.

Scientific Objectives:

Medical–Biological: Development of systems for manned space flight. Effects of space-flight factors on biological specimens including visual (television) observation.

Geophysical: Determination of the composition and intensity of cosmic radiation; measurement of solar X-ray and ultraviolet radiation.

Biological Specimens: Dogs, rats, guinea pigs, mice, insects, plants, fungus cultures, seeds, bacteria, and skin tissue.

SPACESHIP-SATELLITE SERIES

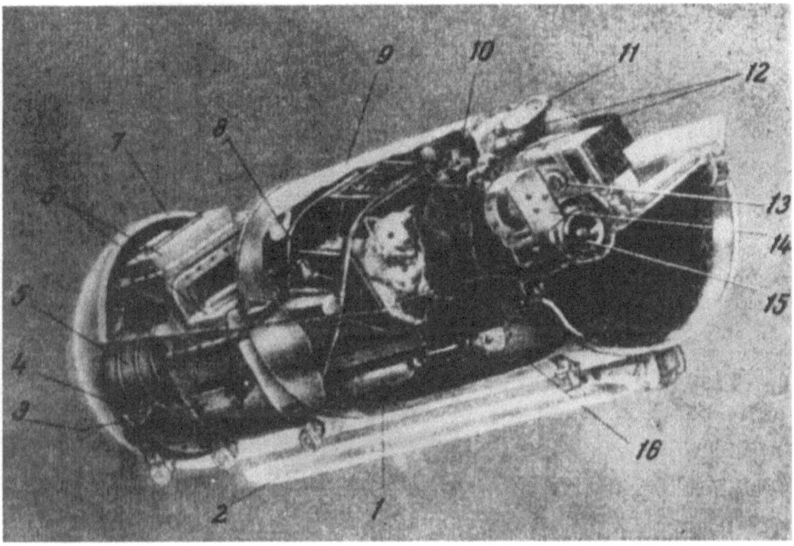

Fig. 2.4(b). Ejectable Animal Capsule

1. Air-supply tank
2. Ejection mechanism
3. Radar unit
4. Special storage battery for heating test tubes with microbes
5. Storage battery
6. Special scientific apparatus
7. Ejection capsule
8. Motion sensor
9. Pressurized capsule for animals
10. Microphone
11. Radar antenna
12. Inhalation and exhalation valves
13. Television camera
14. Mirror
15. Ventilation system
16. Automatic combination feeder

TABLE 2.2

Principal Mission Features

Vehicle	Active Life		Orbital Elements			Transmitter Frequencies, mc	Comments
	Launch Date	Terminal Date	Apogee, km	Perigee, km	Inclination, °		
Spaceship-Satellite 1	5/15/60	Reentered: 9/5/62	369	312	65	19·995	Perfection of all systems ensuring safety of manned flight, including deceleration and separation of capsule before descent from orbit.
Spaceship-Satellite 2	8/19/60	Recovered: 8/20/60 on 18th orbit	339	306	64·95	Not reported	Improvement of systems for man's safety, medical-biological experiment (dogs Belka and Strelka), cosmic radiation, vegetative studies, cytological and radiogenetic studies.
Spaceship-Satellite 3	12/1/60	Burned in atmosphere during reentry: 12/2/60	265	187·3	64·97	19·995	Same as Spaceship-Satellite 2 (dogs, Pchelka and Mushka).
Spaceship-Satellite 4	3/9/61	Recovered: 3/9/61 after 1 orbit	248·8	183·5	64·93	Not reported	Spaceship-Satellite's design for manned flight, medical-biological experiment (dog Chernushka), hematological effects (mice), radiogenetic investigations (bacteria), vegetative studies, cytological and radiogenetic studies.
Spaceship-Satellite 5	3/25/61	Recovered: 3/25/61 after 1 orbit	247	178·1	64·90	Not reported	Same as Spaceship-Satellite 4 (dog Zvezdochka), total flight time, $1^{h}\ 45^{m}$.

SUMMARY

According to news releases appearing in "Pravda" and "Izvestiya" the primary mission of the spaceship-satellite series was to conduct further investigations of the effects of space flight factors on biological specimens, to develop systems for guaranteeing man's vital activity in space, and also to ensure safe return to earth. The first spaceship-satellite, however, reportedly had no biological specimens on board, as the payload consisted of a pressurized capsule weighing about 2500 kg with a load or dummy equivalent to the weight of a man. Instrumentation and power supply, including solar converters and chemical batteries, weighed an additional 1500 kg. Although initial Soviet releases were unclear as to whether actual recovery of the capsule was planned, subsequent reports noted that the pressurized cabin was detached from the satellite on May 19 1960, after completing its mission. It did not fall earthward as planned but, because of a malfunction in the satellite's orientation system, went into a higher orbit with a period of 94·25 minutes, a perigee of 307 km and an apogee of 690 km. Spaceship-Satellite 1 decayed on September 5 1962 although portions of the satellite survived reentry. No detailed descriptions of equipment on board or specific results of this flight have appeared in Soviet technical publications.

Spaceship-Satellite 2 has been termed appropriately a "flying zoo". The pressurized cabin contained an ejectable capsule in which were the dogs, "Belka" and "Strelka", 12 mice, insects, plants, fungus cultures, seeds of corn, wheat, peas, and onions, microbes, and other biological specimens. Outside the ejection unit, but within the pressurized reentry sphere, were 28 mice and two white rats. After completing 17 orbits the cabin reentered the atmosphere and was recovered successfully. The ejection unit with the dogs and other biological specimens was catapulted from the cabin at an altitude of approximately 7000 meters, and recovered by parachute. This flight provided the Soviets with a wealth of information concerning the effect of space flight on physiological, genetic, and cytological reactions of living organisms ranging from cellular to mammalian subjects. Numerous Soviet publications are available presenting exhaustive analyses of the data obtained on this flight. Two television cameras aboard the satellite also provided unique information. The recording of the televised images was synchronized with the telemetry data; this made it possible to compare the direct observations of the animals (dogs) with the data on changes in their physiological functions.

Spaceship-Satellite 3 was essentially a repeat of the second spaceship-satellite. The only differences noted appear to be the Soviet references to improved biotelemetry and a slight variation in instrumentation schemes employed for the cosmic and radiation studies. The reentry sphere, with the dogs "Pchelka" and "Mushka" and all the other biological specimens, failed to descend along the calculated trajectory and burned up while entering the dense layers of the atmosphere. Accordingly, the only biomedical results obtained by the Soviets are the telemetry and television data relating to the psychophysiological reactions of the dogs. All technical publications describing the results of this event relate to only the radiation studies undertaken.

Spaceship-Satellites 4 and 5, which completed one orbit rather than the 17 completed by the second and third spaceship-satellites, were essentially tests of the Vostok life-support and landing systems. In these two flights, the Vostok spacesuit was tested. The suit was put on a rubber mannequin, in which a number of biological specimens from various levels of evolutionary development were placed. The mannequin was situated in the ejectable unit. On Spaceship-Satellite 4 the dog, "Chernushka", with nine rats, guinea pigs, and other biological specimens were placed in the pressurized reentry sphere outside of the ejectable unit. The same system, with the dog "Zvezdochka", was employed in Spaceship-Satellite 5. Thus, both systems of landing, in the cabin or after ejection from the spacecraft, were evaluated during these two flights. Publications relating to the results of these flights refer only to biomedical studies.

POLET SERIES

PAYLOAD CHARACTERISTICS

	Polet 1	*Polet 2*
Launch Date:	November 1 1963	April 12 1964
Weight:	Not reported	Not reported
Size:	Not reported	Not reported

Scientific Objectives: Although scientific instruments were reportedly a part of the Polet payloads, the major mission of this series apparently was to test spacecraft maneuverability.

TABLE 2.3
Principal Mission Features

Vehicle	Active Life		Orbital Elements			Transmitter Frequencies, mc
	Launch Date	Terminal Date	Apogee, km	Perigee, km	Inclination,°	
Polet 1	11/1/63	Silent: 11/3/63 Life Expectancy: Years	Initial: 592 Final: 1437	Initial: 339 Final: 343	Initial: not reported Final: 58·55	19·945
Polet 2	4/12/64	Silent: 4/15/64 Life Expectancy: several years	Initial: not reported Final: 500	Initial: not reported Final: 310	Initial: not reported Final: 58·06	19·895

SUMMARY

To date the Soviets have launched only two vehicles of the Polet series. The first was launched on November 1 1963, the second was placed into orbit on April 12 1964. According to the Soviets, both Polet 1 and Polet 2 were equipped with special guidance equipment and engines to ensure their stabilization and maneuverability in space. Reportedly these vehicles successfully performed multiple maneuvers in different directions changing both the orbital inclinations and other orbital elements. The Soviets also indicated that both vehicles carried scientific instrumentation, although no results of scientific data obtained by this series have been published.

ELECTRON SERIES

Fig. 2.5. Electron 1

Fig. 2.6. Electron 2

PAYLOAD CHARACTERISTICS

Electron 1/2 (Twin Event)

Launch Date: January 30 1964

Weight: < one metric ton

Size: Not reported

Scientific Objectives:
Simultaneous investigation of the radiation belts and associated geophysical phenomena.
Specifically:

1. Inner and outer radiation belts of the earth
2. Low energy charged particles
3. Electron and positive ion density
4. Chemical composition
5. Magnetic field
6. Nuclear component of cosmic radiation
7. Solar shortwave radiation
8. Propagation of radio waves
9. Galactic radio emissions
10. Density of meteoric matter

Electron 3/4 (Twin Event)

Launch Date: July 11 1964

Weight: Not reported

Size: Not reported

Scientific Objectives:
Same as Electron 1/2

TABLE 2.4

Principal Mission Features

Vehicle	Active Life		Orbital Elements			Transmitter Frequencies, mc
	Launch Date	Terminal Date	Apogee, km	Perigee, km	Inclination,°	
Electron 1	1/30/64	Many years	7100	406	61	19·943 19·954 20·005 30·0075 90·0225
Electron 2	1/30/64	Many years	68 200	460	61	Same as Electron 1
Electron 3	7/11/64	Not reported	7040	405	60·52	19·943 19·954 20·005 30·007 90·022
Electron 4	7/11/64	Not reported	66 235	489	60·52	Same as Electron 3

SUMMARY

The first Soviet space launching comprising more than one scientific satellite payload was that of January 30 1964. This system consisted of two heavily instrumented satellites, Electron 1 and Electron 2 (see Figures 2.5 and 2.6), which were placed into substantially different orbits by a single rocket. The purpose of the twin launching was reportedly to undertake simultaneous radiation and related geophysical measurements (using identical equipment) in and on both sides of the radiation belts. These studies were in support of Soviet International Quiet Sun Year (IQSY) activities as well as programs concerned with radiation safety for manned space flight.

The second such system, Electron 3 and 4, launched less than six months later on July 11 1964, had similar orbits, equipment, and objectives. Their launching, however, extended the period of observation and thus provided data relative to the extent of seasonal environmental variations. Because of the comprehensiveness and simultaneity of the Electrons' investigations this series of satellites must be considered as one of the most scientifically significant of the Soviet space program to date. Several Soviet publications are available which extensively describe the various instrumentation schemes on board these vehicles as well as the various scientific results obtained. Specifically, results have been published for studies of the geomagnetic field, spatial and energy distribution of charged particles, soft corpuscular radiation, ion composition, electron density, micrometeorites, and cosmic radio emission.

COSMOS SERIES

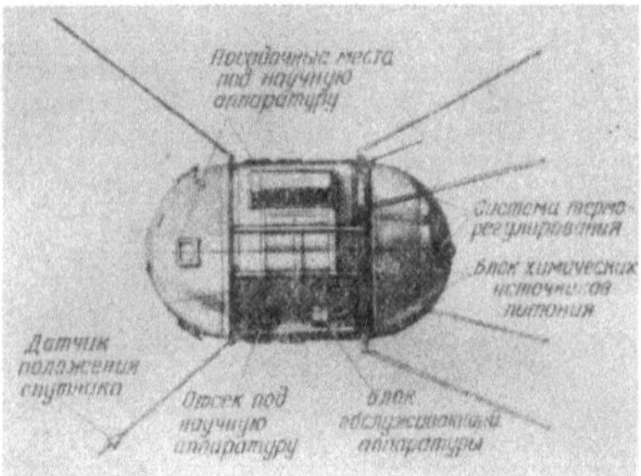

Fig. 2.7. Basic 49° Cosmos Satellite Using Chemical Power Sources

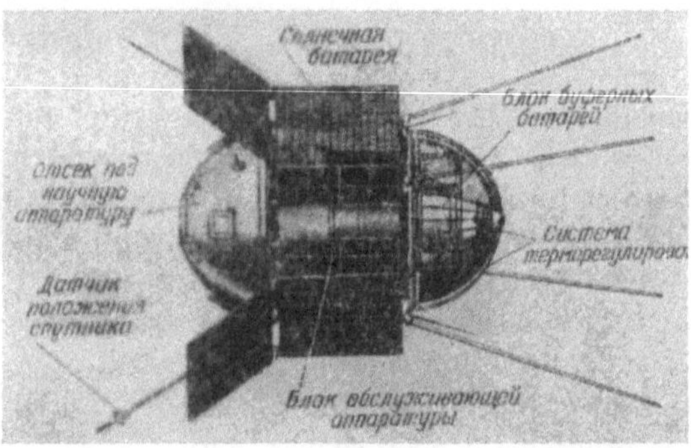

Fig. 2.8. Basic 49° Cosmos Satellite Using Solar Batteries

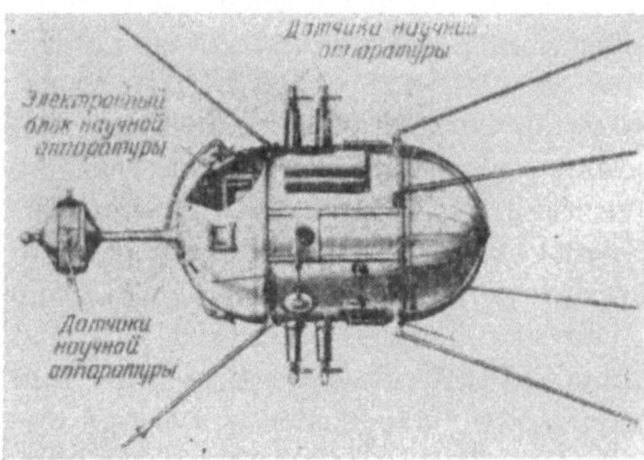

Fig. 2.9. Cosmos Satellite for the Study of the Upper Atmosphere, Solar Radiation, and Micrometeorites

PAYLOAD CHARACTERISTICS

49° Cosmos

Launch Date: 33 satellites launched between March 16 1962 and December 31 1966

Weight: Not reported

Size: Cylindrical with hemispherical top and bottom

Scientific Objectives:

For all Cosmos launchings the Soviets repeat the initial TASS announcement made on March 16 1962 noting that the Cosmos mission included the following scientific investigations:

(1) Charged particle density in connection with radiowave propagation
(2) Corpuscular Streams and low energy particles
(3) Energy spectra of radiation belts
(4) Composition and intensity of primary cosmic rays
(5) Geomagnetic field
(6) UV radiation of the sun and other cosmic bodies
(7) Upper atmosphere
(8) Effect of meteoric material on the structural elements of space vehicles
(9) Distribution and formation of cloud systems.

COSMOS SERIES

Fig. 2.10. Cosmos Satellite for Ionospheric Studies

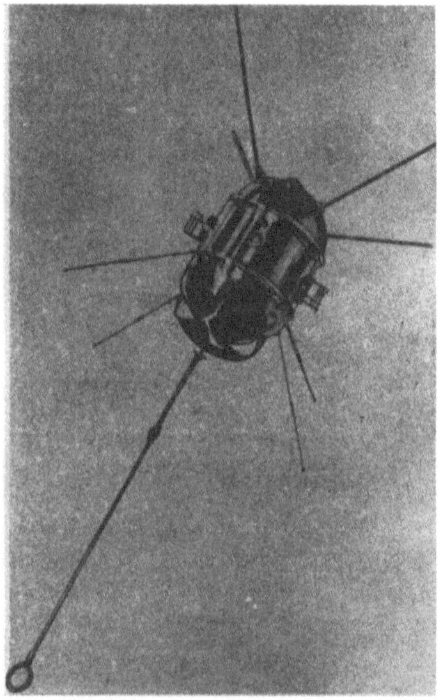

Fig. 2.11. Cosmos Satellite for Geomagnetic Field Studies

Fig. 2.12. Cosmos Satellite for Investigation of Low Energy Particles

COSMOS SERIES

Fig. 2.13. Cosmos 97 for Testing Laser Operation in Space

Fig. 2.14. One Type of 65° Cosmos Satellite (Cosmos 122)—First Soviet Meteorological Satellite

PAYLOAD CHARACTERISTICS

65° Cosmos

Launch Date: 104 satellites through December 31 1966

Weight: Not reported.

Size: Not reported

Scientific Objectives:
Reported to have the same mission as the 49° Cosmos.

TABLE 2.5

Principal Mission Features

Vehicle	Active Life		Orbital Elements			Reported Scientific Studies
	Launch Date	Terminal Date	Apogee, km	Perigee, km	Inclination, °	
Cosmos 1	3/16/62	5/25/62	980	217	49	Electron density, propagation
Cosmos 2	4/6/62	8/20/63	1545·6	211·6	49	Electron density, propagation, ion composition, solar ultraviolet radiation
Cosmos 3	4/24/62	10/17/62	720	229	48·98	Electron density, solar and cosmic radiation, atmospheric density
Cosmos 4	4/26/62	4/29/62*	330	298	65	Cosmic radiation
Cosmos 5	5/28/62	5/2/63	1600	203	49·07	Electron density, solar and cosmic radiation, atmospheric density
Cosmos 6	6/30/62	8/8/62	360	274	49	Cosmic radiation
Cosmos 7	7/28/62	8/1/62	369	210	65	Cosmic radiation, electron density
Cosmos 8	8/18/62	8/17/63	604	256	49	Micrometeorites, atmospheric density, electron density
Cosmos 9	9/27/62	10/1/62	358	301	65	Cosmic radiation
Cosmos 10	10/17/62	10/21/62	380	210	65	Cosmic radiation
Cosmos 11	10/20/62	5/18/64	921	245	49	Ion density, propagation, atmospheric density
Cosmos 12	12/22/62	12/30/62	405	211	65	Cosmic radiation, electron density
Cosmos 13	3/21/63	3/29/63	337	205	64·97	Cosmic radiation
Cosmos 14	4/13/63	8/29/63	512	265	48·95	—
Cosmos 15	4/22/63	4/27/63	371	173	65	Cosmic radiation, electron density
Cosmos 16	4/28/63	5/8/63	401	207	65·02	Cosmic radiation,

* *Reportedly recovered.*

TABLE 2.5 (Continued)
Principal Mission Features

	Active Life		Orbital Elements			
Vehicle	Launch Date	Terminal Date	Apogee, km	Perigee, km	Inclination, °	Reported Scientific Studies
Cosmos 17	5/22/63	6/2/65	788	260	49·03	Cosmic radiation, atmospheric density, ion density, propagation
Cosmos 18	5/24/63	6/2/63	301	209	65·02	Cosmic radiation
Cosmos 19	8/6/63	3/30/64	519	270	49	Cosmic radiation, atmospheric density
Cosmos 20	10/18/63	10/26/63	311	206	65	—
Cosmos 21	11/11/63	11/14/63	229	195	64·83	—
Cosmos 22	11/16/63	11/22/63	394	205	64·94	—
Cosmos 23	12/13/63	3/27/64	613	240	49	—
Cosmos 24	12/19/63	12/28/63	408	211	65	—
Cosmos 25	2/27/64	11/21/64	526	272	49	Cosmic radiation
Cosmos 26	3/18/64	9/28/64	403	271	49	Magnetic field
Cosmos 27	3/27/64	3/28/64	237	192	64·8	—
Cosmos 28	4/4/64	4/12/64	395	209	65	—
Cosmos 29	4/25/64	5/2/64	309	204	65·07	—
Cosmos 30	5/18/64	5/26/64	383·1	206·6	64·94	—
Cosmos 31	6/6/64	10/20/64	508	228	49	—
Cosmos 32	6/10/64	6/18/64	333	209	51·28	—
Cosmos 33	6/23/64	7/1/64	293	209	65	—
Cosmos 34	7/1/64	7/9/64	360	205	64·96	—
Cosmos 35	7/15/64	7/23/64	268	217	51·3	—
Cosmos 36	7/30/64	2/28/65	503	259	49	—
Cosmos 37	8/14/64	8/22/64	300	205	65	—
Cosmos 38	8/18/64	11/8/64	876	210	56·16	—
Cosmos 39	8/18/64	11/17/64	876	210	56·16	—
Cosmos 40	8/18/64	11/18/64	876	210	56·16	—
Cosmos 41	8/22/64	?	39 855	394	64	Cosmic radiation
Cosmos 42	8/22/64	12/19/65	1099	232	49	—
Cosmos 43	8/22/64	12/27/65	1099	232	49	—
Cosmos 44	8/28/64	?	860	618	65	—
Cosmos 45	9/13/64	9/18/64*	327	206	64·9	Terrestrial infrared and ultraviolet radiation
Cosmos 46	9/24/64	10/2/64	271	215	51·3	—
Cosmos 47	10/6/64	10/7/64	413	177	64·77	—
Cosmos 48	10/14/64	10/20/64	295	203	65·07	—
Cosmos 49	10/24/64	8/21/65	490	260	49	Magnetic field
Cosmos 50	10/28/64	11/5/64	241	196	51·3	—
Cosmos 51	12/9/64	11/14/65	554	264	48·8	Luminosity of the stellar background
Cosmos 52	1/11/65	1/19/65	304	205	65	—
Cosmos 53	1/30/65	8/12/66	1192	227	48·8	—
Cosmos 54–56	2/21/65	?	1856	279·7	56·1	—

* *Reportedly recovered.*

TABLE 2.5 (Continued)
Principal Mission Features

Vehicle	Active Life		Orbital Elements			Reported Scientific Studies
	Launch Date	Terminal Date	Apogee, km	Perigee, km	Inclination, °	
Cosmos 57	2/22/65	2/22/65	512	175	64·8	—
Cosmos 58	2/26/65	?	659	581	65	—
Cosmos 59	3/7/65	3/15/65	339	209	65	—
Cosmos 60	3/12/65	3/17/65	287	201	64·7	—
Cosmos 61–63	3/15/65	?	1837	273	56	—
Cosmos 64	3/25/65	4/2/65	271	206	65	—
Cosmos 65	4/17/65	4/25/65*	342	210	65	Terrestrial infrared and ultraviolet radiation
Cosmos 66	5/7/65	5/15/65	291	197	65	—
Cosmos 67	5/25/65	6/2/65	350	207	51·8	—
Cosmos 68	6/15/65	6/23/65	334	205	65	—
Cosmos 69	6/25/65	7/3/65	332	211	65	—
Cosmos 70	7/2/65	?	1154	220	48·8	—
Cosmos 71–75	7/16/65	?	550	550	56·1	—
Cosmos 76	7/23/65	3/16/66	530	261	48·8	—
Cosmos 77	8/3/65	8/11/65	291	200	51·84	—
Cosmos 78	8/14/65	8/22/65	329	206	69	—
Cosmos 79	8/25/65	9/2/65	359	211	64·9	—
Cosmos 80–84	9/3/65	?	≈1500	≈1500	56	—
Cosmos 85	9/9/65	9/17/65	319	212	65	—
Cosmos 86–90	9/18/65	?	1690	1380	56	—
Cosmos 91	9/23/65	10/1/65	342	212	65	—
Cosmos 92	10/16/65	10/24/65	353	212	65	Terrestrial infrared and ultraviolet radiation
Cosmos 93	10/19/65	1/3/66	522	220	48·4	—
Cosmos 94	10/28/65	11/5/65	293	211	65	—
Cosmos 95	11/4/65	1/18/66	521	207	48·4	—
Cosmos 96	11/23/65	12/9/65	310	227	51·9	—
Cosmos 97	11/26/65	?	2100	220	49	Laser experiment
Cosmos 98	11/27/65	12/5/65	570	216	65	—
Cosmos 99	12/10/65	12/18/65	320	199	65	—
Cosmos 100	12/17/65	?	650	650	65	—
Cosmos 101	12/21/65	7/12/66	550	260	49	—
Cosmos 102	12/28/65	1/13/66	278	218	65	—
Cosmos 103	12/28/65	?	600	600	56	—
Cosmos 104	1/7/66	1/15/66	401	204	65	—
Cosmos 105	1/22/66	1/30/66	324	204	65	—
Cosmos 106	1/25/66	?	564	290	48·4	—
Cosmos 107	2/10/66	2/18/66	322	204	65	—
Cosmos 108	2/11/66	?	865	227	48·9	—

* *Reportedly recovered.*

Table 2.5 (Continued)
Principal Mission Features

Vehicle	Active Life		Orbital Elements			Reported Scientific Studies
	Launch Date	Terminal Date	Apogee, km	Perigee, km	Inclination, °	
Cosmos 109	2/19/66	2/27/66	309	209	65	—
Cosmos 110	2/22/66	3/16/66*	904	187	51·9	Medicobiological experiment with two dogs
Cosmos 111	3/1/66	3/3/66	226	191	51·8	—
Cosmos 112	3/17/66	3/25/66	565	214	72	—
Cosmos 113	3/21/66	3/29/66	327	210	65	—
Cosmos 114	4/6/66	4/14/66	374	210	73	—
Cosmos 115	4/20/66	4/28/66	294	190	65	—
Cosmos 116	4/26/66	?	478	294	48·4	—
Cosmos 117	5/6/66	5/14/66	308	207	65	—
Cosmos 118	5/11/66	?	640	640	65	—
Cosmos 119	5/24/66	?	1305	219	48·5	—
Cosmos 120	6/8/66	?	300	200	51·8	—
Cosmos 121	6/17/66	?	354	210	72·9	—
Cosmos 122	6/25/66	?	625	625	65	Meteorological satellite
Cosmos 123	7/8/66	?	529	263	48·8	—
Cosmos 124	7/14/66	8/22/66	303	208	51·8	—
Cosmos 125	7/20/66	?	250	250	65	—
Cosmos 126	7/28/66	?	359	212	51·8	—
Cosmos 127	8/8/66	8/16/66	279	204	51·9	—
Cosmos 128	8/27/66	9/4/66	364	212	65	—
Cosmos 129	10/14/66	?	307	202	65	—
Cosmos 130	10/20/66	10/28/66	340	211	65	—
Cosmos 131	11/12/66	11/20/66	360	205	72·9	—
Cosmos 132	11/19/66	?	280	207	65	—
Cosmos 133	11/28/66	?	232	181	51·9	—
Cosmos 134	12/3/66	?	319	214	65	—
Cosmos 135	12/12/66	?	662	259	48·5	—
Cosmos 136	12/19/66	?	305	198	64·6	—
Cosmos 137	12/21/66	?	1720	230	48·8	—

* *Reportedly recovered.*

SUMMARY

The Cosmos satellite series initiated on March 16 1962 represents the largest Soviet satellite series implemented to date. Since the initial launching an impressive total of 137 satellites have been launched through 1966 with no indications of program termination plans. Based upon the sparse amount of information available concerning this series, it appears that this activity comprises two discernible programs. One includes the launching of smaller, long lifetime payloads into 49°-

inclination orbits for almost routine scientific investigations in the near-earth space environment, principally in or above the ionospheric region. Over 30 vehicles have been launched in this category through 1966. The other program consists of larger short-lived and/or recoverable satellites launched primarily into 65°-inclination orbits considered to have scientific, biomedical, and technical objectives. One hundred and four satellites in this group have been launched since the initiation of the Cosmos program in 1962. Several of these, however, are multiple satellite launchings.

Descriptive information relative to the payload characteristics of the numerous Cosmos satellites is almost nonexistent. Photographs and sketches of Cosmos vehicles available to date are shown in Figures 2.7 to 2.14. With the exception of Cosmos 122 (Figure 2.14) (first Soviet meteorological satellite) all payloads pictured are believed to be of the smaller series.

Following the successful launching of Cosmos 1 on March 16 1962, the Soviets released the nine-item scientific mission noted earlier and have repeated this announcement following almost every subsequent Cosmos launching. Exceptions have included references to laser studies aboard Cosmos 97, biomedical (dog) studies aboard Cosmos 110, and more recently Cosmos 122 meteorological satellite accomplishments.

Compared to the number and frequency of Soviet launchings of this series of satellites it is surprising that so little published information is available. Most of that available relates to studies aboard the 49° Cosmos vehicles. This includes, for example, results of radio propagation and related ion density studies aboard Cosmos 1, 2, 3, 5, 11, 12, 15, and 17, cosmic radiation results from Cosmos 3, 5, 6, 17, and 19, magnetic field data from Cosmos 26 and 49, and reference to micrometeorite studies on Cosmos 8 and stellar ultraviolet measurements on Cosmos 51. Also, the Soviets have published some indirect determinations of atmospheric density using orbital decay data from Cosmos 1, 3, 5, 8, 11, 17, and 19. For 65° Cosmos satellites, published results have been limited to descriptions of cosmic radiation equipment aboard and some results obtained from Cosmos 4, 7, 9, 10, 12, 13, 15, 16, 18, and 41. More recently, results and description of instrumentation associated with visible, ultraviolet, and infrared atmospheric and terrestrial radiation studies aboard Cosmos 45, 65, and 92 have been described. Likewise, the Soviets have released some data, particularly visible and infrared cloud cover photographs from Cosmos 122 and biomedical results from the Cosmos 110 flight.

PROTON SERIES

Fig. 2.15. Schematic View of Proton Satellite
1. Solar panels; 2. Attitude sensors
3. Hermetically sealed casing; 4/5/6. External surface and antennas
7/8. Scientific, electrical, and radio equipment
9. External heat exchanger

PAYLOAD CHARACTERISTICS

Proton 1

Launch Date: July 16 1965

Weight: 12·2 metric tons Size: Not reported

Scientific Objective:
To study ultrahigh-energy cosmic radiation

Proton 2

Launch Date: November 2 1965

Weight: 12·2 metric tons Size: Not reported

Scientific Objective:
To study ultrahigh-energy cosmic radiation

PROTON SERIES

PAYLOAD CHARACTERISTICS

Proton 3

Launch Date: July 6 1966

Weight: > 12 metric tons Size: Not reported

Scientific Objective:
To study ultrahigh-energy cosmic radiation

Fig. 2.16. Proton Scientific Apparatus

1. Electron spectrometer (SEZ–12)
2. Ionization calorimeter; 3. Gamma telescope (GG–1)
4. Low-energy particle charge spectrometer (SEZ–1)

TABLE 2.6
Principal Mission Features

Vehicle	Active Life		Orbital Elements			Transmitter Frequencies, mc
	Launch Date	Terminal Date	Apogee, km	Perigee, km	Inclination, °	
Proton 1	7/16/65	Decayed: 10/11/65	627	190	63·5	19·910
Proton 2	11/2/65	Decayed: 1/2/66	637	191	63·5	19·545
Proton 3	7/6/66	Decayed: 9/16/66	630	190	63·5	19·545

SUMMARY

The introduction of the over 12 metric ton payload Proton series of satellites in 1965 provided the Soviets with unprecedented opportunities for both space exploration and exploitation. The three satellites of this series launched to date have been quite similar (see Figure 2.15) and have concentrated on the study of cosmic ray particles of high and ultra-high-energies of both scientific and manned space flight interests. For these studies the Soviets installed the complex array of scientific apparatus shown in Figure 2.16. The main portion of this apparatus was the ionization calorimeter designed for investigating the energy spectrum and chemical composition of primary cosmic rays in the energy spectrum up to 10^{14} eV. The instrument also permitted the investigation of the nuclear interaction of cosmic particles of ultrahigh-energies up to 10^{12} eV. The electron spectrometer (SEZ–12) attached to the bottom of the configuration (shown in Figure 2.16) measured high-energy electrons whereas the spectrometer (SEZ–1) on top investigated cosmic rays with energies below 30 BeV including those occasionally emitted by the sun. This instrument's sensitivity was sufficient to record small cosmic ray variations associated with solar activity. The gamma telescope (GG–1) also shown in Figure 2.16 recorded high-energy electromagnetic radiation (gamma-quanta) in a wavelength range millions of times shorter than visible light. These instruments aboard the Proton satellite provided data relating to the degree of uniformity of cosmic rays and interstellar matter distribution in the galaxy. The few Soviet papers available to date discussing the Proton satellites contain mostly only preliminary analysis of Proton 1 and Proton 2 data. Reportedly, Proton 3 undertook the same studies but also carried additional equipment for searching for fundamental elementary particles, the so-called "quarks" in primary cosmic rays.

Most significant among the Proton results reported to date were that (1) the proportion of heavy nuclei of primary cosmic rays as compared to protons in the energy range 10^{10}–10^{12} eV remained constant, (2) the intensity of high and ultrahigh energy particles measured directly by the Proton satellites was lower than that determined from ground-based observations, (3) the measurement of high-energy electrons aboard the Proton satellites have, however, revealed an unexpectedly large flux of electrons with energies over 300 MeV. Measured intensity values were almost 10 times greater than expectations based upon earlier stratospheric observations.

MOLNIYA SERIES

Fig. 2.17. Axial View of Molniya 1 Vehicle

Fig. 2.18. Side View of Molniya 1 Vehicle

Fig. 2.19. View Showing Relative Size of Molniya 1 Vehicle

PAYLOAD CHARACTERISTICS

The Molniya 1 Events

Launch Dates:

First Event:
April 23 1965

Second Event:
October 14 1965

Third Event:
April 25 1966

Fourth Event:
October 20 1966

Weight: Not reported

Size: Not reported

Scientific Objectives:

Experimental long-distance two-way multichannel telecommunication, experimental long-distance television transmission, communications satellite system component improvement, data accumulation for use in establishing a commercial communications system utilizing artificial earth satellites, and from the third and fourth Molniya 1 vehicles experimental televising from near-apogee altitudes of observations of the earth's surface for meteorological or other purposes.

TABLE 2.7

Principal Mission Features

Vehicle	Launch Date	Orbital Elements: Apogee, km	Perigee, km	Inclination, °	Period
First Molniya 1	4/23/65				
Initially		39 380	497	65	$11^h 48^m$
After adjustment		39 957	548	65	12^h
Second Molniya 1	10/14/65	40 000	500	65	$11^h 59^m$
Third Molniya 1	4/25/66	39 500	499	64·5	$11^h 50^m$
Fourth Molniya 1	10/20/66	39 700	485	64·9	$11^h 53^m$

NOTE: Expected lifetimes for these vehicles range from 6½ to 14 years. No report of the operating frequencies has been noted.

SUMMARY

The Soviets have launched four experimental communications satellites, all designated Molniya 1 (pronounced "Mol'nee–ah", meaning lightning). All have been launched into essentially the same highly elliptical orbit, the apogees being in the Northern Hemisphere. With their approximate 12–hour period, any of the vehicles can relay communications between points in the Soviet Union during the greater part of every other orbit traverse. During the corresponding part of the intervening traverse, communications can be relayed between the European part of the USSR and North America. The vehicles apparently are essentially identical, except that each of the latter two has been equipped with a television camera. Relay is accomplished by a 40-watt wideband transceiver, having a capacity for one television channel, or the equivalent in multichannel telecommunication.

The first Molniya 1, launched April 23 1965, was maneuvered into a slightly different orbit on May 2 1965 to achieve a more prolonged relay interval over the Soviet Union. The initial television relay was of a three-hour program on the day of launch, from Vladivostok to Moscow. The first series of telephone transmissions between Moscow and Vladivostok via the satellite occurred May 7 1965. A second Molniya 1 was launched October 14 1965, with an announced purpose of permitting experimentation with a multivehicle system. On November 30 1965 an experimental Moscow-to-Paris color television transmission via the Molniya 1 system was conducted. The third Molniya 1 was launched April 25 1966, and on May 18 1966 the Soviets reported its first transmission of television pictures of the earth, taken from heights between 30 000 and 40 000 km. On May 28 1966 a Paris-to-Moscow experimental color television transmission was reported, the first of a series. On October 20 1966 the fourth Molniya 1, also equipped with a television camera, was launched. The Soviets apparently are continuing to experiment with the Molniya 1 system.

VOSTOK SERIES

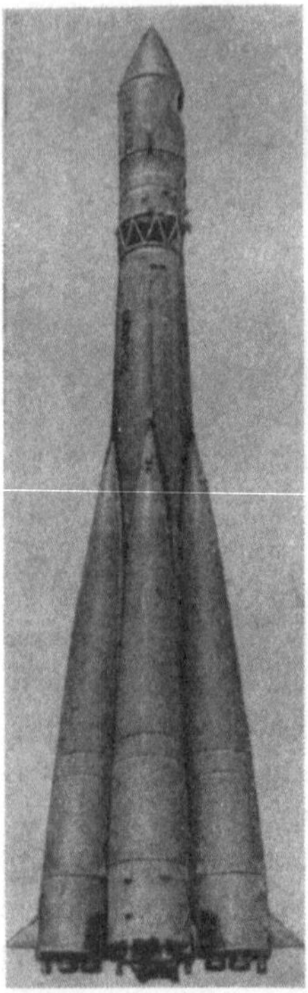

Fig. 2.20. Vostok: Launch Configuration

PAYLOAD CHARACTERISTICS

Vostoks 1–6

Weight:
- Capsule: 4·73 metric tons
- Orbit payload: 6·17 metric tons

Size:
- Capsule: Sphere: 2·3-m diameter
- Orbit payload:
  - Cylinder: 7·36 m, length

Scientific Objectives
- Primary: Manned space flight investigations; test and development of ship systems
- Secondary: Manned geophysical and astronomical studies

Fig. 2.21. Vostok: Orbital Configuration

Fig. 2.22. Vostok: Reentry Sphere

VOSTOK SERIES

Fig. 2.23. Vostok Pilot's Couch and Ejection Seat

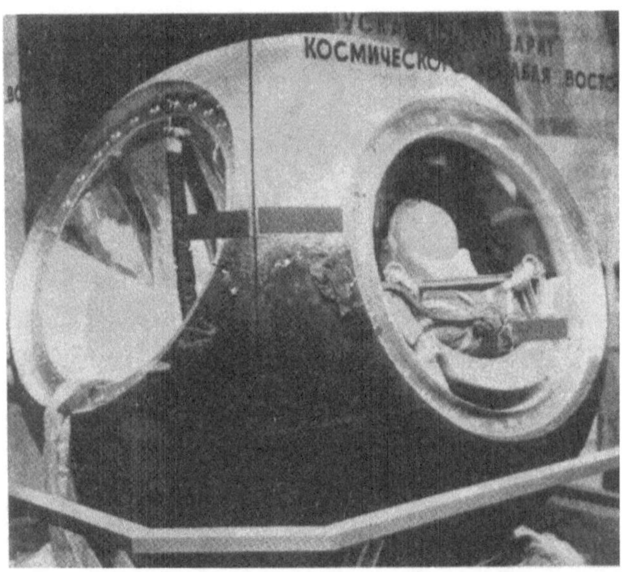

Fig. 2.24. Vostok Cabin and Ejection Seat

TABLE 2.8

Principal Mission Features

Vehicle	Active Life, Moscow Time	Initial Orbital Data: Apogee, km	Perigee, km	Inclination, degrees	Transmitter Frequencies, mc	Comments
Vostok 1	0907^h 12 Apr. 1961– 1055^h 12 Apr. 1961 (1^h 48^m)	302	175	~65	19·103 AM 20·006 AM 143·625 FM 19·995	Pilot: Yu. A. Gagarin; Call Sign: Kedr; landed near Smelkova in Saratovskaya Oblast; First manned space flight; on board: Drosophila, dry seeds, lysogenic bacteria.
Vostok 2	0900^h 6 Aug. 1961– 1018^h 7 Aug. 1961 (25^h 18^m)	244	183	~55	15·765 AM 20·006 AM 143·625 FM 19·995	Pilot: G. S. Titov; Call Sign: Orel; landed near Krasnyy Kut hamlet in Saratovskaya Oblast; on board: Drosophila, dry seeds, lysogenic bacteria.
Vostok 3	1130^h 11 Aug. 1962– 0955^h 15 Aug. 1962 (94^h 25^m)	251	183	~65	20·006 AM 143·625 FM 19·995	Pilot: A. G. Nikolayev; Call Sign: Sokol; landed at 48°02′N–75°45′E; on board: Drosophila, dry seeds, lysogenic bacteria, microspores.
Vostok 4	1102^h 12 Aug. 1962– 1001^h 15 Aug. 1962 (70^h 59^m)	254	180	~65	20·006 AM 143·625 FM 19·990	Pilot: P. R. Popovich; Call Sign: Berkut; landed at 48°10′N–71°51′E; with Vostok 3 constituted the world's first group flight; on board: Drosophila, dry seeds, lysogenic bacteria, microspores, human cancer cells.
Vostok 5	1500^h 14 June 1963– 1406^h 19 June 1963 (119^h 6^m)	235	181	~65	20·006 AM 143·625 FM 19·948	Pilot: V. F. Bykovskiy; Call Sign: Yastreb; landed 540 km north west of the city of Karaganda at 53°N latitude; on board: Drosophila, dry seeds, lysogenic bacteria, microspores, Chlorella.
Vostok 6	1230^h 16 June 1963– 1120^h 19 June 1963 (70^h 50^m)	233	183	~65	20·006 AM 143·625 FM 19·995	Pilot: V. V. Tereshkova; Call Sign: Chayka; landed 620 km northeast of city of Karaganda at 53°N latitude; on board: Drosophila, dry seeds, lysogenic bacteria, microspores, Chlorella, human cancer cells.

SUMMARY

The Vostok series represented a number of firsts for the Soviets. Manned space flight began with the successful one-orbit flight of Yuriy Gagarin in Vostok 1 on April 12 1961. The first group space flight, with two ships in close proximity to each other, was achieved with the orbiting of Vostok 3 with Andriyan Nikolayev on August 11 1962, followed by the insertion of Vostok 4 with Pavel Popovich into an almost identical orbit on August 12 1962. The flight of Vostok 6 on June 16 1963 represented the first venture into space by a woman cosmonaut. Her partner in that group flight, Valeriy Bykovskiy in Vostok 5, set an endurance record for space flight of 119 hours, which lasted until well into the U.S. Gemini series.

Throughout the series, Soviet concern for the neurogenic and vestibular effects of weightlessness were evident. In all flights the reactions of the cosmonauts were monitored by two television cameras, one transmitting a full-face view and the other a view of the face in profile. Commencing with Vostok 3, electroencephalogram, electro-oculogram, and galvanic skin reactions were recorded and telemetered throughout the flights. Extensive vestibular studies were conducted on all flights, particularly with the long duration flights starting with Vostok 2. In spite of this concern with vestibular reactions, Titov and Tereshkova were seriously disoriented during their flights, and Popovich experienced milder symptoms.

Most of the earlier publications relating to the Vostok flights presented rather skimpy details of psychophysiological reactions, training techniques and postflight evaluations. Other than references to the weight of the Vostok, few details were released as to the Vostok spacecraft or life-support systems until a model of the Vostok was displayed at the Soviet Economic Achievements Exhibition in Moscow in April 1965. As illustrated in Figure 2.21, the Vostok ship consists of a 4·73 metric ton spherical capsule 2·3 meters in diameter. The overall length of the orbiting payload is 7·35 meters with a total weight of 6·17 metric tons.

The cosmonauts on the Vostoks had a choice of landing techniques. They could elect to remain in the cabin which descended by parachute, or catapult from the ship at an altitude of seven km, separate from the ejection seat, and descend by personal parachute. Gagarin landed in the capsule of Vostok 1, but apparently all the other Vostok cosmonauts ejected from the capsule and landed apart from the ship. A model of the ejection seat is also shown in Figure 2.23.

In addition to the principal mission of manned space flight investigations, the Vostok series involved numerous other experiments. The cosmonauts' flight program included astronomical and geophysical studies such as observations of constellations, photographs of the sun and the disc of the earth both at daybreak and sunset, the visual and optical observations of the earth's surface. Meteorological and radiation studies of a photographic nature were also conducted. All flights also involved studies of a histological, cytological, radiobiological, or genetic nature. For example, the payloads of Vostoks 3 through 6 included specimens of Drosophila, seeds of higher plants, lysogenic bacteria, microspores, and human cancer cells. Seeds, Drosophila, and lysogenic bacteria were also on board Vostoks 1 and 2.

VOSKHOD SERIES

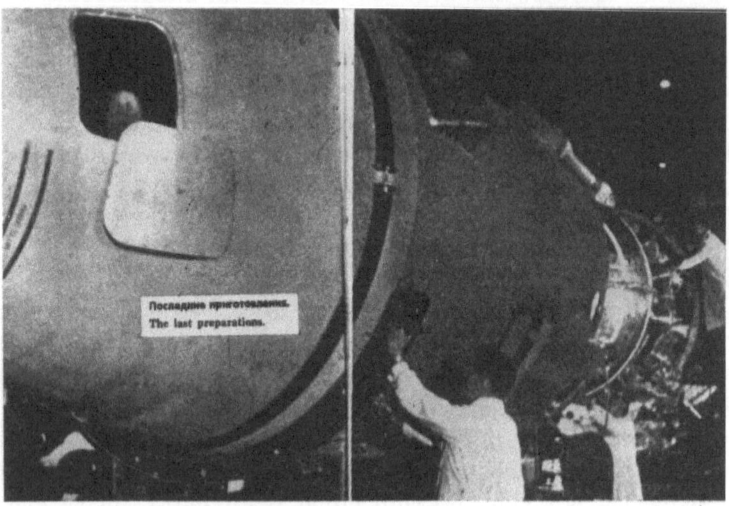

Fig. 2.25. Assembly of Voskhod Spacecraft

VOSKHOD SERIES

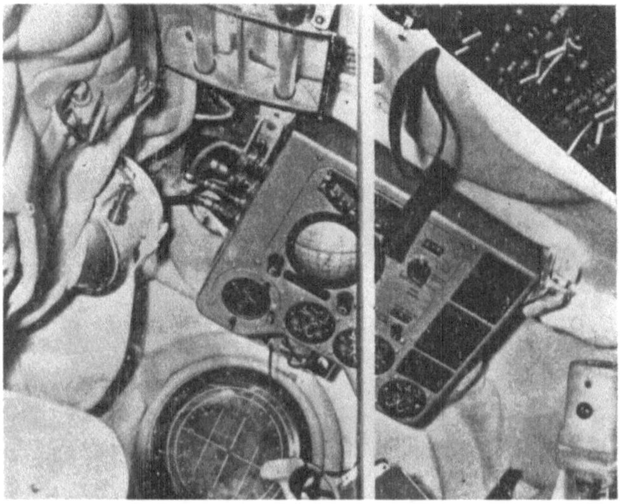

Fig. 2.26. Voskhod Cabin Interior and Instrument Panel

PAYLOAD CHARACTERISTICS

Voskhods I and 2

Launch Date:
- Voskhod 1: October 12 1964
- Voskhod 2: March 18 1965

Weight:
- Voskhod 1: 5320 kg
- Voskhod 2: 5682 kg

Size: Not reported

Scientific Objectives

Medical-Biological: Test new multiseat manned spacecraft, including soft landing; conduct broader medical and biological experiments under space flight conditions; study work capacity and functions of group of cosmonauts; evaluate comfort and reliability of space suit and life-support system and ability of man to work in space outside of spacecraft.

Geophysical: Visual, photometric, and photographic observations of the earth and its atmosphere, polar aurorae, luminescent particles, and earth's horizon and stellar backgrounds.

TABLE 2.9

Principal Mission Features

Vehicle	Active Life, Moscow Time	Initial Orbital Data			Transmitter Frequencies, mc	Comments
		Apogee, km	Perigee, km	Inclination, degrees		
Voskhod 1	1030^{h} 12 Oct. 1964– 1047^{h} 13 Oct. 1964 (24^{h} 17^{m})	409	178	~65	143·625 17·365 18·035 19·974 19·994 19·995	Crew: V. Komarov (pilot), K. Feoktistov (scientist) and B. Yegorov (physician). Call Sign: Rubin. Landed 563 km north of Baikonur, Kazakhstan. First multi-crew flight.
Voskhod 2	1000^{h} 18 Mar. 1965– 1202^{h} 19 Mar. 1965 (26^{h} 2^{m})	495	173	65	143·625 17·365 18·035 19·996 19·994 19·995	Crew/Pilots: P. I. Belyayev and A. A. Leonov; Call Sign: Almaz; landed near Perm, snow-covered wooded area in the foothills of the Urals. First egress into space by A. A. Leonov and 10-minute extravehicular activity.

SUMMARY

The Voskhod 1 flight, with its three-man crew (two of whom were nonpilots), the shirt-sleeve environment with no spacesuits, and the soft landing, represented a new plateau for Soviet space sciences. From the biomedical viewpoint the presence of the physician, Dr. Yegorov, on board was especially significant, for direct observation of physiological reactions, including self-observation, was accomplished. The biomedical program was aimed at a detailed study of the reactions of the central nervous system and the cardiovascular system to space flight stresses. Blood morphology was also studied on samples taken on board. Dr. Yegorov had a system of detachable biosensors available to him to supplement the telemetered data from electro-oculograms, ECG, seismocardiograms, pulse, respiration, and psychomotor performance measurements. Electroencephalograms were obtained directly by Dr. Yegorov as well as telemetered to the ground. Visual acuity of the cosmonauts was also evaluated by noting their ability to identify both natural and man-made objects on the earth's surface.

Numerous geophysical and astronomical observations were conducted by the scientist-cosmonaut K. P. Feoktistov. Results of visual, photometric, and photographic observations of (1) the earth and its atmosphere, (2) polar aurora and luminescent particles, and (3) azimuthal (horizon) and stellar backgrounds for navigation and orientation purposes have been published by the Soviets.

The Soviets claimed that a new orientation system, termed an ion velocity vector plotter was successfully tested on the Voskhod 1 flight. However, no details as to the principle employed or the functioning of this device have been published to date.

The highlight of the Voskhod 2 event was, of course, the egress of Lt. Col. A. Leonov into space and his 10-minute extravehicular activity. Another feature of the flight was the unscheduled manual reentry, some one- and one-half orbits later than originally programmed. This resulted from a failure of the solar-orientation system used in automatic reentry. A new television system, employing a 625 line, 25 frames-per-second scan rate was used on this mission to transmit pictures of Leonov's egress and extravehicular activities.

Published information regarding this mission has been concerned primarily with the physiological reactions of the cosmonauts, the difficulties of working in space in a suit pressurized to 0·4 atmosphere, and the relative advantages of an air-lock system for egress as compared to complete depressurization of the capsule.

Chapter 3

Lunar and Planetary Probes

G. E. Wukelic

The Soviets embarked on an ambitious lunar and planetary rocket exploration program shortly after the inauguration of their artificial earth satellite program. For example, the Soviets announced the successful launching of their first lunar probe on January 2 1959 a little over one year after orbiting Sputnik 1. Their first announced planetary probe occurred some two years later with the launching of a 643-kg payload into a Venus trajectory on February 12 1961.*As of December 1966, the Soviets have announced the successful launching of 13 lunar probes and seven planetary probes having Mars or Venus missions.

The purpose of this chapter is to indicate, based upon a review of the numerous technical and popular Soviet items available, the nature, objectives, and degree of success of each of the announced events. Specifically, available information regarding the payload characteristics (i.e., weight, size, objective(s), and photographic illustrations when available) and principal mission features (i.e., closest approach to moon or planets, operating lifetime, and degree of success) associated with each event of the Luna, Mars, Venera, and Zond series are noted. Also, general summaries of the highlights and significance of these series are provided. For the discussion of the scientific results associated with each of these events the reader is referred to the various technical chapters contained in Part II.

* The so-called "heavy satellite" launching on February 4 1961 was considered by some Western experts to have been the first, but completely unsuccessful, Soviet Venus probe attempt.

LUNA SERIES

Fig. 3.1. Luna 1

PAYLOAD CHARACTERISTICS

Initial Luna Series

Luna 1

Launch Date: January 2 1959 Weight: 361·3 kg Size: Sphere: 120-cm diameter

Scientific Objectives:

Measurement of magnetic field, solar and cosmic radiation, micrometeorites, and composition and density of interplanetary ionized gas in cislunar space. Also, artificial sodium cloud experiment conducted for tracking purposes.

LUNA SERIES

Fig. 3.2. Luna 2

PAYLOAD CHARACTERISTICS

Initial Luna Series

Luna 2

Launch Date: September 12 1959 Weight: 390·2 kg Size: Sphere: 120-cm diameter

Scientific Objectives:

Measurement of magnetic field, solar and cosmic radiation, micrometeorites, and composition and density of interplanetary ionized gas in cislunar space. Also, artificial sodium cloud experiment conducted for tracking purposes. In addition, Luna 2 employed an improved magnetometer and a lunar radar altimeter.

LUNA SERIES

Fig. 3.3. Luna 3

PAYLOAD CHARACTERISTICS

Initial Luna Series

Luna 3

Launch Date: October 4 1959

Weight: 278·5 kg

Size: Cylindrical
  Length: 130 cm
  Width: 120 cm

Scientific Objectives:
  Primary: Photography of the back side of the moon
  Secondary: Measurements of solar and cosmic radiation, micrometeorites, and ion density studies to and near the moon

PAYLOAD CHARACTERISTICS

Soft-Landing Series

Lunas 4–9

Launch Date: 6 launchings between April 2 1963 and January 31 1966. Only Luna 9 successfully soft-landed

Weight: ≈100 kg

Size: Almost spherical, ≈ 58 cm in diameter (excluding antennas)

Scientific Objectives:
  Primary: To soft-land and photograph lunar surface
  Secondary: Radiation measurements near and on the lunar surface

Fig. 3.4. Entire Luna 9 Spacecraft

LUNA SERIES

Fig. 3.5. Soft-Landed Luna 9 Payload

1. The lobe antennas
2. The whip antennas
3. The instrument extension mechanisms
4. The ground-testing device
5. The radiation density meter
6. The TV camera

Fig. 3.6. Soft-Landed Luna 13 Payload

PAYLOAD CHARACTERISTICS
Soft-Landing Series

Luna 13

Launch Date: December 21 1966

Weight: Not reported. Similar to Luna 9 except contained additional surface-testing instruments. Therefore > 100 kg.

Size: Almost spherical like Luna 9, therefore ≈ 58 cm in diameter (excluding antennas)

Scientific Objectives:

Primary: To soft-land and photograph lunar surface.

Secondary: Radiation near and on lunar surface, physical properties of lunar surface

LUNA SERIES

Fig. 3.7. Drawing of Luna 10 Spacecraft

PAYLOAD CHARACTERISTICS

Lunar Orbiter Series

Luna 10 and Luna 11

Launch Date: March 31 1966 and August 24 1966

Weight: Orbited payload 245 kg

Size: Not reported

Scientific Objectives:

Direct: To measure cosmic and lunar-surface gamma radiation, micrometeorites, magnetic field, infrared radiation, positive ion density, and X-ray and fluorescent lunar surface radiation from lunar-orbiting satellite

Indirect: Determination of shape, mass, and gravitation field of the moon from orbital measurements

1. The equipment of the radiomeasurement system
2. Artificial moon satellite
3. The system for separation of the artificial moon satellite
4. Equipment of the astro-orientation system
5. Engine installation

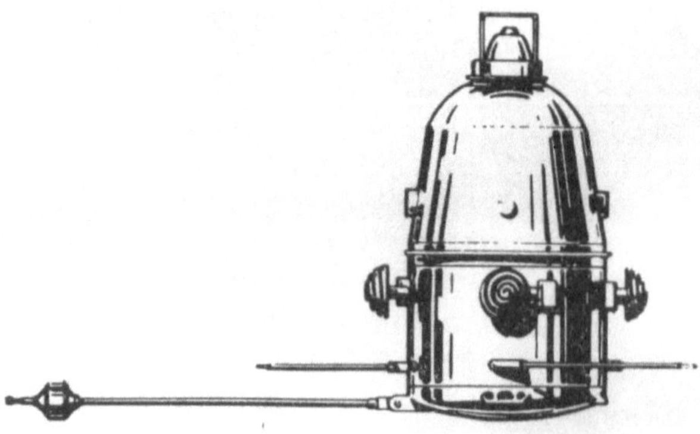

Fig. 3.8. Drawing of Luna 10 Orbiting Payload

Outside Capsule: Antennas; magnetometer (on 1·5-meter extensible boom to remove it from fields due to capsule); meteor particle traps; heat sensor unit; solar plasma instrument

Inside Capsule: Instrument electronics, radiometer for low- and high-energy particle measurements, and gamma spectrometer for detecting moon's radiation

LUNA SERIES

Fig. 3.9. Luna 12 Spacecraft

1. Gas containers for astro-orientation system; 2. Photo-TV equipment; 3. Radiator of temperature control system; 4. Radiometer; 5. Instrument compartment; 6. Chemical batteries; 7. Optical-mechanical unit of astro-orientation system; 8. Antennas; 9. Electronic unit of astro-orientation system; 10. Control engines; 11. Retrorocket system.

PAYLOAD CHARACTERISTICS

Lunar Orbiter Series

Luna 12

Launch Date: October 22 1966 Weight: Not reported Size: Not reported

Scientific Objectives:

Primary: Obtain photographs of lunar surface from a lunar orbit.

Secondary:

Direct: Gamma, X-ray, and corpuscular radiation, micrometeorite, and radio astronomical studies

Indirect: Determination of moon's gravitational field from satellite orbital data

TABLE 3.1

Principal Mission Features

Vehicle/ Launch Date	Total Spacecraft Weight, kg	Closest Approach to the Moon	Comments
		Initial Luna Series	
Luna 1 (Mechta)/ Jan. 2 1959	1472	Came within 5000–6000 km on January 4 1959 (34 hours after launch)	First, at least partially successful, lunar probe to undertake direct scientific measurements in cislunar space. Went into heliocentric orbit on January 7/8 1959. Radio signals ceased on January 5 1959.
Luna 2/ Sep. 12 1959	1511	Impacted lunar surface on September 14 1959 at $00^h\ 02^m\ 24^s$ Moscow time.	More successful lunar mission to undertake direct scientific measurements in cislunar space became silent upon impact.
Luna 3/ Oct. 4 1959	1553	Circumnavigated the moon and came within 7900 km of moon's center on October 6 1959.	Successfully photographed much of the back side of the moon invisible from earth and made scientific measurements en route. Stopped transmitting on November 15 1959, went into orbit around the earth and decayed in April 1960.
		Soft-Landing Series	
Luna 4/ Apr. 2 1963	1422	Came within 8500 km of lunar surface on April 6 1963.	Unsuccessful attempt to soft land instrument package on moon went into heliocentric orbit.
Luna 5/ May 9 1965	1476	Impacted lunar surface on May 12 1965 at 2210 hours.	Unsuccessful lunar soft-landing attempt became silent upon impact.
Luna 6/ June 8 1965	1442	—	Unsuccessful lunar soft-landing attempt failed during mid-course correction; went into heliocentric orbit.
Luna 7/ Oct. 4 1965	1506	Impacted lunar surface on October 8 1965 at 0108 hours.	Unsuccessful lunar soft-landing attempt became silent upon impact.
Luna 8/ Dec. 3 1965	1552	Impacted lunar surface on December 4 1965 at $00^h\ 51^m\ 30^s$.	Somewhat more successful soft lunar landing attempt but became silent upon impact.

Vehicle/ Launch Date	Total Spacecraft Weight, kg	Closest Approach to the Moon	Comments
		Soft-Landing Series (Continued)	
Luna 9/ Jan. 31 1966	1583	Soft landed on lunar surface on February 3 1966 at $21^h 45^m 30^s$.	Achieved first successful soft lunar landing; transmitted photographs of lunar surface until February 6 1966 when power supply was depleted.
Luna 13/ Dec. 21 1966		Soft landed on lunar surface on December 24 1966 at 2101 Moscow time.	Second successful soft lunar landing and transmitted photographs and measurements of the physical characteristics of lunar surface. Had reportedly stopped transmitting by December 31 1966.
		Lunar Orbiter Series	
Luna 10/ Mar. 31 1966	1600	Went into orbit around the moon on April 3 1966 with the following orbital parameters: Period: $2^h 58^m 15^s$ Aposelene: 1017 km Periselene: 350 km Inclination: 71°54″	First artificial satellite of the moon. Transmitted lunar environmental data (no photography). Power supply exhausted May 30 1966. Will probably continue to orbit for several years.
Luna 11/ Aug. 24 1966	1640	Went into orbit around the moon on August 28 1966 with the following orbital parameters: Period: $2^h 58^s$ Aposelene: 1200 km Periselene: 160 km Inclination: 27°	Second successful lunar orbiter. Transmitted some lunar environmental data (no photography). Power supply exhausted October 1 1966.
Luna 12/ Oct. 22 1966		Went into orbit around the moon on October 25 1966 with the following orbital parameters: Period: $3^h 25^m$ Aposelene: 1740 km Periselene: 100 km Inclination: nearly equatorial	Third successful artificial satellite orbiting the moon. Transmitted varying amounts of lunar environmental data (radiation and micrometeorites) including the first Soviet photography of lunar surface from a lunar orbiting satellite. Radio contact terminated January 19 1967.

SUMMARY

The Luna Series, of which the Soviets have announced 13 events through 1966, began with Luna 1 launched January 2 1959. The first three vehicles in the series were launched by the direct-ascent technique. Luna 1 bypassed the moon by 5000–6000 km during an unspecified mission; Luna 2 impacted on the moon on September 14 1959; and Luna 3 was successfully launched on October 4 1959 with the announced objective of photographing the hidden side of the moon. Published photographs confirm that this goal was achieved.

It was not until over three years later that the Soviets resumed their lunar exploration program by launching Luna 4 on April 2 1963 employing for the first time the earth-parking orbit launch technique. Although the specific mission of Luna 4 was never stated, a soft landing of an instrument payload on the moon probably was intended. Luna 4 passed within 8500 km of the moon. Lunas 5, 7, and 8, which were also most likely soft-landing attempts, actually hit the moon and were destroyed on impact. Luna 6 was even less successful, going far afield when the midcourse-correction motor failed to shut off. It was Luna 9, launched on January 31 1966 that achieved the world's first successful soft landing, and transmitted the first photographs directly from the moon's surface. The probe transmitted photofacsimiles and other lunar environmental data for about 72 hours.

Luna 10, launched on March 31 1966, became the first artificial satellite of the moon, and Lunas 11 and 12 were successfully placed in lunar orbit in August and October. All the orbiters transmitted varying amounts of lunar environmental data. In addition, Luna 12 took pictures from orbit, but they were of rather poor quality.

On December 24 1966 Luna 13 became the second Soviet space probe to successfully soft land on the moon. Like Luna 9, this vehicle successfully transmitted photographs of the lunar surface. In addition, Luna 13 was instrumented for determining the density and firmness of the lunar surface.

Although the Soviet lunar exploration program has had its share of difficulties, particularly those encountered in the development of the soft-landing series, the Soviets have succeeded in obtaining several creditable "firsts" with their lunar program to date. Also, unmanned lunar missions undertaken by the Soviets through December 1966 are compatible with interests and requirements associated with a manned lunar landing. Their scheduled date for attempting such a mission, however, has never been announced.

PLANETARY SERIES

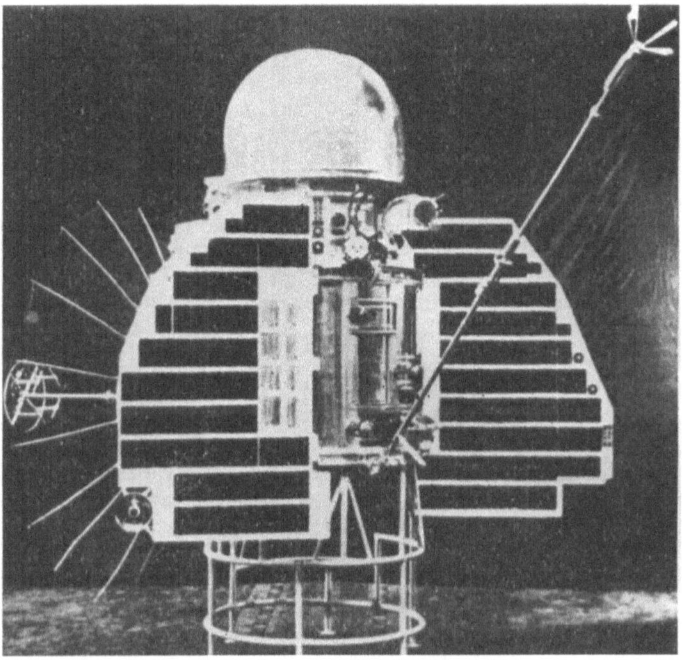

Fig. 3.10. Venera 1

PAYLOAD CHARACTERISTICS—*Venera Series*

Venera 1

Launch Date: February 12 1961 Weight: 643·5 kg

Size: Cylindrical (not including antennas and solar batteries)

Length: 2·035 meters Diameter: 1·050 meters

Scientific Objectives:

Solar and cosmic radiation, micrometeorites, magnetic field, and positive-ion density measurements in interplanetary space and in the vicinity of Venus

Venera 2

Launch Date: November 12 1965

Weight: 963 kg

Size: Similar to Mars 1

Scientific Objectives:

Primary: Photography of Venus

Secondary: Same as Venera 1 mission except included radio astronomical studies

Venera 3

Launch Date: November 16 1965

Weight: 960 kg

Size: Similar to Mars 1 (also contained landing module 90 cm in diameter)

Scientific Objectives:

Primary: Soft-land instrument capsule on surface of Venus.

Secondary: Same as Venera 1 mission except included radio astronomical studies

PLANETARY SERIES

Fig. 3.11. Mars 1

PAYLOAD CHARACTERISTICS

Mars Series

Mars 1

Launch Date: November 1 1962

Weight: 893·5 kg

Size: Cylindrical (excluding solar batteries)
Length: 3·3 meters
Diameter: 1·1 meters

Scientific Objectives:
Primary: Photography of Mars
Secondary: Solar and cosmic radiation, ion composition and density, micrometeorites, magnetic field, and radio astronomical studies in interplanetary space and in the vicinity of Mars. Also contained equipment for detecting life on Mars.

Zond Series

	Zond 1	*Zond 2*	*Zond 3*
Launch Date:	April 2 1964	November 30 1964	July 18 1965
Weight:	Not reported	Not reported	Not reported
Size:	Not reported (similar to Mars 1)	Not reported (similar to Mars 1)	Not reported (similar to Mars 1)
Scientific Objectives:	Not reported (possible Venus mission)	Unidentified scientific measurements toward and in vicinity of Mars	Primary: Test planetary photographic equipment (using moon) Secondary: Magnetic field, solar and cosmic radiation, micrometeorites, optical property and radio astronomical studies

TABLE 3.2

Principal Mission Features

Vehicle	Launch Date	Closest Approach to Planet	Comments
		Venera Series	
Venera 1	Feb. 12 1961	Came within 100 000 km of Venus on May 19–20 1961 and went into heliocentric orbit.	First Soviet planetary probe. Only partially successful as radio contact was lost on February 17–21 1961 a few days after launch and never re-established.
Venera 2	Nov. 12 1965	Came within 24 000 km of Venus on February 27 1966 and went into heliocentric orbit.	Second successful Venus probe had photographic capability. However, radio contact was lost when spacecraft approached flyby position thus terminating mission prior to accomplishment of objective.
Venera 3	Nov. 16 1965	Impacted Venus on March 1 1966 at $9^{h}\,56^{m}$ Moscow time after 106-day flight.	First spacecraft to impact on a planet. Venera 3 had soft-landing instrument capsule and radio contact was lost once again as spacecraft approached the planet thus eliminating possibility of complete mission success.
		Mars Series	
Mars 1	Nov. 1 1962	Was to perform Mars flyby (within 1000–11 000 km) in the latter part of May or early June 1963.	First Soviet planetary probe to Mars. Was more sophisticated than the first Venus probe and contained, among other apparatus, phototelevision equipment for photographing Mars. Never fulfilled mission objective. However, as Soviets reported on March 1963 the radio contact terminated around 106 million km or about one-half the distance to Mars.

Table 3.2 (continued)

Principal Mission Features

Vehicle	Launch Date	Closest Approach to Planet	Comments
		Zond Series	
Zond 1	Apr. 2 1964	Not reported.	Although no specific mission was announced Zond 1 was believed to have been an unsuccessful Venus probe attempt.
Zond 2	Nov. 30 1964	Reportedly passed within 1500 km of Mars in early August 1965.	Zond 2 was successfully launched toward Mars two days after Mariner 4 but suffered a 50% power failure upon launch. Soviet scientists subsequently reported that Zond 2's telemetry system had worked irregularly for months stopping entirely before the probe reached the vicinity of Mars.
Zond 3	July 18 1965	Came within 10 000 km of lunar surface on July 20 1965. Went into heliocentric orbit. Still operating as of March 1966.	Zond 3 was successfully launched into heliocentric orbit with the announced objective of testing planetary photographic equipment. The back side of the moon was successfully photographed with much improved operations over the 1959 Luna 3 mission. The Zond 3 photographs were not transmitted until nine days after they were taken and from a distance of over two million km. Successful retransmittal of these photos three months later from a distance of over 33 million km was reported on October 28 1965.

SUMMARY

As of January 1967, direct planetary exploration by the Soviets has been concerned with only Mars and Venus. The first, at least partially successful, Soviet planetary probe was launched toward Venus on February 12 1961. The next announced probe was Mars 1 on November 1 1962 followed by three Zond probes in 1964–1965 and the twin Venera 2 and 3 launchings to Venus in November 1965.

Nearly all Soviet planetary probes had in-flight difficulties which eliminated any possibility of accomplishing their primary mission objectives. In the case of Venus exploration, for example, radio contact was lost with Venera 1 only 5–7 days after launch and never reestablished. Likewise the somewhat more successful Venera 2 and 3 spacecrafts, which were programmed for Venus flyby photography and soft-landing missions, respectively, both suffered complete telemetry loss as they approached the planet. The other successfully launched Venus probe attempt, Zond 1, in April 1964 is also believed to have undergone communications failure sometime before its approach to Venus in July-August 1964.

Soviet exploration of Mars has been similarly disappointing. The sophisticated Mars 1 probe successfully launched in November 1962 on a Martian photographic flyby mission suffered a telemetry loss in March 1963 some three months prior to reaching Mars. The only other announced Soviet Mars probe was Zond 2 which lost 50% of its power at launch and the remainder before it reached the vicinity of Mars. The most successful of the Soviet planetary series appears to have been Zond 3 which unfortunately had no designated planetary assignment but rather represented only a planetary systems test. The probe successfully rephotographed the back side of the moon and made repeated transmissions of these photographs from increasing distances in space evidently to evaluate planetary-probe photographic equipment performance capabilities.

The Soviets have published photographs for only the Venera 1 and Mars 1 spacecrafts as shown above. They have indicated, however, that subsequent Venera and Zond launchings have basically employed the Mars 1 spacecraft with minor modifications according to the mission objectives.

Although, as noted previously, all Soviet probes to date have failed to accomplish their primary objective of undertaking measurements and photography in the vicinity of Mars and Venus, they have provided varying amounts of scientific data as to the nature of the interplanetary medium. Likewise, the number of attempts and the size and sophistication of the payloads launched thus far reflect the seriousness of the Soviet planetary exploration program.

Part II

Soviet Rocket, Satellite, and Space Probe Investigations (Excluding Biomedical Studies)

Part II contains a summary of the type of physical or environmental measurements undertaken aboard Soviet rockets, satellites, and space probes between 1957–1966 and the nature of results obtained. The discussion is based upon the perusal and condensation of the numerous Soviet technical literature items available in the U.S. in either the English translation or the original Russian. Since the Handbook is meant to be a comprehensive reference tool for scientists, engineers, and students interested in space research and having limited access to Soviet literature, references to and comparison with U.S. space science accomplishments have been kept to a minimum.

Based upon the content of available Soviet literature, Part II has been divided into the following chapters:

4. Micrometeorites and Meteoric Dust.
5. Chemical Composition of the Upper Atmosphere and Interplanetary Space.

6. Optical Phenomena in the Upper Atmosphere.
7. Physical Properties of the Upper Atmosphere.
8. Magnetic Fields.
9. Ionospheric Electron Density Studies.
10. Solar and Cosmic Electromagnetic and Charged-Particle Radiations.
11. Astronomical Investigations Above the Terrestrial Atmosphere.
12. Technical and Scientific Studies Aboard Manned Satellites.

Discussion within each of the above chapters is divided into five sections. The first is an introduction describing the nature and history of Soviet activity in the area (including identification of the specific Soviet space vehicles thus far associated with the investigation). The second contains a brief description of the methodology and instrumentation (including photographs when available) used in the measurement of the parameter. The third is a summary of major published results and conclusions. The fourth provides a synopsis of closely related research. And lastly, a reference list of the more significant Soviet publications in the field is given. The references are arranged alphabetically by senior author permitting rapid identification of the total works of any particular researcher. Also, existing translations of referenced items are noted in parentheses following the original Russian reference. References to U.S. items are given separately in footnotes as they are mentioned in the text.

Chapter 4

Micrometeorites and Meteoric Dust

G. E. Wukelic

INTRODUCTION

Since initiating direct measurements of micrometeorites aboard geophysical rockets in May 1957, the Soviets have been quite active in instrumenting their space research vehicles for investigations of this parameter. The specific geophysical rockets, artificial earth satellites, and space probes launched between 1957–1966 that have been associated with micrometeorite studies are noted in Table 4.1. No literature was identified which discussed micrometeorite studies aboard Soviet geophysical rockets since 1959. Also, the type and results of such studies aboard the Cosmos satellite series have not been reported by the Soviets. And finally, Soviet literature is not clear as to the number of Soviet lunar and planetary probes involved in these studies.

TABLE 4.1

Soviet Space Vehicles Associated With Micrometeorite Studies (1957–1966)

Rockets		Satellites		Space Probes	
Geophysical Rockets	(km)	*Initial Sputnik Series*		*Lunar Probes*	
May 24 1957	212	Sputnik 3	May 15 1958	Luna 1	Jan. 2 1959
Aug. 25 1957	212			Luna 2	Sep. 12 1959
Aug. 31 1957	212	*Electron Series*		Luna 3	Oct. 4 1959
Sep. 9 1957	212	Electron 1/2	Jan. 30 1964	Luna 10	Mar. 31 1966
		Electron 3/4	July 11 1964	Luna 11	Aug. 24 1966
Feb. 21 1958	473			Luna 12	Oct. 22 1966
Aug. 2 1958	212	*Cosmos Series*			
Aug. 13 1958	212			*Planetary Probes*	
Aug. 27 1958	451	Cosmos 8	Aug. 18 1962	Venera 1	Feb. 12 1961
Sep. 19 1958	473			Mars 1	Nov. 1 1962
Oct. 4 1958	110			Zond 2	Nov. 30 1964
Oct. 10 1958	110			Zond 3	July 18 1965
Oct. 31 1958	473			Venera 2	Nov. 12 1965
Dec. 23 1958	110			Venera 3	Nov. 16 1965
Dec. 25 1958	110				
July 2 1959	212				
July 10 1959	212				

INSTRUMENTATION

For the direct determination of the quantity of micrometeorite particles in the upper atmosphere and interplanetary space the Soviets appear to rely on the popular acoustic or microphone-sensing technique. The sensors or detectors installed aboard their geophysical rockets, satellites, and space probes consist of a piezoelectric crystal attached to a metallic plate which generates an electrical signal upon particle impact. The determination of both the number and mass of the impacting particle is possible providing the proper calibration and velocity assumptions are made.

Few Soviet technical papers are available discussing micrometeorite instrumentation schemes. Of those available, two[6, 9] describe the acoustical method (including calibration procedures) and one[1] examines the possibility of using specially designed secondary-electron multipliers for recording micrometeorites with a mass less than 10^{-10} g.

A diagram illustrating the construction of the acoustic ballistic micrometeorite detector used on geophysical rockets and Sputnik 3 is shown in Figure 4.1[9]. Also, two views of the early micrometeorite equipment used on Sputnik 3 are shown in Figures 4.2–4.3[2, 10].

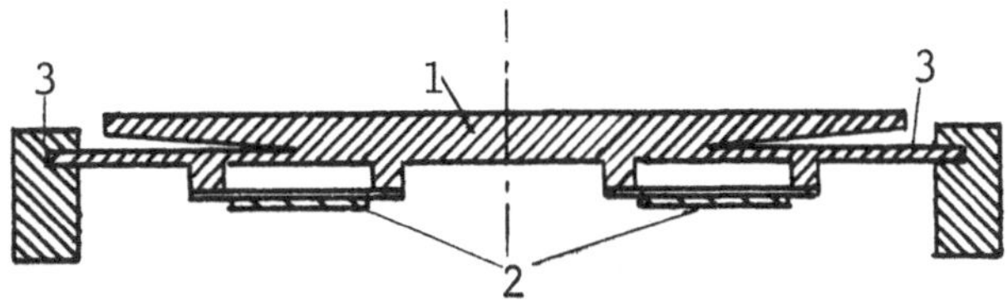

Fig. 4.1. Acoustic (Ballistic) Micrometeorite Sensor[9]

1. Plate
2. Ammonium phosphate piezoelements
3. Flat spring

Fig. 4.2. Type of Micrometeorite Sensor Used on Geophysical Rockets and Sputnik 3[10]

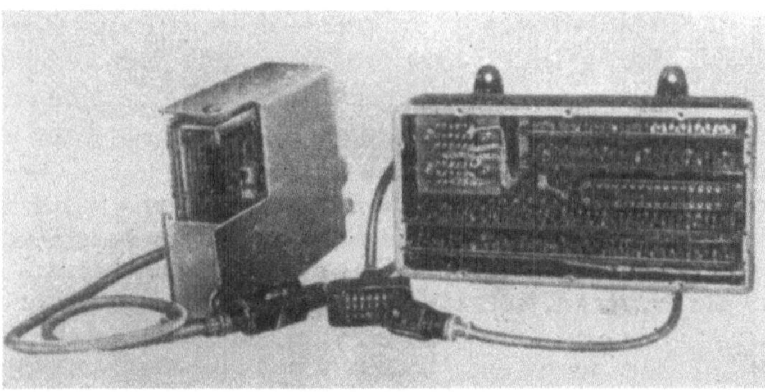

Fig. 4.3. View of Entire Sputnik 3 Micrometeorite Apparatus[2]

Geophysical Rockets

The Soviets indicate that four detectors, occupying a total area of approximately 900 cm^2, were mounted on the perimeter of the nose cones of their early geophysical rocket flights. Piezoelectric (barium titanate or ammonium phosphate) transducers were used having a linear sensitivity range of 10^{-1} to 10^3 g cm/sec. Assuming a mean particle velocity of 40 km/sec, the sensors were reportedly capable of measuring the energy of particles with a mass $>10^{-9}$ g. Assuming a particle velocity of 15 km/sec, the mass sensitivity is reduced to 10^{-8} g.

Artificial Earth Satellites

For micrometeorite studies aboard artificial earth satellites, the Soviets indicated that such studies were made initially by Sputnik 3, using detectors having characteristics similar to those used aboard geophysical rockets. The major differences were that the four sensing elements used on Sputnik 3 were larger (comprising a total area of ·34 m^2) and were further graduated to record particles in the mass ranges from 8×10^{-9} to $2{\cdot}7 \times 10^{-8}$ g, $2{\cdot}7 \times 10^{-8}$ to $1{\cdot}5 \times 10^{-7}$ g, $1{\cdot}5 \times 10^{-7}$ to $5{\cdot}6 \times 10^{-6}$ and $>5{\cdot}6 \times 10^{-6}$ g. A telemetry signal was triggered after 32 impacts were accumulated within the smallest mass range, after the accumulation of 16 and four impacts, respectively, within the next two ranges, and after each impact within the last range. The Soviets also indicated that Sputnik 3 equipment was somewhat sensitive to impacts occurring on the satellite's surface.

Soviet announcements have indicated that satellites of the Cosmos series are involved in "studies of the effect of meteoric matter on the structural elements of space vehicles." Information pertaining to the Cosmos satellites involved and the associated instrumentation schemes is not available, except that Cosmos 8 has specifically been so identified, reportedly employing the previously described acoustic technique. One illustration of a Cosmos satellite reportedly instrumented for micrometeorite studies is shown in Part I, Figure 2.9.

Other Soviet satellites associated with micrometeorite studies belong to the Electron satellite series. Publications to date have associated Electrons 1, 2, and 4 with ballistic piezoelectric detectors having a sensitive area of 0·03 m^2 and capable of registering particles with a mass of 10^{-8} g–10^{-9} g dep nding upon particle velocity assumed.

Lunar and Planetary Probes

For their initial Luna series, the Soviets have reported that equipment installed on the first space rocket was calibrated to record particles with masses ranging between $2{\cdot}5 \times 10^{-9}$ to $1{\cdot}5 \times 10^{-8}$ g, $1{\cdot}5 \times 10^{-8}$ to 2×10^{-7} g, and heavier than 2×10^{-7} g. A telemetry signal was emitted after the accumulation of 15 impacts in the smallest mass range, four impacts in the intermediate mass range, and after each impact in the largest mass range. Luna 2 was calibrated to record meteoric particles within three mass ranges: 2×10^{-9} to 6×10^{-9} g, 6×10^{-9} to $1{\cdot}5 \times 10^{-8}$ g, and heavier than $1{\cdot}5 \times 10^{-8}$ g. The equipment for such studies aboard Luna 3 was calibrated to record particles in the following ranges: 10^{-9} to 3×10^{-9} g, 3×10^{-9} to 8×10^{-9} g, and particles having a mass larger than 8×10^{-9} g. On Lunas 2 and 3 the instrumentation was designed so that no accumulation of impacts occurred, i.e., each impact was registered separately. Also unlike Sputnik 3, equipment on the Luna series was insensitive to impacts on the skin of the container. And finally whereas the sensors on Lunas 1 and 2 occupied a total area of $0{\cdot}2$ m^2, Luna 3 sensors occupied a total area of $0{\cdot}1$ m^2. More recently the Soviets have associated their Luna orbiting series (Lunas 10, 11, and 12) with micrometeorite measurements. Specifically, the first artificial lunar satellite, Luna 10, employed piezoelectric sensors having a total area of $1{\cdot}2$ m^2 and a sensitivity to micrometeorite particles with masses equal to or greater than 7×10^{-8} g (assuming particle velocity of 15 km/sec).

For their planetary probes, the Soviets have associated Veneras 1, 2, and 3, Mars 1, and Zonds 2 and 3 with micrometeorite investigations. To date, however, only micrometeorite equipment aboard Mars 1 has been described. This consisted of piezoelectric sensors, in this case located on the back side of the solar cells. The area sensitive to micrometeorite impacts on Mars 1 was $1{\cdot}5$ m^2. The apparatus recorded micrometeorite impacts with masses of 10^{-7} g and larger.

SUMMARY OF RESULTS

As is the case for several space research topics, the Soviets have not published extensively in the micrometeorite area. To date few technical papers have appeared containing experimental results and these for the most part relate to investigations made during the IGY. Also, few Soviet researchers have been involved in the publication of research results in this area. Although the topic is treated in almost all Soviet national reports to COSPAR[5], most technical papers in this area have been authored by T. N. Nazarova and her associates at the Institute of Geochemistry and Analytical Chemistry of the USSR Academy of Sciences. It is also noteworthy that in reporting the experimental results some confusion has resulted because of the differences in instrument calibration techniques and assumptions regarding the mean micrometeorite velocities. Both the U.S. and USSR researchers initially assumed a 40 km/sec average particle velocity but the U.S. soon changed to 30 km/sec. The Soviets have, however, used 15 km/sec in more recent publications. Also, in the case of Sputnik 3, a 70 km/sec particle velocity was assumed and for one shower encounter with Electron 2, a 61 km/sec velocity. Thus, caution must be exercised when comparing reported U.S. and USSR particle flux rates (particles/m^2/sec). Also noteworthy is the fact that all Soviet investigations reported to date refer to the momentum-sensitive, acoustic or microphone system; whereas U.S. investigations have involved photomultiplier and penetration or fracture-measuring techniques as well as the acoustic technique.

Major conclusions drawn by Soviet investigators from micrometeorite data obtained to date include:

1. The flux of sporadic meteoric matter near the surface of the earth is not constant but varies with time and height. Flux measurements to date have varied between 10^{-2} to 10^{-5} particles/m^2/sec. Also, in several cases near the earth and at large distances from it, individual showers of micrometeorites have been encountered. In such showers the measured flux has varied between 10^{-1} to 10^{-4} particles/m^2/sec.
2. Maximum number of recorded impacts appear to be particles in the mass ranges from 10^{-9} to 10^{-8} g and smaller.
3. Meteoric danger to manned space flight is less than originally estimated.
4. Meteoric (dust) belts or zones having different densities appear to exist around the earth and most likely exist around other planetary bodies.
5. Measured meteoric flux in the vicinity of the moon exceeds the mean interplanetary space impact rate by two orders of magnitude. This increased flux is considered to be of lunar origin.

All reported data obtained from direct micrometeorite measurements aboard Soviet space vehicles since 1957 are summarized in Table 4.2.

TABLE 4.2

Results of Micrometeorite Studies Aboard Soviet Rockets, Satellites, and Space Probes (1957–1966)

Vehicle and Date	Height, km	Area, meter^2	Exposure, $m^2 \cdot sec$	Mass-Sensitivity, grams	Velocity Assumed, km/sec	Micrometeorite Flux, particle/m^2/sec	References (Date)
				Geophysical Rockets			
May 24 1957	100–200	4	536	10^{-8}	15	0·06	Nazarova (1961)[33], (1962)[27], (1965)[29]
Aug. 25 1957	100–200	4	592	10^{-8}	15	0·05	
Feb. 21 1958	150–300	0·09	—	10^{-9}	40	50	Krasovskiy (1958)[10]
	150–300	0·09	—	10^{-9}	40	31	Nazarova (1958)[21]
	126–297	4	340	10^{-8}	15	0·75	Nazarova (1961)[33], (1962)[27], (1965)[29]
Average of above three rocket experiments	200	—	—	10^{-8}	15	0·3	Moroz (1962)[18]
				Artificial Earth Satellites			
Sputnik 3 May 15 1958	Average	0·34	—	10^{-9}–10^{-8}	40	$1{\cdot}7 \times 10^{-3}$	Komissarov, *et al.* (1958)[9] and
	500–600	0·34	—	10^{-9}–10^{-8}	40	9	
	1300–1500	0·34	—	10^{-9}–10^{-8}	40	10	Blagonravov (COSPAR 1960)[5]
	1700–1880	0·34	—	10^{-9}–10^{-8}	40	22	
	400–1880	0·34	—	2×10^{-8}	40	5–10	Krasovskiy (1959)[11]
	Average	0·084	—	10^{-9}	40	0·2	Nazarova (1958)[21]
	500–600	0·084	—	10^{-9}	40	38 ± 10	
	1300–1500	0·084	—	10^{-9}	40	40 ± 18	
	1700–1880	0·084	—	10^{-9}	40	90 ± 34	
	400–1880	0·34	—	8×10^{-9}–$2{\cdot}7 \times 10^{-8}$	40	4–11	Nazarova (1960)[22–25]
	400–1880	0·34	6×10^3	6×10^{-8}–2×10^{-7}	15	7	Nazarova (1961)[33], (1962)[27], (1965)[29]
	400–2000	0·34	—	5×10^{-10}	70	4–11	Nazarova (1961)[26]

TABLE 4.2 (Continued)

Results of Micrometeorite Studies Aboard Soviet Rockets, Satellites, and Space Probes (1957–1966)

Vehicle and Date	Height, km	Area, meter²	Exposure, $m^2 \cdot sec$	Mass-Sensitivity, grams	Velocity Assumed, km/sec	Micrometeorite Flux, particle/m²/sec	References (Date)
May 16–17 1958	400–1880	0·34	—	8×10^{-9}–3×10^{-8}	40	5×10^{-3}	Krasovskiy (1959)[11]
	Average	—	—	8×10^{-9}	40	2×10^{-3}	Moroz (1962)[18]
				6×10^{-8}	15	$1{\cdot}2 \times 10^{-2}$	
	400–1880	0·34	—	8×10^{-9}–$2{\cdot}7 \times 10^{-8}$	40	5×10^{-4}	Nazarova (1960)[22–25]
	400–2000	0·34	6×10^3	6×10^{-8}–2×10^{-7}	15	5×10^{-4}	Nazarova (1961)[26, 33], (1962)[27], (1965)[29]
May 18–26 1958	400–1880	0·34	—	8×10^{-9}–3×10^{-8}	40	$< 10^{-4}$	Krasovskiy (1959)[11] and Nazarova (1960)[22–25]
	400–700	0·34	6×10^3	6×10^{-8}–2×10^{-7}	15	$< 10^{-4}$	Nazarova (1961)[33], (1962)[27], (1965)[29]
Electron 1/2							
Jan. 30–31 1964	400–7130	0·03	$1{\cdot}6 \times 10^3$	$\geqslant 1{\cdot}3 \times 10^{-9}$	61	$1{\cdot}1 \times 10^{-1}$	Nazarova and Rybakov (1965)[34],
Feb. 11–13 1964	400–7130	0·03	$4{\cdot}5 \times 10^3$	$\geqslant 2 \times 10^{-8}$	15	$2{\cdot}4 \times 10^{-3}$	Blagonravov (COSPAR
Feb. 23–25 1964	400–7130	0·03	$4{\cdot}8 \times 10^3$	$\geqslant 2 \times 10^{-8}$	15	$5{\cdot}8 \times 10^{-3}$	1965)[5], and
Feb. 3–Mar. 10 1964	405–7046	0·03	5×10^4	$\geqslant 1{\cdot}3 \times 10^{-9}$	15	$7{\cdot}5 \times 10^{-4}$	Nazarova (1965)[30]
Electron 3/4							
July 11–Sept. 1 1964	405–7046	0·03	10^5	$6{\cdot}5 \times 10^{-8}$–2×10^{-8}	15	8×10^{-5}	Nazarova (1966)[31]
				Lunar and Planetary Probes			
Luna 1		—	—	10^{-9}	40	$<2 \times 10^{-4}$	Krasovskiy (1959)[11]
Jan. 2 1959	180 000	—	—	$2{\cdot}5 \times 10^{-9}$	40	$<2 \times 10^{-3}$	Mirtov (1961)[17]
	(average)	—	—	$1{\cdot}4 \times 10^{-8}$	15	$<3 \times 10^{-3}$	
		0·2	—	$2{\cdot}5 \times 10^{-9}$–$1{\cdot}5 \times 10^{-8}$	40	$<2 \times 10^{-3}$	Nazarova (1960)[22, 23, 25]
		—	—	$1{\cdot}5 \times 10^{-8}$–$2{\cdot}0 \times 10^{-7}$	40	$<5 \times 10^{-4}$	
		—	—	$>2 \times 10^{-7}$	40	$<10^{-4}$	

Vehicle and Date	Height, km	Area, meter^2	Exposure, m$^2 \cdot$ sec	Mass-Sensitivity, grams	Velocity Assumed, km/sec	Micrometeorite Flux, particle/m^2/sec	References (Date)
	2000–360 000	0·2	$4{\cdot}2 \times 10^3$ [33]	2×10^{-8}–10^{-7}	15	$<2 \times 10^{-3}$	Nazarova (1961)[33], (1962)[27], (1965)[29]
		—	$7{\cdot}2 \times 10^3$ [27, 29]	10^{-7}–10^{-6}	15	$<5 \times 10^{-4}$	
		—	—	$>10^{-6}$	15	$<10^{-4}$	
Luna 2		—	—	10^{-8}	40	$\sim 2 \times 10^{-4}$	Krasovskiy (1959)[11] and Moroz (1962)[18]
Sep. 12 1959	180 000	—	—	$2{\cdot}5 \times 10^{-9}$	40	9×10^{-5}	
		—	—	$1{\cdot}4 \times 10^{-8}$	15	$1{\cdot}5 \times 10^{-4}$	
		0·2	—	2×10^{-9}–6×10^{-9}	40	$<5 \times 10^{-5}$	Nazarova (1960)[22, 23, 25]
		—	—	6×10^{-9}–$1{\cdot}5 \times 10^{-8}$	40	$<5 \times 10^{-5}$	
		—	—	$>1{\cdot}5 \times 10^{-8}$	40	9×10^{-5}	
	2000–360 000	0·2	$2{\cdot}2 \times 10^5$	10^{-8}–4×10^{-8}	15	$<5 \times 10^{-5}$	Nazarova (1961)[33], (1962)[27], (1965)[29]
		—	—	4×10^{-8}–4×10^{-7}	15	4×10^{-5} [33] 5×10^{-5} [27]	
		—	—	$>10^{-7}$	15	9×10^{-5}	
		0·2	—	10^{-8}	15	9×10^{-4}	Nazarova (1961)[26]
Luna 3	300 000	—	—	10^{-9}	40	$2{\cdot}8 \times 10^{-3}$	Moroz (1962)[18]
Oct. 4–18 1959				6×10^{-8}	15	$1{\cdot}8 \times 10^{-3}$	
		0·1	—	10^{-9}–3×10^{-9}	40	4×10^{-4}	Nazarova (1960)[22, 23, 25]
		—	—	3×10^{-9}–8×10^{-9}	40	2×10^{-3}	
		—	—	$>8 \times 10^{-9}$	40	4×10^{-4}	
	102 000–470 000	0·1	$2{\cdot}3 \times 10^5$	7×10^{-9}–2×10^{-8}	15	4×10^{-4}	Nazarova (1961)[33], (1962)[27], (1964)[29]
		—	—	2×10^{-8}–6×10^{-8}	15	2×10^{-3}	
		—	—	$>6 \times 10^{-8}$	15	4×10^{-4}	
		0·1	—	10^{-8}	15	1×10^{-3}	Nazarova (1961)[26]
Lunas 1, 2, 3	—	—	—	10^{-7}–10^{-9}	—	2×10^{-3}–5×10^{-5}	Mirtov (1961)[17]
Luna 10 Apr. 3–May 12 1966	355–1050 (above moon's surface)	1·2	—	$>7 \times 10^{-8}$	15	4×10^{-3}	Nazarova, *et al.* (1966)[35]

TABLE 4.2 (Continued)

Results of Micrometeorite Studies Aboard Soviet Rockets, Satellites, and Space Probes (1957–1966)

Vehicle and Date	Height, km	Area, meter^2	Exposure, m$^2 \cdot$ sec	Mass-Sensitivity, grams	Velocity Assumed, km/sec	Micrometeorite Flux, particle/m^2/sec	References (Date)
Mars 1							
Nov. 1 1962	6600–42 000	1·5	—	$>10^{-7}$	—	7×10^{-3} (stream) 10^{-5} (sporadic)	Nazarova (1961)[26], (1965)[29, 30]
Nov. 15–Dec. 31 1962	$<23 \times 10^6$	—	—	—	—	10^{-5} (sporadic)	Nazarova, *et al.* (1963)[32]
Dec. 31 1962–Jan. 30 1963	23×10^6–45×10^6	—	—	—	—	$4{\cdot}5 \times 10^{-3}$	Blagonravov (COSPAR 1963)[5]
Zond 3							
July 18 1965	26×10^6	1·5	$2{\cdot}4 \times 10^7$	$>10^{-7}$	15	7×10^{-5}	Nazarova (1966)[31]
Venera 2							
Nov. 12 1965–Jan. 26 1966	28×10^6	1·5	$9{\cdot}6 \times 10^6$	$>10^{-7}$	15	3×10^{-5}	Nazarova (1966)[31]

The table identifies the vehicle, height, sensor area, exposure, mass sensitivity, assumed particle velocity, and reference associated with each micrometeorite flux value reported in Soviet technical publications.

RELATED WORK

Several Soviet technical publications treat topics related to direct micrometeorite investigations, viz., geophysical interpretation of results, micrometeorite effects, and theoretical and laboratory studies.

Five of these discuss the nature of the meteoric dust envelop of the earth[8, 16, 18, 27, 36]. In the earliest paper, B. A. Mirtov[16] considers dust particles of terrestrial origin to be negligible above 80 to 90 km and meteoric particles of cosmic origin to form a scattering layer above this height. He considers the thickness of the scattering layer to be dependent upon the total flux of incoming cosmic particles. Further, Mirtov advances the hypothesis of the micrometeorite origin of the green-line luminosity of the night airglow. The other four papers associate rocket and satellite data obtained to date with the possible origin, density, and nature of micrometeorite dust belt(s) surrounding the earth[8, 18, 27, 36]. From a comparison of U.S. and USSR data, T. N. Nazarova[27] concludes that direct micrometeorite observations definitely indicate a zone of increased density of micrometeorite particles near the earth at a distance of 100–300 km from the surface. V. I. Moroz[18] theorizes that there are actually three zones or belts around the earth which appear to have a constant meteoric density but vary abruptly from zone to zone. These three postulated zones are: Zone 1, 100–400 km, $N \approx 0{\cdot}1\text{–}1\ m^{-2}\ sec^{-1}$; Zone 2, 400 km $- 2R_e$, $N \approx 10^{-4}$ to $10^{-2}\ m^{-2}\ sec^{-1}$; Zone 3, $>2R_e$, $N \approx 10^{-6}$ to $10^{-4}\ m^{-2}\ sec^{-1}$, where R_e is the radius of the earth and N the number of particles. Mechanisms to explain the origin of these trapped dust zones have been proposed by Moroz[18], Ruskol[36], and more recently by Katasev and Kulikova[8]. The latter authors report the results of a computer analysis of the dimensions and densities of the capture zones for particles with different masses and velocities. The authors conclude that a similar increase in meteoric flux should be detectable in the vicinity of other planets having rather dense atmospheres. It is interesting to note, however, that several recent U.S. publications have questioned the existence, nature, and density of these belts which have been hypothesized in the U.S. as well as in the USSR.

In comparison to U.S. literature, few Soviet papers are available which treat the effect of micrometeorite particles upon space vehicles. Earlier theoretical treatments of the general problem of high velocity solid particle impact, directly applicable to the meteoric hazard problem, have been published by M. A. Lavrent'yev[12] and K. P. Stanyukovich[37]. Another paper which discusses the change in the albedo of Sputnik 1, by I. M. Yatsunskiy and O. V. Gurko[39], identifies micrometeorites among several environmental factors contributing to the reduction of the reflecting properties of the satellite.

Since 1964, an increasing number of papers have appeared describing results of theoretical and experimental studies in the Soviet Union aimed at determining meteoric puncture, penetration, and collision probabilities in space [3, 4, 14, 20, 38]. Of particular interest to space-science investigations is an item concerned with evaluating the effect of micrometeorite particles on telescope optical surfaces beyond the earth's atmosphere[7]. In this paper the author concludes that the effect caused by extraterrestrial meteoric particles on the optical surfaces of instruments is insignificant, becoming perceptible only after a period of about 10 years. Similarly one paper describes a technique using a multistage linear synchronous accelerator to simulate the collision of meteoric bodies with materials in space in the velocity range 30–70 km/sec[19]. Another contains an interesting analysis of the increased meteor danger to spacecraft due to meteoric concentrations in the anti-radiants of meteor streams[15]. And lastly one paper considers the structure and density of meteoroids themselves, a topic which is basic to the understanding of interaction of satellites and space probes with solid interplanetary material[13]. From a theoretical analysis of the heat, deformation, and fragmentation associated with ablating meteoroids, the authors conclude that an extremely loose structure should not be assumed for meteoroids.

References

1. A. I. Akishin and L. I. Tseplyayev, "Secondary-Electron Multiplier for Recording Micrometeors", Geomagnetizm i Aeronomiya, 1964, Vol 4, No. 1, pp 202–205 (Geomagnetism and Aeronomy, pp 157–160).

2. S. G. Aleksandrov and R. Ye. Fedorov, "Soviet Satellites and spaceships", Publishing House of the Academy of Sciences, USSR, Moscow, Second Edition (enlarged and revised), 1961, 439 pp (JPRS: 15 805).

3. E. I. Andriankin, "The Penetration of Barriers by Meteorites", Kosmicheskiye Issledovaniya, 1966, Vol 4, No. 2, pp 280–290 (HT–66–244, pp 165–185).

4. E. I. Andriankin and Yu. S. Stepanov, "Impact Penetration Depth of Meteoric Particles", Iskusstvennyye Sputniki Zemli, 1964, No. 15, pp 44–52 (Planetary and Space Science, 1963, Vol 11, pp 1365–1374).

5. USSR National Reports to COSPAR, 1960–1966, presented by A. A. Blagonravov (In English).

6. M. A. Isakovich and N. A. Roi, "An Acoustical Method for the Measurement of the Mechanical Parameters of Meteorites", Iskusstvennyye Sputniki Zemli, 1958, No. 2, p 81–82 (Artificial Earth Satellites, pp 105–107).
7. I. D. Karachentsev, "Effects of Micrometeoric Particles on the Optical Surfaces of Telescopes Beyond the Earth's Atmosphere", Iskusstvennyye Sputniki Zemli, 1964, No. 15, pp 57–65 (Artificial Earth Satellites, pp 58–67).
8. L. A. Katasev and I. V. Kulikova, "On Meteoric Satellites of the Earth", 7th COSPAR, Vienna, Austria, May 1966 (In English).
9. O. D. Komissarov, T. N. Nazarova, L. N. Neugodov, S. M. Poloskov, and L. Z. Rusakov, "Micrometeorite Studies Using Rockets and Satellites", Iskusstvennyye Sputniki Zemli, 1958, No. 2, pp 54–58 (Artificial Earth Satellites, pp 68–74).
10. V. I. Krasovskiy, "Exploration of the Upper Atmosphere With the Help of the Third Soviet Sputnik", 9th International Astronautical Congress, August 1958, Academic Press, pp 614–625.
11. V. I. Krasovskiy, "Investigations Made by Soviet Sputniks and Cosmic Rockets", 14th Annual ARS Meeting, Washington, D.C., November 16–20 1959 (In English).
12. M. A. Lavrent'yev, "The Problem of Piercing at Cosmic Velocities", Iskusstvennyye Sputniki Zemli, 1959, No. 3, pp 61–65 (Artificial Earth Satellites, pp 85–91).
13. V. N. Lebedinets and Yu. I. Porthiagin, "On the Ablation of Meteoroids in the Earth's Atmosphere", 7th COSPAR, Vienna, Austria, May 1966 (In English).
14. L. B. Livanov, "Comparison of Collision Effects Produced by Meteorites Against Surfaces of Various Metals", Kosmicheskiye Issledovaniya, 1965, Vol 3, No. 4, pp 659–660 (TT–65–1262, pp 266–269).
15. P. V. Makovetskiy, "Meteor Danger in the Antiradiant of a Meteor Stream", Kosmicheskiye Issledovaniya, 1966, Vol 4, No. 3, pp 493–495.
16. B. A. Mirtov, "Meteoric Matter and Some Geophysical Problems of the Upper Atmosphere", Iskusstvennyye Sputniki Zemli, 1960, No. 4, pp 118–134 (Artificial Earth Satellites, pp 334–357).
17. B. A. Mirtov, "Rockets, Satellites, and Investigations of the Upper Atmosphere", Priroda, 1961, No. 10, pp 23–31 (TT 61–217).
18. V. I. Moroz, "On a 'Dust Envelop' of the Earth", Iskusstvennyye Sputniki Zemli, 1962, No. 12, pp 151–158 (Planetary and Space Science, 1963, Vol 11, pp 387–394).
19. A. K. Mukhamedzhanov, "A Method for Accelerating Microparticles to Meteoric Velocities", Iskusstvennyye Sputniki Zemli, 1964, No. 15, pp 53–56 (Artificial Earth Satellites, pp 54–57).
20. A. K. Mukhamedzhanov, "The Penetration of a Thin Shield by a Meteorite", Kosmicheskiye Issledovaniya, 1966, Vol 4, No. 2, pp 291-295 (HT-66-244, pp 186-194).
21. T. N. Nazarova, "Rocket and Satellite Investigations of Meteors", CSAGI Meeting, Moscow, 1958 (In English).
22. T. N. Nazarova, "The Results of Studies of Meteoric Dust by Means of Sputnik III and Space Rockets", Space Research I, Proceedings of the First International Space Science Symposium, Nice, France, January 11–16 1960, North-Holland Publishing Company, Amsterdam, 1960, pp 1059–1062.
23. T. N. Nazarova, "Results of Exploring Meteoric Matter With Instrumentation of Sputnik III and Space Probes", 11th International Astronautical Congress, Stockholm, Sweden, August 15–20 1960 (In English).
24. T. N. Nazarova, "Study of Meteoric Particles Through Instruments on the Third Soviet Artificial Satellite", Iskusstvennyye Sputniki Zemli, 1960, No. 4, pp 165–170 (Planetary and Space Science, 1961, Vol 8, pp 82–85).
25. T. N. Nazarova, "Results of the Investigation of Meteoric Substance With the Help of Instruments Mounted in Space Rockets", Iskusstvennyye Sputniki Zemli, 1960, No. 5, pp 38–40 (Artificial Earth Satellites, pp 524–527).

26. T. N. Nazarova, "Rocket and Satellite Meteoric Dust Investigations", 12th International Astronautical Congress, Washington, D.C., October 1961 (In English).

27. T. N. Nazarova, "Investigation of Meteoric Dust by Means of Rockets and Artificial Earth Satellites", Iskusstvennyye Sputniki Zemli, 1962, No. 12, pp 141–144 (Planetary and Space Science, 1963, Vol 11, pp 305–309).

28. T. N. Nazarova, "Preliminary Results of Investigations of Meteoric Matter Along the Trajectory of the Mars I Probe Flight", Space Research IV, Proceedings of the Fourth International Space Science Symposium, Warsaw, Poland, June 4–10 1963, North-Holland Publishing Company, Amsterdam, 1964, pp 921–924.

29. T. N. Nazarova, "Soviet Devices to Measure Meteor Impacts", Geofizicheskiy Byulleten', 1964, No. 14, pp 89–91 (JPRS: 32 734).

30. T. N. Nazarova, "Investigation of Meteoric Dust Using Rockets and Satellites", Symposium on Meteor Orbits and Dust, Cambridge, Massachusetts, August 9–13 1965 (In English).

31. T. N. Nazarova, "Investigation of Meteoric Bodies by Means of Rockets and Satellites", Kosmicheskiye Issledovaniya, 1966, Vol 4, No. 6, pp 900–909.

32. T. N. Nazarova, A. K. Bektabegov, and O. D. Komissarov, "Preliminary Results of the Investigation of Meteoric Matter Along the Flight of the Mars 1 Interplanetary Station", Kosmicheskiye Issledovaniya, 1963, Vol 1, No. 1, pp 169–171 (TT–63–968, pp 302–307).

33. T. N. Nazarova and V. I. Moroz, "The Research of Micrometeorites With the Help of Rockets and Satellites", Space Research II, Proceedings of the Second International Space Science Symposium, Florence, Italy, April 10–14 1961, North-Holland Publishing Company, Amsterdam, 1961, pp 639–644.

34. T. N. Nazarova and A. K. Rybakov, "Meteoric Dust Measured on the Electron 2 Satellite", Space Research VI, Proceedings of the Sixth International Space Science Symposium, Mar Del Plata, Argentina, May 10–21 1965, North-Holland Publishing Company, Amsterdam, 1966, pp 946–951.

35. T. N. Nazarova, A. K. Rybakov, and G. D. Komissarov, "Preliminary Results of Investigation of Solid Interplanetary Matter in the Neighborhood of the Moon", Doklady Akademii Nauk SSSR, 1966, Vol 170, No. 3, pp 578-579 (JPRS: 38 798) and Kosmicheskiye Issledovaniya, 1966, Vol 4, No. 6, pp 910–911.

36. E. L. Ruskol, "On the Origin of the Concentration of Interplanetary Dust Around the Earth", Iskusstvennyye Sputniki Zemli, 1962, No. 12, pp 145–150 (Planetary and Space Science, 1963, Vol 11, pp 311–316).

37. K. P. Stanyukovich, "Elements of the Theory of the Impact of Solid Bodies with High (Cosmic) Velocities", Iskusstvennyye Sputniki Zemli, 1960, No. 4, pp 86-117 (Artificial Earth Satellites, pp 292–333).

38. Yu. S. Stepanov, "Certain Questions Regarding the Puncture on Collision With Meteoric Particles", Kosmicheskiye Issledovaniya, 1965, Vol 3, No. 6, pp 903–916 (TT–65–1985, pp 154–176).

39. I. M. Yatsunskiy and O. V. Gurko, "Change of the Albedo of the First Artificial Earth Satellites as a Result of the Action of External Factors", Iskusstvennyye Sputniki Zemli, 1960, No. 5, pp 71–73 (Artificial Earth Satellites, pp 573–576).

Chapter 5

Chemical Composition of the Upper Atmosphere and Interplanetary Space

G. E. Wukelic and R. A. Duffee

INTRODUCTION

The direct investigation of the gaseous composition of the upper atmosphere and interplanetary space has been another research area of prime Soviet concern. Following early studies aboard balloons, the Soviets developed a unique (detachable) glass-flask sampling technique for use on their meteorological and geophysical rockets. Although this represented an improvement over other existing flask sampling techniques, it was soon to be replaced by the more sophisticated radio-frequency mass spectrometer technique previously adapted for rocket ion composition studies in the U.S. in the early 1950's. Also, as in the U.S., current Soviet investigations are emphasizing studies of the neutral composition of the upper atmosphere. As noted in Table 5.1, the first Soviet mass spectrometer studies were undertaken aboard the geophysical rocket of September 9 1957, although glass-flask sampling reportedly began as early as 1949. RF mass spectrometer studies have been associated with satellite payloads aboard Sputnik 3, the four Electron satellites, and possibly several of the Cosmos satellites. And lastly, ion trap studies aboard Soviet lunar and planetary probes have also provided direct information regarding the positive ion concentration in the upper atmosphere and interplanetary space.

TABLE 5.1

Soviet Space Vehicles Associated with Composition Studies of the Upper Atmosphere and Interplanetary Space (1957–1966)

Rockets	Satellites	Space Probes (Related Ion-Trap Studies)
Meteorological Rockets	*Initial Sputnik Series*	*Lunar Probes*
Glass flask sampling beginning in 1949	Sputnik 3 May 15 1958	Luna 1 Jan. 2 1959
		Luna 2 Sep. 12 1959
	Electron Series	Luna 3 Oct. 4 1959
Geophysical Rockets	Electron 1/2 Jan. 30 1964	
Glass flask sampling (km)	Electron 3/4 July 11 1964	*Planetary Probes*
		Venera 1 Feb. 12 1961
Aug. 25 1957 212	*Cosmos Series*	Mars 1 Nov. 1 1962
Aug. 31 1957 212	Cosmos 2 (ion-trap studies)	Zond 3 July 18 1965
Radio-frequency mass spectrometer		Venera 2 Nov. 12 1965
		Venera 3 Nov. 16 1965
Sep. 9 1957 212		
Feb. 21 1958 473		
Aug. 2 1958 212		
Aug. 13 1958 212		
Aug. 27 1958 451		
Sep. 19 1958 473		
Oct. 31 1958 473		
July 2 1959 212		
July 10 1959 212		
July 14 1959 212		
July 22 1959 195		
June 15 1960 212		
June 24 1960 212		
Sep. 16 1960 212		
Sep. 22 1960 212		
Sep. 23 1960 212		
Nov. 15 1961 430		
Oct. 18 1962 500		
Oct. 25 1963 430		
Dec. 24 1963 422		

INSTRUMENTATION

Flask Sampling

Rocket investigation of the chemical composition of the upper atmosphere reportedly began in the USSR using the classical flask sampling technique in 1949. In an effort to minimize rocket exhaust contamination, the Soviets developed special automatic containers which detached themselves from the rocket at a specific point in the trajectory and descended separately by means of parachutes. The container, which is shown in Figure 5.1, consisted of a metal cylinder about three meters

Fig. 5.1. The Automatic Container[68]

long and 40 cm in diameter. It had separate compartments for the parachute, the instruments, and the power supply. Two variations of the open sampling compartment are shown in Figures 5.2 and 5.3. Inside both versions, the Soviets used (nonoxidizing) glass sampling flasks (3000 cc and 900 cc) rather than metal flasks such as were employed in the West. Because of these sampling improvements and the use of more sophisticated sample analysis techniques, the Soviets claimed the development of a system superior to that used in the West. Reportedly the sampling program started in 1949, but most published results are for the 1951–1957 interval.

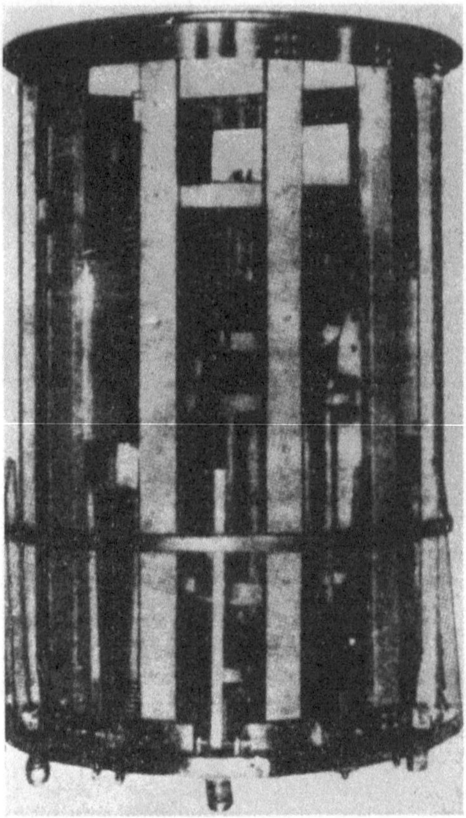

Fig. 5.2. The Open Compartment of the Container With the Bulbs for Taking Samples at Altitudes Below 100 km[68]

Radio-Frequency Mass Spectrometry

The classical flask sampling method described above, which provided valuable information regarding the "normal" composition of the atmosphere below 100 km, proved completely inadequate for dealing with the problem of investigating the unstable radicals and molecules as well as the ions in the upper atmosphere. This requirement resulted in the development in the U.S. between 1952–1956 of a special type (Bennett) radio-frequency mass spectrometer for use aboard vertical research rockets. Subsequently, V. G. Istomin, in the USSR, used a RF mass spectrometer aboard a geophysical rocket in 1957. His apparatus was based on the same principle as Townsend's (U.S.) apparatus, but differed primarily in the presence of "one layer" grids in the analyzer and ion

Fig. 5.3. Compartment of the Container With Projecting Bulbs for Taking Samples at Altitudes Above 100 km[68] (Before and after sampling the flaps are closed.)

source. This mass spectrometer which was also used in early ion composition satellite studies was designated the RMS (see Figure 5.5), the prefix RMS being simply an acronym for radio-frequency mass spectrometer. Automation and miniaturization improvements by the Soviets led to the development of the MKh spectrometer series, used for both neutral and ion composition studies aboard later rocket and satellite flights. Thus far four versions of this series have been developed and employed for upper atmosphere studies (see Figures 5.6–5.9). In this series, from MKh-6401 through MKh-6407, the MKh prefix indicates the equipment is for chemical analysis. The first digit notes the ion separation principle of the apparatus—the six indicating that it is radio frequency. The second digit identifies the area of application—the four indicating that the apparatus is for a special-purpose. The last two digits are model numbers only.*

* "Radio-Frequency Mass Spectrometry", Library of Congress, AID Report P–64–39, June 11 1964.

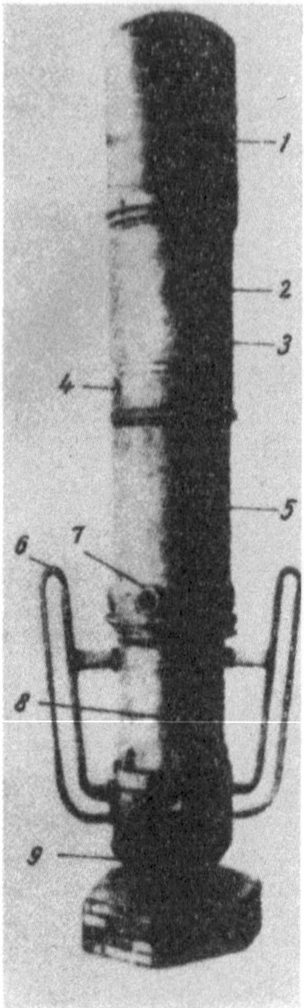

Fig. 5.4. Automatic Container With RF Mass Spectrometer[68]

Fig. 5.5. RMC–1 (RMS–1) Radio-Frequency Mass Spectrometer[34]

Fig. 5.6. MX–6401 (MKh–6401) Radio-Frequency Mass Spectrometer[75]

Fig. 5.7. MX–6403 (MKh–6403) Radio-Frequency Mass Spectrometer[42]

Fig. 5.8. MX–6405 (MKh–6405) Radio-Frequency Mass Spectrometer[90]

A description and comparison of all RF mass spectrometers associated with Soviet rocket and satellite ion and neutral composition studies to date are contained in Table 5.2.

Geophysical Rockets

In their rocket spectroscopy studies up to around 200 km, once again, in contrast to the U.S. method, the Soviets mounted their RF mass spectrometers inside detachable cylinders similar to those employed in their glass sampling technique (see Figure 5.4). More recently, rocket composition studies up to 500 km have been accomplished in the USSR by using so-called high-altitude automatic geophysical stations (VGAS)

Fig. 5.9. MX–6407M (MKh–6407M) Radio-Frequency Mass Spectrometer[73]

which are detachable spherical containers with automatic orientation systems (see Figure 1.11 in Chapter 1). In October 1962 the Soviets employed a gas-discharge device containing several air capsules which permitted controlled amounts of oxygen and nitrogen to be expelled near the rocket mass spectrometer[19]. The program objective was to study the reaction between the atmospheric O^+ and N^+ ions and the O_2 and N_2 molecules discharged from the capsules.

Between 1957–1963, the Soviets associated some 20 geophysical rockets with either or both ion and neutral composition studies. All versions of the RF mass spectrometer shown in Figures 5.5–5.9 except MKh–6407 have been associated with these rocket flights. According to the Soviets, the MKh–6407 is planned for both rocket and satellite use.

TABLE 5.2

Comparison of Characteristics of Radio-Frequency Mass Spectrometers Used for Atmospheric Composition Studies in the Soviet Union[31–32, 34, 63, 68, 70–73, 75, 85–86, 90]

Characteristic	Spectrometer				
	RMS–1	MKh–6401	MKh–6403	MKh–6405	MKh–6407M
Associated Vehicles	Geophysical rockets and Sputnik 3	Geophysical rockets	Geophysical rockets	Geophysical rockets and Electron satellites	Planned for rocket and satellite studies
Weight, kg	6·5	3·3	2·0	2·0	2·5
Power Requirement, watts	25–30	5·3 and 6	3·2 and 4	4·0	3–5
Scan Time, sec	1·7	3·0	3·0	3·0	2·5
Mass Range, amu	6–48 8–63	1–4/12–56	1–4 12–60	1–2/12–36 4–34	1–4 12–48
Resolving Power, $\frac{M}{\Delta M}$	20–30	50	50	20	5–20
Sensitivity, ions/cm³	—	$1{\cdot}6 \times 10^{8}$	$1{\cdot}6 \times 10^{7}$	10	—
mm Hg	—	5×10^{-9}	5×10^{-10}	1×10^{-10}	$1\text{–}3 \times 10^{-11}$
Number of Cycles	5–7	5–9/4–7	5–9/4–7	5–9/4–7	—
Dimensions, mm	—	—	—	$l = 300$, $3 = 50$	—
Measuring unit	—	$210 \times 90 \times 70$	—	—	$120 \times 70 \times 40$
Analyzer (without ion source)	—	$l = 240$, d = 50	—	—	—
Ion Source	—	$l = 140$, d = 50	—	—	—

Artificial Earth Satellites

The radio-frequency mass spectrometer RMS–1 shown in Figure 5.5 was initially installed on Sputnik 3. In this experiment the spectrometer was designed to record only positive ions. The equipment started functioning as soon as the protective cone was jettisoned. During the period of its intermittent operation, from May 15–25 1958, about 15 000 mass spectra of positive ions were obtained for altitudes between 225–980 km.

For composition studies aboard the Electron satellites, the Soviets have employed the MKh–6405 model radio-frequency mass spectrometer shown in Figure 5.8. A special feature of this instrument was the use of a variable relay which made it insensitive to the background and thus prevented the recording of spurious mass peaks. Although some variations existed between the Electron 1 and Electron 2 apparatus in general, the range of the instruments was from 1–34 amu and their resolution about 20 in the region $M = 20$. Sensitivity was about 10 ions cm^{-3} when making ion investigations and about 10^{-10} mm Hg when analyzing neutral composition.

Ion-Trap Measurements

Numerous Soviet rocket, satellite, and space probes have been instrumented with ion traps for ion-density determinations. Such investigations, which also provide information of significance to positive ion composition studies, are discussed in Chapters 9 and 10.

SUMMARY OF RESULTS

Direct Flask Sampling

Between 1949 and 1957, the Soviets expended a large amount of effort on direct flask sampling of the upper atmosphere using vertical research rockets. The samples were collected in the middle latitudes of the European territory of the USSR and usually in the spring and summer and during the morning hours. Table 5.3 contains the results of the analyses of these samples for which the experimental accuracy is not in doubt[20, 66, 68].

Based upon these and similar data, the Soviets have drawn the following conclusions regarding the composition of the atmosphere up to 100 km.

1. There is no appreciable separation between oxygen and nitrogen up to 95 km.
2. The quantity of the heaviest gas, argon, at altitudes of 85–98 km is less than at the earth's surface.
3. Although the decrease in the argon content with height is slight, it can nevertheless be concluded that diffusive separation does occur at altitudes of about 100 km.
4. There remains some disagreement between theory and measurement.

TABLE 5.3

Results of Flask Sampling Studies Aboard Soviet Rockets

Height, km	Composition of Sample, volume per cent			Height, km	Composition of Sample, volume per cent		
	O_2	N_2	A		O_2	N_2	A
65	19·0	80	0·91	82–85	24·5	74	0·77
75–80	21·0	78	0·93	82–85	20·5	78	0·79
75–80	21·0	78	0·93	82–85	19·0	80	0·91
80	21·5	78	weak	85	21·0	78	0·86
80	19·0	80	0·86	85	21·0	78	0·90
80	22·0	77	0·87	85	21·0	78	0·88
80	23·0	76	0·90	95	21·5	77	0·76

Radio-Frequency Mass Spectrometry

By using various types of RF mass spectrometers aboard their research rockets and artificial earth satellites (see Figures 5.5–5.9), the Soviets have attempted to determine the vertical profiles and the time and spatial distribution of ionized and neutral components of the upper atmosphere. However, as in the U.S., these measurements are spread out in both time and location thus making further investigation necessary if complete understanding of the normal and disturbed conditions of the upper atmosphere is to be obtained. Most of the Soviet research has been performed by a few Soviet investigators at the Institute of Applied Geophysics (IPG) of the Academy of Sciences of the USSR, namely, V. G. Istomin[31–44, 47], A. A. Pokhunkov[74–81], and B. A. Mirtov[66–68, 70]. Theoretical assessments of the mechanisms and processes influencing atmospheric composition, in an effort to explain experimental results, have been provided chiefly by A. D. Danilov[4–15, 17, 19], G. S. Ivanov-Kholodny[48–53], S. P. Yatsenko[88–89], and V. V. Rybin[84] of the same facility. These authors have contributed over 50 of the 90 Soviet items published on this space topic. Several papers relating to mass spectrometer improvements and applications have been published by Soviet researchers at the Central Aerological Observatory[3, 54, 58, 62, 64]. Most of these are in association with meteorological rocket studies of the lower atmosphere (principally ozone and water vapor concentration) in progress at this facility.

Summaries of the major Soviet results of rocket and satellite studies of the ion and neutral composition of the upper atmosphere are presented below. Specific results as they are associated with vehicles and references are contained in Table 5.4.

Neutral Composition

As in early U.S. rocket neutral composition studies, some misinterpretations have occurred in early USSR work, particularly that reported by A. A. Pokhunkov. In both cases improper estimates for the extent of oxygen recombination inside the mass spectrometers were the cause. The result was that early experimental determinations of the quantity of molecular oxygen in the upper atmosphere were on the high side and correspondingly, low estimates resulted for the atomic oxygen-molecular oxygen ratio. However, subsequent improvements in the mass spectrometers and data processing techniques in both the U.S. and USSR have minimized the effect.

By using various modified versions of a mass spectrometer in geophysical rockets between 1959–1961, the Soviets have recorded and determined the relative concentrations of the following atmospheric gases: O, N, O_2, N_2, A, and CO_2 between 90–430 km. He and H_2 were not detected and the Soviets have therefore concluded that their concentration must be less than 10^7 molecules per cm^3. Molecular nitrogen is the predominant component up to 210 km and perhaps dominates composition up to 280 km or higher. Like molecular oxygen, molecular nitrogen changes very little between 100–200 km and appears to undergo no diurnal or seasonal variations. Atomic nitrogen, on the other hand, does not occur in appreciable amounts between 100–200 km, being somewhat less than 2% of molecular nitrogen. The ratio of atomic oxygen to molecular oxygen begins to increase around 100 km and continues up to 200 km. Daytime atomic oxygen values between 120–160 km were higher than at night. Correspondingly a discernible diurnal variation in the atomic oxygen-molecular nitrogen ratio in the 100–200 km region is attributable to the morning increase of atomic oxygen. And finally, in almost all experiments, diffusive separation (based upon the change in the argon-molecular nitrogen ratio) was found to occur around 100–120 km.

TABLE 5.4

Results of Atmospheric Composition Studies Aboard Soviet Rockets and Satellites (1957–1966)

Date of Investigation	Time of Day	Mass Spectrometer Employed	Height Range, km	Type of Measurement	Major Results Reported	References
					Geophysical Rockets	
Sep. 9 1957	Evening	RMS-1	105–206	Ion composition	Altogether 70 spectra were recorded. During this flight, three ions were found in the ionosphere, with mass numbers 16, 30, and 32 (O^+, NO^+, and O_2^+). Between 105–190 km the concentration of O^+ and O_2^+ did not exceed 20 per cent of the concentration of NO^+ which predominated at all altitudes (105–206 km). O^+ ions were recorded between 190–206 km and their concentration was observed to increase in this range.	Blagonravov (1960–1966)[2] Istomin (1958)[31], (1961)[40], Istomin and Pokhunkov (1963)[17], Mirtov (1961)[68]
Feb. 21 1958	Morning	RMS-1	105–206	Ion composition	In all three ascents the results obtained agreed with those obtained previously (both in the USSR and the U.S.). The main ion in the lower layers of the ionosphere was NO^+, while O^+ appeared in measurable amounts from 150–160 km and increased rapidly with height; O_2^+ was present in appreciable concentrations mainly in the lower layers of the ionosphere (100–150 km). The concentration of O_2^+ had a tendency to decrease during the day. Also detected traces of Fe^+ and, possibly, Si^+ or N_2^+.	Istomin (1961)[37, 40, 42] Khostikov (1964)[55], Mirtov (1961)[68]
Aug. 2 1958	Morning	RMS-1	100–210	Ion composition		
Aug. 13 1958	Morning	RMS-1	100–211	Ion composition		
July 14 1959	Morning	MKh-6401	92–203	Neutral composition	In these flights the gases of H, H_2, N, O, OH, H_2O, N_2, O_2, A, CO_2, and N_2O were recorded. The character of the recordings of H_2O, OH, H, and H_2 indicated that these gases were, however, stray gases from the container or rocket and the only true atmospheric gases recorded were O, N, O_2, N_2, A, and CO_2 or possibly N_2O which has the same mass number. Data showed diffusive separation between argon and molecular nitrogen above 100 km. Also, the data obtained concerning the ions of the light gases (hydrogen and helium) during the July 22 flight showed that the concentration of hydrogen and helium ions could not exceed 10^3 ions/cm^3.	Istomin (1961)[40–41], Mirtov (1961)[68], Pokhunkov (1960)[74], (1961)[75], (1962)[77]
July 22 1959	Morning	MKh-6401	91–211	Neutral and ion composition		
July 15 1960	Morning	MKh-6403	92–206	Ion composition	In all, slightly more than 100 spectra were obtained. The mass spectrometer registered ions of Mg^+, Ca^+, Fe^+, N_2^+, or possibly Si^+ as well as the usual positive ions of NO^+ and O_2^+. The uncharacteristic ions of Mg^+, Ca^+, N_2^+, and Fe^+ were considered to be of meteoric origin.	Istomin (1961)[36–37, 41–42], (1963)[17]

TABLE 5.4 (Continued)

Results of Atmospheric Composition Studies Aboard Soviet Rockets and Satellites (1957–1966)

Date of Investigation	Time of Day	Mass Spectrometer Employed	Height Range, km	Type of Measurement	Major Results Reported	References
Sep .23 1960	Midnight	MKh-6403	100–210	Neutral composition	Fifty mass spectra were obtained during the ascent of the container and 51 during descent. Principal components once again were N_2, O, and O_2. Up to 210 km, N_2 was the predominant component. N content did not exceed 2 per cent of N_2. O increased with height until 210 km equalled 65 per cent of N_2 concentration. O_2 unchanged at 100 km and at 210 km equalled 14 per cent of N_2 concentration. Results showed at least in the middle and circumpolar latitudes that the change in the argon-nitrogen ratio with height indicates that diffusive separation begins at about 100–110 km. Concentration of H_2 must be less than 3×10^7 molecules/cm^3 and He less than 6×10^7 molecules/cm^3 at altitudes over 100 km.	Pokhunkov (1962)[77–78], (1963)[76]
Nov. 15 1961	Evening	MKh-6401 (with improved sensitivity	130–430	Neutral and ion composition	During the experiment 85 neutral and ionized gas spectra were obtained. Ions of N^+, O^+, NO^+, O_2^+ were registered. Also some peaks were due to gases released from the rocket surface but these were easily distinguishable. At heights above 210 km, O^+ ions were the predominant component. For the first time helium ions were recorded having a density equal to $1{\cdot}7 \times 10^3$ ions/cm^3 at 370 km and increased linearly to 7×10^3 ions/cm^3 at 430 km. As for the neutral components, O, O_2, N_2, and A were recorded and their relative concentrations determined. Daytime O values were higher than at night between 130–160 km. Confirmed once again diffusive separation around 100 km.	Danilov and Pokhunkov (1966)[13] Pokhunkov (1963)[76, 81], (1964)[79]
Various Geo-physical Rockets 1959–1961	Various	Various	100–430	Neutral and ion composition	Diffusive separation of A and N_2 gases occurred above 105–110 km. N_2 considered main component up to 280 km. Quantity of N up to 210 km was $<2\%$ of N_2. O existed above 100 km and was the predominant component above 280 km. O_2 existed above 100 km and its altitude distribution corresponded to diffusive separation between O_2 and N_2. The limiting values of the concentrations of minor atmospheric gases (NO, H_2O, OH, H_1, H_2, and He) were determined. Value of NO did not exceed $0{\cdot}1\%$ of the concentration of N_2 in the 130–180 km range. $H_2O<6\%$, and	Pokhunkov (1965)[82]

Date of Investigation	Time of Day	Mass Spectrometer Employed	Height Range, km	Type of Measurement	Major Results Reported	References
					$OH < 6 \times 10^{-3}$ % of the total atmospheric pressure. The concentration of $H_1 < 10^8$ particles/cm^3, H_2 above 100 km$< 3 \times 10^7$ particles/cm^3 and He^+ in the 100–210 km range $< 10^2$ ions/cm^3. He^+ also detected above 370 km. Magnesium oxide of meteoric origin was detected at 100–130 km.	
Various Geophysical Rockets 1959–1961	Various	Various	100–210	Neutral composition	U.S. and USSR neutral composition results compared. Based on USSR results (1959–1961) it is concluded that diffusive separation between A and N_2 occurred at 105–120 km; N_2 main constituent up to 200 km; ratio of O/N_2 depended upon time of day; N content less than 5%; and the mean molecular weight decreased from 28 at 100 km to 22–23 at 200 km.	Danilov (1965)[14], (1966)[16]
Oct. 18 1962	Morning	MKh-6403 (also air discharge device)	150–508	Ion composition, ion exchange studies	Recorded ions of O^+, H_2O^+, NO^+, and O_2^+. H_2O^+ from container surface. Variation of O^+, NO^+ with height determined. The air discharge device permitted controlled amounts of air to be released from capsules near the mass spectrometer so reactions between the O^+ and N^+ ions and the discharged O_2 and N_2 could be recorded. Spectra obtained showed O_2^+ and some traces of NO^+ ions. N_2^+ ions were not recorded. Comparison of results obtained with Western and laboratory experiments shows considerable disagreement.	Danilov and Yatsenko (1964)[19]
				Artificial Earth Satellites		
Sputnik 3 (May 15–25 1958)	Mainly in the morning	RMS-1	225–980	Ion composition	During the period May 15–25 1958, 15 000 spectra of positive ions were recorded. The predominant being that of atomic oxygen (O^+), the next in intensity being that of atomic nitrogen (N^+). Near perigee (225–250 km) ions of N_2^+, O_2^+, and NO^+ were observed. The strongest of the three was NO^+. Also all spectra contained a weak trace of the oxygen isotope $(O^{18})^+$ thought to be both satellite-induced contamination as well as of atmospheric origin. Studies revealed that the composition at these altitudes varied as a function of height. The relative content of NO^+, O_2^+, and N_2^+ (relative to O^+) decreased with increased height, and the relative content of the light ion N^+ increased with increased height. Beginning at about 500 km the atmosphere became atomic. Also a dependence of composition on latitude was established.	Fedynskiy (1962)[20], Istomin (1958)[31], (1959)[33], (1960)[35], (1961)[37–38], Istomin and Pokhunkov (1963)[47], Khvostikov (1964)[55], Mirtov (1961)[68]

TABLE 5.4 (Continued)

Results of Atmospheric Composition Studies Aboard Soviet Rockets and Satellites (1957–1966)

Date of Investigation	Time of Day	Mass Spectrometer Employed	Height Range, km	Type of Measurement	Major Results Reported	References
Electron 1/2 (Jan. 30 1964)	—	MKh-6405	400–3000	Ion and neutral composition	Ion composition studies aboard Electrons 1, 2, 3, and 4 established that in the period of solar activity minimum the concentration of molecular ions NO^+, O_2^+, and N_2^+ at heights from 400–500 km was at least one order of magnitude lower than in the period of the maximum. It was also established that at all heights under investigation helium ions constituted a small portion of the over-all ion concentration and that nowhere were they predominant ions. The equality of concentration of O^+ and H^+ ions was reached at heights of the order of 900 km. Thereafter H^+ predominated.	Blagonravov (1959–1966)[2], Istomin (1965)[44], Zarkhin (1965)[90]
Electron 3/4 (July 11 1964)	—	MKh-6405	400-3000	Ion and neutral composition		
Electron 2	—	MKh-6405	400–1000	Ion and neutral composition	Data indicated that concentration of major ion components (O^+, H^+, and N^+) strongly dependent on time. In the region 400–500 km the relative ion concentration of H^+ varied from 8–10% during morning and evening hours and from 2–3% in the afternoon. At 900 km (morning) relative concentration of O^+ was 20% and increased to 70–80% during the day. Both O^+ and H^+ reached extreme values two to three hours after local noon. N^+ on the other hand increased steadily from 3–6% in the region of satellite perigee and from 1–10% at 900–1000 km. It is postulated that diurnal N^+ variations are obscured by latitude effects. The He^+ varied during the day from 0·3–3% at 400 km to 8–10% at 100 km.	Istomin (1966)[46]
Electron 2 (Jan. 30–Apr. 1 1964)	—	MKh-6405	410–2750	Ion and neutral composition	For daylight values on February 21 1964 between 425–1300 km (at latitudes of 25°, 38°, 60°, and 56° N), H^+ and O^+ were the predominant ions. N^+ and He^+ were detected but were much less intense. Afternoon measurements between February 10–16 1964 at 440–1800 km showed O^+ and H^+ remained predominant ions at all altitudes and were equal at ≈900 km. No molecular ions detected by Electron satellites. He^+ ions were not predominant at any altitude according to Electron 2 findings. With an increase in altitude the "oxygen" ionosphere changes directly into a "hydrogen" ionosphere or protonosphere.	Istomin (1965)[45]

Ion Composition

In general ion composition data obtained from initial rocket experiments (1957–1959) in middle latitudes of the European part of the USSR were quite similar to early U.S. data even though experiments were undertaken at different times and locations and utilized mass spectrometers having different characteristics. Also, as in the early U.S. rocket composition studies, much erroneous data were introduced by the rocket and measuring equipment. A major difference in the results was the quantity of negative ions (particularly NO_2^-) detected in early U.S. rocket studies. The Soviet studies showed that the lower ionosphere is composed mainly of positive ions of mass number 30 (nitric oxide, NO^+) except in the middle of the day, when considerable quantities of O_2^+ are found at 100–120 km. As the height increases, O^+ ions (which appear initially around 150 km) rapidly replace the NO^+ and O_2^+ ions and become predominant at altitudes above 200 km. The distribution and origin of the minor constituents such as N_2^+, H_2O^+, Fe^+, and Si^+ were not resolved during these early studies. Also, hydrogen, helium, and hydroxyl ions were not detected.

Subsequent rocket ion composition studies employing improved mass spectrometers (1960–1962) confirmed the predominant presence of NO^+, O_2^+, and O^+ in the upper atmosphere and identified the existence of N^+. Most significant was the frequent recording of the uncharacteristic heavy ions—Mg^+, Ca^+, Fe^+, and Si^+, detected originally in 1958, which the Soviets ascribed to extraterrestrial—possibly meteoric—origin. This led to the Soviet conclusion that meteors play an important role in the ionization of the lower region of the ionosphere. The existence of these metallic ions have been confirmed by recent U.S. rocket studies and conclusions regarding their source and lower ionization effects parallel those of the USSR. Some results relating to the variation of O^+ and NO^+ with height, determined from more recent rocket flights, are in considerable disagreement with U.S. experiments, however.

From their 1961 geophysical rocket series, the Soviets reported for their initial detection of helium ions a density less than 10^2 ions/cm^3 between 100–130 km and 10^3 ions/cm^3 between 370–430 km. Also, based upon results of ion-trap measurements aboard Cosmos 2, several Soviet scientists concluded that a region of helium ion predominance (as claimed in the U.S.) existed between 520–650 km. However, considering the small amount of helium ions subsequently detected by their Electron satellites, between 400–3000 km, most Soviet investigators currently postulate that He^+ ions at all heights add little to the hydrogen-oxygen ionosphere and that the terrestrial atmosphere is void of a heliosphere.

With the exception of the above helium results very little in quantity or significance has been published regarding Soviet satellite composition studies. Few satellite results reported to date (Sputnik 3 and Electron satellites) have shown O^+, H^+, and N^+ to be the predominant ions and to exhibit diurnal, altitude, latitude, and solar dependence. Molecular N_2^+ and O_2^+ ions and NO^+ ions were recorded but were found to decrease rapidly with height.

RELATED RESEARCH

Examples of related Soviet space research include the results of frequent rocket investigations of the vertical distribution of atmospheric ozone[57, 61, 65, 83, 87] and water vapor[21–22] published in recent years. Also, several papers contain interesting discussions of disturbances induced in the gaseous medium during the flight of a satellite[67, 88], estimates of the amount of neutral hydrogen in space[59–60], and the determination of correction coefficients for mass spectrometers[80]. Also closely related are the numerous publications concerning ion density studies aboard rockets, satellites and space probes principally by Gringauz, *et al.*[1, 23–30]. Results of the latter are discussed in Chapters 9 and 10.

References

1. V. V. Afonin, T. K. Breus, G. L. Gdalevich, B. N. Gorozhankin, K. I. Gringauz, and R. Ye. Rybchinskiy, "Brief Survey of the Results of Physical Experiments on the Satellite 'Cosmos 2' in the Ionosphere", Trudy Vsesoyuznoy Konferentsii po Fizike Kosmicheskogo Prostranstva, Izdatel'stvo "Nauka", Moscow, June 10–16 1965, pp 151–167 (NASA TT F–389, pp 202–224).

2. USSR National Reports to COSPAR, 1960–1966, presented by A. A. Blagonravov (In English).

3. A. F. Chizhov, "Use of the Theory of Free Molecular Flow for Processing Data Obtained With a Quadrupole Mass Filter", Trudy Tsentral'noy Aerologicheskoy Observatorii, 1965, No. 61, pp 44–49.

4. A. D. Danilov, "About the Formation of NO^+ in the Upper Atmosphere", Iskusstvennyye Sputniki Zemli, 1960, No. 5, pp 60–65 (Artificial Earth Satellites, pp 556–563).

5. A. D. Danilov, "Molecular Ions in the Upper Atmosphere", Doklady Akademi Nauk SSSR, 1961, Vol 137, No. 5, pp 1098–1101 (NASA TT F–8014).

6. A. D. Danilov, "Formation of $0_2{}^+$ Ions in the Upper Atmosphere", Iskusstvennyye Sputniki Zemli, 1961, No. 7, pp 56–59 (Planetary and Space Science, Pergamon Press, 1962, Vol 9, pp 175–178).

7. A. D. Danilov, "Formation of Molecular Ions in the Upper Atmosphere", Iskusstvennyye Sputniki Zemli, 1961, No. 8, pp 72–76 (Artificial Earth Satellites, pp 233–237).
8. A. D. Danilov, "The Excitation Mechanism of the Red Oxygen Line in the Night Glow", Iskusstvennyye Sputniki Zemli, 1961, No. 8, pp 77–80 (Planetary and Space Science, Pergamon Press, 1962, Vol 9, pp 499–502).
9. A. D. Danilov, "Molecular Nitrogen in the Upper Atmosphere", Iskusstvennyye Sputniki Zemli, 1961, No. 10, pp 98–101 (Planetary and Space Science, Pergamon Press, 1962, Vol 9, pp 877–879).
10. A. D. Danilov, "Some Problems Connected With the Recombination and Ionization Processes in the Earth's Atmosphere", Iskusstvennyye Sputniki Zemli, 1963, No. 15, pp 38–43 (Artificial Earth Satellites, pp 39–44); Space Research III, Proceedings of the Third International Space Science Symposium, Washington, D.C., May 2–8 1962, North-Holland Publishing Company, Amsterdam, 1963, pp 274–281.
11. A. D. Danilov, "Ion Exchange Processes in the Upper Atmosphere", Iskusstvennyye Sputniki Zemli, 1963, No. 17, pp 19–30 (Artificial Earth Satellites, pp 18–29).
12. A. D. Danilov, "Argon Ionization in the Upper Atmosphere", Kosmicheskiye Issledovaniya, 1963, Vol 1, No. 2, pp 256–260 (TT–64–42, pp 108–115).
13. A. D. Danilov, "The Formation of Ions in the Ionosphere", Kosmicheskiye Issledovaniya, 1964, Vol 2, No. 6, pp 865–880 (TT–64–1316, pp 75–101).
14. A. D. Danilov, "On the Atmosphere Composition in the Range of 100 to 200 km", Space Research VI, Proceedings of the Sixth International Space Science Symposium, Mar Del Plata, Argentina, May 11–19 1965, North-Holland Publishing Company, Amsterdam, 1966, pp 395–400; Trudy Vsesoyuznoy Konferentsii po Fizike Kosmicheskogo Prostranstva, Izdatel'stvo "Nauka", Moscow, June 10–16 1965, pp 48–50 (NASA TT F–389, pp 62–65).
15. A. D. Danilov, "Rates of the Main Ion-Molecular Processes in the Ionosphere", Seventh International Space Science Symposium, Vienna, Austria, May 1966 (In English).
16. A. D. Danilov, "Experimental Studies of the Neutral Composition of the Atmosphere in the Altitude Range 100–200 km", Kosmicheskiye Issledovaniya, 1966, Vol 4, No. 1, pp 47–65 (TT–66–76, pp 71–101).
17. A. D. Danilov, V. G. Istomin, and S. M. Poloskov, "Ionospheric Composition Investigated by Rockets and Sputniks, and Physical Processes Determining the Structure of the Ionosphere", Space Research II, Proceedings of the Second International Space Science Symposium, Florence, Italy, April 10–14 1961, North-Holland Publishing Company, Amsterdam, 1961, pp 993–1001.
18. A. D. Danilov and A. A. Pokhunkov, "Mass Spectrometric Investigation of the Interaction of Atmospheric Ions With Molecules of Rocket Gas Release", Seventh International Space Science Symposium, Vienna, Austria, May 1966 (In English).
19. A. D. Danilov and S. P. Yatsenko, "Experimental Investigation of the Constants of Ionospheric Ion Exchange Processes", Kosmicheskiye Issledovaniya, 1964, Vol 2, No. 2, pp 276–279 (TT–64–547, pp 163–169).
20. A. V. Fedynskiy, "Certain Data on the Composition of Upper Atmospheric Layers", Trudy Tsentral'noy Aerologicheskoy Observatorii, 1962, No. 42, pp 5–19 (Library of Congress, AID P–63–36).
21. A. V. Fedynskiy, "Possible Method of Measuring Water Vapor Concentration at High Altitudes", Trudy Ts–tral'noy Aerologicheskoy Observatorii, 1965, No. 61, pp 59–67.
22. A. V. Fedynskiy, S. P. Perov, and A. F. Chizhov, "Some Results of Measuring Concentration of Water Vapor, Atomic Oxygen, and Total Density in the Mesosphere by Thermal Sensors", Seventh International Space Science Symposium, Vienna, Austria, May 1966 (In English).
23. K. I. Gringauz, "The Structure of the Earth's Ionized Gas Envelope Based on

Results of Direct Measurements in the USSR of Charged Particle Local Concentrations", Space Research II, Proceedings of the Second International Space Science Symposium, Florence, Italy, April 10–14 1961, North-Holland Publishing Company, Amsterdam, 1961, pp 574–592.

24. K. I. Gringauz, "The Structure of the Earth's Ionized Gas Envelop According to Data of Direct Measurements of Local Concentrations of Charged Particles Made in the USSR", Iskusstvennyye Sputniki Zemli, 1962, No. 12, pp 105–118 (Artificial Earth Satellites, pp 114–130).

25. K. I. Gringauz, V. V. Bezrukikh, L. S. Musatov, E. K. Solomatina, and R. E. Rybchinsky, "Some Results of Measurements Conducted by Means of the Charged Particle Trap on the Electron 2 Satellite", Space Research VI, Proceedings of the Sixth International Space Science Symposium, Mar Del Plata, Argentina, May 11–19 1965, North-Holland Publishing Company, Amsterdam, 1966, pp 850–861.

26. K. I. Gringauz, V. V. Bezrukikh, and V. D. Ozerov, "Results of Measurements of Positive Ion Concentrations in the Ionosphere by Means of Ion Traps on the Third Soviet Earth Satellite", Iskusstvennyye Sputniki Zemli, 1961, No. 6, pp 63–100 (Artificial Earth Satellites, pp 77–121).

27. K. I. Gringauz, B. N. Gorozhankin, G. L. Gdalevich, V. V. Afonin, R. E. Rybchinsky, and N. M. Shutte, "The Technique and Some Results of Experiments Conducted on the Satellite Cosmos 2 by Means of Langmuir Probes and Ion Traps of Honeycomb Type", Space Research V, Proceedings of the Fifth International Space Science Symposium, Florence, Italy, May 12–16 1964, North-Holland Publishing Company, Amsterdam, 1965, pp 733–750.

28. K. I. Gringauz, B. N. Gorozhankin, N. M. Shutte, and G. L. Gdalevich, "Altitude Variations of Charged Particles and Ion Composition in the Outer Ionosphere Since the Solar Activity Maximum According to Data Collected by Ion Traps on the Cosmos 2 Satellite", Space Research IV, Proceedings of the Fourth International Space Science Symposium, Warsaw, Poland, June 4–10 1963, North-Holland Publishing Company, Amsterdam, 1964, pp 473–479.

29. K. I. Gringauz, B. N. Gorozhankin, N. M. Shutte, and G. L. Gdalevich, "Some Experiments Carried Out Aboard the Satellite Cosmos 2", Astronautyka, 1963, No. 4, pp 2–4.

30. K. I. Gringauz, B. N. Gorozhankin, N. M. Shutte, and G. L. Gdalevich, "Height Distribution of Charged Particles in the Ionosphere Layers According to Experimental Data From the Ion Traps on the Cosmos 2 Satellite", Doklady Akademii Nauk SSSR, 1963, Vol 151, No. 3, pp 560–563 (TT–63–20853 and TT–64–71154).

31. V. G. Istomin, "Investigation of the Ion Composition of the Earth's Atmosphere by Rockets and Satellites", CSAGI Meeting, Moscow, 1958; Iskusstvennyye Sputniki Zemli, 1958, No. 2, p 32 (Artificial Earth Satellites, pp 40–44).

32. V. G. Istomin, "Grids for Radio-Frequency Mass Spectrometers", Pribory i Tekhnika Eksperimenta, 1958, No. 2, p 111 (Instruments and Experimental Techniques, pp 306–307).

33. V. G. Istomin, "Mass Spectrometer Measurements of the Ionic Composition of the Upper Atmosphere by the Third Artificial Satellite", Doklady Akademii Nauk SSSR, 1959, Vol 129, No. 1, pp 81–84 (Doklady of the Academy of Sciences of the USSR, Earth Sciences Sections, pp 1129–1132).

34. V. G. Istomin, "Radio-Frequency Mass Spectrometer for Investigation of Ionic Composition of the Upper Atmosphere", Iskusstvennyye Sputniki Zemli, 1959, No. 3, pp 98–112 (Artificial Earth Satellites, pp 137–160).

35. V. G. Istomin, "Some Results of Measurement of Mass Spectra of Positive Ions on the Third Soviet Artificial Earth Satellite", Iskusstvennyye Sputniki Zemli, 1960, No. 4, pp 171–183 (Artificial Earth Satellites, pp 411–429).

36. V. G. Istomin, "Magnesium and Calcium Ions in the Earth's Upper Atmosphere", Doklady Akademii Nauk SSSR, 1961, Vol 136, No. 5, pp 1066–1068

(Doklady of the Academy of Sciences of the USSR, Earth Sciences Sections, pp 169–172).

37. V. G. Istomin, "Nitrogen Ions in the Upper Atmosphere and the Ionization of the Region at Night", Doklady Akademii Nauk SSSR, 1961, Vol 137, No. 5, pp 1102–1105 (Doklady of the Academy of Sciences of the USSR, Earth Sciences Sections, pp 411–414).

38. V. G. Istomin, "Mass Spectrometric Measurements of Gas Composition of the Earth Atmosphere by Means of Rockets and Satellites", Geomagnetizm i Aeronomiya, 1961, Vol 1, No. 3, pp 359–368 (Geomagnetism and Aeronomy, pp 321–328).

39. V. G. Istomin, "Concentration-Height Distribution of Positive Ions According to Data of Mass-Spectrometric Measurements on the Third Satellite", Iskusstvennyye Sputniki Zemli, 1961, No. 6, pp 127–131 (Planetary and Space Science, Pergamon Press, 1961, Vol 8, pp 179–182).

40. V. G. Istomin, "Investigations of the Ionic Composition of Earth Atmosphere on Geophysical Rockets in 1957–1959", Iskusstvennyye Sputniki Zemli, 1961, No. 7, pp 64–77 (Planetary and Space Science, Pergamon Press, 1962, Vol 9, pp 179–193).

41. V. G. Istomin, "Absolute Concentrations of Ionic Components of Earth's Atmosphere at Altitudes From 100 to 200 km", Iskusstvennyye Sputniki Zemli, 1961, No. 11, pp 94–97 (Planetary and Space Science, 1963, Vol 11, pp 169–172).

42. V. G. Istomin, "Ions of Extraterrestrial Origin in Earth's Ionosphere", Iskusstvennyye Sputniki Zemli, 1961, No. 11, pp 98–107 (Planetary and Space Science, 1963, Vol 11, pp 173–181); Space Research III, Proceedings of the Third International Space Science Symposium, Washington, D.C., May 2–8 1962, North-Holland Publishing Company, Amsterdam, 1963, pp 209–220.

43. V. G. Istomin, "Upper Atmospheric Detection of 0^+ Ions With Higher Than Thermal Energy", Kosmicheskiye Issledovaniya, 1963, Vol 1, No. 2, pp 261–266 (TT-64-42, pp 116–127).

44. V. G. Istomin, "Measurements of Ion Composition in the Outer Ionosphere With Mass Spectrometers Installed on Electron Satellites", abstract only, Sixth International Space Science Symposium, Mar Del Plata, Argentina, May 11–19 1965 (In English).

45. V. G. Istomin, "Composition of the Outer Ionosphere of the Earth Based on Measurement Data From the Electron Satellites", Trudy Vsesoyuznoy Konferentsii po Fizike Kosmicheskogo Prostranstva, Izdatel'stvo "Nauka", Moscow, June 10–16 1965, pp 192–202 (NASA TT F-389, pp 259–272).

46. V. G. Istomin, "Ion Composition Variations During a Day at Heights of 400–1000 km According to Data of Measurements From the Electron 2 Satellite", Seventh International Space Science Symposium, Vienna, Austria, May 1966 (In English).

47. V. G. Istomin and A. A. Pokhunkov, "Mass Spectrometer Measurements of Atmospheric Composition in the USSR", Space Research III, Proceedings of the Third International Space Science Symposium, Washington, D.C., May 2–8 1962, North-Holland Publishing Company, Amsterdam, 1963, pp 117–131.

48. G. S. Ivanov-Kholodnyy, "On the Rate of Ionization and Recombination Processes in the Ionosphere", Space Research III, Proceedings of the Third International Space Science Symposium, Washington, D.C., May 2–8 1962, North-Holland Publishing Company, Amsterdam, 1963, pp 303–309.

49. G. S. Ivanov-Kholodnyy, "Intensity of Short-Wave Solar Radiation and the Rate of Ionization and Recombination Processes in the Ionosphere", Geomagnetizm i Aeronomiya, 1962, Vol 2, No. 3, pp 377–407 (Geomagnetism and Aeronomy, pp 315–336).

50. G. S. Ivanov-Kholodnyy, "Ion Composition and Effective Recombination Coefficient Variations of the Ionosphere", Seventh International Space Science Symposium, Vienna, Austria, May 1966 (In English).

51. G. S. Ivanov-Kholodnyy, "Anomaly in the Composition of Generated Ions and

Certain Phenomena in the Lower Atmosphere", Doklady Akademii Nauk SSSR, 1966, Vol 170, No. 4, pp 831–834.

52. G. S. Ivanov-Kholodnyy, "Intensity of Ion Formation at Altitudes of 100–300 km", Geomagnetizm i Aeronomiya, 1966, Vol 6, No. 2, pp 382–385.

53. G. S. Ivanov-Kholodnyy and A. D. Danilov, "Variations in Atmospheric Ion Composition at Altitudes of 100–200 km", Trudy Vsesoyuznoy Konferentsii po Fizike Kosmicheskogo Prostranstva, Izdatel'stvo "Nauka", Moscow, June 10–16 1965, pp 217–219 (NASA TT F-389, pp 291–294); Kosmicheskiye Issledovaniya, 1966, Vol 4, No. 3, pp 439–453.

54. A. I. Ivanovskiy, "Aerodynamics of Manometers and Mass Spectrometers in Rockets and Earth Satellites", Trudy Vsesoyuznoy Konferentsii po Fizike Kosmicheskogo Prostranstva, Izdatel'stvo "Nauka", Moscow, June 10–16 1965, pp 56–61 (NASA TT F-389, pp 74–80).

55. I. A. Khvostikov, "The Upper Layers of the Atmosphere", Gidrometeorologicheskoy Izdatel'stvo, Leningrad, 1964, 514 pp (NASA TT F-315).

56. V. N. Konashenok," Photochemical Equilibrium and Ionic Composition of the Upper Layers of the Atmosphere", Kosmicheskiye Issledovaniya, 1966, Vol 4, No. 1 pp 89–94 (TT-66-76, pp 144–151).

57. L. A. Kudryavtseva, "Results of Measuring by Means of a Rocket the Vertical Distribution of Atmospheric Ozone", Trudy Tsentral'noy Aerologicheskoy Observatorii, 1960, No. 37, pp 24–29.

58. G. N. Kudryavtsev, G. N. Levina, and V. P. Lepekhina, "Certain Characteristics and Potentials of a Small Mass Spectrometer for Use During Flight", Trudy Tsentral'noy Aerologicheskoy Observatorii, 1965, No. 61, pp 93–99.

59. V. G. Kurt, "Neutral Hydrogen in the Vicinity of the Earth and in Interplanetary Space", Uspekhi Fizicheskikh Nauk, 1963, Vol 81, No. 2, pp 249–270 (Soviet Physics Vol 6, No. 5, pp 701–714).

60. V. G. Kurt, "Problem of the Total Amount of Neutral Hydrogen in the Upper Atmosphere of the Earth", Kosmicheskiye Issledovaniya, 1966, Vol 4, No. 1, pp 111–115 (TT-66-76, pp 180–188); Trudy Vsesoyuznoy Konferentsii po Fizike Kosmicheskogo Prostranstva, Izdatel'stvo "Nauka", Moscow, June 10–16 1965, p 51 (NASA TT F-389, p 66).

61. A. A. L'vova, A. Ye. Mikirov, and S. M. Poloskov, "Rocket Determinations of the Ozone Profile Above the Level of Maximal Concentration", Seventh International Space Science Symposium, Vienna, Austria, May 1966 (In English).

62. G. V. Malyarova and G. M. Marty'nkevich, "Application of the Molecular System of Overlapping for the Study of Factors Affecting the Accuracy of Analyzing the Neutral Atmospheric Components by Mass Spectrometry", Trudy Tsentral'noy Aerologicheskoy Observatorii, 1965, No. 61, pp 50–54.

63. G. M. Marty'nkevich, "Mass Spectrometry Methods for Studying the Molecular (Neutral) Composition of the Upper Atmosphere", Trudy Tsentral'noy Aerologicheskoy Observatorii, 1963, No. 46, pp 63–75.

64. G. M. Marty'nkevich, "Mass-Spectrometric Approach to the Study of the Neutral and Charged Components of the Upper Atmosphere", Trudy Tsentral'noy Aerologicheskoy Observatorii, 1965, No. 61, pp 18–27.

65. A. Ye. Mikirov, "Determining the Ozone Concentration at Altitudes of 44–102 km Based on Data Obtained From Nocturnal Launches of Geophysical Rockets", Trudy Vsesoyuznoy Konferentsii po Fizike Kosmicheskogo Prostranstva, Izdatel'stvo "Nauka", Moscow, June 10–16 1965, p 65 (NASA TT F-389, p 74); Geomagnetizm i Aeronomiya, 1965, Vol 5, No. 6, pp 1120–1123 (Geomagnetism and Aeronomy, pp 882–884).

66. B. A. Mirtov, "Rocket Investigations of the Composition of the Atmosphere at High Altitudes", Uspekhi Fizicheskikh Nauk, 1957, Vol 63, No. 1b, pp 181–196 (Advances in Physical Sciences, pp 243–265).

67. B. A. Mirtov, "Disturbances Occurring in the Gaseous Medium During the Flight of a Satellite", Iskusstvennyye Sputniki Zemli, 1958, No. 2, pp 17–25 (Artificial Earth Satellites, pp 20–31).
68. B. A. Mirtov, "Gas Composition of Earth's Atmosphere and Methods of Its Analysis", Publishing House of the Academy of Sciences, USSR, Moscow, 1961, 262 pp (NASA TT F-145).
69. B. A. Mirtov, "Photochemical Equilibrium of Atmospheric Oxygen at Altitudes Above 150 km", Geomagnetizm i Aeronomiya, 1966, Vol 6, No. 2, pp 292–297.
70. B. A. Mirtov and V. G. Istomin, "Investigation of the Ion Composition of the Ionized Layers of the Atmosphere", Uspekhi Fizicheskikh Nauk, 1957, Vol 63, No. 1b, pp 227–238 (Advances in Physical Sciences, pp 305–320).
71. V. A. Pavlenko, A. Ye, Rafal'son, and M. D. Shutov, "A Series of Miniature Mass Spectrometers for Investigating the Composition of Neutral and Ionized Gases in the Upper Layers of the Atmosphere", Kosmicheskiye Issledovaniya, 1963, Vol 1, No. 2, pp 287–295 (TT-64-1316, pp 162–174).
72. V. A. Pavlenko, A. Ye. Rafal'son, M. Ye. Slutskiy, G. A. Tsveyman, and M. D. Shutov, "Radio-Frequency Mass Spectrometer for the Analysis of the Ionic and Molecular Composition of the Upper Layers of the Atmosphere", Pribory i Tekhnika Eksperimenta, 1960, No. 6, pp 89–95 (Instruments and Experimental Techniques, pp 948–953).
73. V. A. Pavlenko, B. I. Zarkhin, A. Ye. Rafal'son, and M. Ye. Slutskiy, "Highly Sensitive Radio-Frequency Mass Spectrometer for Investigating Ion and Neutral Composition of the Upper Atmosphere", Kosmicheskiye Issledovaniya, 1966, Vol 4, No. 3, pp 457–463.
74. A. A. Pokhunkov, "A Study of the Neutral Composition of the Upper Atmosphere at Altitudes Above 100 km", Izvestiya Akademii Nauk SSSR, Seriya Geofizicheskaya, 1960, No. 11, pp 1649–1657 (Bulletin [Izvestiya] Academy of Sciences, USSR, Geophysics Series, 1960, No. 11, pp 1099–1105).
75. A. A. Pokhunkov, "Mass-Spectrometric Investigations of Structural Parameters of Earth Atmosphere at Altitudes From 100–210 km", Iskusstvennyye Sputniki Zemli, 1961, No. 7, pp 89–100 (NRL 852; Artificial Earth Satellites, pp 88–100).
76. A. A. Pokhunkov, "Gravitational Separation, Composition, and the Structural Parameters of the Atmosphere at Altitudes Above 100 km", Space Research III, Proceedings of the Third International Space Science Symposium, Washington, D.C., May 2–8 1962, North-Holland Publishing Company, Amsterdam, 1963, pp 132–142.
77. A. A. Pokhunkov, "Variation in the Mean Molecular Weight of the Night Air at Heights From 100 to 210 km Shown by Mass Spectrometry", Iskusstvennyye Sputniki Zemli, 1962, No. 12, pp 133–140 (Artificial Earth Satellites, pp 145–153).
78. A. A. Pokhunkov, "Gravitational Separation, Composition, and Structural Constants of the Night Atmosphere at Altitudes From 100 to 210 km", Iskusstvennyye Sputniki Zemli, 1962, No. 13, pp 110–118 (Artificial Earth Satellites, pp 116–126).
79. A. A. Pokhunkov, "Mass-Spectrometric Measurements of the Distribution of Helium, Nitrogen, and Argon in the Earth's Atmosphere Higher than 130 km", Space Research IV, Proceedings of the Fourth International Space Science Symposium, Warsaw, Poland, June 4–10 1963, North-Holland Publishing Company, Amsterdam, 1964, pp 325–332.
80. A. A. Pokhunkov, "The Possibility of Using a Radio-Frequency Mass Spectrometer for the Measurement of Absolute Concentrations of Atomic Gases in the Upper Atmosphere", Geomagnetizm i Aeronomiya, 1963, Vol 3, No. 2, pp 252–261 (Geomagnetism and Aeronomy, pp 203–210).
81. A. A. Pokhunkov, "Mass Spectrometric Measurements of He^+, N^+, O^+, NO^+, and O_2^+ Ion Distribution in the Earth's Atmosphere up to 430 km", Kosmicheskiye Issledovaniya, 1963, Vol 1, No. 2, pp 267–270 (TT-64-1316, pp 128–134).
82. A. A. Pokhunkov, "Mass-Spectrometric Measurements of the Neutral Compo-

sition of the Earth's Atmosphere at Altitudes of 100–430 km", V Razdel Programmy MGG: Ionosfera, Sbornik Statey, 1965, No. 14, pp 26–40.

83. S. M. Poloskov, A. A. L'vova, and A. Ye. Mikirov, "Rocket Measurements of Ozone Profiles Above the Level of Maximum Concentration", Seventh International Space Science Symposium, Vienna, Austria, May 1966 (In English).

84. V. V. Rybin, "On Collisional Interactions With Neutral Particles at the Ionosphere Altitudes, Where the Frequency of Electron-Ion Collision is High", Seventh International Space Science Symposium, Vienna, Austria, May 1966 (In English).

85. M. Ya. Shcherbakova, "Optimum Analyzer Designs for a Radio-Frequency Mass Spectrometer With Two and Three Drift Spaces", Kosmicheskiye Issledovaniya, 1965, Vol 3, No. 2, pp 309–314 (TT–65–414, pp 213–220).

86. M. Ye. Slutskiy, B. I. Zarkhin, and M. A. Pushkina, "Miniature Wide-Band Electrometer Amplifier", Kosmicheskiye Issledovaniya, 1963, Vol 1, No. 2, pp 296–302 (TT–64–42, pp 175–185).

87. A. V. Yakovleva, L. A. Kudryavtseva, A. S. Britayev, V. F. Gerasev, V. P. Kachalov, A. P. Kuznetsov, N. A. Pavlenko, and V. A. Iozenas, "A Spectrometric Investigation of the Ozone Layer Up to 60 km Altitude", Iskusstvennyye Sputniki Zemli, 1962, No. 14, pp 57–68 (Artificial Earth Satellites, pp 56–68).

88. S. P. Yatsenko, "Ionization of Gases Carried Into the Upper Layers of the Atmosphere by Satellite", Iskusstvennyye Sputniki Zemli, 1961, No. 7, pp 60–63 (Artificial Earth Satellites, pp 57–62).

89. S. P. Yatsenko, "Chemical Processes in the Ionosphere", V Razdel Programmy MGG: Ionosfera, Sbornik Statey, 1965, No. 14, pp 21–25.

90. B. I. Zarkhin, V. G. Istomin, A. Ye. Rafal'son, and M. Ye. Slutskiy, "Radio-Frequency Mass Spectrometer for Electron Satellite", Kosmicheskiye Issledovaniya, 1965, Vol 3, No. 5, pp 768–781 (TT–65–1702, pp 167–189).

Chapter 6

Optical Phenomena in the Upper Atmosphere

G. E. Wukelic

INTRODUCTION

Numerous aircraft and rocket photometric, spectrophotometric, and spectroscopic measurements of night, twilight, and, more recently, the day airglow have been undertaken in the U.S. since the early 1950's. This work initially in the visible region of the spectrum has been successfully extended to include ultraviolet airglow and aurora measurements. In addition to these studies, the U.S. has recently made rocket observations of daytime aurora and obtained photographs of the airglow from aboard manned satellites.

The USSR on the other hand, although historically quite active in almost all phases of theoretical and ground-based airglow and aurora studies, does not appear to have pursued rocket and satellite airglow and aurora studies with comparable enthusiasm. This is not to say that the Soviets are without such activities. For example, as can be seen in Table 6.1, the Soviets have undertaken numerous day sky brightness studies aboard both meteorological and geophysical rockets and photometric day, twilight, and night airglow measurements (including ultraviolet studies) aboard several geophysical rockets since 1960. More recently the Soviets have reported on spectrophotometric and photographic observations made aboard several manned and unmanned satellites.

TABLE 6.1

Soviet Space Vehicles Associated With Studies of Optical Phenomena in the Upper Atmosphere (1957–1966)

Rockets		Satellites	Space Probes
Meteorological Rockets		Cosmos 45, 65, and 92	Luna 1 and Luna 2
Day sky brightness studies		Night and dayglow	Creation of
beginning in 1958		(visible and ultra-	artificial
Geophysical Rockets		violet studies)	sodium cloud
Day Sky Brightness		Optical studies aboard	for tracking
October 4 1958	(110 km)	Vostoks and Voskhods	purposes
October 10 1958	(110 km)	are discussed in	
December 23 1958	(110 km)	Chapter 12	
December 25 1958	(110 km)		
July 21 1959 a.m.	(110 km)		
July 21 1959 p.m.	(110 km)		
Night Airglow			
June 24 1960	(212 km)		
September 22 1960	(210 km)		
September 23 1960	(212 km) ?		
September 24 1960	(200 km)		
Day and Night Airglow (Ultraviolet Studies)			
June 6 1963	(515 km)		
June 18 1963	(510 km)		
October 25 1963	(430 km)		
December 24 1963	(422 km)		
Artificial Airglow			
September 19 1958	(473 km)		
June 15 1960	(212 km)		

As in the U.S., the objective of such rocket and satellite studies is to determine height, latitude, and seasonal or diurnal variations in intensity of known airglow emissions and, more recently in ultraviolet studies, to measure airglow and aurora emissions not detectable by ground-based observing techniques. Also, in the case of aurora, these studies include the measurement of the charged particle flux (electrons) most closely associated with the auroral excitation mechanism.

INSTRUMENTATION

Rocket Techniques[1–2, 16–18, 25–27, 30–31]

For rocket studies of the day sky brightness the Soviets have utilized two techniques. The earlier technique involved the determination of the day sky brightness values from photographic film exposed during meteorological rocket flights for orientation purposes. The meteorological rocket camera system consisted of five cameras—four with

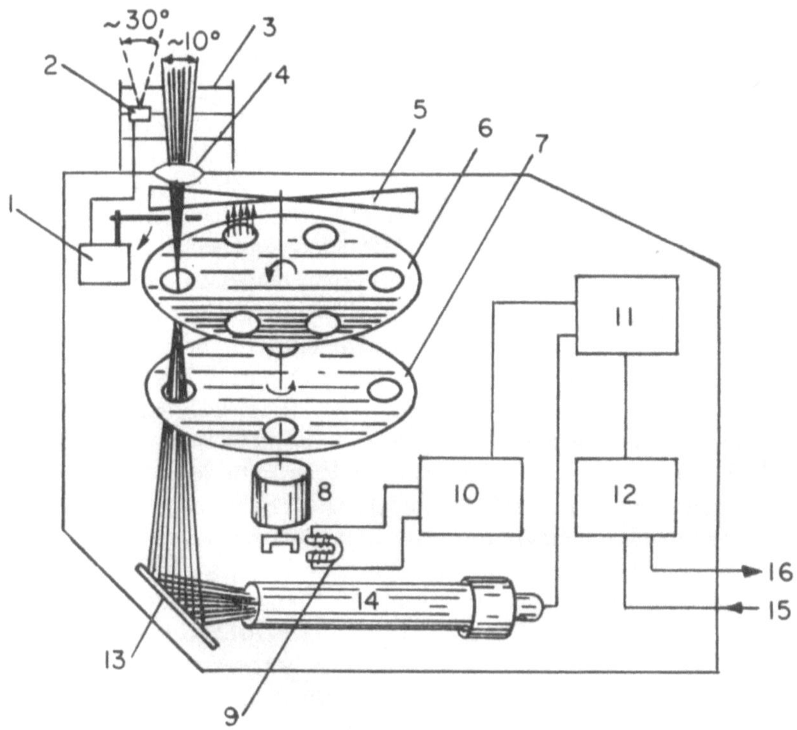

Fig. 6.1. Block Diagram of a Geophysical Rocket Photoelectric Photometer[17]

1. Relay for protective plate
2. Protective resistance
3. Entrance cap
4. Objective
5. Modulating disk
6. Disk with filters and light source
7. Disk with diaphragms
8. Motor
9. Generator
10. Synchronous detector
11. Amplifier
12. Output stage
13. Mirror
14. Photomultiplier (FEU-25)
15. Feed
16. Output to telemetry

mutually perpendicular axes oriented perpendicular to the longitudinal axis of the rocket and the fifth directed vertically upwards. The latter thus continuously photographed the zenith during ascent. A photograph was taken every five seconds. Exposure times were $\frac{1}{20}$ to $\frac{1}{50}$ sec and the camera utilized a PO–4 objective lens with a 35-mm focal length. The system was sensitive to all wavelengths from 4000–6500 Å. The other technique developed at the Institute of Applied Geophysics (IPG) involved the use of several photoelectric photometer designs. The photometer used for geophysical rocket flights to around 100 km had a field of view of about 10°. A block diagram of this instrument is shown in Figure 6.1. This design incorporated a quartz lens, a standard light source, light filters, attenuating diaphragms, a spherical mirror, and a FEU–25 photomultiplier. As can be seen in the figure, the design is such that the filters, an open window, the standard source, and the diaphragms could automatically be introduced into the path of the incident beam of light. By using the filters UFS–1, SS–8, ZhZS–1, OS–14, and ZhS–4 either individually or in combination, the instrument has measured intensities in the 3450, 4150, 5300, and 5850 Å wavelength regions. A complete measurement cycle comprised 24 individual measurements requiring six seconds. Each individual measurement took 0·2 sec. These features enabled the instrument to attain a sensitivity of $0{\cdot}9 \times 10^{-8}$ stilb or approximately 450 Rayleighs (R)*. The introduction of attenuating diaphragms into the design enabled brightness to be measured over a range of $0{\cdot}5 \times 10^{-5}$ to $0{\cdot}5 \times 10^{-8}$ stilb or similarly 250 kR to 250 R. For geophysical rocket sky brightness studies to 500 km, the Soviets developed the FIR–3 photoelectric photometer which could measure small light fluxes in the near ultraviolet, the visible, and the near infrared. The photometer contained photomultipliers (FEU 35 and 62) which provided a working spectral sensitivity range from 1700–11 000 Å. The FIR–3 (see Figure 6.2) operated in the 500 kR to 5 R brightness range. Up to 13 glass filters mounted on a rotating disk were used for spectral range selectivity. The photometer had a measurement cycle of 4–5 sec.

Little information describing Soviet instrumentation utilized in rocket nightglow photometry is available. For the much publicized geophysical rocket flight of September 23 1960 to 200 km, apparently two photoelectric photometers were utilized having sensitivities in the 2600–6000 Å and 6000–11 000 Å ranges. One contained a FEU–VEI–3 photomultiplier (from the All-Union Electrotechnical Institute) with an oxygen-cesium cathode while the other had a FEU–R photomultiplier with an antimony-cesium cathode. Photometer orientation during this

* Rayleigh = 10^6 photon/cm^2 (column) sec $\approx 6 \times 10^{-7}$ apostilb or $\approx 2 \times 10^{-11}$ stilb.

flight was determined to within ±2° between 65–100 km, to within ±10° between 100–130 km, and ±20° between 130–190 km. The field of view of the photometer was 6°. Reportedly, simultaneous measurements (maximum interval 1.6 sec) could be undertaken in all six spectral regions under investigation.

In their 1963 geophysical rocket studies of the ultraviolet airglow, the Soviets employed two SFM–1, nitrogen-oxide-filled, Geiger photon

Fig. 6.2. FIR–3 Photoelectric Photometer for Rocket Studies in the Range 1700–11 000Å[18]

counters with lithium fluoride windows sensitive in the 1050–1340 Å area. A supplementary calcium fluoride filter was placed in front of one of the counters to improve spectral sensitivity for atomic oxygen studies around 1300 Å. Measurement of the L_α intensity at 1216 Å was done with the LiF windows only. The external appearances of both the sensing and electronic units of this instrument are shown in Figure 6.3.

Satellite Technique[6, 13–15]

For measuring nightglow and scattered ultraviolet radiation (on the day side of the earth) aboard Cosmos 45, 65, and 92, the Soviets employed, with slight modification among flights, the equipment shown schematically in Figure 6.4. This equipment consisted of an ultraviolet spectrophotometer and a colorimeter. The ultraviolet spectrophotometer was a double monochromator which operated in the 2250–3100 Å region. Incoming light passed first through the quartz window, then through the spectrophotometer inlet slit, and onto a concave diffraction grating (600 lines/mm) having a focal length of 125 mm. The diffraction grating was oriented for first-order reflection of the maximum amount of light at wavelengths around 2500 Å. A spectrum with a dispersion of 67 Å/mm was received at the image plane of the inlet slit, where a

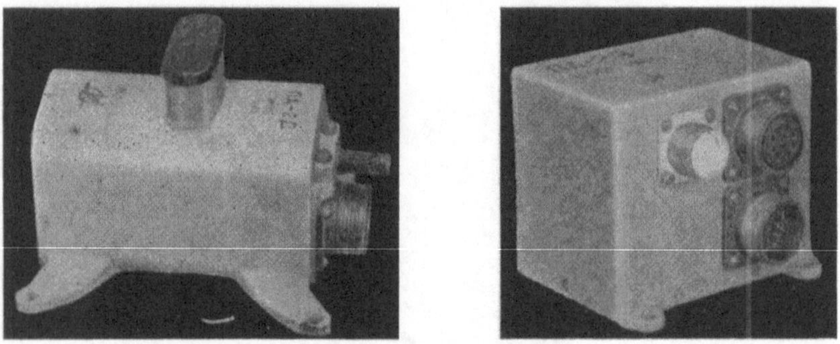

a b

Fig. 6.3. General Appearance of the Instrument for Rocket Ultraviolet Studies (1050–1340 Å)[1]
a. The sensor module containing two SFM–1 (geiger photon) counters
b. Electronic module

movable outlet slit separated a narrow band of wavelengths from the spectrum. The recording cycle took about 30 sec. From the outlet slit, the light was collimated before hitting a flat diffraction grating having 2400 lines/mm. The linear dispersion of the monochromator with the flat diffraction grating was equal in magnitude to that of the concave grating but opposite in sign. Thus, regardless of wavelength, the light hit the same spot on the photomultiplier photocathode. The latter was antimony-cesium on quartz having an amplification factor of around 10^7. Double monochromatization was reportedly employed because of the need for maximum reduction of scattered light. For the same purpose, the spectrophotometer components were blackened, and the different parts of the optical system were photoinsulated from each other as much as possible. Instrument resolution was reported as 15 Å with a 20 km field of view.

The colorimeter for measuring nightglow intensities had a shutter and a disc with light filters arranged along one axis. The shutter modulated the light flux with a frequency of 500 ± 10 cps. The disc with the light filters was rotated once every 12·5 sec. The filters used differed among flights. For example, on Cosmos 45, a ZhS–4 light filter was used to cut off ultraviolet radiation, a UFS–1 for the 2500–4000 Å range, and two

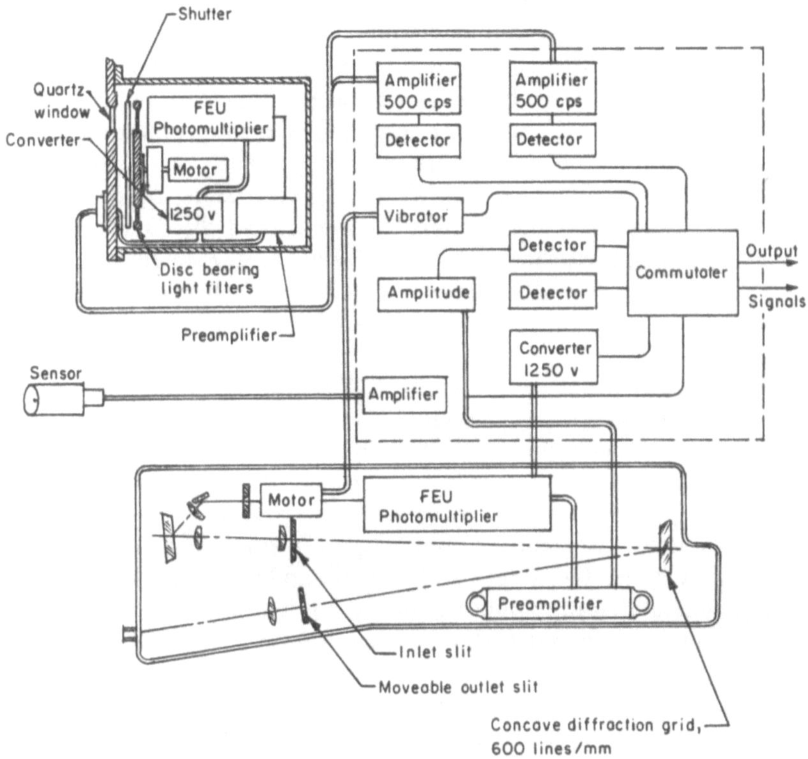

Fig. 6.4. Block diagram of Satellite Visible-UV Photometric Equipment[15]

narrow-band light filters discerned airglow emissions at 5577 and 3914 Å. Three additional filters were used on Cosmos 92. These included the ZhS–18, UFS–3, and a KS–11. The ZhS–18 in combination with the ZhS–4 permitted measurements at 3950–5000 Å and 5000–6000 Å. The UFS–3 covered the 3150–3950 Å range and in combination with the UFS–1 permitted observation in the 2500–3150 Å range. The open window provided for measurements in the 1600–2500 Å region after the UFS–1 and ZhS–4 readings were subtracted. The electrical components of the colorimeter were similar to those used by the ultraviolet spectrophotometer. The colorimeter field of view was 120 km.

SUMMARY OF RESULTS

The Soviets have obtained estimates of the day sky brightness from utilizing meteorological rocket photographs obtained initially for orientation purposes and from direct photometric studies aboard at least six geophysical rockets (during the period 1958–1960). Published results of day sky brightness values obtained from meteorological rocket data have appeared in only one Central Aerological Observatory publication, by L. A. Biryukova (1959)[2]. This work, which apparently was inspired by the earlier results obtained by Western investigators Miley, *et al.* (1953), and Berg (1955), concerning the intensity of the dayglow component, concluded that the total brightness of the day sky in the visible region of the spectrum decreases by a factor of 10^3 between 0–60 km. At 60 km the determined value corresponded to a brightness of over 10^6 R. All results of geophysical rocket investigations of the daytime sky brightness have been authored by A. Ye. Mikirov of the Institute of Applied Geophysics[16, 18–19]. Although the Soviets have noted, particularly in their national reports to COSPAR, that studies were undertaken in the 1700–11 000 Å range, results published to date cover the 3200–3800 Å and 3450–5850 Å spectral intervals only. (Some results of balloon infrared daytime sky brightness studies up to 30 km have, however, been published by N. M. Gopshteyn and V. I. Kishpil'[5].) These papers conclude in general that the brightness of the sunlit atmosphere varies with altitude and latitude. Data show higher intensity values for middle latitudes than for northern latitudes. Also a uniform decrease in sky brightness has been observed above 120 km and at 450 km equals approximately 500 R. The latter value, according to the Soviets, is much higher than can be assigned to Rayleigh scattering. The nature of the excessive glow has not yet been unambiguously established; however, most Soviet publications favor the hypothesis that it is due primarily to scattering by an aerosol layer existing between 80–500 km rather than to intrinsic airglow emissions. A detailed summary of published rocket day sky brightness results are contained in Table 6.2.

For rocket night airglow investigations all items published to date have been by T. M. Tarasova[24–31] of the Institute of Applied Geophysics. These papers, all fairly repetitious, discuss the height distribution and the intensities of the nightglow emissions as measured aboard the geophysical rocket flight of September 1960. In general, published

TABLE 6.2

Results of Optical Studies Aboard Soviet Rockets and Satellites (1957–1966)

Type of Measurement	Instrument Carrier/Dates	Altitude, km	Spectral Region Investigated, Å	Instrumentation/ Technique	Major Results Reported	References
				Rockets		
Day sky brightness	Various meteorological rockets since 1958	$\leqslant 60$	4000–6500	Derived from exposure data obtained by cameras employed aboard meteorological rockets for orientation.	The absolute brightness of the day sky at the zenith and at 60 km corresponded to approximately the limit of the film sensitivity, $>10^6$ R. The author concluded therefore that the brightness of the day sky in the visible region decreases by a factor of 10^3 between 0 and 60 km.	L. A. Biryukova (1959)[2]
Day sky brightness	Two unspecified geophysical rockets	$\leqslant 100$	3450–5850	Photoelectric photometer	Day sky brightness was found to be higher in middle latitudes of the USSR than in northern latitudes. The author assigned this variation to a larger number of scattering aerosols at middle latitudes. The author noted further that experimental data indicated the existence of an aerosol layer around the earth between approximately 80–100 km and perhaps higher.	A. Ye. Mikirov (1962)[16–17]
Day sky brightness	Geophysical rocket June 6 1963 Geophysical rocket June 18 1963	$\leqslant 450$	3200–3800	FIR-3 photoelectric photometer (spectral sensitivity 1700–11 000 Å).	The author reported that the brightness of the atmosphere illuminated by the sun decreases monotonically, starting from an altitude of 120 km; at an altitude of about 450 km it was equal to approximately 500 R. A comparison of the measured values and the values calculated for an absolutely pure atmosphere indicated the presence of a large excess of energy brightness in the daytime sky, apparently caused by aerosol scattering of light. The aerosol scattering factor and the scattering density were evaluated. It was concluded that the maximum concentration of the aerosol layer is at an altitude of about 80 km and that it extends to altitudes above 500 km.	A. Ye. Mikirov (1965)[18]
Day sky brightness	Three unspecified geophysical rockets	$\leqslant 450$	3450–5300	FIR photoelectric photometer	Author concluded that day sky brightness at 70–100 km is generated by molecular and aerosol constituents and at altitudes >100 km by aerosols alone. Data indicated variation of mean particle size with height—with most of the large particles localized in the 120–160 km region with the maximum at 120 km ($\lambda = \cdot 53\mu$). Density of the aerosol substance at 70–450 km was calculated. Maximum density at 120 km equalled 4×10^{-14} g/cm^3. The 5577 Å emission does not contribute significantly to the measured 5300 Å radiation.	A. Ye. Mikirov (1966)[19]

TABLE 6.2 (Continued)

Results of Optical Studies Aboard Soviet Rockets and Satellites (1957–1966)

Type of Measurement	Instrument Carrier/Dates	Altitude, km	Spectral Region Investigated, Å	Instrumentation/ Technique	Major Results Reported	References
Night Airglow (height deter-minations)	Undesignated geophysical rocket in 1960 (most likely September 23 1960)	<200	8650 9100–10 700 5577 5300	Photoelectric photometer	First published results of USSR rocket airglow studies. The author reported detecting layers of hydroxyl emission (73–100 km), O_2 emission (74–100 km), emission of line λ5577 (90; 136–154 km) and glow of continuous atmospheric background of terrestrial origin (64–110 km).	T. M. Tarasova (1961)[24]
Night Airglow (height determina-tions)	Geophysical rocket September 23 1960	65–200	O: 5577 6300 Na: 5893 OH: 9100–10 700 O_2: 8650 Continuous spectrum: 5300	Photoelectric photometer with interference filter of half-width $\Delta\lambda = 280$ Å	Papers contain first detailed results of Soviet rocket airglow investigations. All radiations detected appeared in the $74 < h < 130$ km region. Data obtained indicated the mid-latitude appearance of nightglow emissions having the following intensities and altitude distributions:	T. M. Tarasova (1963)[25–26] T. M. Tarasova and V. A. Slepova (1964)[31]

Intensity (in Rayleighs)

Emission, Å	Altitude 0 ( )	65 (↑)	130 (↓)	180 (↑)
9100–10 700 Relative	14	16	18	1
8650 Relative	21	22·5	25	2
5300 Total	220	185	72	75
/1 Å	1·2 R/Å	1 R/Å	0·4 R/Å	0·4 R/Å
5577 Absolute	240	200	210·5	0
5893 Absolute	270	300	200	90
6300 Absolute	250	300	130	80–100

Layer Height, km

	Maximum	Lower	Upper
9100–10 700	78 ±2	73 ±2	110 ±10
8650	81 ±2	74 ±2	110 ±10
5300	64–180	<64	>150
5577	90	82 ±2	100
5893	80–85	70	>200
6300	?	<130	>200

Type of Measurement	Instrument Carrier/Dates	Altitude, km	Spectral Region Investigated, Å	Instrumentation/ Technique	Major Results Reported	References
Night Airglow (height determinations)	Geophysical rocket September 1960	65–200	6300	Measurements at 6300 Å were carried out by two photometers, one a FEU-VEI-3 photomultiplier utilized an oxygen-cesium cathode, the other an FEU-R photomultiplier used an antimony-cesium cathode.	Analysis of theoretical and experimental (rocket) data permitted the preliminary conclusion that atomic oxygen glow (λ6300) is not localized in the form of a thin layer but is distributed over a great depth of the atmosphere from about 100 km up to heights of more than 1000 km.	T. M. Tarasova (1964)[27], (1966)[30]
Night Airglow (height determinations)	Geophysical rocket September 1960	65–200	5893 6300	Photoelectric photometer	A comparison of sodium (5893 Å) and atomic oxygen radiation (at 6300 Å) revealed an unexpectedly close spatial correlation. The ratio of the total light fluxes and the intensity ratio of the λ 5893/λ 6300 emissions remained constant during rocket ascent and descent. Thus laws governing the height distribution for sodium and atomic oxygen were found to be the same. Also the identification of a significant portion of sodium luminescence (∼30%) at high altitude makes it imperative that the existing concepts regarding the agent responsible for exciting sodium atoms be re-examined. Data obtained supported the conclusion that the lower boundary of the 6300 Å emission is <200 km while the upper boundary for Na luminescence is located >200 km.	T. M. Tarasova (1965)[28], (1966)[29]
Night and Day Airglow (ultraviolet studies) Twilight/ Dayglow Twilight/ Dayglow Nightglow	3 geophysical rockets June 6 1963 (sun 5° above horizon) October 25 1963 (sun 3° above horizon) December 24 1963 (sun 17° below horizon)	∼500	L_α (1216) OI triplet (1302, 1304, 1306)	All three flights employed two SFM-1 Geiger photon counters (filled with nitrogen oxide) having lithium fluoride windows sensitive in the relatively narrow spectrum range of 1050–1340 Å. One of the photometers in each experiment had an additional calcium fluoride filter with known transmissivity. The two photometers were installed in the automatic high altitude geophysical station (VGAS) on all three launchings.	The absolute intensity value for 1216 Å recorded on the June 6 1963 rocket flight was 3·2 kilorayleighs (kR), on October 25 1963, 4 kR, and on December 24 1963, 0·8 kR. Incorporating a correction factor for container gas emission and rotation increases the value to 8 kR for the October flight and 1·5 kR for the December flight. For the OI triplet, the maximum intensity was recorded between 180–200 km. The intensity became constant at 400 km. Data differed slightly during ascending and descending trajectories and according to the probe's axis. For example, at 120 km the intensity at the zenith was 10 times that at the nadir. Also, both the zenith distance and height of the sun affected the maximum airglow intensity. In interpreting their results, Soviet researchers claim that even though their results differ by as much as 2–3 times they must be considered to be in general agreement with Western UV data because of differences in instrumental errors and sun elevations.	S. I. Babichenko *et. al.* (1965)[1] S. A. Kaplan, *et al.* (1965)[7] S. A. Kaplan, *et al.* (1965)[9, 10] V. V. Katyushina, (1965)[11] V. V. Katyushina, *et al.* (1965)[12]

TABLE 6.2 (Continued)

Results of Optical Studies Aboard Soviet Rockets and Satellites (1957–1966)

Type of Measurement	Instrument Carrier/Dates	Altitude, km	Spectral Region Investigated, Å	Instrumentation/ Technique	Major Results Reported	References
				Artificial Satellites		
Day and Night Glow (Visible and UV Studies)	Cosmos 45 September 1964	206–327				A. I. Lebedinskiy, *et al.* (1965)[15]
Day Airglow			2250–3100	Ultraviolet spectrophotometer with a field of view of 20 km and a resolving power of 6 Å. Measurement cycle took 40 sec and resolving power on the earth's surface was ≈ 25 km.	Spectrum of radiation reflected from the earth's atmosphere in the near-ultraviolet was determined primarily by Rayleigh scattering of the solar radiation and ozone absorption. At small zenith distances of the Sun, a considerable portion of the outgoing atmospheric radiation in the wavelength of ~3000 Å was caused by reflection from the clouds, which was revealed by a correlation of observed intensities in the 3000 Å, 3100 Å, 6000 Å, and 8000 Å region. The ozone distribution in the earth's atmosphere was found to exhibit local variations noticeable from the changes in the ultraviolet spectrum and from the intensity of the ozone infrared band in the wavelength 9·6 μ.	
Night Airglow			1700–6000 3900–6000 2400–3900 3914 5577	Colorimeter had three wide band filters and two narrow band filters. The passband of the latter was ≈ 100 Å The light filters were changed every 2·5 sec.	From the results of the colorimeter measurements the cross sections of the earth's night side were constructed and latitude variations and local inhomogeneities noted. In the ultraviolet interval 2400 Å to 3900 Å, relative changes in the light flux were smaller than in the visible portion of the spectrum. Also, measurements indicated that illumination from stars and zodiacal light was relatively small. A comparison of these readings with ground observations led to the conclusion that night airglow radiation at 2500–3000 Å is small and at 3200–4000 Å does not exceed stellar and zodiacal light glow. Measurements at 1700–2500 Å indicated that no night airglow radiation exists in this region. Thus it was concluded that measurements throughout the entire range 1700–4000 Å confirmed the absence of high-energy excitation processes in the night sky.	

Type of Measurement	Instrument Carrier/Dates	Altitude. km	Spectral Region Investigated, Å	Instrumentation/ Technique	Major Results Reported	References
Day Airglow (ultraviolet studies, UV reflected from terrestrial atmosphere)	Cosmos 65 April 1965	210–342	2250–3070	Ultraviolet spectrophotometer oriented 7° toward the nadir. The resolution was 15 Å and effective field of view was $2{\cdot}5 \times 10^{-3}$ sterad. Took one minute to record the spectrum. During instrument's operation (∼55 orbits) about 2500 spectra were obtained.	The UV intensity varied with zenith distance, latitude, and local differences believed caused by variation in ozone concentration. Clouds also had a detectable affect particularly at longer wavelengths. Spectra of the atmosphere and the sun obtained under same conditions had close similarity. However, the decrease in atmospheric intensity with decreasing wavelength was greater than in the solar spectrum. Results were compared to U.S. rocket and satellite data.	V. A. Iozenas, *et al.* (1966)[6] V. A. Krasnopol'skiy. *et. al.* (1966)[13]
Night Airglow (intensity variations in visible and UV emissions)	Cosmos 92 October 1965	212–353	2850–6100 5000–6000 3950–6000 1650–6000 2500–3450 3150–3950 5577 3914	Colorimeter with eight filters, field of view ≈150 km. One revolution of the light filters made each 12·5 sec.	The Soviets indicated that night airglow studies aboard Cosmos 92 confirmed the existence of a correlation of the night airglow in different intervals in the visible and ultraviolet regions. Analysis of the mean results was performed by using mean climatic cloudiness magnitudes which at large distance vary comparatively little. The average intensity of the night airglow in the region 1650–2500 Å gradually reduced from ∼0·5 Rayleigh/Å during early orbits to ∼0·05 Rayleigh/Å beginning with the fifth orbit. Cosmos 92 results showed that the almost complete absence of the night airglow in the hard ultraviolet revealed earlier (Cosmos 45) does not always take place. The average radiation intensity in the interval 2500–3150 Å decreased from ∼1·6 Rayleigh/Å to about 0·6 Rayleigh/Å between the first and fifth orbits. Preliminary results yielded in this region on the average 0·45 Rayleigh/Å. Average intensities of the night airglow in the visible region and in the soft ultraviolet were close to the results of ground-based measurements, and during early orbits were approximately two times higher than later ones. The ratio of average intensities on the first and remaining orbits gradually decreased with the increase in wavelength from ∼10 in the region from 1650–2500 Å to ∼1 in the region 5000–6000 Å. The Soviets also concluded that there are occasional night airglows which manifest themselves on a global scale.	V. A. Krasnopol'skiy, *et al.* (1966)[14]

results suggest that at mid-latitude the continuous nightglow background luminescence (5300 Å) has no distinct stratification, while that of molecular oxygen (8640 Å), hydroxyl (9100–10 700 Å), and atomic oxygen (5577 Å) are localized in narrow layers. The atomic oxygen emission at 6300 Å is not localized but rather is distributed over a region of the atmosphere from about 100 km up to 1000 km. Likewise, sodium glow at 5893 Å revealed an unexpectedly close spatial correlation with the 6300 Å emission. However, Soviet data suggest that the lower boundary of the 6300 Å emission is <200 km, while the upper boundary of the Na (5893 Å) glow is located somewhere above 200 km. The derived values for the height distribution and intensities for all the nightglow emissions investigated aboard Soviet geophysical rockets reported to date are shown in Table 6.2.

Results of rocket ultraviolet airglow investigations are few and appeared first in 1965[1, 7, 9–12]. The publications refer to instruments, results, and interpretation of results associated with three geophysical rocket flights in 1963. Two of the rockets investigated the twilight ultraviolet, having been launched just before dawn, and the other the nightglow, having been launched when the sun was 17° below the horizon. The results of UV airglow studies of the L_α (1216 Å) and the OI triplet radiation at around 1300 Å as reported by the Soviets are noted in Table 6.2. In general the Soviets conclude that, considering instrumental errors and the difference in solar elevations during U.S. and USSR investigations, their results do not differ significantly from those obtained in the West. Soviet measurements, however, have been made to over 500 km whereas corresponding U.S. rocket studies have been considerably lower.

More recently (1965–1966) results of visible and ultraviolet studies undertaken aboard Cosmos satellites 45, 65, and 92 have been reported by A. I. Lebedinskiy, *et al.*, from Moscow State University[6, 13–15]. As can be seen in Table 6.2, these studies have been concerned primarily with spectrophotometric measurements of scattered ultraviolet (between 2250–3100 Å) on the day side of the earth and photometric measurements of both visible and ultraviolet nightglow emissions between 1650–6100 Å. The scattered ultraviolet studies have shown that the UV intensity (between 200–350 km) varies with zenith distance, latitude, and local concentrations of ozone. Clouds were also found to have a detectable effect particularly at longer wavelengths. Satellite studies undertaken on the night side of the earth have provided data on the nature of latitude and local variations of the nightglow in the visible and ultraviolet regions as well as on the illumination component associated with stellar and zodiacal light. Initial measurements aboard Cosmos 45 suggested an absence of hard ultraviolet airglow radiation at 1700–2500 Å. However, subsequent measurements aboard Cosmos 92 detected a slight radiation in this region. For the most part, average recorded intensities of the nightglow in the visible and soft ultraviolet region have paralleled the results obtained from ground-based airglow observations. Additional details of these studies are likewise contained in Table 6.2.

The few published results of optical observations made by Soviet cosmonauts aboard the manned Vostok and Voskhod satellites are discussed in Chapter 12.

RELATED RESEARCH

Another area of space research closely related to atmospheric optical studies is the creation of artificial airglows by ejecting photochemically active materials such as sodium, barium, lithium, cesium, nitric oxide, and ethylene into the upper atmosphere from high altitude research rockets. This chemical release technique, which has been used frequently in the U.S. (since around 1956) for determining the physical and chemical properties of the upper atmosphere and, in particular, in studies of upper atmospheric winds, has also been used in the USSR, but to a much more limited extent. For example, to date the Soviets have associated only two geophysical rockets with such artificial cloud studies and results have been reported for only the sodium cloud experiment of the September 19 1958 geophysical rocket flight to 473 km[22]. The Soviets have, however, shown originality in the successful employment of this technique in space for assisting in the optical tracking of their two initial moon probes, Lunas 1 and 2[8, 21, 23].

Other items of related interest include Soviet published efforts to associate meteoric dust to airglow and zodiacal light, such as those by Mirtov[20] and Divari[4], and to use rocket and satellite composition studies to calculate airglow emission intensities, e.g., Danilov[3]. Also, numerous Soviet papers discuss the association of charged particles to the creation of the aurora. The latter are treated, however, in chapter 10.

References

1. S. I. Babichenko, I. P. Karpinskiy, S. A. Kaplan, V. V. Katyushina, L. N. Krylov, V. G. Kurt, R. M. Pustovayt, and A. V. Shifrin, "Investigation of Scattered Ultraviolet Radiation in the Upper Atmosphere of the Earth. 1. Apparatus", Kosmicheskiye Issledovaniya, 1965, Vol 3, No. 2, pp 237–243 (TT–65–414, pp 91–102).

2. L. A. Biryukova, "Experiment in Determining the Brightness of the Sky Up to an Altitude of 60 Kilometers", Trudy Tsentral'noy Aerologicheskoy Observatorii, 1959, No. 25, pp 77–84 (JPRS: 7489).

3. A. D. Danilov, "The Mechanism of the Stimulation of the Red Oxygen Line in the Airglow", Iskusstvennyye Sputniki Zemli, 1961, No. 8, pp 77–80 (Artificial Earth Satellites, pp 238–241).

4. N. B. Divari, "Contribution of the Circumterrestrial Dust Cloud to the Brightness of Zodiacal Light and the F-Corona", Astronomicheskiy Zhurnal, 1965, Vol 42, No. 3, pp 645–652 (Soviet Astronomy AJ, Vol 9, No. 3, pp 493–499).

5. N. M. Gopshteyn and V. I. Kishpil', "The Daylight Luminescence of the Upper Layers of the Terrestrial Atmosphere in the 1.25 μ Region", Kosmicheskiye Issledovaniya, 1964, Vol 2, No. 4, pp 619–622 (Planetary and Space Science, 1965, Vol 13, pp 457–460).

6. V. A. Iozenas, V. A. Krasnopol'skiy, and A. I. Lebedinskiy, "The Earth's Ultraviolet Spectrum Deduced From Cosmos 65", Paper presented at the Seventh International Space Science Symposium, Vienna, Austria, May 1966 (In English).

7. S. A. Kaplan, V. V. Katyushina, and V. G. Kurt, "Intensity Measurements of Scattered Ultraviolet Radiation (1216 Å and 1300 Å) in the Upper Atmosphere", Space Research V, Proceedings of the Fifth International Space Science Symposium, Florence, Italy, May 12–16 1964, North-Holland Publishing Company, Amsterdam, 1965, pp 595–611.

8. S. A. Kaplan and V. G. Kurt, "Expansion of a Sodium Cloud in Interplanetary Space", Astronomicheskiy Zhurnal, 1960, Vol 37, No. 3, pp 536–542 (Soviet Astronomy AJ, 1960, Vol 4, No. 3, pp 508–514).

9. S. A. Kaplan and V. G. Kurt, "Ultraviolet Radiation Scattering (λ1300 Å) in the Upper Atmosphere of the Earth", Trudy Vsesoyuznoy Konferentsii po Fizike Kosmicheskogo Prostranstva, Izdatel'stvo "Nauka", Moscow, June 10–16 1965, p 112 (NASA TT F–389, pp 152–153).

10. S. A. Kaplan and V. G. Kurt, "Investigation of Scattered Ultraviolet Radiation in the Upper Atmosphere of the Earth 5. Interpretation of Observations of the OI Triplet (λ1300 Å) in the Upper Atmosphere", Kosmicheskiye Issledovaniya, 1965, Vol 3, No. 2, pp 256–261 (TT–65–414, pp 124–134).

11. V. V. Katyushina, "Investigation of the Scattered Ultraviolet Radiation in the Upper Atmosphere of the Earth. 3. Measurements of the Intensity of the Upper Atmosphere Glow in the Lines of the OI Triplet ($\lambda \sim$ 1300 Å) at Altitudes of 100–500 km", Kosmicheskiye Issledovaniya, 1965, Vol 3, No. 2, pp 248–251 (TT–65–414, pp 110–116).

12. V. V. Katyushina and V. G. Kurt, "Investigation of Scattered Ultraviolet Radiation in the Upper Atmosphere of the Earth. 2. Measurements of Scattered Lα Radiation in the Upper Atmosphere at Altitudes up to 500 km", Kosmicheskiye Issledovaniya, 1965, Vol 3, No. 2, pp 243–247 (TT–65–414, pp 102–110).

13. V. A. Krasnopol'skiy, A. P. Kuznetsov, and A. I. Lebedinskiy, "Measurements of the Ultraviolet Spectrum of the Earth Made by the Cosmos 65", Geomagnetizm i Aeronomiya, 1966, Vol 6, No. 2, pp 185–189 (Geomagnetism and Aeronomy, pp 145–148).

14. V. A. Krasnopol'skiy and A. I. Lebedinskiy, "Measurements of Night Airglow From Cosmos 92 Satellite", Paper presented at the Seventh International Space Science Symposium, Vienna, Austria, May 1966 (In English).

15. A. I. Lebedinskiy, V. A. Krasnopol'skiy, A. P. Kuznetsov, and V. A. Iozenas, "Investigation of Terrestrial Atmospheric Radiation in the Visible and Ultraviolet Regions", Trudy Vsesoyuznoy Konferentsii po Fizike Kosmicheskogo Prostranstva, Izdatel'stvo "Nauka", Moscow, June 10–16 1965, pp 78–88 (NASA TT F–389, pp 105–119).

16. A. Ye. Mikirov, "The Aerosol Layer in the Upper Atmosphere", Doklady Akademii Nauk SSSR, 1962, Vol 142, No. 3, pp 587–588 (Doklady of the Academy of Sciences of the USSR, Earth Science Sections, pp 3–4).

17. A. Ye. Mikirov, "A Study of the Day Sky Brightness and the Scattering Coefficient of the Upper Atmosphere", Iskusstvennyye Sputniki Zemli, 1962, No. 13, pp 97–106 (Artificial Earth Satellites, pp 102–112).

18. A. Ye. Mikirov, "Investigations of Atmospheric Brightness at Altitudes of 120–450 km", Kosmicheskiye Issledovaniya, 1965, Vol 3, No. 2, pp 284–296 (TT–65–414, pp 170–191).

19. A. Ye. Mikirov, "Mean Particle Size at 70–450 km Heights", Paper presented at the Seventh International Space Science Symposium, Vienna, Austria, May 1966 (In English).

20. B. A. Mirtov, "Meteoric Matter and Some Problems of Geophysics of the High Layers of the Atmosphere", Iskusstvennyye Sputniki Zemli, 1960, No. 4, pp 118–143 (Artificial Earth Satellites, pp 334–357).

21. I. S. Shklovskiy, "An Artificial Comet as a Method for Optical Tracking of Cosmic Rockets", Iskusstvennyye Sputniki Zemli, 1960, No. 4, pp 195–204 (Artificial Earth Satellites, pp 445–468).

22. I. S. Shklovskiy and V. G. Kurt, "The Determination of the Density of the Atmosphere at an Altitude of 430 Kilometers by the Sodium Vapor Diffusion Method", Iskusstvennyye Sputniki Zemli, 1959, No. 3, pp 66–76 (Artificial Earth Satellites, pp 92–107).

23. I. S. Shklovskiy, V. F. Yesipov, V. G. Kurt, V. I. Moroz, and P. V. Shcheglov, "An Artificial Comet", Astronomicheskiy Zhurnal, 1959, Vol 36, No. 6, pp 1073–1077 (Soviet Astronomy AJ, pp 986–991).

24. T. M. Tarasova, "Direct Measurements of the Night Airglow", Astronomicheskiy Tsirkulyar, May 30 1961, No. 222, pp 31–32.

25. T. M. Tarasova, "Night Sky Emission-Line Intensity Distribution With Respect to Height", Space Research III, Proceedings of the Third International Space Science Symposium, Washington, D.C., May 2–8 1962, North-Holland Publishing Company, Amsterdam, 1963, pp 162–172 (In English).

26. T. M. Tarasova, "Direct Measurement of the Night Airglow in the $\lambda = 8640$ Å Spectral Region", Iskusstvennyye Sputniki Zemli, 1963, No. 13, pp 107–109 (Artificial Earth Satellites, pp 113–115).

27. T. M. Tarasova, "On Airglow of the Atmosphere Above 200 km", Space Research IV, Proceedings of the Fourth International Space Science Symposium, Warsaw, Poland, June 4–10 1963, North-Holland Publishing Company, Amsterdam, 1964, pp 235–243 (In English).

28. T. M. Tarasova, "Eigen Radiation of the Atmosphere", Trudy Vsesoyuznoy Konferentsii po Fizike Kosmicheskogo Prostranstva, Izdatel'stvo "Nauka", Moscow, June 10–16 1965, pp 89–90.

29. T. M. Tarasova, "On the Space Correlation of Night Sky Emission", Paper presented at the Seventh International Space Science Symposium, Vienna, Austria, May 1966 (In English).

30. T. M. Tarasova, "The Nightglow of the Atmosphere in the λ6300 Å Region", Kosmicheskiye Issledovaniya, 1966, Vol 4, No. 2, pp 242–248 (HT–66–244, pp 100–112).

31. T. M. Tarasova and V. A. Slepova, "Altitude Distribution of the Radiation Intensity of the Main Emission Lines of the Night Sky", Geomagnetizm i Aeronomiya 1964, Vol 4, No. 2, pp 321–327 (Geomagnetism and Aeronomy, pp 250–254).

Chapter 7

Physical Properties of the Upper Atmosphere

G. E. Wukelic

INTRODUCTION

As in the U.S., rocket and satellite studies of the physical properties of the upper atmosphere have occupied a major role in Soviet upper atmosphere and space research. These studies started initially on a regular basis in the USSR with meteorological rocket determinations of pressure, temperature, and density to around 80–90 km in 1951. Beginning around 1957, Soviet studies became more sophisticated with the introduction of a variety of geophysical rockets which permitted irregular determination of upper atmosphere properties to altitudes now in excess of 500 km. A discussion of both the meteorological and geophysical rockets utilized in these investigations and descriptions of the associated payloads involved are contained in Part I, Chapter 1.

The total number of over 1000 Soviet meteorological and around 50 geophysical rockets launched during the last 10 years is rather small in comparison to the several thousand U.S. vertical research rockets launched during the same interval. Likewise, the U.S. has geographically extended its rocket sounding capability through various cooperative programs with other countries throughout the world. On the other hand, the USSR has relied on the portable nature of their meteorological rockets to acquire an almost world-wide sounding capability. Thus far

TABLE 7.1

Soviet Rockets and Satellites Associated With Studies of the Physical Properties of the Upper Atmosphere (1957–1966)

Direct Studies	Indirect Studies
	Rockets
Meteorological Rockets	*Geophysical Rockets*
All Soviet meteorological rockets measure temperature and pressure (density computed) in the upper atmosphere < 100 km	Determination of temperature from mass spectrometer (composition) studies: July 14 1959 Sep. 23 1960 July 22 1959 Nov. 15 1961 Determination of density from artificial sodium cloud studies: Sep. 19 1958
Geophysical Rockets	
Principally the determination of density and temperature from *in situ* pressure measurements Aug. 25 1957 June 15 1960 Aug. 31 1957 June 24 1960 Sep. 9 1957 Sep. 6 1960 Feb. 21 1958 Sep. 16 1960 Aug. 2 1958 Sep. 19 1960 Aug. 13 1958 Sep. 21 1960 Aug. 27 1958 Sep. 22 1960 Sep. 19 1958 Oct. 18 1962 Oct. 4 1958 June 6 1963 Oct. 10 1958 June 18 1963 Oct. 31 1958 Dec. 23 1958 Dec. 25 1958 July 2 1959 July 10 1959	
	Satellites
Sputnik 3 (density and temperature determinations from direct pressure measurements)	Orbital data from numerous USSR and U.S. satellites have been used by the Soviets to determine density values.

Soviet rocket launchings have occurred in the central portion of the USSR, the Arctic and Antarctic regions, and, to a lesser extent from research ships in the North and South Pacific Ocean.

With the exception of pressure gauge determinations of density aboard Sputnik 3, satellite studies have been principally indirect and involve the determination of density variations from the analysis of orbital decay data. Also, some published research effort has been associated with developing techniques for interpreting atmospheric property variations from mass spectrometer and thermal radiation studies aboard both rockets and satellites. Specific Soviet rocket and satellite vehicles that have been associated with either direct or indirect studies of the physical properties of the upper atmosphere are noted in Table 7.1.

INSTRUMENTATION

Meteorological Rockets[4, 6, 8–9, 27, 29, 59]

Almost all meteorological rockets launched in the USSR are equipped with resistance thermometers for measuring atmospheric temperature. Normally four such thermometers (comprised of tungsten filaments 40 μ in diameter) are mounted in the center section of the thin (25-mm diameter, 740 mm long) boom or spire projecting from the top of the meteorological rocket nose cone (see Figure 1.6). In addition, the temperature of the boom is determined by an additional thermometer wrapped around it. Four bolometers are located at the base of the boom for measuring the intensity of the radiant energy affecting the thermometer readings. The Soviets indicate that the standard deviation of individual temperature measurements acquired in this manner does not exceed $\pm 5°$ up to 40 km, at an altitude of 50 km it amounts to $\pm 10°$, and increases to approximately $\pm 20°$ at altitudes of 70 km or higher. The maximum altitude for the application of this method in its present form is reportedly limited to around 75–80 km. In their routine use of this method, the Soviets continuously compare results of temperature measurements with values derived from measured pressure values.

The measurement of pressure aboard Soviet meteorological rockets is accomplished by using ordinary membrane (diaphragm) transducers sensitive to the pressure range 760 to 5 mm Hg and two types of hot-wire (Pirani) gauges for the pressure range 5·0 to 5×10^{-3} mm Hg. The first type of hot-wire gauge is a tungsten filament (13 μ in diameter) pressure indicator sensitive in the range 5·0 to 0·1 mm Hg. The second of the hot-wire gauges is a helical tungsten filament (with a helix diameter of 60 μ) sensitive to the range 0·3 to 5×10^{-3} mm Hg. The pressure range measured by the diaphragm gauges is divided into three sub-ranges, and a separate gauge is used for each range. Also resistance thermometers are used to check the temperature of the pressure gauges during flight. The actual arrangement of the pressure gauges in the nose cone of Soviet meteorological rockets is shown in Figure 1.6. The Soviets indicate that the standard deviation of a single pressure measurement is equal to ±4 per cent. Analogous to the temperature procedure, the measured pressure values are checked against calculated values derived from direct temperature measurement.

Usually, atmospheric densities at meteorological rocket altitudes are computed from measured values of temperature and pressure by use of the equation of state. However, the Soviets have been conducting experiments to measure density directly aboard meteorological rockets, using a so-called ion densimeter. In this technique lithium ions are introduced into the unperturbed region just ahead of the rocket nose cone. These lithium ions move under the influence of an electric field to the receiver. The amount of atmospheric retardation, which is proportional to atmospheric density, determines the magnitude of the ion current. According to Soviet reports, measured density values using this technique thus far have exceeded accepted density values by about 20–30%. However, further improvements in this technique are reportedly possible. Currently the Soviets are considering the possibility of using resistance thermometers aboard meteorological rockets for directly determining atmospheric density at altitudes between 60–80 km. Reports indicate that theoretical analyses and experiments performed to date show this method to be feasible to heights of at least 85–90 km.

Geophysical Rockets[27, 47, 49–50, 52–53]

Direct measurement of physical properties of the upper atmosphere aboard Soviet geophysical rockets has been limited to pressure studies from which density and temperature values are calculated using the barometric formula. As noted and described in considerable detail in Chapter 1, the Soviets employ two types of instrumented geophysical rocket payloads (nonoriented cylindrical container, Figure 1.10, and an automatically oriented spherical container [VGAS], Figure 1.11, for making direct soundings of the upper atmosphere to altitudes up to 500 km). Both containers are jettisoned from the parent rocket to minimize measurement contamination and are parachute-recovered.

For making pressure measurements aboard both types of containers, the Soviets use varying numbers and types of pressure gauges. In earlier (1957–1958) soundings using the nonoriented container, the Soviets employed magnetic (electric-discharge) gauges sensitive in the range 5×10^{-2} to 1×10^{-5} mm Hg and thermal (ionization) gauges operating in the range 8×10^{-1} to 5×10^{-2} mm Hg. The sealed gauges were mounted in the environment-exposed portion of the cylindrical instrument container (Figure 1.10), and activated by breaking a special seal at the required height. Pressure measurements aboard the spherical high altitude geophysical automatic station (VGAS) during the 1962–1963 period (the last studies reported to date) involved as many as five pressure gauges per flight. In these experiments three magnetic gauges (2×10^{-2} to 3×10^{-6} mm Hg) and two thermal (ionization) gauges (10^{-5} to 10^{-9} mm Hg) were uniformly distributed around the spherical container.

Artificial Satellites[3, 27, 29, 47]

Although numerous satellites have been utilized in the indirect determination of upper atmosphere density, the Soviets have associated only Sputnik 3 with direct measurements of the physical properties of the upper atmosphere. This satellite, launched in 1958, carried two ionization gauges and one magnetic (electric-discharge) gauge designed for *in situ* pressure measurement from which other properties were determined. The magnetic gauge pictured in Figure 7.1 was designed to measure pressure in the range 10^{-5} to 10^{-7} mm Hg. The ionization gauges, shown in Figure 7.2, were sensitive in the pressure range 10^{-7}

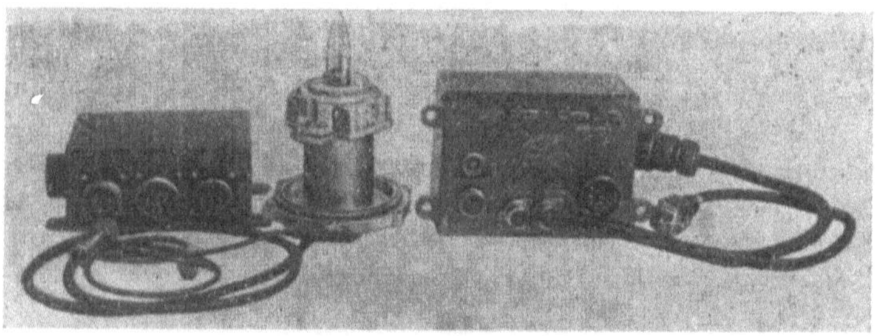

Fig. 7.1. Magnetic Pressure Gauge Aboard Sputnik 3[3]

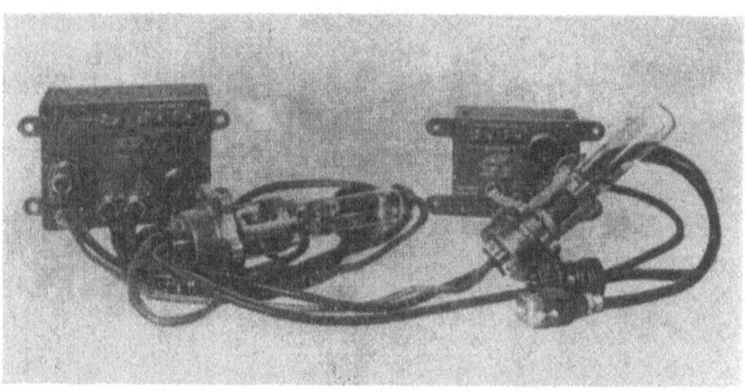

Fig. 7.2. Ionization Pressure Gauges Aboard Sputnik 3[3]

to 10^{-9} mm Hg. Figures 7.3 and 7.4 show schematic drawings of both type gauges.

Stabilization of the emission current in the ionization gauge (Figure 7.4) was accomplished by an additional flat grid (4) placed between the anode grid (1) and the cathode (3). The cathode was made of a

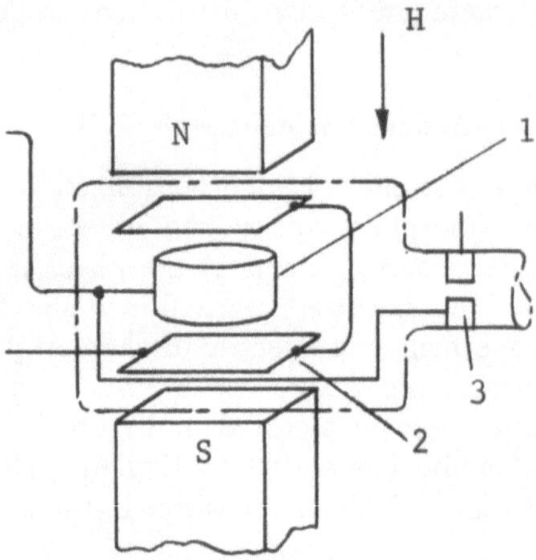

Fig. 7.3. Schematic Drawing of Sputnik 3 Magnetic Pressure Gauge[47]

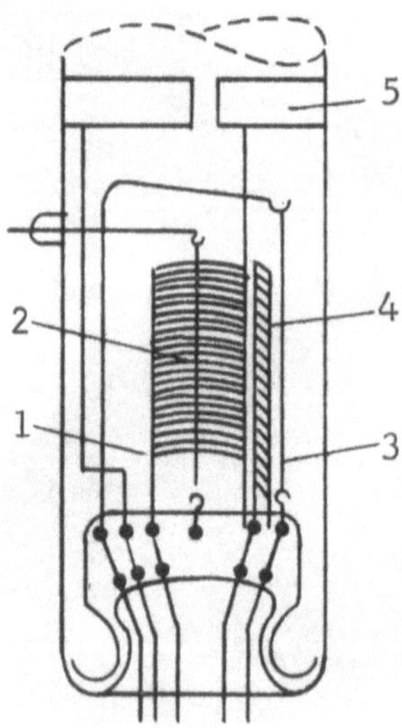

Fig. 7.4. Schematic Drawing of Sputnik 3 Ionization Pressure Gauge[47]

tungsten wire 0·05 mm in diameter and arranged asymmetrically. The cylindrical anode was made of nickel wire 0·2 mm in diameter. The collector (2) was made of tungsten wire 0·1 mm. The gauge contained traps (5) at a high potential for preventing penetration of positive ions from the ambient medium.

In the magnetic gauge (Figure 7·3) the anode was designed as a ring (1) attached to a rod electrode, while the cathode was in the form of plates (2). The penetration of ambient positive ions was once again prevented from entering the gauge by the employment of highly charged traps (3) connected to the anode. The entire assembly was placed inside a strong magnetic field to increase the discharge current and to maintain a continuous discharge at low pressure.

SUMMARY OF RESULTS

Direct Studies

Although general rocket information is annually supplied to the Committee on Space Research (COSPAR) by the Soviets, surprisingly little has been published or exchanged in the way of Soviet rocket data. As shown in Table 7.2, results of meteorological rocket determinations of temperature, pressure, and density published to date have been limited to infrequent analysis of older data obtained primarily during the International Geophysical Year (IGY) and International Geophysical Co-operation (IGC) periods. Most of this work has been done by I. A. Khvostikov, Yu. A. Bragin, Ye. G. Shvidkovskiy, P. P. Alekseyev, M. N. Izakov, G. A. Kokin, G. I. Golyshev, S. P. Perov, A. I. Ivanovskiy, and A. M. Kasatkin of the Central Aerological Observatory (TsAO). Papers by these authors consider types of thermal stratification, characteristics of diurnal, annual, and geographical temperature, pressure, and density variations, and explanations for anomalies noted in rocket soundings below 100 km. Similarly the characteristics of stratospheric winds and circulation with height, latitude, and season are frequently discussed[1, 3–6, 13, 27–28, 53, 56, 66, 68–70, 75, 78]. Several recent papers describe Soviet attempts to determine density directly using resistance thermometers and lithium-ion densimeters aboard meteorological rockets[8, 9, 59]. According to Soviet reports, most meteorological rocket data are in good agreement with calculated values and U.S. rocket results. These claims are supported somewhat by the similarity of Soviet standard atmospheres (SA–48 and SA–60) and recent U.S. reference atmospheres for the 0 to 80 km region.

TABLE 7.2

Results of Soviet Rocket and Satellite Studies of the Physical Properties of the Upper Atmosphere (1957–1966)

Vehicle/ Date of Investigation	Atmospheric Property Determined	Altitude, km	Technique	Results	References
				Rockets (*Direct Measurements*)	
Meteorological Rockets (Middle latitudes of the USSR, Arctic and Antarctic regions, in the tropics, and in middle latitudes of southern hemisphere)	Direct temperature and pressure measurements Density calculated from measured pressure and temperature values	<100	Temperature measurements made by tungsten resistance thermometers. Pressure measurements made by (diaphragm) resistance pressure gauges (range 760 to 5 mm Hg) and hot wire (Pirani) gauges (range $5–5 \times 10^{-3}$ mm Hg)	Regular meteorological rocket soundings of the upper atmosphere have been in progress in the USSR since 1951. Since 1957 some 1000 rockets have been launched for measuring thermodynamic properties (temperature and pressure) and dynamic properties (wind velocity and direction) below 100 km. General information concerning these launchings has been routinely reported by the Soviets but publication or exchange of USSR rocket data has been infrequent. To date, limited data available have been for the IGY and IGC periods (1957–1960). Soviet publications have emphasized the extent of annual and geographical temperature variations, stratospheric temperature stratification, sudden variation in polar temperatures, comparisons of physical properties over oceans and continents, and seasonal data on stratospheric winds.	Anonymous (1963)[1], Trudy Central Aerological Observatory (1964)[2], Alekseyev, *et al.* (1957)[1], (1962)[5], Dubentsov (1961)[13], Khvostikov (1964)[27], Khvostikov, *et al.* (1963)[28], Mikhnevich and Khvostikov (1957)[53], Perov (1965)[56], Ryazantseva and Khvostikov (1965)[66], Shvidkovskiy (1958)[68], (1959)[69], (1960)[70], Yakovkin and Korkina (1961)[75], Zhanturov (1965)[78]
	Wind velocity and direction determined by several techniques	—	Determinations of velocity and direction of winds have been made by employing principally the parachute tracking technique but artificial smoke cloud grenades, and dipole reflectors (metallized chaff) techniques have also been referred to upon occasion.		
Meteorological Rockets (Middle latitudes)	Density	50–70	Determination of density directly using ion densimeter in which lithium ions are introduced into the atmosphere ahead of the rocket	Density values determined by this technique were within 5% of those calculated from temperature and pressure values measured at the same time using standard measurement technique. Reportedly, further improvements in densimeter technique should eliminate all errors.	Bragin (1962)[8], (1964)[9]
Meteorological Rockets	Density	—	Development of a method to measure upper atmospheric density by using standard resistance thermometers	Soviets have demonstrated that resistance thermometers can be used aboard meteorological rockets for measuring density directly. Density data resulting from two experiments (spring and autumn) between 60–80 km are presented.	Perov (1965)[56]

Vehicle/ Date of Investigation	Atmospheric Property Determined	Altitude, km	Technique	Results	References
Geophysical Rockets (Before and during 1957)	Pressure (direct measurement) Temperature (computed from pressure)	57–100	Thermal (Pirani) and magnetic pressure gauges (8×10^{-1} to 2×10^{-5} mm Hg)		Mikhnevich (1957)[46], Mikhnevich and Khvostikov (1957)[53]

Altitude, km	Pressure, mm Hg	Temperature (determined from pressure), °K
57·5	$3{\cdot}15 \times 10^{-1}$	303
60·0	$2{\cdot}39 \times 10^{-1}$	282
62·5	$1{\cdot}71 \times 10^{-1}$	263
65·0	$1{\cdot}23 \times 10^{-1}$	245
67·5	$8{\cdot}75 \times 10^{-2}$	237
70·0	$6{\cdot}05 \times 10^{-2}$	221
72·5	$4{\cdot}14 \times 10^{-2}$	204
75·0	$2{\cdot}69 \times 10^{-2}$	187
77·5	$1{\cdot}69 \times 10^{-2}$	172
80·0	$1{\cdot}01 \times 10^{-2}$	161
82·5	$5{\cdot}90 \times 10^{-3}$	154
85·0	$3{\cdot}35 \times 10^{-3}$	154
87·5	$1{\cdot}93 \times 10^{-3}$	156
90·0	$1{\cdot}15 \times 10^{-3}$	163
92·5	$6{\cdot}90 \times 10^{-4}$	173
95·0	$4{\cdot}30 \times 10^{-4}$	181
97·5	$2{\cdot}77 \times 10^{-4}$	192
100·0	$1{\cdot}81 \times 10^{-4}$	208

Vehicle/ Date of Investigation	Atmospheric Property Determined	Altitude, km	Technique	Results	References
Various Geophysical Rockets (1957–1958)	Density determined from *in situ* pressure measurements	60–260	Magnetic (10^{-5} to 10^{-7} mm Hg) and ionization (10^{-5} to 10^{-9} mm Hg) pressure gauges	Density of the Atmosphere from Measurements Using Rockets (1957–1958)	Mikhnevich (1958)[47], Mikhnevich and Khvostikov (1957)[53]

Altitude, km	Density, g/cm³	Altitude, km	Density, g/cm³
60	$3{\cdot}9 \times 10^{-7}$	170	$6{\cdot}4 \times 10^{-13}$
70	$1{\cdot}36 \times 10^{-7}$	180	$4{\cdot}4 \times 10^{-13}$
80	$2{\cdot}9 \times 10^{-8}$	190	$3{\cdot}3 \times 10^{-13}$
90	$3{\cdot}3 \times 10^{-9}$	200	$2{\cdot}7 \times 10^{-13}$
100	$4{\cdot}0 \times 10^{-10}$	210	$2{\cdot}0 \times 10^{-13}$
110	$9{\cdot}8 \times 10^{-11}$	220	$1{\cdot}6 \times 10^{-13}$
120	$2{\cdot}2 \times 10^{-11}$	230	$1{\cdot}25 \times 10^{-13}$
130	$7{\cdot}4 \times 10^{-12}$	240	$1{\cdot}1 \times 10^{-13}$
140	$3{\cdot}2 \times 10^{-12}$	250	$9{\cdot}0 \times 10^{-14}$
150	$1{\cdot}7 \times 10^{-12}$	260	$6{\cdot}9 \times 10^{-14}$
160	$9{\cdot}5 \times 10^{-13}$		

TABLE 7.2 (Continued)

Results of Soviet Rocket and Satellite Studies of the Physical Properties of the Upper Atmosphere (1957–1966)

Vehicle/ Date of Investigation	Atmospheric Property Determined	Altitude, km	Technique	Results	References
Geophysical Rockets February 21 1958 August 27 1958	Density determined from *in situ* pressure measurements	100–200	Magnetic and ionization pressure gauges. Magnetic gauge sensitivity 10^{-5} to 10^{-7} mm Hg August 27 two magnetic, one ionization	A comparison of 1958 data with 1962–1963 data showed that the density below 200 km did not vary much with solar activity. However, a diurnal effect was observed below 200 km. The value of density above 200 km appreciably decreased from 1958 to 1963 with the decrease of solar activity.	Mikhnevich (1965)[49], Mikhnevich, *et al.* (1965)[52]
October 18 1962	Density determined from *in situ* pressure measurements	100–350	Ionization gauge on VGAS (10^{-5} to 10^{-9} mm Hg)		
June 6 1963	Density determined from *in situ* pressure measurements	100–180	Two magnetic gauges 2×10^{-2} to 3×10^{-6} mm Hg and one ionization gauge 10^{-5} to 10^{-9} mm Hg on VGAS		
June 18 1963	Density determined from *in situ* pressure measurements	50–380	Three magnetic gauges 2×10^{-2} to 3×10^{-6} mm Hg; two ionization gauges 10^{-5} to 10^{-9} mm Hg on VGAS		
Geophysical Rockets June 6 1963 and June 18 1963	Calculated temperature and density from measured pressure values	$\leqslant 500$	VGAS Ionization and magnetic gauges	Comparison of density values with those obtained by Sputnik 3 in 1958 showed that at 300 km the density decreased by a factor of 10 over 1958 values due to decreased solar activity. Correspondingly temperature at 300 km during the two periods changed from 1159° K (1958) to 775° K (1963). Unsuspected was that between 150–170 km the temperature did not increase monotonically but rather an inversion was observed. Inversion considered to be temporary condition possibly due to inversion in molecular weight but requires further study.	Mikhnevich (1965)[50]

Vehicle/ Date of Investigation	Atmospheric Property Determined	Altitude, km	Technique	Results	References
				Rockets (Artificial Sodium Cloud Technique)	
Geophysical Rocket September 19 1958	Density	430	Determination of atmospheric density from studies of artificial sodium cloud diffusion	First approximation of density at 430 km from this method equalled $4{\cdot}7 \times 10^{-15}$ g/cm^3. More rigorous calculations gave a value of $6{\cdot}7 \times 10^{-15}$ g/cm^3. These values reportedly agreed well with density values determined by satellite drag technique.	Shklovskiy and Kurt (1959)[67]
				Rockets (Mass Spectrometer Technique)	
Geophysical Rockets				Maximum error in temperature determination reportedly less than 10%. The following temperature values were determined:	Pokhunkov (1961),[57] (1962)[58], (1963)[59], (1966)[60]
July 14 1958 (morning)	Temperature	100–210	Temperature values from mass spectrometer characterization of the variation of the relative pressure of two gases (Ar and N_2) or changes in the ratio of ion currents Ar/N_2.	h, km: 100 110 120 130 140 150 160 170 T, °K: 230 390 580 670 750 810 870 930	
July 22 1959 (morning)				180 190 200 210 960 990 1020 1050	
September 23 1960 (00h 56m)	Temperature	100–200	Temperature values from mass spectrometer characterization of the variation of the relative pressure of two gases (Ar and N_2) or changes in the ratio of ion currents Ar/N_2.	h, km: 100 110 120 130 140 150 160 170 T, °K: 215 265 325 395 490 600 715 785 180 190 200 210 825 860 895 925	
November 15 1961 (16h00m)	Temperature	130–325	Temperature values from mass spectrometer characterization of the variation of the relative pressure of two gases (Ar and N_2) or changes in the ratio of ion currents Ar/N_2.	h, km: 130 140 150 160 170 180 190 T, °K: 425 480 600 700 1000 1070 1150 200 220 240 260 280 300 325 1200 1260 1300 1360 1410 1440 1470	
				Artificial Earth Satellites (Direct Measurements)	
Sputnik 3 May 1958	Density determined from *in situ* pressure measurements	260–355	Two ionization gauges and one magnetic gauge	At 260 km the density equalled 10^{-13} g/cm^3 and at 355 km 9×10^{-15}g/cm^3	Krasovskiy (1958 IAF)[36]

TABLE 7.2 (Continued)

Results of Soviet Rocket and Satellite Studies of the Physical Properties of the Upper Atmosphere (1957–1966)

Vehicle/ Date of Investigation	Atmospheric Property Determined	Altitude, km	Technique	Results	References
Sputnik 3 May 16 1958 (57° N–65° N) (1300–1900h local time)	Pressure measured directly (temperature and density determined from pressure measurements)	225–500	Two ionization gauges and one magnetic gauge	Physical properties between 225 and 500 km as determined from Sputnik 3 pressure measurements are noted below. These values were reportedly in fair agreement with density values determined from the orbital drag of various U.S. and USSR satellites. The pressure and density values obtained were however 1·5–10 times larger than values proposed by early atmospheric models.	Mikhnevich, *et al.* (1959)[51]

Height, km	P, mm Hg	ρ, g/cm^3	T, °K
225	$6{\cdot}25 \times 10^{-7}$	$2{\cdot}12 \times 10^{-13}$	936
230	5·54	1·79	938
235	4·92	1·70	941
240	4·40	1·42	946
245	3·94	1·25	952
250	3·54	1·10	958
255	3·17	$9{\cdot}73 \times 10^{-14}$	964
260	2·88	8·66	971
265	2·60	7·77	979
270	2·35	6·83	987
275	2·14	6·10	996
280	1·95	5·44	1005
285	1·78	4·87	1015
290	1·62	4·36	1026
295	1·49	3·93	1037
300	1·37	3·53	1048
305	1·29	3·26	1059
310	1·17	2·90	1072
315	1·008	2·63	1084
320	1·00	2·39	1097
325	$9{\cdot}28 \times 10^{-8}$	2·17	1110
330	8·62	1·98	1124
335	8·06	1·82	1138
340	7·52	1·66	1153
345	7·46	1·52	1169

Vehicle/ Date of Investigation	Atmospheric Property Determined	Altitude, km	Technique	Results	References

Height, km	P, mm Hg	ρ, g/cm^3	T, °K
350	6·58	1·40	1185
355	6·18	1·29	1200
360	5·82	1·19	1219
365	5·53	1·10	1237
370	5·19	1·02	1257
375	4·90	$9{\cdot}41 \times 10^{-15}$	1276
380	4·64	8·72	1295
385	4·44	8·24	1305
390	4·19	7·56	1335
395	3·98	7·07	1353
400	3·79	6·60	1373
405	3·60	6·16	1393
410	3·46	5·78	1417
415	3·30	5·41	1440
420	3·17	5·09	1465
425	3·04	4·79	1489
430	2·93	4·51	1514
435	2·82	4·25	1539
440	2·72	4·03	1563
445	2·62	3·80	1589
450	2·53	3·60	1614
455	2·44	3·40	1643
460	2·37	3·23	1675
465	2·30	3·06	1709
470	2·25	2·92	1745
475	2·19	2·79	1781
480	2·13	2·65	1810
485	2·08	2·53	1845
490	2·02	2·42	1880
495	1·98	2·31	1917
500	1·94	2·21	1953

Artificial Earth Satellites (Indirect Studies)

Vehicle/ Date of Investigation	Atmospheric Property Determined	Altitude, km	Technique	Results	References
Sputniks 1 and 2 (Also Sputnik 1 rocket) (1957)	Density	225	Density determination from satellite drag	Average density from Sputnik 1 and rocket carrier data equalled $2{\cdot}9 \times 10^{-13}$ g/cm^3. Density from Sputnik 2 data equalled $4{\cdot}1 \times 10^{-13}$ g/cm^3. The difference resulted because Sputnik 1 represented nighttime values whereas Sputnik 2 data were for the sunlit zone.	El'yasberg (1958)[14]

TABLE 7.2 (Continued)

Results of Soviet Rocket and Satellite Studies of the Physical Properties of the Upper Atmosphere (1957–1966)

Vehicle/ Date of Investigation	Atmospheric Property Determined	Altitude, km	Technique	Results	References
Sputniks 1–3 Sputnik 1 Rocket U.S. Explorer Satellites (1957–1958)	Density	228–368	Density determination from satellite drag	At a height of 228 km (Soviet satellite data) the density was between 2·4 to $3{\cdot}2 \times 10^{-12}$ g/cm^3 whereas at 368 km (U.S. satellite data) the density was calculated to be between 1·37 to $1{\cdot}51 \times 10^{-14}$ g/cm^3. Diurnal variation at mean latitudes amounted to 20–30%.	Lidov (1958)[36]
Sputnik 3 (1958)	Density	246	Density determination from satellite drag	Calculated values varied between 1·61 to $1{\cdot}94 \times 10^{-13}$ g/cm^3 with an average of $1{\cdot}72 \times 10^{-13}$ g/cm^3. Since observations were for only several hours, small variations were not ascribed to a solar influence.	El'yasberg and Yastrebov (1960)[15]
Sputnik 1 (Rocket carrier) and Sputnik 2 Sputnik 3 (Rocket carrier) (1957–1958)	Density	$\sim$226	Density determination from satellite drag	Most calculated values increased at noon and decreased at midnight. There was also a regular eight-hour variation which could be caused by tidal action. Also some data showed maxima at midnight and minima at noon which could not be ascribed to either source.	Kolegov (1960)[32]
Sputnik 3 (1957) Explorer 1 (1958) Explorer 8 (1960) Explorer 11 (1961)	Density	180–200	Density determination from satellite drag	Results showed a diurnal effect beginning at 180–200 km and increasing with height. Initial results indicated that variations in density occurred two to three days after corresponding variations in solar activity.	Fominov (1963)[16–17]
U.S. and Soviet Satellites (1957–1960)	Density	>200	Density determination from satellite drag	World-wide review of satellite drag determinations of density variations due to diurnal, monthly, semiannual, solar activity, geomagnetic, and latitude effects. Concluded that density latitude dependence found in U.S. and USSR rocket experiments was probably due to diurnal and geomagnetic variations.	Mikhnevich (1963)[48]
Cosmos 3 (April–May 1962) Cosmos 5 (June–November 1962)	Density	>200	Density determined from satellite drag	For altitudes of $\sim$200 km the average density determinations were $5{\cdot}8 \times 10^{-13}$ g/cm^3 (for H = 30 km), $4{\cdot}7 \times 10^{-13}$ g/cm^3 (for H = 45 km) and $4{\cdot}1 \times 10^{-13}$ g/cm^3 (for H = 60 km).	Marov (1963)[40]
Cosmos 1, Cosmos 2, Cosmos 3, Cosmos 5 (April 1962–August 1963) Cosmos 8 Spaceship-Satellite 1 (May 1960–May 1963)	Density	180–300	Density determination from satellite drag	For altitudes of $\sim$200 km, the average density values were $1{\cdot}9 \times 10^{-13}$ g/cm^3 (H = 26 km) at 230 km, $6{\cdot}5 \times 10^{-14}$ g/cm^3 (H = 41 km). At 300 km a density value of $\sim 1{\cdot}5 \times 10^{-14}$ g/cm^3 (H = 60 km) was determined.	Marov (1964)[41–42]

Vehicle/ Date of Investigation	Atmospheric Property Determined	Altitude, km	Technique	Results	References
16 Soviet Satellites Including: 14 Cosmos Spaceship-Satellite 1 Sputnik 3 (1958–1964)	Density	180–300	Density determination from satellite drag	Extensive information was obtained on the distribution of density for average diurnal, maximum daytime, and minimum nighttime conditions during solar activity minimum. A correlation of density with decimetric solar radiation was confirmed during the above-mentioned period and diurnal variations were recorded which ranged from approximately 50–60% near 200 km to a factor of 2 at 300 km. On the basis of the distribution obtained and correlation with the solar decimetric wavelength flux a tentative prediction of atmospheric density between 180 and 300 km for the next few years is presented.	Marov (1965)[44–45]
				Comparison of 1964 and 1958 data (for a period of maximum solar activity) showed a decrease in average density of a factor of 2 at 200 km to 3·5–4·0 near 300 km.	Marov (1965)[43]
Sputnik 3 (November 1958– February 1960)	Density	180–200	Density determination from satellite drag	Diurnal variations in density values noted maximum variation for 200 km was $\sim 2\times10^{-13}$ g/cm^3. During the daytime a variation of $\sim 0{\cdot}6\times10^{-13}$ g/cm^3 was found. The average day to night change between 205–206 km was $\sim 1{\cdot}5\times10^{-13}$ g/cm^3.	Kozlov (1965)[74]
Spaceship-Satellite 1 (June 1960– December 1963) Cosmos 2 (April 1962– March 1963) Cosmos 11 (October 1962 May 1963)	Density	200–300	Density determination from satellite drag	Diurnal fluctuations of atmospheric density occurred at altitudes of 200–300 km during 1960–1963. Ratio max/min = 1·9 at an altitude of about 300 km (Spaceship-Satellite 1). At an altitude of about 250 km, max/min = 1·5 (Cosmos 11). At an altitude of about 200 km, max/min = 1·1 and 1·2 (Cosmos 2). The results obtained for low satellites were reportedly in good agreement with Western results. Also supported earlier U.S. and Soviet conclusion that near solar activity minimum, the amplitude of the diurnal atmospheric density variations at 200 km remain the same as in the period of near maximum activity.	Slovokhotova (1965)[72]
Soviet and U.S. Satellites (1957–1963)	Wind velocity	170–360	Wind velocity (or index of circulation) determinations from satellite orbital data	A mean circulation index of −0·4 was determined for the case when the upper atmosphere assumed to have the same velocity as solid earth. Technique believed to be more representative than those involving noctilucent clouds and ionospheric drift observations since the wind values are averaged over a rather long perigee segment and random local variations are eliminated. Tables showing satellite, date, perigee, and perigee latitude, and circulation index are presented.	Trubnikov (1965)[74]

Direct measurements of the physical properties of the upper atmosphere above 100 km in the USSR have been rather infrequent. Since 1957, only 20 some geophysical rockets and one satellite, Sputnik 3, have been associated with such studies. In both cases, the previously described technique of determining density from *in situ* pressure measurements has been the principal means for obtaining such data between 100–500 km. These measurements aboard Soviet rockets and satellites, have been accomplished principally by V. V. Mikhnevich and her associates at the Institute of Applied Geophysics (IPG). No mention of any Soviet interest in or attempt to use the falling sphere density determination technique, commonly employed in the U.S., was noted in available Soviet literature.

As in the U.S., direct structural property data are too few and errors too large to permit accurate determination of the fine structure of the upper atmosphere. Soviet results reported to date (see Table 7.2) are extremely sparse and are limited basically to comparisons of data obtained during geophysical rocket experiments in 1957–1963 and Sputnik 3 in 1958. In general the Soviets found little if any density dependence upon solar activity below 200 km, although a diurnal effect, below and above 200 km, was discernible. Also, density above 200 km showed a heavy dependence upon solar activity. For example, the density at 300 km was found to decrease by a factor of 10 between 1958 and 1963. As for latitude dependence, the Soviets now believe that density variations ascribed to this effect in early U.S. and USSR rocket experiments, were probably due to diurnal and geomagnetic effects. Also, density results obtained from the earlier (1957–1958) geophysical rocket studies were lower than corresponding U.S. results and differed considerably from more recent satellite data[47–52, 62].

Direct measurements of temperature in the Soviet Union were undertaken in several early (1958) rocket experiments to altitudes around 110 km. Measurements above this altitude, however, have been limited to infrequent attempts to determine electron temperatures aboard geophysical rockets and satellites. With the exception of the brief reference to Cosmos 2 (Langmuir probe type) investigations, most Soviet publications discuss only U.S. results[10]. Also a single geophysical rocket experiment to determine density up to 430 km using the artificial sodium cloud (rocket-release) technique was apparently successfully performed by Shklovskiy and Kurt in 1958[67].

More recently attempts have been made to deduce upper atmospheric properties, particularly temperature, from rocket mass spectrometer composition studies. In the USSR, this work was suggested and tried by A. A. Pokhunkov[57–60] of the Institute of Applied Geophysics. He used results of mass spectrometer data (changes in the ion current ratio for A and N_2) from four geophysical rocket experiments (1959–1961) to determine temperature profiles between 100–325 km. According to Pokhunkov, the maximum error in these determinations was less than 10%. The results of the few temperature determinations reported to date using this technique are shown in Table 7.2. Use of this technique for density and pressure has also been studied[11–12, 73].

One of the interesting Soviet claims is the identification of a possible temperature inversion in the 150–170 km region of the upper atmosphere. According to the Soviets, determination of the nature and cause of this inversion requires further study[50].

From *in situ* pressure gauge measurements aboard Sputnik 3, the Soviets determined density values between 225–500 km, which reportedly were in fair agreement with indirect determinations made in connection with orbital evolution data from early U.S. and USSR satellites. Calculations based on these data constituted the tentative standard atmosphere for the 225–500 km region submitted by the Soviets to the Committee on Space Research (COSPAR) in 1961. These values were, however, 1–10 times larger than values contained in earlier atmosphere models (see Table 7.2). Also, the Soviet model for temperature has been frequently criticized since it depicts an unacceptable increase in the temperature gradient from ≈ 1°K/km at 250 km to ≈ 7°K/km at 500 km.[35, 51]

Indirect Studies (Satellite Drag)

One of the earliest papers considering the possibility of determining atmospheric density from an analysis of the drag encountered by artificial satellites was written by I. M. Yatsunskiy in 1957[76]. Subsequent papers by El'yasberg[14] and Lidov[36] in 1958 contained initial Soviet attempts to determine density values from the change in periods of Sputniks 1, 2 and 3 and early U.S. Explorer satellites. These studies detected the existence of diurnal (daytime and nighttime) variations in density amounting to as much as 20–30% at 225 km. Follow-on studies using the same satellite decay data by El'yasberg[15], Kolegov[32], and Fominov[16–17] in 1960–1963, confirmed the diurnal or day-and-night effect and concluded the possible existence of a solar activity influence (previously noted in the West). In addition, Kolegov[32] postulated that a discernible eight-hour density variation was possibly the result of tidal action.

Since 1962, most of the Soviet work in this area appears routine and is limited to that of M. Ya. Marov of the Institute of Physics of the Atmosphere (IFA) of the USSR Academy of Sciences. His publications, which contain the results of calculating some 150 values of the $\rho\sqrt{H}$ product (which relates density ρ to atmospheric scale height H) from Cosmos satellites' orbital data, emphasize the determination of maximum (day), minimum (night), and mean density values between 180–300 km[40–45]. During these recent studies, the correlation between the change in secular satellite acceleration and solar radio emission at 10·7 cm (also originally detected in the U.S.) was confirmed for periods of intermediate and minimum solar activity[41–42]. Using this established solar dependence, Marov has attempted to make tentative predictions of atmospheric density (in the region 180–300 km) for the next few years[44–45].

Little consideration has been given in available Soviet literature to annual, semiannual, and geomagnetic variations of upper atmospheric density and temperature currently receiving emphasis in the West. Also, few papers attempting to refine structural values or improve atmospheric models based upon satellite-drag data have appeared[24–26, 48, 79]. One interesting paper considers the effect of solar radiation pressure on density estimates which can limit the applicability of the satellite-drag technique. In this paper the authors conclude that solar radiation pressure could produce a deceleration equivalent to that caused by atmospheric density on the order of 10^{-16} g/cm^3.[52]

Standard Atmosphere Activities

One of the major contributions of rocket and satellite research is in the preparation of standard, reference, or model atmospheres. Such models, which express the variation of upper atmosphere properties with height for a variety of scientific and engineering purposes, have been periodically prepared in the U.S. and West (NACA, ARDC, ICAO, COESA)* since about 1947. A similar Soviet model titled "Table of Tentative Standard Atmosphere, 1960" (TSA–60)[75] was submitted to the Committee on Space Research (COSPAR) for consideration in the preparation of CIRA 1961†. This report makes reference to an

* NACA—National Advisory Committee for Aeronautics, now NASA
ARDC—Air Research and Development Command
ICAO—International Civil Aviation Organization
COESA—Committee on Extension to the Standard Atmosphere.

† CIRA 1961, COSPAR International Reference Atmosphere, 1961, North-Holland Publishing Company, Amsterdam, 1961, 177 pp.

earlier Soviet 1948 standard atmosphere designated SA–48 (All-Union State Standard 4401–1948). Also Izakov[25] makes passing reference to a more recent standard atmosphere, SA–62, published in Moscow in 1963. The TSA–60 contains tables of mean values (to five significant figures) for temperature, pressure, density, velocity of sound, coefficient of viscosity, acceleration of gravity, and mean free path for altitudes from –2 km to 200 km, in 20-meter height intervals. Curves showing the variation of the same parameters with height to 200 km are also included. As for most reference or standard atmospheres, TSA–60 agrees in general but differs in detail from other existing models. The Soviets noted that at altitudes up to 25 km, TSA–60 coincides closely with their 1948 standard atmosphere but differs in that the variation of acceleration of gravity with altitude was considered in the 1960 determination of the mean values of pressure and density. The usefulness of TSA–60 is somewhat limited in that it reflects only the average condition of the upper atmosphere and does not include supporting models (for various latitudes, seasons, and solar activity conditions) as are contained in recent U.S. reference atmospheres.

In addition to TSA–60, the Soviets submitted a so-called "Tentative Atmosphere for the Region 225–500 km" for COSPAR consideration. The latter contained the results of Sputnik 3 data as originally published by V. V. Mikhnevich, et al., in 1959[51]. Both tables are included in their entirety in CIRA 1961, pages 101–155. In contrast, Soviet acknowledgment in the more recent CIRA 65* is limited to three curves (Figures 1, 3, and 4) contained in Part I and passing references in Part III to density determinations from satellite drag[40] and geophysical rockets[49]. The Soviet curves in Part I, all for the 0–80 km region, show (1) the annual mean temperature for the northern hemisphere plus temperature deviation for various departure probabilities, (2) the annual mean density for the northern hemisphere and curves of extreme densities, and (3) the annual pressure for the northern hemisphere and curves of extreme pressures. These data were originally contained in a report by the Central Aero-Hydrodynamic Institute of the State Committee on Aviation Technology of the USSR[1]. Although the major contributor to the formulation of Soviet models is probably M. N. Izakov, other Soviet scientists participating in the COSPAR international reference atmosphere working group (IV) include V. F. Arkhangel'skiy, V. A. Bugayev, S. M. Poloskov, and Ye. G. Shvidkovskiy.

* CIRA 1965, COSPAR International Reference Atmosphere, North-Holland Publishing Company, Amsterdam, 1965, 313 pp.

RELATED RESEARCH

Several related Soviet papers have appeared, mostly from the Central Aerological Observatory (TsAO), considering the possibilities and limitations of determining upper atmospheric density and temperature by employing mass spectrometers, thermometers, and pressure gauges on rockets and satellites [20, 22, 23, 54–56]. Other papers treat instrumentation performance and sensor and media interactions[19–21,30–31, 64–65, 71, 77].

In addition to direct and indirect determination of atmospheric temperature, pressure, and density, the Soviets have also made reference to rocket and satellite studies of upper atmosphere winds. Although few results of these determinations have been published to date, the Soviets have referred to fairly extensive use of the parachute tracking method[6], isolated experiments using artificial smoke and/or sodium clouds[61, 67], and more recently radar tracking of dipole reflectors released from rockets[6]. Also, Soviet efforts to determine wind velocity or index of circulation at altitudes of 170–360 km from analyses of U.S. and USSR satellite orbital data have recently been reported. According to this report, the Soviets consider this method to be more representative than results obtained from observations of luminous clouds and ionospheric drifts[74].

Recently Soviet papers have noted the practicality of determining the vertical temperature and water vapor profiles from the interpretation of rocket and satellite thermal radiation data[6–7, 33, 37–39]. Such studies are considered in more detail in Chapter 16 relating to satellite meteorology. The Soviets have also made reference to aerodynamic re-entry studies undertaken in connection with several geophysical rocket flights, but no published results of these studies have been identified[6].

The reader is reminded that the closely allied topic of atmospheric chemical composition studies is considered in Chapter 5.

References

1. "Distribution of Basic Thermodynamic Parameters of the Atmosphere Over the Northern Hemisphere", Supplement II to Table of Provisional Standard Atmosphere (TSA–60), Bureau of Scientific Information, USSR, 1963.

2. Trudy Tsentral'noy Aerologicheskoy Observatorii, 1964, No. 52, and No. 56.

3. S. G. Aleksandrov and R. Ye. Fedorov, "Soviet Satellites and Spaceships", Izdatel'stvo Akademii Nauk SSSR, Moscow, 1961, 440 pp (TT–62–121).

4. P. P. Alekseyev, Ye. A. Besyadovskiy, G. I. Golyshev, M. N. Izakov, A. M. Kasatkin, G. A. Kokin, I. S. Livshits, N. D. Masanova, and Ye. G. Shvidkovskiy, "Rocket Investigations of the Atmosphere", Meteorologiya i Gidrologiya, 1957, No. 8, pp 2–13 (SCL–T–189).

5. P. P. Alekseyev, Ye. A. Besyadovskiy, L. A. Biryukova, G. I. Golyshev, A. I. Ivanovskiy, M. N. Izakov, G. A. Kokin, Yu. V. Kurilova, N. S. Livshits, A. A. Petrov, B. G. Rozhdestvenskiy, N. V. Solov'yev, K. Ye. Speranskiy, I. A. Khvostikov, Ye. G. Shvidkovskiy, and I. A. Shcherba, "Study of the Upper Layers of the Atmosphere With the Aid of Meteorological Rockets", Trudy Vsesoyuznogo Nauchnogo Meteorologicheskogo Soveschaniya, T.I.L., 1962, pp 91–103.

6. USSR National Reports to COSPAR, 1960–1966, presented by A. A. Blagonravov (In English).

7. V. G. Boldyrev, L. I. Korprova, and M. S. Malkevich, "The Role of Vertical Temperature and Humidity Profiles During the Determination of the Earth's Surface Temperature From Outgoing Radiation", Izvestiya Akademii Nauk SSSR, Fizika, Atmosfery i Okeana, 1965, Vol 1, No. 7, pp 703–714.

8. Yu. A. Bragin, "Measurement of Gas Density by the Determination of Lithium-Ion Range", Trudy Tsentral'noy Aerologicheskoy Observatorii, 1962, No. 42, pp 116–118.

9. Yu. A. Bragin, "Measurement of Atmospheric Density in Between 50–70 km", Kosmicheskiye Issledovaniya, 1964, Vol 2, No. 6, pp 917–919 (TT–64–1316, pp 164–169).

10. T. K. Breus and G. L. Gdalevich, "Electron and Ion Temperatures in the Ionosphere (Summary)", Trudy Vsesoyuznoy Konferentsii po Fizike Kosmicheskogo Prostranstva, Izdatel'stvo "Nauka", Moscow, June 10–16 1965, pp 189–192 (NASA TT F–389, pp 254–258); Kosmicheskiye Issledovaniya, 1965, Vol 3, No. 6, pp 877–889 (TT–65–1985, pp 108–133).

11. A. F. Chizhov, "The Measurement of the Temperature of the Free Atmosphere Taking Into Account the Recombination of Atoms", Trudy Tsentral'noy Aerologicheskoy Observatorii, 1963, No. 46, pp 76–84 (NASA TT F–215, 12 pp).

12. A. F. Chizhov, "Use of the Theory of Free Molecular Flow for Processing Data Obtained With a Quadrupole Mass Filter", Trudy Tsentral'noy Aerologicheskoy Observatorii, 1965, No. 61, pp 44–49.

13. V. R. Dubentsov, "Main Features of the Temperature Distribution in the Atmosphere in Different Seasons", Meteorologiya i Gidrologiya, 1961, No. 9, pp 3–12 (JPRS: 11 845).

14. P. Ye. El'yasberg, "Determination of the Density of the Outer Atmosphere From Secular Variations of the Orbital Elements of the First Two Artificial Satellites", Iskusstvennyye Sputniki Zemli, 1958, No. 1, pp 21–24 (Artificial Earth Satellites, pp 25–29).

15. P. Ye. El'yasberg and V. D. Yastrebov, "Determination of Upper-Atmosphere Density From the Results of Measurements of the Flight of the Third Soviet Artificial Earth Satellite", Iskusstvennyye Sputniki Zemli, 1960, No. 4, pp 18–30 (Artificial Earth Satellites, pp 202–219).

16. A. M. Fominov, "The Determination of Some Parameters of the Earth's Atmosphere From the Motion of Satellites", Astronomicheskiy Tsirkulyar, 1963, No. 255, pp 1–6 (Royal Aircraft Establishment Translation No. 1059).

17. A. M. Fominov, "The Motion of an Artificial Earth Satellite in a Nonspherical Atmosphere", Byulleten' Instituta Teoreticheskoy Astronomiy, 1963, Vol 9, No. 3 (106), pp 185–203 (Royal Aircraft Establishment Translation No. 1066).

18. A. M. Fominov, "Determination of the Parameters Characterizing the Nonsphericity of the Earth's Atmosphere From Changes in Artificial-Satellite Orbit Elements", Byulleten' Instituta Teoreticheskoy Astronomiy, 1964, Vol 9, No. 7, pp 499–521.

19. A. I. Ivanovskiy, "Problems Related to the Conductivity of Cylindrical Pipes in Free Molecular Flow", Trudy Tsentral'noy Aerologicheskoy Observatorii, 1965, No. 61, pp 7–17.

20. A. I. Ivanovskiy, "Aerodynamics of Manometers and Mass Spectrometers Carried on Rockets and Satellites", Trudy Vsesoyuznoy Konferentsii po Fizike Kosmicheskogo Prostranstva, Izdatel'stvo "Nauka", Moscow, June 10–16 1965, pp 56–61 (NASA TT F–389, pp 74–80).

21. A. I. Ivanovskiy and A. I. Repnev, "Spatial Distribution of Mass, Impulse, and Energy Fluxes Behind a Small Aperture", Trudy Tsentral'noy Aerologicheskoy Observatorii, 1962, No. 40, pp 72–76.

22. M. N. Izakov, "Measurements of the Atmospheric Density by an Instrumented Nonoriented Satellite", Kosmicheskiye Issledovaniya, 1963, Vol 1, No. 1, pp 156–158 (TT–64–90, pp 285–301).

23. M. N. Izakov, "Measurements of Atmospheric Temperature by an Instrumented Satellite", Kosmicheskiye Issledovaniya, 1963, Vol 1, No. 1, pp 159–168 (TT–64–90, pp 279–284).

24. M. N. Izakov, "Proposal for a New Addition of the COSPAR International Reference Atmosphere (CIRA–64) and Further Progress in Studying the Structure of the Upper Atmosphere", paper presented at the Sixth International Space Science Symposium, Mar del Plata, Argentina, May 1965 (In English).

25. M. N. Izakov, "Accounting for Variability in the Coefficient of Aerodynamic Resistance in Determining Atmospheric Density on the Basis of Artificial-Satellite Deceleration", Kosmicheskiye Issledovaniya, 1965, Vol 3, No. 2, pp 297–308 (TT–65–414, pp 192–212).

26. M. N. Izakov, "Some Problems in the Study of Structure of the Upper Atmosphere", Trudy Vsesoyuznoy Konferentsii po Fizike Kosmicheskogo Prostrantstva, Izdatel'stvo "Nauka", Moscow, June 10–16 1965, pp 30–39 (NASA TT F–389, pp 30–49).

27. I. A. Khvostikov, "The Upper Layers of the Atmosphere", Gidrometeorologicheskoye Izdatel'stvo, Leningrad, 1964, 606 pp (NASA TT F–315).

28. I. A. Khvostikov, M. N. Izakov, G. A. Kokin, Yu. V. Kurilova, and N. S. Livshits, "Studies of the Stratosphere With the Aid of Meteorological Rockets in the USSR", Meteorologiya i Gidrologiya, 1963, No. 1, pp 3–8 (JPRS: 18 306); presented at the Third International Space Science Symposium, Washington D.C., April 1962.

29. A. A. Kmito, "Methods for Investigation of Atmosphere by Means of Rockets and Satellites", Gidrometeoizdat, Leningrad, 1966, 366 pp.

30. G. A. Kokin, "Some Physical Aspects of Rocket Sounding", Trudy Tsentral'noy Aerologicheskoy Observatorii, 1965, No. 61, pp 3–6.

31. G. A. Kokin, "Determination of the Angle-of-Attack of a Meteorological Rocket in a Free Molecular Field", Trudy Tsentral'noy Aerologicheskoy Observatorii, 1965, No. 61, pp 55–58.

32. G. A. Kolegov, "Variations in Density of the Upper Atmosphere From Data Obtained From the Measured Period of Rotation of Artificial Earth Satellites", Iskusstvennyye Sputniki Zemli, 1960, No. 4, pp 31–34 (Artificial Earth Satellites, pp 220–224).

33. L. I. Koprova and M. S. Malkevich, "The Thermal Emission of a Spherical Atmosphere", Kosmicheskiye Issledovaniya, 1964, Vol 2, No. 6, pp 881–900 (TT–64–1316, pp 102–134).

34. N. N. Kozlov, "Determination of Atmospheric Density From the Observed Deceleration of the Third Soviet Satellite", Vestnik Moskovskogo Universitet, Seriya Fizika, Astronomiya, 1965, No. 5, pp 10–13.

35. V. I. Krasovskiy, "Exploration of the Upper Atmosphere by the Help of the Third Soviet Satellite", 9th International Astronautical Congress, Amsterdam, 1958, Springer-Verlag, 1959, pp 614–625 (In English).

36. M. L. Lidov, "Determination of the Density of the Atmosphere From the Observed Deceleration of the First Artificial Satellite", Iskusstvennyye Sputniki Zemli, 1958, No. 1, pp 9–20 (Artificial Earth Satellites, pp 10–24).
37. A. F. Maklakov and V. S. Khakhalin, "Modern Techniques for Investigating the Atmosphere", Gidrometeoizdat, 1964, 132 pp (JPRS: 30 574).
38. M. S. Malkevich and V. I. Tatarskiy, "Determination of Vertical Temperature Profile of the Atmosphere From Outgoing Radiation in the CO_2 Absorption Band", Kosmicheskiye Issledovaniya, 1965, Vol 3, No. 3, pp 444–456 (TT–65–828, pp 170–190).
39. M. S. Malkevich and V. I. Tatarskiy, "Determining Temperature and Moisture of the Earth's Atmosphere Based on Satellite Measurements of the Earth's Atmosphere", Trudy Vsesoyuznoy Konferentsii po Fizike Kosmicheskogo Prostranstva, Izdatel'stvo "Nauka", Moscow, June 10–16 1965, pp 104–111 (NASA TT F–389, pp 140–152).
40. M. Ya. Marov, "Study of the Upper Atmosphere by Means of Cosmos 3 and Cosmos 5 Satellites. 4. Density of the Upper Atmosphere at Altitudes of 200–230 km", Kosmicheskiye Issledovaniya, 1963, Vol 1, No. 1, pp 143–146 (TT–64–90, pp 256–261).
41. M. Ya. Marov, "Density of the Upper Atmosphere From the Drag of Soviet Satellites", Space Research V, Proceedings of the Fifth International Space Science Symposium, Florence, Italy, May 12–16 1964, North-Holland Publishing Company, Amsterdam, 1965, pp 1140–1149.
42. M. Ya. Marov, "The Density of the Upper Atmosphere During the Year of Minimum Solar Activity", Kosmicheskiye Issledovaniya, 1964, Vol 2, No. 6, pp 909–916 (TT–64–1316, pp 148–163).
43. M. Ya. Marov, "The Density of the Upper Atmosphere", paper presented at the Symposium on Aeronomy, Cambridge, Massachusetts, August 1965, 34 pp (In English).
44. M. Ya. Marov, "Dynamic Nature of Atmospheric Density at Altitudes of 200–300 km", Trudy Vsesoyuznoy Konferentsii po Fizike Kosmicheskogo Prostranstva, Izdatel'stvo "Nauka", Moscow, June 10–16 1965, pp 41–48 (NASA TT F–389, pp 53–61).
45. M. Ya. Marov, "Some Additional Data on the Atmospheric Density at Altitudes of 200–300 km", Space Research VI, Proceedings of the Sixth International Space Science Symposium, Mar del Plata, Argentina, May 11–19 1965, North-Holland Publishing Company, Amsterdam, 1966, pp 386–392.
46. V. V. Mikhnevich, "The Measurement of Pressure in the Upper Atmosphere", Uspekhi Fizicheskikh Nauk, 1957, Vol 63, No. 1b, pp 197–204 ("The Russian Literature of Satellites", Part II, pp 87–94).
47. V. V. Mikhnevich, "Preliminary Results of a Determination of the Density of the Atmosphere Above 100 km", Iskusstvennyye Sputniki Zemli, 1958, No. 2, pp 26–31 (Artificial Earth Satellites, pp 32–39).
48. V. V. Mikhnevich, "Variations of the Atmospheric Density at Elevations in Excess of 200 km", Iskusstvennyye Sputniki Zemli, 1963, No. 17, pp 31–41 (Artificial Earth Satellites, pp 30–42; Planetary and Space Science, Pergamon Press, 1964, Vol 21, pp 1111–1120).
49. V. V. Mikhnevich, "Atmospheric Density at Heights of 100–350 km", Space Research V, Proceedings of the Fifth International Space Science Symposium, Florence, Italy, May 12–16 1964, North-Holland Publishing Company, Amsterdam, 1965, pp 1112–1123.
50. V. V. Mikhnevich, "Density and Temperature of the Atmosphere From Measurement Results on VGAS in 1963", Sixth International Space Sciences Symposium, Mar del Plata, Argentina, May 11–19 1965; Trudy Vsesoyuznoy Konferentsii po Fizike Kosmicheskogo Prostranstva, Izdatel'stvo "Nauka", Moscow, June 10–16 1965, pp 23–34 (NASA TT F–389, pp 25–34).

51. V. V. Mikhnevich, B. S. Danilin, A. I. Repnev, and V. A. Sokolov, "Certain Results in the Determination of the Structural Parameter of the Atmosphere With the Aid of Sputnik 3", Iskusstvennyye Sputniki Zemli, 1959, No. 3, pp 84–97 (Artificial Earth Satellites, pp 119–136).

52. V. V. Mikhnevich, Ye. N. Golubev, and Yu. N. Parfianovich, "Preliminary Results of Determination of Density of the Atmosphere of June 18 1963", Kosmicheskiye Issledovaniya, 1965, Vol 3, No. 3, pp 457–468 (TT–65–828, pp 191–207).

53. V. V. Mikhnevich and I. A. Khvostikov, "Study of the Upper Layers of the Atmosphere", Izvestiya Akademii Nauk SSSR, Seriya Geofizicheskaya, 1957, No. 11, pp 1393–1409 (Bulletin [Izvestiya] of the Academy of Sciences of the USSR, Geophysics Series, pp 88–107).

54. S. P. Perov, "Method for Determining Structural Parameters of the Atmosphere With a Rocket Thermometer", Candidate dissertation in Physics and Mathematics, Vechernyaya Moskva, 1965, No. 40, p 4.

55. S. P. Perov, "Measuring the Coefficient of Air Accommodation on Metallic Threads", Trudy Tsentral'noy Aerologicheskoy Observatorii, 1965, No. 61, pp 68–85.

56. S. P. Perov, "Atmospheric Density Determination With a Rocketborne Resistance Thermometer", Trudy Tsentral'noy Aerologicheskoy Observatorii, 1965, No. 61, pp 86–92 (JPRS: 33 679).

57. A. A. Pokhunkov, "Mass-Spectrometric Investigations of Structural Parameters of the Earth's Atmosphere at Altitudes From 100 to 210 km", Iskusstvennyye Sputniki Zemli, 1961, No. 7, pp 89–100 (Artificial Earth Satellites, pp 88–100; Planetary and Space Science, Pergamon Press, 1962, Vol 9, pp 269–279).

58. A. A. Pokhunkov, "Gravitational Separation, Composition and Structural Parameters of the Night Atmosphere at Altitudes From 100 to 210 km", Iskusstvennyye Sputniki Zemli, 1962, No. 13, pp 110–118 (Artificial Earth Satellites, pp 116–126; Planetary and Space Science, Pergamon Press, 1963, Vol 11, pp 441–449).

59. A. A. Pokhunkov, "The Distribution of Helium, Nitrogen, and Argon in the Earth's Atmosphere up to a Height of 430 km", Kosmicheskiye Issledovaniya, 1963, Vol 1, No. 1, pp 147–155 (TT–63–968, pp 262–278).

60. A. A. Pokhunkov, "Measurements of the Upper Atmosphere Temperature by Mass-Spectrometric Methods", paper presented at the Seventh International Space Science Symposium, Vienna, Austria, 1966 (In English).

61. S. M. Poloskov and B. A. Mirtov, "The Study of the Upper Atmosphere Through the Use of Rockets at the Academy of Science of the USSR", International Congress on Guided Missiles and Rockets, Paris, France, 1956.

62. S. M. Poloskov, "Upper Atmosphere Structure Parameters According to Investigation Data Obtained on Rockets and Satellites in the USSR During the IGY", Space Research, Proceedings of the First International Space Science Symposium, Nice, France, January 11–16 1960, North-Holland Publishing Company, Amsterdam, 1960, pp 95–116.

63. V. V. Radziyevskiy and A. V. Artem'yev, "Effect of Solar Radiation Pressure on the Motion of Artificial Earth Satellites", Astronomicheskiy Zhurnal, 1961, Vol 38, No. 5, pp 994–996 (Soviet Astronomy AJ, 1962, Vol 5, No. 5, pp 758–759).

64. A. I. Repnev, "Interpretation of the Readings of an Instrument Moving at High Velocity in the Upper Layers of the Atmosphere", Trudy Vsesoyuznoye Nauchnoye Meteorologicheskoye Soveshchaniye, 1st, Leningrad, 1961, Pribory i Metody Nablyudeniy, 1963, Vol 9, pp 132–138.

65. A. I. Repnev, "Additional Terms in Hydrodynamic Equations for a Case of Radioactive Recombination in Two Temperature Gases", Trudy Tsentral'noy Aerologicheskoy Observatorii, 1965, No. 61, pp 100–106.

66. L. A. Ryazantseva and I. A. Khvostikov, "Processes in the Stratosphere According to Rocket Sounding Data Meteorological Researches", Rezul'taty Issledovaniy po Mezhdunarodnim Geofizicheskim Proektam, 1965, Razdel II, No. 9, pp 58–63.

67. I. S. Shklovskiy and V. G. Kurt, "The Determination of the Density of the Atmosphere at an Altitude of 430 Kilometers by the Sodium Vapor Diffusion Method". Iskusstvennyye Sputniki Zemli, 1959, No. 3, pp 66–76 (Artificial Earth Satellites, pp 92–107).

68. Ye. G. Shvidkovskiy, "Some Results of Measurements of Thermodynamic Parameters of the Stratosphere Obtained With the Aid of Meteorological Rockets", Iskusstvennyye Sputniki Zemli, 1958, No. 2, pp 10–16 (Artificial Earth Satellites, pp 12–19).

69. Ye. G. Shvidkovskiy, "Rocket Investigations of the Upper Layers of the Atmosphere", Trudy Tsentral'noy Aerologicheskoy Observatorii, 1959, No. 26.

70. Ye. G. Shvidkovskiy, "Rocket Meteorological Measurements", Trudy Tsentral'noy Aerologicheskoy Observatorii, 1960, No. 29.

71. Ye. G. Shvidkovskiy, A. I. Ivanovskiy, and A. I. Repnev, "Some Properties of Binary Two-Temperature Gas", Paper presented at the Seventh International Space Science Symposium, Vienna, Austria, 1966 (In English).

72. N. P. Slovokhotova, "Variations in Atmospheric Density in 1960–1963 as Derived From Measurements of the Orbits of Artificial Earth Satellites", Kosmicheskiye Issledovaniya, 1965, Vol 3, No. 3, pp 469–472 (TT–65–828, pp 208–214).

73. N. A. Sokova and A. F. Chizhov, "Use of the Mass Spectrometric Technique for Determining Partial Densities and Temperatures of the Components of the Upper Atmosphere", Trudy Tsentral'noy Aerologicheskoy Observatorii, 1965, No. 61, pp 28–43.

74. B. N. Trubnikov, "Utilization of Artificial Earth Satellite Orbital Data to Determine the Wind Velocity in the Atmosphere", Trudy Vsesoyuznoy Konferentsii po Fizike Kosmicheskogo Prostranstva, Izdatel'stvo "Nauka", Moscow, June 10–16 1965, pp 51–56 (NASA TT F–389, pp 66–73).

75. M. V. Yakovkin and A. I. Korkina, "Table of Tentative Standard Atmosphere, 1960", CIRA 1961, COSPAR International Reference Atmosphere, 1961, North-Holland Publishing Company, Amsterdam, 1961, pp 103–151.

76. I. M. Yatsunskiy, "The Effect of Geophysical Factors on the Motion of a Satellite", Uspekhi Fizicheskikh Nauk, 1957, Vol 63, No. 1a (Advances in Physical Science, pp 89–106).

77. V. M. Zaletayev, "Temperature Field of Elements of Thin-Walled Satellite Surfaces in Radiant Heat Transfer", Kosmicheskiye Issledovaniya, 1966, Vol 4, No. 1, pp 116–127 (TT–66–76, pp 189–204).

78. R. S. Zhanturov, "Semidiurnal Variations of Pressure", Trudy Tsentral'noy Aerologicheskoy Observatorii, 1965, No. 61, pp 107–119.

79. R. S. Zhantuarov, "Diurnal Variations in Density, Pressure and Temperature in the Atmosphere", Trudy Vsesoyuznoy Konferentsii po Fizike Kosmicheskogo Prostranstva, Izdatel'stvo "Nauka", Moscow, June 10–16 1965, pp 39–42 (NASA TT F-389, pp 49–51).

Chapter 8

Magnetic Fields

N. A. Frazier

INTRODUCTION

The Soviets carried a long-standing interest in the earth's magnetic field into the space age. Beginning *circa* 1930 and continuing to the present, plans have been drafted, implemented, and revised. The general ground magnetic survey of the USSR was undertaken and more than 23000 USSR field determinations were completed between 1931–1943. Several magnetic observatories were organized in the Arctic as well as in the more southerly latitudes of USSR. Magnetic field observations from drifting ice islands were undertaken. Pioneer efforts in magnetic prospecting from aircraft were initiated in the late 1930's. The nonmagnetic research schooner "Zarya" commenced operations during the International Geophysical Year (IGY), and the use of magnetometers towed by steel-hulled ships has been under investigation for several years.

The first magnetic field measurements from artificial earth satellites were made with the USSR's third satellite (Sputnik 3), launched May

15 1958, as a part of the IGY and the World Magnetic Survey program. Also, the Soviets planned to include magnetic field studies as a part of their IGY rocket program*, but execution of such studies has not been noted in reports of Soviet IGY or other USSR rocket investigations. The absence of this work in the USSR's magnetic field research was unexpected in view of (1) the early belief that variations in the magnetic field observed at the earth's surface arise in part from electrical currents in the ionosphere, and (2) the data gap between satellite and aircraft altitudes. U.S. rocket work on this subject started about two years after E. H. Vestine's suggestion in 1947 to use rocket-borne magnetometers to investigate these currents.

The Soviets have reported that magnetic field instrumentation was

TABLE 8.1

Soviet Space Vehicles Instrumented for Magnetic Field Measurements (1957–1966)

Vehicle	Instrument
	Geophysical Rockets
None reported	None reported
	Artificial Earth Satellites
Initial Sputnik Series	
Sputnik 3	Three-component magnetometer
Electron Series	
Electron 2	Two three-component magnetometers
Electron 3	Coil for measuring field pulsations
Electron 4	Two three-component magnetometers
Cosmos Series	
Cosmos 26	Two proton magnetometers
Cosmos 49	Two proton magnetometers
	Lunar and Planetary Probes
Luna Series	
Luna 1	Three-component magnetometer
Luna 2	Three-component magnetometer
Luna 10	Three-component magnetometer
Venera Series	
Venera 1	Three-component magnetometer Variometer with two parallel detectors
Venera 2	Not described
Venera 3	Not described
Zond Series	
Zond 3	Not described
Mars Series	
Mars 1	Single-component magnetometer

* USSR Rocket and Satellite Program submitted on June 10 1957 by Academician I. P. Bardin, President, Soviet IGY Committee, to Dr. L. V. Berkner, CSAGI Reporter for Rockets and Satellites (from Annals of the IGY, Vol 6, "Manual of Rockets and Satellites", Pergamon Press, Ltd., 1958).

on board six earth satellites, three lunar probes, and five planetary probes. Details are shown in Table 8.1.

INSTRUMENTATION

Statistically, at least (see Table 8.1), the Soviets have favored the three-component metal core magnetometers, having indicated this type of instrument was on seven of their nine vehicles associated with total field objectives*. Magnetometers with different ranges and sensitivities have been used depending upon the space region and field to be investigated. Two proton magnetometers were used on each of two earth satellites (Cosmos 26 and 49), a variometer on one space probe, and a coil for measuring field pulsations on one satellite. Two fluxgate instruments were used on each of two satellites (Electrons 2 and 4).

Descriptions of instrumentation from terse statements that magnetometers were on board a particular vehicle (e.g., Veneras 2 and 3) to an occasional and rather complete treatment of instrumentation on other spacecraft. Some of these characteristics are given in Table 8.2 with reference to literature containing additional details.

SUMMARY OF RESULTS

Scalar values of the total field within the earth's magnetosphere have been derived from seven Soviet vehicles—five earth satellites and two lunar probes. Limited amounts of data have been reported on magnetic field pulsations in the 1–10 cps and 30–300 cps bands from Electron 3 and limited variometer data from Venera 1, a Venus probe. Selected results from these vehicles as extracted from Soviet literature are given in Table 8.2.

The majority of Soviet results were reported by members of the Institute of Terrestrial Magnetism, Ionosphere, and Propagation of Radio Waves of the USSR Academy of Sciences. The Soviet journals "Geomagnetizm i Aeronomiya", "Doklady Akademii Nauk SSSR", "Kosmicheskiye Issledovaniya", and the former "Iskusstvennyye Sputniki Zemli" represent the main periodicals containing magnetic field investigations. More timely or annual activities often were presented in the reports of international organizations and meetings, particularly the Committee on Space Research (COSPAR).

* Photographs of some Soviet space vehicles with magnetic field instrumentation are given in Chapters of Part I: Sputnik 3, Electron 2, and Cosmos (numbers unspecified) in Chapter 2: Lunas 1, 2, and 10, Venera 1, and Mars 1 in Chapter 3.

TABLE 8.2

Results of Magnetic Field Studies Aboard Soviet Space Vehicles (1957–1966)

Vehicle/Launch/ Orbital Data	Instrumentation/Investigated Region	Results	References
		Geophysical Rockets	
No launchings noted			
		Earth Satellite Vehicles	
Initial Sputnik Series			
Sputnik 3/ May 15 1958/ Apogee—1881 km, Perigee—226 km, Inclination—65°	Scalar values of total field vector and orientation angles of satellite with respect to total field vector were measured. The magnetometer (Model SG-45) was a relative instrument with three magnetically saturated cores wound with primary and secondary coils. Two of the sensors used to provide data needed for determining orientation and precession, and rotation of the satellite. Details of the orientation platform, magnetometer circuitry, analytical derivations, and magnetic deviations are given in References 15, 33–36, 60.	Total magnetic field intensity versus time elapsed from an undefined time origin have been given for about six passes of Sputnik 3 over USSR. These data are presented on small-scale plots each containing curves of scalar values as measured on board the vehicle. These curves modulate about curves of scalar values corrected for rotation and precession of the satellite and about curves of computed field values along the satellite's paths. Field intensity increased from the time origin when the satellite was near perigee altitude and then decreased as the satellite increased in altitude and passed beyond the USSR[33–35].	Beletskiy, *et al.* (1961)[15] Dolginov, *et al.* (1963)[25], (1960)[31], (1963)[33–34], (1962)[35] (1960)[36] Zonov (1959)[60]
		Measured field values were generally less than computed values to the east of 55° E in USSR and greater westward of 55° E. Values of the former ranged up to about 250γ and the latter to about 100γ[25].	
	Ground tests of the magnetometer over 25 days indicated a zero drift of 135 γ over the first 21 days (average about 5 γ/day) and 12 γ/day over the last four days of the test. Drift was also checked in flight by comparing measured field values and theoretical values at points in orbits with similar coordinates. Zero point of the Sputnik 3 magnetometer was obtained from comparison with a proton magnetometer[35].	Comparisons were made of the signs of the differences in field intensity between two ground stations and satellite values on quiet and disturbed (May 15 and June 1) days. Differences were of the same sign at one location for both disturbed days (satellite at 250 km and 510–600 km). The observed differences were believed more likely attributable to errors in determining magnetic deviation as opposed to location errors or the influence of an F_2 homogeneous current system[35].	
		The problem of the currents induced by the translational motion of a satellite relative to the magnetic field and the change in speed of rotation of a satellite about its own axis due to eddy currents, and the disturbing forces exerted by the magnetic field on a satellite were treated by Zonov.[60] An extensive treatment of the orientation and rotation of Sputnik 3 and associated analytical derivations have been given by Beletskiy and Zonov[15].	
	The magnetic deviations are derived as a function of the orientation with respect to the magnetic field vectors. Average errors in determining the absolute values of magnetic deviations is estimated at 120 γ[35].		

Vehicle/Launch/ Orbital Data	Instrumentation/Investigated Region	Results	References
	Magnetic field measurements were made within the period May 15 to June 15 1958. Majority of the data were obtained over Soviet territory. Total path length for which data over USSR obtained was 80 000 km at altitudes of 225–800 km. Sparse data from higher altitudes up to 1880 km obtained only over the Southern Hemisphere. The satellite transited near or over the maximum of the East Siberian magnetic anomaly on several passes.		
Electron Series			
Electron 2/ Jan. 30 1964/ Perigee—460 km, Apogee—68 200 km, Inclination to earth's equator—61°, Inclination of orbital plane to the plane of the ecliptic—around 50°.	Electrons 2 and 4 each had two three-component fluxgate magnetometers, one of which had a range of $\pm 1200\ \gamma$ and a sensitivity of $20\ \gamma$. Electron 2's other instrument had a range of $\pm 120\ \gamma$ and Electron 4's second magnetometer had a range of $\pm 240\ \gamma$; both had a sensitivity of $2\ \gamma$. The more sensitive instruments made measurements between 6·5–11·6 R_e. Error in measurement of field strength using the more sensitive instrument was not more than ± 3 to $4\ \gamma$. Three components could be interrogated almost simultaneously every two minutes in one mode and every eight minutes in the second mode. Electron 2 magnetometers are described in most detail in Reference 10 and to a lesser degree in Reference 26.	From Electron 2 data, measured minus theoretical values (ΔT) of the total field were characteristically negative on outbound orbits between 3 R_e and 5–5·5 R_e and on inbound orbits from 7·5 or 6 to 3 R_e. In the remaining portions of the orbit ΔT was characteristically positive[26–27, 56–58]. Average maximum value of the negative differences was 100–150 γ at about 3 R_e. Qualitative assessment gave argument to explaining these ΔT values at about 3–5 R_e, which were about 10% of the dipole field at these distances, in terms of trapped protons and electrons[26–27]. The positive differences ranged from 7–70 γ (20 to 200% of the dipole field) depending on magnetic activity. ΔT averaged over each of the three-hour periods when vehicle was between 7 and 11·7 R_e showed close correlation between the K_p indices of magnetic activity over the lifetime of the investigations.[56–57] On quiet-day orbits at distances of 7–11 R_e, the diurnal variation in ΔT was within the limits of 12 γ and was attributable to deformation of the field by solar wind. After two sudden commencements were noted on the ground (February 12–13 and 20–21 1964), a smooth increase in positive ΔT over 15–20 minutes was recorded at 7–11 R_e without abrupt field change observed	Aleksanyan, *et al.* (1966)[10] Dolginov, *et al.* (1966)[26], (1965)[27] Gringauz, *et al.* (1965)[37] Vernov, *et al.* (1965)[55] Yeroshenko (1966)[56] (1965)[57] Yeroshenko, *et al.* (1965)[58]
Electron 3/ July 11 1964/ Perigee—405 km, Apogee—7040 km, Inclination to earth's equator—61°			

TABLE 8.2 (Continued)

Results of Magnetic Field Studies Aboard Soviet Space Vehicles (1957–1966)

Vehicle/Launch/ Orbital Data	Instrumentation/Investigated Region	Results	References
Electron 4/ July 11 1964/ Perigee—459 km, Apogee—66 235 km, Inclination to earth's equator—61°	Data were obtained from Electron 2 during 40 orbits January 30 to May 5–6 1964 at 3–11·6 R_e. Last information on magnetic field received June 1 1964. Apogee positions of orbit changed from near the morning side to the night side of the earth during the investigations. Initially the craft emerged from the night side on outbound orbits with the inbound approach on the morning-day side, passed through a stage when out-bound orbits began on the evening side and inbound orbits terminated on the morning side, and, finally, outbound orbits began on evening side and inbound orbits ended on the night-morning side.	during the initial phase of the storms. This change in satellite values of ΔT during the main phase did not coincide in sign with ΔH on the ground at low and mean latitudes which is interpreted as indicating the source of the main phase was at distances less than 7–11 R_e. (On February 12 an increase in positive ions—notable change from negative to positive currents from charged particle traps on Electron 2—occurred about three hours before sudden commencement was observed on the ground and three hours before ΔT increased aboard the satellite.)[37] Sharp short-time negative ΔT changes were correlated with auroral activity. Results from Electron 2 indicated the magnetosphere boundary to be at greater distances than the Electron 2 apogee during the vehicle's lifetime[56–57].	
		The magnetosphere boundary was higher than Electron 4 during magnetically quiet periods. The vehicle did cross the magnetosphere boundary during the storm of July 17–20 1964. This crossing was indicated by irregularity of the field and departure from sinusoidal nature of the readings from all three sensors. Scalar value amplitudes underwent irregular changes of 8–35 γ. The distance of the boundary changed on outbound orbits from 9·6 R_e on the first day of the storm to 9·3 R_e on the second day and then increased to 10·4 R_e.[56]	
	Electron 3 was equipped with a coil with a ferrite core for measuring magnetic field pulsations. Signals from the coil were fed to two amplifying channels with passbands of 1–10 cps and 30–300 cps. Number of impulses with amplitudes exceeding $\sim$1, 5, and 25 γ were recorded on both channels between interrogations. Six interrogation channels were used, the coarser channels with a threshold of $\gtrsim 25\ \gamma$ and the more sensitive interrogation channels having threshold values $\geqslant 1\ \gamma$[55].	On an average $\Delta T = 0$ at 5 R_e and 26° S on Electron 2's outbound orbits and at 6 R_e and 58° S on inbound orbits. For inbound orbits $\Delta T = 0$ showed a systematic change from 7·5 R_e and 59° S in late January 1964 to 5 R_e and 50° S in late April 1964. The regions of negative ΔT for outbound and inbound orbits are approximately 160° in longitude apart indicating a ringlike character of the effect. The regions of $\Delta T = 0$ showed a dependence on magnetic activity with the magnitude of R_e and ϕ for $\Delta T = 0$ varying inversely with activity index Kp.[26–27] During three moderate magnetic disturbances, measured values from Electron 2 were 300–400 γ less than calculated values over an interval of about 0·2 to 0·5 R_e at a distance of about 3·5 R_e. The main phase of one of these storms (February 20–21) developed 15 hours after this occurrence on the satellite. During the main phase of one storm, $\Delta T = 0$ was depressed to 4·5 R_e during the inbound orbit whereas $\Delta T = 0$ generally occurred at 6–7 R_e. During another storm, $\Delta T = 0$ was depressed to 3·5 R_e during the main phase development.[26–27] In the 1965 report to COSPAR, combined Electron 2 and 4 data showed ΔT values a t10 R_e to be negative on the dayside with increasing magnitudes nearer to noon meridian and toward higher geomagnetic latitudes. With exceptions, positive ΔT occurred between 24^h and 6^h at the lower geomagnetic latitudes.	

Vehicle/Launch/ Orbital Data	Instrumentation/Investigated Region	Results	References
		In Electron 3 magnetic field pulsation studies, more than seven impulses were recorded every two minutes on the more sensitive interrogation channels. As a rule, the coarser channels did not record any impulses between interrogations. The number of pulsations with an amplitude of $\gtrsim 5\ \gamma$ between 1–10 cps was generally greater than in the 30–300 cps. Increase in intensity of pulsations at 1–10 cps as a rule was not accompanied by an increase in 30–300 cps region.[55]	
Cosmos Series			
Cosmos 26/ March 18 1964/ Perigee—271 km, Apogee—403 km, Inclination 49°	Two proton magnetometers with sensors mounted at right angles were on board each vehicle. Ground tests of one of the instruments indicated a measurement accuracy of 2–3 γ. Description of the instruments is given in Reference 22.	The orbit of Cosmos 49 was such that the satellite every 77th orbit more or less "repeated" its flight in nearly identical conditions but with a shift in local time of transit by two hours. The terrestrial equator was crossed either near noon or near midnight hours. Differences in altitude, Δh, on "repeated" orbits was positive in night-time portion of the orbit and negative in day-time portion, where $\Delta h > 0$ if $\Delta h_i - h_{i+77} > 0$.[21]	Dolginov, *et al.* (1966)[21–22]
Cosmos 49/ October 24 1964/ Perigee—260 km, Apogee—490 km, Inclination—49°	Scalar values of magnetic field were measured between latitudes of $\pm 49°$. Cosmos 26 operated from March 18–30 1964, and Cosmos 49 from October 20 to November 26 1964, between altitudes of 270–403 km and 270–490 km, respectively.	In the USSR 1965 report to COSPAR a chart of the residual field at 400 km from two days of Cosmos 49 observations and a chart of the Brazilian anomaly at a height of 350 [?] km from Cosmos 49 and Cosmos 26 data are given. Three positive closures are indicated in the residual field—two at 40° N (one centered over central U.S. and the other over central Siberia) and the third over south central Australia. A negative closure is centered at 40° N over the Pacific, and negative values which may close south of Africa at about 5° E and 45° S.	
		Otherwise, results to date are chiefly those of statistical treatment of the differences from Cosmos 49 of $\delta T = \Delta T_i - \Delta T_{i+77}$, where ΔT_i is the difference between measured and computed strength of the magnetic field on transit i and ΔT_{i+77} is the analogous value on the "repeated" orbit, i.e., the i+77th orbit. Analysis to date has indicated maximum δT values (35–70 γ) to have been more prominent near equatorial latitudes and more frequently observed at noon or midnight hours local time and often had a positive sign during midnight hours and a negative sign during noon hours. North and south of the equator these maximum δT values may be negative at night and positive during daytime. Statistical data are to be examined in terms of sources of magnetic influence at heights exceeding apogee and at heights traversed by Cosmos 49 in the F_2 layer. Also the effect of errors on the distribution of δT from "repeated" orbits, orbit determinations, and electrojets are yet to be evaluated.[21]	

TABLE 8.2 (Continued)
Results of Magnetic Field Studies Aboard Soviet Space Vehicles (1957–1966)

Vehicle/Launch/ Orbital Data	Instrumentation/Investigated Region	Results	References
		Lunar and Planetary Probes	
Luna Series			
Luna 1/ January 2 1959/ Closest approach to moon—6000–8000 km on January 5 1959	Scalar values of total magnetic field determined using three-component magnetometer with magnetically saturated pickups. Magnetic effects of vehicle equipment did not exceed 70 γ and were known. Field values reported between ~14 700 km and 36 000 km from the center of the earth.	Measured values of the magnetic field decreased more rapidly with distance than calculated dipole field. The measured field decreased to a value of 400 γ at the distance of 20 800 km, then increased to 800 γ at 22 000 km, and with some deviations then decreased thereafter. This anomalous distribution of field values examined in terms of external field sources associated with the motion of charged particles in the earth's field and considered to be attributable to a change in current density associated with a change in energy density of particles, and the change in the position of the vehicle relative to the maximum of the current density[14, 23].	Antsilevich (1961)[13] Antsilevich, *et al.* (1960)[14] Dolginov, *et al.* (1959)[23], (1963)[25] (1960)[31]
		A sudden commencement (10 γ) and a D_{st} variation (20 γ) occurred about 5 hours[25] or 6·5 hours[14] prior to Luna 1 measurements above. Luna 1 and surface data on magnetic field were considered not to be in contradiction to the suggestion that anomalous distribution of magnetic field values from Luna 1 may have been due to a ring current at geocentric distances of 25 750–26 800 km in the equatorial plane[13–14].	
Luna 2/ September 2 1959/ Impacted moon—September 14 1959	Total intensity determined using three-component magnetometer with magnetically saturated sensors.	For Luna 2 out to distances of 50 000 km, the smoothed curve of the differences between the measured and calculated field showed negative values (measured < calculated) of about 130 γ at 18 000 km decreasing to zero difference between 30 000 and 35 000 km.[24]	Ben'kova, *et al.* (1962)[18] Dolginov, *et al.* (1963)[24–25], (1961)[28], (1960)[31–32] Shevnin (1961)[52]
	Magnetic field measurements began ~18 000 km from the center of the earth. Last measurement was ~50 km from the moon. Curves have been given of scalar magnitudes between geocentric distances of 18 000 and 95 000 km[24–25, 32], and between 4000 km and 50 km from the moon's surface.[25] Tables of field values between 5150 and 1795 km from the center of the moon are given in Reference 32. Readings are given of components along X-, Y-, and Z-axes fixed in vehicle between 75 000 and 200 000 km from the earth[32], between geocentric distances of 45 000 and 60 000 km[28], from 40 000 to 33 000 km from the moon[28], and from 4000 to 50 km from the moon[28].	On the basis of Luna 1, Luna 2, and U.S. work magnetically active regions in the interval 2–10 R_e were indicated or suggested at 2·5–3·5 R_e, 6 R_e, and 9 R_e.[18, 24, 52] The current system of the 2·5–3·5 R_e region is apparently localized and during magnetic disturbances is probably responsible for the main phase of a magnetic storm[24].	
		If the moon has a magnetic field, Luna 2 instrumentation indicated that the field would be less than about 50 γ and would have a dipole magnetic moment less than 1/10 000 of the magnetic moment of the earth[25, 32].	

Vehicle/Launch/ Orbital Data	Instrumentation/Investigated Region	Results	References
Luna 10 March 31 1966/ April 6 1966: Periselene—350 km, Aposelene—1017 km, Period—$2^h58^m15^s$, Inclination—71° 54′. May 30 1966: Periselene—~378 km, Aposelene—~985 km, Period—$2^h58^m03^s$, Inclination—72° 2′.	Three-component magnetometer used to determine scalar field values from the field components measured parallel and perpendicular to the axis of rotation of the vehicle. The range and sensitivity threshold in each direction were 50 γ and 1 γ, respectively. Absolute and relative errors in determining the resultant magnetic field were estimated to be $\pm 10\ \gamma$ and $\pm 5\ \gamma$, respectively. Luna 10 went into lunar orbit on April 3 1966. On May 30 1966 power supply was exhausted and communications with the vehicle ceased. Magnetic field near the earth-visible side of the moon measured between April 3 and May 4 1966. Eleven mean values of total intensity given with corresponding relationship between position of moon in orbit and earth-sun line[30].	During the Luna 10 observations, magnetic field values were in the following ranges: total intensity 23–40 γ and field components parallel and perpendicular to vehicle's axis of rotation 18–38 and 12–16 γ, respectively. During full moon on April 5 and May 4, mean values of the total field were 28 γ and 40 γ, respectively, and 33 and 35 γ during new moon on April 20. Along other selected portions of the moon's orbit, values of 32, 26, 31, 30, 24, 23, and 32 γ are reported. Magnetic field near moon four to six times greater than interplanetary field. This difference should be attributed to effects of the moon. No field of dipole nature was detected. Magnetic tail of the earth was not detected and if present at 60 R_e during observations its magnitude was less than 5 γ. A correlation was indicated between the index of magnetic activity on earth and the magnetic field values near the moon[30, 59].	Dolginov, *et al.* (1966)[30] Zhuzgov, *et al.* (1966)[59]
Venera Series			
Venera 1/ February 12 1961 Perihelion—106×10^6 km, Aphelion—151×10^6 km	Three-component magnetometer to study magnetic field in the vicinity of Venus and a variometer for measurement of magnetic fields in flight. Variometer was a two-channel instrument with parallel-mounted detectors that measured the same field component. Sensitivity threshold of variometer was 2 γ and the measurement range was from 0–50 γ. Magnetic variometer data for February 12 1961 at distances of 165 000 to 175 000 km from earth and for February 17 1961 when the probe was $1{\cdot}9 \times 10^6$ km from earth. No results reported on magnetic field in the vicinity of Venus.	At 165 000 to 175 000 km from the earth when the vehicle was in the region of the magnetosphere boundary, both variometer channels indicated field changes to be generally $\pm 4\ \gamma$. Changes of the H and D components on the ground for the same time had similar values over the 35-minute observation period.[29] Variometer readings over a 22-minute period at the distance of $1{\cdot}9 \times 10^6$ km showed virtually no changes.[29]	Dolginov, *et al.* (1962)[29]

TABLE 8.2 (Continued)

Results of Magnetic Field Studies Aboard Soviet Space Vehicles (1957–1966)

Vehicle/Launch/ Orbital Data	Instrumentation/Investigated Region	Results	References
Venera 2/ November 12 1965	Not noted.	No results noted.	
Venera 3/ November 19 1965	Not noted.	No results noted.	
Zond Series			
Zond 3/ July 18 1965	Not noted.	No results noted.	
Mars Series			
Mars 1/ November 1 1962	One-component magnetometer with 2 γ sensitivity. Position and times of observations not indicated.	Values of field component usually were on the order of 3-4 γ with some values between 6 and 9 γ[20].	USSR National Reports to COSPAR (1960–1966)[20]

Magnetic Field of the Earth and Earth's Vicinity

The bulk of reported total field data has come from Electron 2 and Cosmos 49, launched in January 1964 and October 1964, respectively. Measurements between geocentric distances of 3–11.6 R_e were made on Electron 2 but the vehicle during its lifetime did not detect the magnetosphere boundary. Cosmos 49, a part of the World Magnetic Survey program (as was Cosmos 26 and the introductory and preliminary Sputnik 3) measured the field between the altitudes of 270 and 490 km. Sputnik 3 results were of marginal value. Electron 4 is the only Soviet earth satellite that has detected the magnetosphere boundary. This crossing, which occurred during a magnetic storm, was indicated by field irregularities and departure from the normal sinusoidal pattern of readouts from sensors of all three field components.

The Soviet treatment of magnetic field measurements in the magnetosphere has stressed the determination and distribution of $\Delta T = T_{\text{measured}} - T_{\text{computed}}$ values of the earth's field, where $T=$ the total scalar field. Analysis of the ΔT results from the viewpoint of the physical mechanisms involved has not been equally stressed to date, but Western, particularly U.S., results and analyses of this type have been duly studied by the Soviets.

On the basis of their lunar probes, Lunas 1 and 2, and U.S. data, magnetically active regions between 2 and 10 R_e have been indicated: 2·6 to 3·5 R_e, 6 R_e, and 9 R_e. The current system at 2·6 to 3·5 R_e appeared to be weakly localized and these radiation belt particles in motion are a principal cause of the main phase of magnetic storms.

From 7 R_e to 11·6 R_e, Electron 2 ΔT values were characteristically positive, ranging from 7 to 70 γ. Three-hour averages of ΔT varied no more than 12 γ and were attributed to deformation of the field by solar wind. ΔT was negative in the region 3 to 5 R_e or 3 to 7·5 R_e, dependent upon whether the vehicle was approaching or leaving earth. The average $\Delta T_{\max}$ in this case was 100–150 γ at 3–5 R_e, which is probably explainable in terms of trapped protons and electrons. ΔT equalled zero on an average at about 5 R_e and 26° S latitude on outbound orbits, and at 6 R_e and 59° S on inbound orbits. Geocentric distances and latitudes of $\Delta T=0$ showed a systematic seasonal change as the orbit changed during the lifetime of the satellite. Also values of R_e and ϕ (latitude) of $\Delta T=0$ exhibited some inverse relation to K_p.

The orbit of Cosmos 49 was such that every 77th orbit the satellite "repeated" its flight path but with a shift in local time of transit of two hours. This property of the orbit is being used to investigate

$\delta T = \Delta T_i - \Delta T_{i+77}$, where ΔT_i and ΔT_{i+77} are the differences between measured and computed values on the ith and the "repeated" i+77th orbit, respectively. Maximum δT values (35–70 γ) from work to date occurred more frequently at noon or midnight hours local time near equatorial latitudes, often with positive sign near midnight hours and a negative sign around noon hours.

Magnetic Fields in the Vicinity of the Moon

Magnetic field measurements on Luna 2 (January 1959) to within 50 km of the lunar surface showed that if the moon had a field, its value would be less than about 50 γ. The next and latest measurements in the vicinity of the moon were from Luna 10, a lunar orbiter with periselene and aposelene of 378 and 1017 km, respectively. Measurements were made above the side of the moon visible from earth. Field values in the range of 23 to 40 γ were obtained and no lunar field of a dipole nature was noted. The magnetic field values near the moon were 4–6 times greater than the interplanetary field values and this difference is attributed to the effects of the moon. The earth's magnetotail was not detected at 60 R_e and, if present during observations, its magnitude was less than 5 γ.

Planetary and Interplanetary Magnetic Fields

Magnetic field instrumentation has been on board four Soviet planetary probes—three with a Venus and one with a Mars mission—and on Zond 3, whose mission was to test photographic transmissions from lunar and planetary distances. Magnetic field instrumentation on two of the Venus vehicles and on Zond 3 has not been described. Essentially no results have been reported from these probes.

RELATED RESEARCH

A definition of "related studies" within the scope of this chapter could reasonably include consideration of world magnetic charts, use of these charts by spherical harmonics and integral methods from potential theory for comparison of measured and computed field values at various geocentric distances, and electrical currents in the vicinity of the earth. Soviet literature on these topics has been sampled, but an all-inclusive presentation of results contained therein relating to the broader aspects and manifestations of the main field and its secular change has not been attempted. Similarly, no consideration is given to instruments, observations, and studies involved in recording and analyzing ground-based data of either the earth's main field or transient variations.

World Magnetic Charts

World magnetic charts are compiled in the USSR by IZMIRAN* every five years, with the latest two being epoch 1955[11] and epoch 1960[2]. Work on epoch 1965 is well underway, if not completed [54].

Data from Soviet expeditionary ships to the Antarctic, from the continent of Antarctica, and from the Soviet nonmagnetic ship "Zarya" have changed appreciably the 1960 over the 1955 charts especially for high southern latitudes of the Southern Hemisphere[2]. Zarya data have been treated from several approaches for the Red Sea, Black Sea, Azov Sea, South and North Atlantic, and the Indian Ocean, in several Soviet papers.

Zarya data from the Atlantic Ocean between 45° N and 35° S and the Indian Ocean were used by the Soviets as a basis for comparing U.S. and British charts, epoch 1955, for these ocean areas. Resultant discrepancies reported show the British charts on the whole to be somewhat superior to those of the U.S.[40].

One of the earlier Soviet post World War II spherical analyses[9] of the geomagnetic field was in 1947 for epoch 1945. Circa 1960 the number of Soviet papers on potential theory as applied to the earth's magnetic field began increasing[1–8, 16–17, 19, 41–51]. In most of these, the method of spherical harmonics is applied. With one exception[9], when charts have been the source of initial data in these analyses, the Soviets have consistently, to the degree noted, used IZMIRAN or British charts but on occasion have used U.S. derived coefficients for comparison in satellite work. The majority of these analyses has been done by members of IZMIRAN (or its predecessors) and the Institute of Mathematics, Siberian Division of the USSR Academy of Sciences.

Magnetic field measurements from Sputnik 3 (in 1958) were chiefly over the Soviet Union. Ben'kova and Tyurmina[17] listed the expansion coefficients used to compute the field at Sputnik 3 altitudes over the USSR for determining differences $\Delta T = T_{\text{measured}} - T_{\text{computed}}$[25].

With initial data from USSR Z-chart 1950 and assuming field sources to be entirely within the earth, coefficients g_n^m and h_n^m to $n = m = 6$ were

* Institute of Terrestrial Magnetism, Ionosphere, and Propagation of Radio Waves of the USSR Academy of Sciences.

corrected to 1958 epoch. Differences between initial and computed field data were then used to compute Δg_n^m for determining new coefficients $g_n'^m = \Delta g_n^m + g_n^m$ through $n = 9$ and $m = 6$. (Distribution of differences were assumed to be symmetrical about 180° longitude and Δh_n^m corrections were identically equal to 0.) Representativeness of the field was then analyzed[17].

Adam, *et al.* have computed at least nine sets[1–4, 8] of expansion coefficients for all or parts of the earth's main field and have examined the results from a variety of viewpoints[1–8]. Most of the computations were to $n = m = 6$, e.g., when using as initial data the IZMIRAN 1955 and British Admiralty 1955 charts[1]. These coefficients have been used to synthesize the X-, Y-, Z- elements and the total field at 12 altitudes between 0 and 15 000 km[6]. Integral methods have also been applied to calculate the Z-field at seven altitudes between 0 and 10 000 km[49]. One analysis of secular change for seven epochs between 1932 and 1958 used coefficients to $n = m = 4$, but sixth order terms have also been used for this purpose[5].

Coefficients from Adam's, *et al.*[8], spherical analysis of IZMIRAN 1960 1:10 000 000 charts* of D, H, and Z elements after reduction to epoch 1964 were used to compute the field in the ΔT determinations from Cosmos 49 magnetic field measurements[21]. The number of points used by Adam, *et al.*, was 2330 with field values taken from the chart using a $5° \times 5°$ grid. Coefficients to $n = m = 6$ were computed and listed for the total, inner, and external field. For two pairs of Cosmos 49 orbits the Soviets also compared the differences between measured and computed magnetic field values as calculated from Adam's, *et al.*[8], coefficients and from coefficients derived in the U.S. by Cain, *et al.*†. Curves of this comparison show some agreement in form but differ in values up to about 200 γ[21].

Adam, *et al.*[1], found best agreement in zonal harmonics from four sets of coefficients which they state strengthens the assumption that the external part of the earth's field is caused by electromagnetic phenomena in radiation belts. In later analyses[4] using the coefficients[1] based on IZMIRAN charts, Adam, *et al.*, found that only in a small area of South Africa was the outer field more than two per cent of the inner field. Otherwise between the latitudes of 80° N and 60° S the outer field averaged about 0·6 per cent of the inner field.

From U.S. data (Explorer 6 and Pioneer 5) and on the basis of mag-

* Ben'kova, *et al.*[16], have reported on the accuracy of the analysis of these 1960 charts using the coefficients reported by Adam, *et al.*[8].

† Journal of Geophysical Research, 1965, Vol 70, No. 15, pp 3647–3674.

netic field[13, 14, 23, 53] and charged particle trap data[38, 39] from Luna 1, Shevnin[52] concluded that an extra-ionospheric current system probably occurs at least in two radiation zones—one in the region of 4 R_e (outer) and the second in the region of 9 R_e (outermost). Ben'kova and Tyurmina[18] computed the field for an equatorial current at 3·5 and 9 R_e for three different latitudes of intersection of the current surface with the ionosphere. They concurred with Shevnin that extra-ionospheric current probably flows at 3-5 R_e and 9–10 R_e (or third zone), adding that different distances for current detection can be ascribed to the fact that the current is dominant in different zones at different times and, particularly, to the displacement of the zones.

References

1. N. V. Adam, N. P. Ben'kova, V. P. Orlov, N. K. Osipov, and L. O. Tyurmina, "Spherical Analysis of the Main Geomagnetic Field for the Epochs 1955 and 1958", Geomagnetizm i Aeronomiya, 1962, Vol 2, No. 5, pp 949–962 (Geomagnetism and Aeronomy, pp 785–796).

2. N. V. Adam, N. P. Ben'kova, V. P. Orlov, N. K. Osipov, and L. O. Tyurmina, "Spherical Analysis of the Constant Geomagnetic Field for the Epochs 1955 and 1958, II", Geomagnetizm i Aeronomiya, 1963, Vol 3, No. 1, pp 121–126 (Geomagnetism and Aeronomy, pp 96–101).

3. N. V. Adam, N. P. Ben'kova, V. P. Orlov, N. K. Osipov, and L. O. Tyurmina, "Spherical Analysis of the Main Geomagnetic Field and Secular Variations", Geomagnetizm i Aeronomiya, 1963, Vol 3, No. 2, pp 336–353 (Geomagnetism and Aeronomy, pp 271–285).

4. N. V. Adam, N. P. Ben'kova, V. P. Orlov, N. K. Osipov, and L. O. Tyurmina, "Synthesis of the Geomagnetic Field From Spherical Harmonic Coefficients", Geomagnetizm i Aeronomiya, 1964, Vol 4, No. 1, pp 151–160 (Geomagnetism and Aeronomy, pp 113–120).

5. N. V. Adam, N. P. Ben'kova, V. P. Orlov, N. K. Osipov, and L. O. Tyurmina, "Analytical Representation of Secular Variation", Geomagnetizm i Aeronomiya, 1964, Vol 4, No. 4, pp 748–752 (Geomagnetism and Aeronomy, pp 584–593).

6. N. V. Adam, N. P. Ben'kova, V. P. Orlov, N. K. Osipov, and L. O. Tyurmina, "Vertical Distribution of the Geomagnetic Field", Geomagnetizm i Aeronomiya, 1964, Vol 4, No. 4, pp 753–761 (Geomagnetism and Aeronomy, pp 587–593).

7. N. V. Adam, N. P. Ben'kova, and L. O. Tyurmina, "Computations of the Geomagnetic Field at Great Heights", Geomagnetizm i Aeronomiya, 1965, Vol 5, No. 3, pp 522–528 (Geomagnetism and Aeronomy, pp 402–407).

8. N. V. Adam, N. K. Osipov, L. O. Tyurmina, and A. P. Shlyakhtina, "Spherical Harmonic Analysis of World Magnetic Charts for the 1960 Epoch", Geomagnetizm i Aeronomiya, 1964, Vol 4, No. 6, pp 1130–1131 (Geomagnetism and Aeronomy, pp 878–879).

9. V. I. Afanas'eva, "Spherical Harmonic Analysis for 1945", Izvestiya Akademii Nauk SSSR, Seriya Geograficheskaya i Geofizicheskaya, 1947, Vol 11, No. 1, pp 55–60.

10. L. M. Aleksanyan, Ye. G. Yeroshenko, L. N. Zhuzgov, and U. V. Fastovskiy, "Magnetometric Equipment Aboard the 'Electron 2' Space Station", Kosmicheskiye Issledovaniya, 1966, Vol 4, No. 2, pp 302–310 (Cosmic Research, pp 208–223).

11. L. I. Al'tshuler, B. D. Vints, K. A. Mal'tseva, Z. S. Chuguyan, and A. P. Shlyakhtina, "World Magnetic Charts for Epoch 1955", Trudy Nauchno Issledovatelskogo Instituta Zemnogo Magnetizma, 1955, No. 11 (21), pp 222–236.

12. L. I. Al'tshuler, K. A. Mal'tseva, Z. S. Chuguyan, and A. P. Shlyakhtina, "World Magnetic Maps for the Epoch, 1960", Geomagnetizm i Aeronomiya, 1964, Vol 4, No. 4, pp 773–780 (Geomagnetism and Aeronomy, pp 601–606).

13. M. G. Antsilevich, "Geomagnetic Field Variations on 9, 10, 24, 25 August, 2 September 1959, and 11 March 1960", Geomagnetizm i Aeronomiya, 1961, Vol 1, No. 3, pp 320–325 (Geomagnetism and Aeronomy, pp 286–291).

14. M. G. Antsilevich and A. D. Shevnin, "On the Problem of the Geomagnetic Observations Made by the First Soviet Cosmic Rocket", Doklady Akademii Nauk SSSR, 1960, Vol 135, No. 12, pp 298–300 (Doklady of the Academy of Sciences USSR, Earth Sciences Sections, pp 1244–1245).

15. V. V. Beletskiy and Yu. V. Zonov, "Rotation and Orientation of the Third Soviet Satellite", Iskusstvennyye Sputniki Zemli, 1961, No. 7, pp 32–55 (Artificial Earth Satellites, pp 29–52).

16. N. P. Ben'kova, N. V. Adam, and L. O. Tyurmiňa, "Precision of Analysis of the World Magnetic Charts of Epoch 1960", Geomagnetizm i Aeronomiya, 1966, Vol 6, No. 1, pp 179–181 (Geomagnetism and Aeronomy, pp 142–144).

17. N. P. Ben'kova and L. O. Tyurmina, "Analytical Representation of the Geomagnetic Field Over the Territory of the Soviet Union for the 1958 Epoch", Geomagnetizm i Aeronomiya, 1961, Vol 1, No. 1, pp 87–96 (Geomagnetism and Aeronomy, pp 81–96).

18. N. P. Ben'kova and L. O. Tyurmina, "The Magnetic Field of the Equatorial Ring (Current)", Geomagnetizm i Aeronomiya, 1962, Vol 2, No. 4, pp 635–641 (Geomagnetism and Aeronomy, pp 528–534).

19. N. P. Ben'kova, T. L. Vinnikova, and L. O. Tyurmina, "Geomagnetic Eccentric Dipole", Geomagnetizm i Aeronomiya, 1964, Vol 4, No. 5, pp 917–923 (Geomagnetism and Aeronomy, pp 713–717).

20. USSR National Reports to COSPAR, 1960–1966, presented by A. A. Blagonravov (In English).

21. Sh. Sh. Dolginov, L. V. Konovalova, V. I. Nalivayko, L. D. Tyurmina, M. M. Chinchevoy, and M. N. Fatkulin, "About Effects of the Outer Sources of the Magnetic Field on the Heights of the F_2 Layer—I", Geomagnetizm i Aeronomiya, 1966, Vol 6, No. 6, pp 969–983.

22. Sh. Sh. Dolginov, V. I. Nalivayko, A. V. Tyurmina, and M. N. Chinchevoy, "Experiments Under the Program of the World Magnetic Survey", Trudy Vsesoyuznoy Konferentsii po Fizike, Kosmicheskogo Prostranstva, Izdatel'stvo "Nauka", Moscow, June 10–16, 1965, pp 606–614 (NASA TT F-389, pp 806–817).

23. Sh. Sh. Dolginov and N. V. Pushkov, "Results of Measurements of Magnetic Field of the Earth With Cosmic Rockets", Doklady Akademii Nauk SSSR, 1959, Vol 12, No. 1, pp 77–80 (Doklady of the Academy of Sciences, USSR, Earth Science Sections, pp 1126–1128).

24. Sh. Sh. Dolginov and N. V. Pushkov, "On Some Results of Earth's Magnetic Field Investigations in Outer Space", Space Research III, Proceedings of the Third International Space Science Symposium, Washington, D.C., May 2–8 1962, North-Holland Publishing Company, Amsterdam, 1963, pp 331–341.

25. Sh. Sh. Dolginov and N. V. Pushkov, "Study of the Magnetic Field in Outer Space", Kosmicheskiye Issledovaniya, 1963, Vol 1, No. 1, pp 55–97 (Cosmic Research, pp 89–171).

26. Sh. Sh. Dolginov, Ye. G. Yeroshenko, and L. N. Zhuzgov, "A Survey of the Earth's Magnetosphere in the Region of the Radiation Belt ($3R_E - 6\ R_E$) in February–April 1964 on the Electron 2 Satellite", Space Research VI, Proceedings of the Sixth International Space Science Symposium, Mar Del Plata, Argentina, May 11–19 1965, North-Holland Publishing Company, Amsterdam, 1966, pp 790–809.

27. Sh. Sh. Dolginov, Ye. G. Yeroshenko, and L. N. Zhuzgov, Investigation of the Earth's Magnetosphere in the Radiation Zone Region (3–6 R_e) February–April, 1964", Trudy Vsesoyuznoy Konferentsii po Fizike, Kosmicheskogo Prostranstva, Izdatel'stvo "Nauka", Moscow, June 10–16 1965, pp 342–356 (NASA TT F-389, pp 464–482).

28. Sh. Sh. Dolginov, Ye. G. Yeroshenko, L. N. Zhuzgov, and N. V. Pushkov "Investigation of the Magnetic Field of the Moon", Geomagnetizm i Aeronomiya, 1961, Vol 1, No. 1, pp 21–29 (Geomagnetism and Aeronomy, pp 18–26).

29. Sh. Sh. Dolginov, Ye. G. Yeroshenko, L. N. Zhuzgov, and N. V. Pushkov, "Magnetic Measurements by the Venus Automatic Interplanetary Station", Geomagnetizm i Aeronomiya, 1962, Vol 2, No. 1, pp 38–40 (Geomagnetism and Aeronomy, pp 28–30).

30. Sh. Sh. Dolginov, Ye. G. Yeroshenko, L. N. Zhuzgov, and N. V. Pushkov, "Magnetic Field Measurement in the Moon's Neighborhood at the Luna 10 Satellite", Doklady Akademii Nauk SSSR, 1966, Vol 170, No. 3, pp 574–577.

31. Sh. Sh. Dolginov, Ye. G. Yeroshenko, L. N. Zhuzgov, N. V. Pushkov, and, L. O. Tyurmina, "Measuring the Magnetic Field of the Earth and Moon by Means of Sputnik III and Space Rockets I and II", Space Research I, Proceedings of the First International Space Science Symposium, Nice, France, January 11–16 1960, North-Holland Publishing Company, Amsterdam, 1960, pp 863–869.

32. Sh. Sh. Dolginov, Ye. G. Yeroshenko, L. N. Zhuzgov, and L. O. Tyurmina, "Magnetic Measurements With the Second Cosmic Rocket", Iskusstvennyye Sputniki Zemli, 1960, No. 5, pp 16–23 (Artificial Earth Satellites, pp 490–502).

33. Sh. Sh. Dolginov, L. N. Zhuzgov, N. V. Pushkov, L. O. Tyurmina, and I. V. Fryazinov, "Some Results of the Constant Geomagnetic Field Measurements Carried Out From Sputnik III Over the Territory of the USSR", 12th International Astronautical Congress, Washington, D.C., 1961, Vol 2, Academic Press, 1963, pp 939–950 (In English).

34. Sh. Sh. Dolginov, L. N. Zhuzgov, N. V. Pushkov, L. O. Tyurmina, and I. V. Fryazinov, "Some Results of the Constant Geomagnetic Field Measurements Carried Out From Sputnik III Over the Territory of the USSR", Space Research III, Proceedings of the Third International Space Science Symposium, Washington, D.C., May 2–8 1962, North-Holland Publishing Company, Amsterdam, 1963, pp 342–354.

35. Sh. Sh. Dolginov, L. N. Zhuzgov, N. V. Pushkov, L. O. Tyurmina, and I. V. Fryazinov, "Some Results of Measuring the Constant Magnetic Field of the Earth With the Third Artificial Sputnik of the Earth Above the Territory of the USSR", Geomagnetizm i Aeronomiya, 1962, Vol 2, No. 6, pp 1061–1075 (Geomagnetism and Aeronomy, pp 877–889).

36. Sh. Sh. Dolginov, L. N. Zhuzgov, and V. A. Selyutin, "Magnetometers in the Third Soviet Earth Satellite", Iskusstvennyye Sputniki Zemli, 1960, No. 4, pp 135–160 (Artificial Earth Satellites, pp 358–396).

37. K. I. Gringauz, Sh. Sh. Dolginov, V. V. Bezrukikh, Ye. G. Yeroshenko, L. N. Zhuzgov, L. S. Musatov, E. K. Solomatina, and U. V. Fastovskiy, "Comparison of Simultaneous Measurements of the Magnetic Field and Positive Ion Streams Within the Earth's Magnetosphere, Performed on the Satellite 'Electron 2' ", Trudy Vsesoyuznoy Konferentsii po Fizike, Kosmicheskogo Prostranstva, Izdatel'stvo "Nauka", Moscow, June 10–16 1965, pp 336–341 (NASA TT F-389, pp 457–463).

38. K. I. Gringauz, V. G. Kurt, V. I. Moroz, and I. S. Shklovskiy, "Ionized Gas and Fast Electrons in the Earth's Neighborhood and in Interstellar Space", Doklady Akademii Nauk SSSR, 1960, Vol 132, No. 5, pp 1062–1066 (Doklady of the Academy of Sciences USSR, Earth Sciences Sections, pp 580–583).

39. K. I. Gringauz and S. M. Rytov, "Relationship Between Charged Particle Measurements Made by Means of Traps on Soviet Space Rockets and Explorer VI and Pioneer V Data on Magnetic Field", Doklady Akademii Nauk SSSR, 1960, Vol 135, No. 1, pp 48–51 (Doklady of the Academy of Sciences, Soviet Physics Sections, 1961, Vol 5, No. 6, pp 1225–1228).

40. M. M. Ivanov, "Accuracy of World Magnetic Charts for Ocean Areas", Geomagnetizm i Aeronomiya, 1961, Vol 1, No. 1, pp 104–110 (Geomagnetism and Aeronomy, pp 97–103).

41. Yu. D. Kalinin, "Expression in Spherical Coordinates of the Potential of a Magnetic Field Situated at Any Point in Space and Having a Magnetic Moment of Any Direction", Geomagnetizm i Aeronomiya, 1963, Vol 3, No. 6, pp 1089–1093 (Geomagnetism and Aeronomy, pp 876–879).

42. N. K. Osipov, "Computer Synthesis of Geomagnetic Data", Geomagnetizm i Aeronomiya, 1961, Vol 1, No. 3, pp 432–435 (Geomagnetism and Aeronomy, pp 387–389).

43. N. K. Osipov, "Algorithms for Use in Preparing Analytical Representatives of Observed Geomagnetic Fields on Digital Computers", Geomagnetizm i Aeronomiya, 1962, Vol 2, No. 5, pp 963–971 (Geomagnetism and Aeronomy, pp 797–804).

44. N. K. Osipov, "The Energy Spectrum of the Inner Part of the Constant Geomagnetic Field in the Epochs 1955 and 1958", Geomagnetizm i Aeronomiya, 1963, Vol 3, No. 1, pp 174–176 (Geomagnetism and Aeronomy, pp 141–143).

45. N. K. Osipov, "The Outer Part of the Earth's Main Field", Geomagnetizm i Aeronomiya, 1963, Vol 3, No. 2, pp 354–361 (Geomagnetism and Aeronomy, pp 286–292).

46. N. K. Osipov, "Secular Variations of the Energy Spectrum of the Inner Part of the Earth's Main Magnetic Field", Geomagnetizm i Aeronomiya, 1963, Vol 3, No. 2, pp 381–382 (Geomagnetism and Aeronomy, pp 312–313).

47. N. K. Osipov, "Analytical Representation of the Inner Part of the Potential of the Constant Geomagnetic Field From 1820 to 1960", Geomagnetizm i Aeronomiya, 1965, Vol 5, No. 1, pp 197–199 (Geomagnetism and Aeronomy, pp 150–152).

48. K. Pepushoy, "Spherical Analysis of the Geomagnetic Field for the Epoch 1955", Geomagnetizm i Aeronomiya, 1962, Vol 2, No. 1, pp 161–166 (Geomagnetism and Aeronomy, pp 138–143).

49. I. M. Pudovkin, "The Spatial Structure of the Geomagnetic Field and Some Aspects of the Inner Structure of the Earth, I", Geomagnetizm i Aeronomiya, 1964, Vol 4, No. 2, pp 376–389 (Geomagnetism and Aeronomy, pp 292–302).

50. I. M. Pudovkin, "The Spatial Structure of the Geomagnetic Field and Some Aspects of the Inner Structure of the Earth, II", Geomagnetizm i Aeronomiya, 1964, Vol 4, No. 3, pp 539–547 (Geomagnetism and Aeronomy, pp 427–433).

51. I. M. Pudovkin, "The Spatial Structure of the Geomagnetic Field and Some. Aspects of the Inner Structure of the Earth, III", Geomagnetizm i Aeronomiya, 1964, Vol 4, No. 4, pp 762–772 (Geomagnetism and Aeronomy, pp 594–600).

52. A. D. Shevnin, "The Extra Ionospheric Current System", Geomagnetizm i Aeronomiya, 1961, Vol 1, No. 2, pp 178–193 (Geomagnetism and Aeronomy, pp 160–171).

53. A. D. Shevnin, "The Magnetic Field of the Ring Current", Geomagnetizm i Aeronomiya, 1963, Vol 3, No. 2, pp 213–222 (Geomagnetism and Aeronomy, pp 172–179).

54. N. M. Sisakyan, "Soviet Achievements in Geophysics 1965", Vestnik Akademii Nauk SSSR, 1966, No. 3, pp 43–47.

55. S. N. Vernov, A. Ye. Chudakov, P. V. Vakulov, Ye. V. Gorchakov, Yu. I. Logachev, A. G. Nikolayev, I. A. Rubinshteyn, Ye. N. Sosnovets, and M. V. Ternovskaya, "Pulsations of the Magnetic Field According to the Measurements on the Satellite 'Elektron 3' ", Trudy Vsesoyuznoy Konferentsii po Fizike Kosmicheskogo Prostranstva, Izdatel'stvo "Nauka", Moscow, June 10–16 1965, pp 433–434 (NASA TT F–389, pp 587–588).

56. Ye. G. Yeroshenko, "Investigation of the Earth's Magnetosphere at Distances 7–11·6 R_E by the Electron Satellites", Space Research VI, Proceedings of the Sixth International Space Science Symposium, Mar Del Plata, Argentina, May 11–19 1965, North-Holland Publishing Company, Amsterdam, 1966, pp 629–641.

57. Ye. G. Yeroshenko, "Study of the Earth's Magnetosphere at a Distance of 7–11·7 R_e on the 'Electron' Satellites", Trudy Vsesoyuznoy Konferentsii po Fizike Kosmicheskogo Prostranstva, Izdatel'stvo "Nauka", Moscow, June 10–16 1965, pp 356–367 (NASA TT F–389, pp 483–489).
58. Ye. G. Yeroshenko, Sh. Sh. Dolginov, L. N. Zhuzgov, U. V. Fastovsky, and L. M. Aleksanyan, "Magnetic Investigations on the Electron 2 Satellite", Space Research V, Proceedings of the Fifth International Space Science Symposium, Florence, Italy, May 12–16 1964, North-Holland Publishing Company, Amsterdam, 1965, pp 980–984.
59. L. N. Zhuzgov, Sh. Sh. Dolginov, and Ye. G. Yeroshenko, "Investigation of the Magnetic Field From the Satellite Luna 10", Kosmicheskiye Issledovaniya, 1966, Vol 4, No. 6, pp 880–897.
60. Yu. V. Zonov, "On the Problem of the Interaction Between a Satellite and the Earth's Magnetic Field", Iskusstvennyye Sputniki Zemli, 1959, No. 3, pp 118–124 (Artificial Earth Satellites, pp 169–182).

Chapter 9

Ionospheric Electron Density Studies

R. C. Behn

INTRODUCTION

Although there had been some earlier sporadic activity, the cohesive Soviet study of ionospheric electron densities by means of geophysical rockets began in 1957. In the same year studies utilizing artificial earth satellites were initiated. To date, over 20 geophysical rockets and 12 satellites have been associated with reported experiments.

Table 9.1 identifies the vehicles and the associated observational techniques which have been utilized by the Soviets to make these studies. As shown in the table, some direct sensing of electron densities has been done, principally using ion traps. These instruments, as well as the data with regard to both positive ions and electrons which have been obtained by their use, are discussed in Chapter 10. For convenience, however, the vehicles associated with positive ion studies, as well as with electron temperature measurements, are included in the table.

TABLE 9.1

Soviet Rockets and Satellites Associated With Ionospheric Charged-Particle Studies (1957–1966)

Vehicle			Parameter Measured and Method											
			Positive Ions		Electrons									
			Concentration	Temperature	Density									Temperature
					Direct				Indirect					
Launch Date	Time of Day	Alti-tude, km	Ion Trap	Honeycomb Ion Trap	Technique Unspecified	Antenna Probe	Ion Trap	Langmuir Probe	Dispersion Interferometry	Doppler Frequency Difference	Faraday Rotation	Joint Amplitude/ Phase Analysis	Radio Rise and Set	Presumably Langmuir Probe
Geophysical Rockets														
May 16 1957	0618	212	—	—	—	—	—	—	×	—	—	—	—	—
Aug. 25 1957	0627	212	×	—	—	—	—	—	×	—	—	—	—	—
Aug. 31 1957	Morning	212	×	—	—	—	—	—	×	—	—	—	—	—
Sep. 9 1957	1954	212	×	—	—	—	—	—	×	—	—	—	—	—
Feb. 21 1958	1140	473	×	—	—	—	—	—	×	—	×	—	—	×
Aug. 2 1958	0947	212	—	—	—	×	—	—	—	—	—	—	—	—
Aug. 27 1958	0806	451	×	—	—	—	—	—	×	—	×	—	—	×
Sep. 19 1958	Night	473	×	—	—	—	—	—	×	—	—	—	—	×
Oct. 4 1958	Day	110	×	—	—	—	—	—	—	—	—	—	—	—
Oct. 10 1958	Day	110	×	—	—	—	—	—	—	—	—	—	—	—
Oct. 31 1958	1554	473	×	—	—	—	—	—	×	—	×	—	—	×
Dec. 23 1958	?	110	×	—	—	—	—	—	—	—	—	—	—	—
Dec. 25 1958	Day	110	×	—	—	—	—	—	—	—	—	—	—	—
July 2 1959	0740	212	×	—	—	×	—	—	×	—	—	—	—	×
July 10 1959	Morning	212	×	—	—	—	—	—	×	—	—	—	—	×
July 14 1959	0740	212	—	—	—	—	—	—	×	—	—	—	—	—
July 21 1959	Morning	110	×	—	—	—	—	—	—	—	—	—	—	—
July 21 1959	Evening	110	×	—	—	—	—	—	—	—	—	—	—	—
July 22 1959	0514	195	—	—	—	—	—	—	×	—	—	—	—	—
June 15 1960	0643	212	—	—	—	—	—	—	×	—	—	—	—	—
June 24 1960	?	212	—	—	×	—	—	—	—	—	—	—	—	—
Nov. 15 1961	Afternoon	430	—	—	—	—	—	—	×	—	—	—	—	—
Oct. 18 1962	Afternoon	500	—	—	—	—	—	—	×	—	—	—	—	—

Vehicle			Parameter Measured and Method											
			Positive Ions		Electrons									
			Concentration	Temperature	Density									Temperature
					Direct				Indirect					
Launch Date	Time of Day	Altitude, km	Ion Trap	Honeycomb Ion Trap	Technique Unspecified	Antenna Probe	Ion Trap	Langmuir Probe	Dispersion Interferometry	Doppler Frequency Difference	Faraday Rotation	Joint Amplitude/ Phase Analysis	Radio Rise and Set	Presumably Langmuir Probe
June 6 1963	Morning	515	—	—	—	—	—	—	×	—	—	—	—	—
June 18 1963	?	510	—	—	×	—	—	—	—	—	—	—	—	—
Sep. 20 1965	0400	500	—	—	—	—	—	×	×	—	—	—	—	×
Oct. 1 1965	0400	500	—	—	—	—	—	—	×	—	—	—	—	×
Oct. 13 1966	Morning	500	—	—	—	—	—	—	×	—	—	—	—	×
Artificial Earth Satellites														
Sputnik 1/Oct. 4 1957			—	—	—	—	—	—	—	—	—	—	×	—
Sputnik 2/Nov. 3 1957			—	—	—	—	—	—	—	—	—	—	×	—
Sputnik 3/May 15 1958			×	—	—	—	—	—	—	—	—	—	—	—
Cosmos 1/Mar. 16 1962			—	—	—	—	—	—	×	×	×	×	—	—
Cosmos 2/Apr. 6 1962			×	×	—	—	×	—	×	×	×	×	—	×
Cosmos 3/Apr. 24 1962			×	—	—	—	—	—	—	—	—	—	—	—
Cosmos 5/May 28 1962			×	—	—	—	—	—	—	—	—	—	—	—
Cosmos 11/Oct. 20 1962			—	—	—	—	—	—	×	×	—	—	—	—
Cosmos 12/Dec. 22 1962			×	—	—	—	×	—	—	—	—	—	—	—
Cosmos 15/Apr. 22 1963			×	—	—	—	×	—	—	—	—	—	—	—
Cosmos 17/May 22 1963			—	—	—	—	—	—	×	—	—	—	—	—
Electron 1/Jan. 30 1964			—	—	—	—	—	—	×	×	—	—	—	—
Electron 2/Jan. 30 1964			×	—	—	—	×	—	—	—	—	—	—	—
Electron 3/July 11 1964			—	—	—	—	—	—	×	×	—	—	—	—
Electron 4/July 11 1964			×	—	—	—	×	—	—	—	—	—	—	—

INSTRUMENTATION AND OBSERVATIONAL TECHNIQUES

Dispersion Interferometry Technique[27, 35–36, 47]

As illustrated in Table 9.1, all indirect Soviet ionospheric electron density studies which have been conducted utilizing geophysical rockets have made use of the dispersion interferometry technique, nearly to the exclusion of other indirect methods. In addition, the technique has been used with most of the satellites with which Soviet indirect studies have been made. The Soviets attribute the origin of the technique to a 1937 proposal by L. I. Mandel'shtam and N. D. Papaleksi[1], which proposal stemmed from a phase probe method first used in studying the ionosphere during a 1936 solar eclipse[61]. Basically, the method makes use of the frequency dispersion of coherent RF transmissions introduced by the ionosphere. The analysis, however, neglects electron collision and geomagnetic effects. Two ultrashortwave coherent signals on multiply related frequencies are transmitted by a vehicle in flight, and a continuous record is made of their amplitudes and phase difference as received. Vehicle position tracking data are simultaneously obtained. Electron densities are determined from their relation to the computed space rate of change of phase difference along the transmission path. The theoretical analysis is available in the Soviet literature.

Two frequencies, 144 mc and 48 mc, were used by the Soviets until 1958, when the capability of the transmitter was extended to provide transmission on 24 mc. Thereafter a second phase difference, between 144 mc and 24 mc, was measured in some experiments to supplement the first. A view of the transmitter, with housing removed, as described in 1961, appears in Figure 9.1. The successively higher frequencies are produced by frequency multiplication of the 24-mc quartz-crystal controlled master oscillator frequency, and output powers are 15 watts, 8 watts, and 3 watts, for the 144-mc, 48-mc, and 24-mc channels, repectively. Two dipole antennas are used, one for 144 mc and 48 mc, the second for 24 mc.

A view of a ground station appears in Figure 9.2. Each antenna assembly shown consists of two electrically independent antennas having mutually perpendicular linear polarizations to isolate polarization changes in signal levels from other amplitude fluctuations, and to prevent interruptions due to polarization nulls. The half power beam width of the 144-mc antenna in the plane of the supporting poles is 25°, and 35° in the orthogonal plane. For the 48-mc antenna these values are 35° and 50°, respectively. The antenna array for the 24-mc transmissions

Fig. 9.1. Vehicle Transmitter Assembly[36]

1. 24-mc master oscillator
2. 144-mc channel components
3. 48-mc and 24-mc channel components

Fig. 9.2. Ground Receiving Station[36]

1. 144-mc antenna
2. 48-mc antenna
3. Apparatus van

(not shown) is comprised of four half-wave dipoles suspended one-quarter wave length above the ground between four masts. Parallel dipoles are in phase and respond to similarly polarized signals. The radiation pattern, of a vertically directed axially symmetric form, has a directional gain of about 4.

Fig. 9.3. Receiving Equipment[47]

The receiving equipment, illustrated in Figure 9.3, consists of a three-channel superheterodyne capable of simultaneous measurement of changes in the phase differences of two sets of frequencies. Each channel consists of two IF stages with appropriate frequency multipliers between stages, the final output frequency to the phase detectors being 8·4 mc. The paired voltages are added and detected by the phase detectors, the resulting interference frequency signals being fed through cathode followers to the recording devices. In addition to the interference frequencies, the second IF frequency pairs, prior to frequency multiplication, are fed to a cathode ray oscilloscope to provide Lissajous figure indications of variations in phase differences, by a method attributed in the Soviet literature to Ye. Ya. Shchegolev, *et al.*[65].

The recording equipment is illustrated in Figure 9.4. The interference frequency signals, receiver input voltage signals, and time-

Fig. 9.4. Recording Equipment[47]
1. Loop oscillograph
2. Lissajous figure recorder
3. Photorecorder

marking pulses are applied to a loop oscillograph, being recorded on 35-mm motion picture film. Four oscillographs working in unison are used. Sample oscillograms for various sections of a trajectory, the rocket velocities and electron densities differing, are presented in Figure 9.5. Complementary calculations of electron densities are made from the phase difference and input voltage curves.

A number of potential sources of error exist in the accumulation of data by this technique. Included are Doppler frequency shifts due to vehicle motion, instrumental error in the receiver, primarily in the heterodyning process, and inaccuracies introduced by the recording equipment. Comparison of records of transmissions received simultaneously at a number of physically separated ground stations have shown, however, that results usually vary by no more than 5 per cent of the measurement value.

Doppler Frequency Difference Technique[6, 17, 20, 50, 52–53, 64, 66, 71]

A second principal indirect method employed by the Soviets when utilizing a space vehicle to study ionospheric electron densities is the

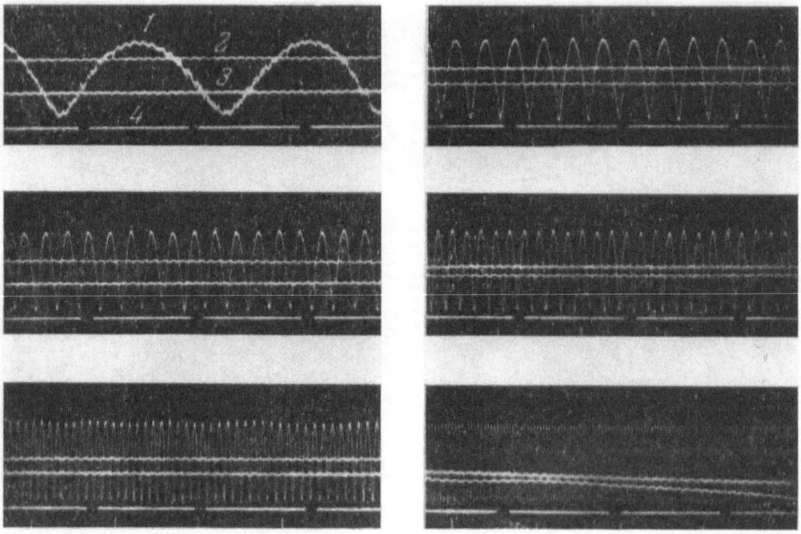

Fig. 9.5. Sample Oscillograms[35]

1. Phase difference
2–3. Receiver input voltages
4. Time markers (0·25 sec/interval)

Doppler frequency difference technique, first proposed[3] in 1958 by Ya. L. Al'pert, of the Institute of Terrestrial Magnetism, Ionosphere, and Radio Wave Propagation, of the Soviet Academy of Sciences. The theory has been discussed and extended in a number of later papers[4–5, 10, 19] by Al'pert and others, and the method was first used in March and April 1962 in conjunction with Cosmos 1 and 2. By measuring the difference in Doppler shifts of two harmonically related frequencies transmitted from a moving vehicle, i.e., the reduced phase (or frequency) difference, the effect of the medium is attained in terms of refraction angle, total electron content along the transmission path and electron density at the vehicle. Electron collision and geomagnetic effects are neglected. The data analysis technique, which assumes a horizontally quasi-homogeneous ionosphere, provides the mean electron densities of orbital segments corresponding to the measurement time intervals (e.g., 40 km and 5 seconds).Orbital data are required for the computations. The analysis is further extended to provide gradients of the electron densities and to permit determinations of the linear extent of inhomogeneities intervening along the propagation path. It is of interest to point out that in a 1965 article, K. I. Gringauz, *et al.*[33], of the Radio Engineering Institute of the Soviet Academy of Sciences, Moscow, take exception to Al'pert's neglecting gradient and nonstationary effects in his determinations of electron densities, and allege other shortcomings in his analysis. In his rebuttal[9], Al'pert rejects their arguments.

Observations by this method* have been made at Moscow, Novosibirsk, Sverdlovsk, and in the Crimea, utilizing signals from Cosmos vehicles 1, 2, and 11, and from Electron 1. The "Mayak" (Beacon) transmitter aboard each vehicle emitted at 20·0050 mc and 90·0225 mc and on an additional frequency, 30·0075 mc, in the case of Electron 1. An external view of this equipment appears in Figure 9.6. Although other type antennas have been used, signals have been received at the ground principally either on vertical quarter-wave dipoles having a metallized ground plane, or on crossed half-wave dipoles (turnstile antennas), capable of use as independent crossed horizontal antennas, or in a circularly polarized mode.

Figure 9.7 is a general view of the receiver (center rack) and associated phase recording equipment. The receiver is a two-channel superheterodyne, each channel of which has two IF stages. Appropriate frequency multipliers produce in each channel a second-stage output of 75 kc, required by the phase recorder used. This device permits the determination of the phase difference of the two signals. It is a dual instrument, capable of simultaneously processing two pairs of signals and recording both outputs on one 36-mm moving film. Several sample phase difference records appear in Figure 9.8. Recording signals include three reference spots producing the three horizontal traces, the distance between each adjacent pair representing a phase shift of 180°. The vertical lines indicate one-second intervals. The measuring spot moves vertically as the phase between a pair of received signals changes, producing the inclined and other (anomalous) traces on the moving film.

Tests have indicated that the accuracy of the determination of the difference in Doppler frequencies is greater than 1/20, and possibly

* The principal source of the description of the ground installation contained herein is Reference 11.

Fig. 9.6. The "Mayak" (Beacon) Transmitter[22]

Fig. 9.7. Receiving Apparatus[73]

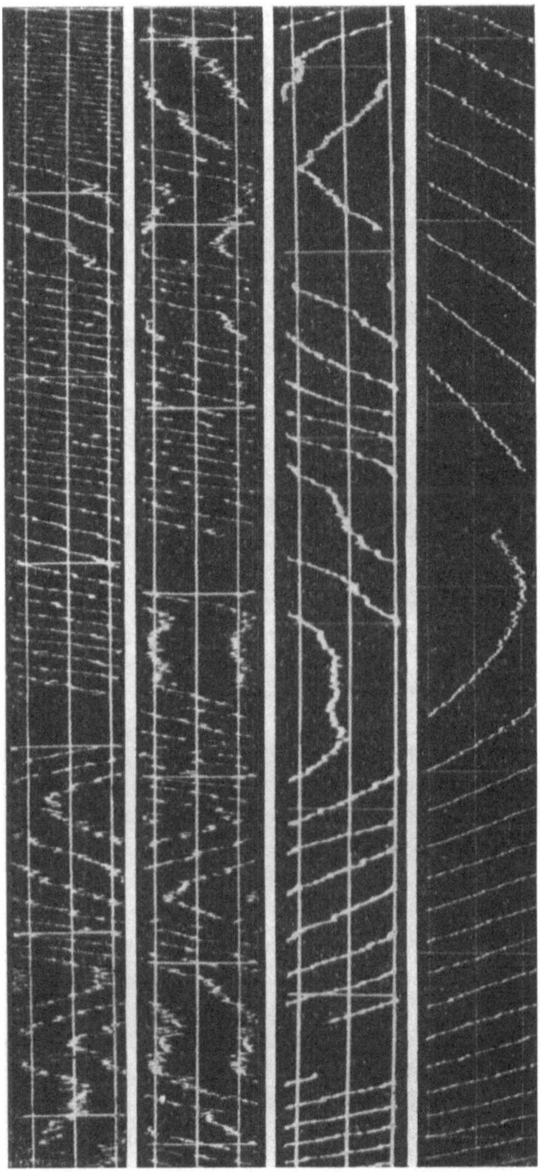

Fig. 9.8. Doppler Phase Difference Photo-oscillograms[52]

as precise as 1/50 cycle/second. For strictly vertical motion of a vehicle, an accuracy of 1/10 cycle/second corresponds to a minimum determinable electron concentration of about 2×10^3 cm^{-3} over a distance of 1 km, and 2×10^2 cm^{-3} over a 10-km path.

Other Techniques

Other vehicle-related indirect techniques to study ionospheric electron densities which the Soviets have used or discussed include the radio-rise-and-set method, which was used with Sputnik 1[12–13], and presumably with other early satellites; Faraday (rotational Doppler) effect measurements, either simply[34, 56, 59], or in relation to other methods[49, 50, 53, 71]; and specialized signal amplitude fluctuation studies[43, 68].

A 1965 paper[46] describes a technique for measuring electron densities and collision frequencies in the lower ionosphere by recording the input impedance at 50 kc of a miniature rocket-borne antenna. Telemetered data consist of two voltages having known functional relationships to the antenna resistance and reactance. The paper develops relationships for these antenna parameters as functions of the electron density, collision frequency, gyrofrequency, and the radiation frequency. Other plasma parameters are neglected. Two related papers[25, 45] have been noted.

SUMMARY OF RESULTS

Table 9.2 provides a summarization of the results of Soviet experiments utilizing rockets and satellites which have been associated with the measurement of ionospheric electron density parameters, except for the results of those experiments involving the use of ion traps. The latter are discussed in Chapter 10. The table categorizes the data with respect to direct measurements (made by rockets only), and indirect measurements, the latter being considered separately with regard to experiments involving the use of rockets and those utilizing satellites. All rockets were launched from middle latitudes of the USSR.

Few Soviet experiments are reported in which direct measurements of ionospheric electron densities were made. Two rocket observations, nearly a year apart (1958–1959), have provided consistent D-region data obtained with a low frequency impedance probe. No well-defined maximum of electron density was observed[45–46]. In 1965 a Langmuir probe recorded electron densities on the descent of a rocket from 480 km. The variation of values with height was consistent with that indicated by data obtained by dispersion interferometry, but the magnitudes were uniformly lower. The Soviets provide no indication as to which set of results is favored[31].

TABLE 9.2

Results of Soviet Rocket and Satellite Studies (Except Ion-Trap) of Ionospheric Electron Densities (1957–1966)

Vehicle/Experiment	Altitude, km	Measurement Technique	Results	References
		Direct Measurements		
Geophysical Rockets				
Aug. 2 1958/0947 LT July 2 1959/0740 LT	212 212	Antenna probe (50-kc impedance type)	Electron densities increase monotonically from $2{\cdot}7 \times 10$ to 8×10^3 cm^{-3} in the altitude range 70–90 km (1958) and from 5×10 to 7×10^3 cm^{-3} in the altitude range 80–92 km (1959). No well-defined maximum was found in the D region. The profiles agree with those determined by VLF techniques.	Krasnushkin, *et al.* (1962)[45], (1965)[46]
Sep. 20 1965/0400 LT	500	Langmuir probe	Electron density profile obtained during rocket's descent in the approximate range 120–480 km; profile contour agreed very closely with that obtained simultaneously by dispersion interferometry, but values were consistently lower. A minor max near 120 km indicated on the interferometric, is not evident on the probe profile.	Gringauz, *et al.* (1966)[31]
		Indirect Measurements		
Geophysical Rockets				
May 16 1957/0618 LT Aug. 25 1957/0627 LT Sep. 9 1957/1954 LT Feb. 21 1958/1140 LT	212 212 212 473	Dispersion interferometry (144 mc and 48 mc) Dispersion interferometry (144 mc and 48 mc) Dispersion interferometry (144 mc and 48 mc) Dispersion interferometry (144 mc and 48 mc); also Faraday rotation (144 mc, 48 mc, and 24 mc)	Two 1957 electron density profiles show a minor but well-defined max near 130 km, but according to Gringauz no well-defined E layer is indicated. The 1958 (ascent) profile shows no E-region max; above the F-layer main max the decrease in electron density in the first 100 km is 12·6%, and the decrease is about $1{\cdot}8$ to 1×10^6 cm^{-3} from 290 to 473 km; the rate of change of integral electron content near the trajectory peak was about 5×10^9 cm^{-2} sec^{-1}. Ionosonde F-layer-max heights exceeded the interferometer data by 50–150 km. The discrepancy is not reconciled.	Gringauz (1958)[27] (1961)[28], (1962)[29] Gringauz, *et al.* (1961)[35], (1962)[30] Krasovskiy (1958)[47] Zhekulin (1958)[74]
Aug. 27 1958/0806 LT Oct. 31 1958/1554 LT	451 473	Dispersion interferometry (144 mc and 48 mc); also Faraday rotation (144 mc, 48 mc, and 24 mc)	Profiles (made on ascent) show no E-region max; the F maxima are both near 325 km, but densities differ, as do density decreases topside (13·4% in the first 100 km and about $1{\cdot}1$ to $0{\cdot}7 \times 10^6$ cm^{-3} from 325 to 440 km for August 27; and 16·1% in the first 100 km and about $2{\cdot}9$ to $1{\cdot}4 \times 10^6$ cm^{-3} from 325 to 430 km for October 31). For the rocket of August 27, the rate of change of integral electron content near the trajectory peak was about 5×10^9 cm^{-2} sec^{-1}; Faraday rotation data acquired at three sites agreed closely, but values were generally somewhat less than those of the interferometer data.	Gringauz (1961)[28] (1962)[29] Gringauz, *et al.* (1960)[34], (1961)[35], (1962)[30] Gorozhankin, *et al.* (1965)[26] Rudakov (1964)[63]

TABLE 9.2 (Continued)

Results of Soviet Rocket and Satellite Studies (Except Ion-Trap) of Ionospheric Electron Densities (1957–1966)

Vehicle/Experiment	Altitude, km	Measurement Technique	Results	References
July 14 1959/0740 LT July 22 1959/0514 LT June 15 1960/0643 LT	212 195 212	Dispersion interferometry (144 mc, 48 mc, and 24 mc)	Profiles (made on ascent) to just below 200 km show roughly monotonic increases with small maxima near 125 km. On July 14 1959, in an illuminated ionosphere, electron densities increased from 1 to 3×10^5 cm^{-3} from 100 to 190 km, while on July 22, with weak illumination, the profile is nearly vertical, near $0{\cdot}4 \times 10^5$ cm^{-3}.	Rudakov (1961)[62]
Nov. 15 1961/Afternoon	430	Dispersion interferometry	Profile to just above 400 km shows minor max at about 120 km, rapid rise (5×10^4 to 6×10^5 cm^{-3} from 160 to 235 km) to F region max, with a decrease topside to $1{\cdot}5 \times 10^5$ just above 400 km.	Gorozhankin, *et al.* (1965)[26] Rudakov (1964)[63]
Oct. 18 1962/Afternoon June 6 1963/Morning	500 515	Dispersion interferometry (144 mc and 48 mc)	Profiles (made on descent) of both years to about 450 km show nearly monotonic increases to the F region max (October 18 1962—6×10^5 cm^{-3} near 220 km; June 6 1963—about 3×10^5 near 250 km); the 1963 profile has a distinct minor maximum near 125 km. Topside, densities decrease to about 8×10^4 cm^{-3} near 450 km in both cases. Comparison with earlier profiles (August 27 1958 and October 31 1958) indicates that the principal ionization maximum in quiet-sun periods is several times less dense and 50–100 km lower than in periods near the solar activity maximum; also the topside density decreases more rapidly.	Gorozhankin, *et al.* (1965)[26] Rudakov (1964)[63]
Sep. 20 1965/0400 LT Oct. 1 1965/0400 LT	500 500	Dispersion interferometry (144 mc and 48 mc)	Profiles (made on descent) of both flights to about 430 km show nearly monotonic increases to the F region max (September 20—$1{\cdot}5 \times 10^5$ cm^{-3} near 225 km; October 1—3×10^5 near 210 km); the September profile has a distinct minor max of about 5×10^4 cm^{-3} near 120 km; densities near 430 km were about 4·5 and $5{\cdot}8 \times 10^4$ cm^{-3}, respectively. Values above 430 km were found to be subject to considerable error caused by fluctuations in time of the total electron content, measured to be 10^9 cm^{-2} sec^{-1}.	Gringauz, *et al.* (1966)[31]

Vehicle/Experiment	Altitude, km	Measurement Technique	Results	References
Artificial Earth Satellites				
Sputnik 1 (Observations made at six sites between latitudes 20–45° N, October 5–7 1957, local times between 0740–0940)	500–650	Radio-rise-and-set (40 mc)	Of about 600 observations, only about 10% provided usable data, from which the mean values for the height of the base and max density level of the F region were found to be 200 and 320 km, respectively; the average max electron density about $1{\cdot}8\times10^{6}$ cm^{-3}; and the index for the assumed exponential decrease in electron density topside approximately $3{\cdot}5\times10^{-3}$ km^{-1}. Pointing out that this yields values above 500–600 km which are too low, electron densities were nevertheless computed out to 3100 km as follows: Altitude, km: 200, 320, 400, 500 Electron density, cm^{-3}: 10^5, $1{\cdot}8\times10^6$, $1{\cdot}4\times10^6$, 7×10^5 Altitude, km: 1150, 1800, 2450, 3100 Electron density, cm^{-3}: 10^5, 10^4, 10^3, 10^2	Al'pert (1958)[2] Al'pert, *et al.* (1958)[12, 13]
Explorer 7, Discoverer 36, Cosmos 1 and Cosmos 2 (Observations at Gor'kiy, December 25 1961–April 28 1962)	No data reported	Signal intensity fluctuation analysis (20 mc)	The diurnal incidence of ionospheric small-scale inhomogeneities was investigated by subjectively categorizing the induced individual signal fluctuations as to intensity, and statistically analyzing the assigned indices. No variation from winter to spring was noted. Occurrence was principally at night between 2000–0600, the maximum being near 2200–2300 hours. Intensities varied from day to day, but not in periods of 3–4 hours. Non-uniformities occurred in frequently well-delineated regions extending from 10's to 100's of km. Localization occurred most frequently in a quiet ionosphere, or by day. Persistence was usually at least 3–4 hours. Height and dimension results for inhomogeneities were obtained by analysis of simultaneous signal intensity fluctuations recorded at three points several hundred meters apart. One 1600-hour observation and 43 in the interval 1900–0700 hours, Moscow time, were analyzed. Most frequent occurrence was in the height range 280–360 km and within ±70 km of the F-region maximum. Quasi-periodic variations in height of 50–100 km were noted during several observation periods. The inhomogeneities ranged up to ~2 km in extent, with the max frequency of occurrence being near 1 km. No significant variation with height was noted, but some increase in the smaller dimensions occurred during periods of increased magnetic activity. The inhomogeneities were found to be elongated along the geomagnetic field.	Korobkov, *et al.* (1965)[43] Yerukhimov (1962)[68], (1965)[70] Yerukhimov, *et al.* (1965)[72]

TABLE 9.2 (Continued)

Results of Soviet Rocket and Satellite Studies (Except Ion-Trap) of Ionospheric Electron Densities (1957–1966)

Vehicle/Experiment	Altitude, km	Measurement Technique	Results	References
Explorer 7 (Observations aboard ship on about 50 days between May 13–August 28 1962 in region 18° S–41° N, 138° E–205° E)	No data reported	Subjective signal intensity fluctuation analysis (19·992 mc)	Individual signal fluctuation records were subjectively categorized as to intensity, and the assigned indices statistically analyzed; the indices are implicitly indicative of the incidence of electron density inhomogeneities. No Longitudinal dependence of fluctuation intensities was found, but mean intensities reached maxima near the equator and 30° N; no significant diurnal variation was evident; in equatorial regions the correlation between fluctuation intensity and degree of diffuse ionospheric reflections (also subjectively categorized) was 0·6; little correlation of fluctuation intensities with the geomagnetic k-index was noted. Estimates of the dimension of inhomogeneities assumed to be near the F_2 max were 0·4–1·5 km, and 1 km for those assumed to be in the E_s layer; regions containing inhomogeneities extended for several hundred km, separated by homogeneous regions extending 400–600 km.	Korobkov, *et al.* (1965)[43] Yerukhimov (1965)[70]
Explorer 7 (Observations at Gor'kiy, January 24–March 20 1962)	656–920	Faraday rotation (19·995 mc)	Eighty observations of total electron content to satellite altitude were made, but many night recordings were disturbed by fluctuations caused by electron density inhomogeneities. Data from selected observations are as follows:	Mityakova (1964)[56] Yerukhimov, *et al.* (1965)[72]

Date, 1962	Time, LT	Satellite Height km	Total Electron Content $\times 10^{-13}$, cm^{-2}
June 24	1307	678	1·97
June 28	1314	677	2·3
June 26	1404	710	2·23
Mar. 20	1927	920	1·23
Mar. 5	2350	735	0·8
Feb. 26	0123	656	1·05

Vehicle/Experiment	Altitude, km	Measurement Technique	Results	References
Cosmos 1 and Cosmos 2 (Observations between March 23 and April 23 1962)	252–450	Doppler frequency difference (20·005 mc and 90·0225 mc)	This paper presented what were probably the earliest published results from data obtained by the Doppler frequency technique. The preliminary analysis indicated that the range of electron densities along the orbit was 0·4 to $7{\cdot}0\times10^5$ cm^{-3}, and the range of electron density gradients was 1 to 5×10^3 cm^{-3} km^{-1}. The dimensions of inhomogeneities varied between 100–500 km, with max frequencies of occurrence being in the ranges 2–6 km and 100–130 km. Consideration of data on the variations of the angle of refraction indicated horizontal electron density gradients had to be further considered in evaluating measurements made by the Doppler technique.	Al'pert, *et al.* (1963)[10]
Cosmos 1 (Observations in the Crimea, March 17–23 1962) Cosmos 2 (Observations in the Crimea, April 6–19 1962)	220–850	Doppler frequency difference (20·005 mc and 90·0225 mc) also Faraday rotation (20·005 mc)	Data obtained by the Doppler frequency difference technique for 9 of 15 observations of Cosmos 1 and 48 of 60 observations of Cosmos 2, for a total of 57, were integrally analyzed. Observation times ranged from 1000 to 2200 hours, Moscow daylight time. The total electron content to satellite height ranged from $4{\cdot}3\times10^{13}$ cm^{-2} (1312 hours, 820 km altitude) to $0{\cdot}1\times10^{13}$ cm^{-2} (2100 hours, 240 km altitude) except for four sessions after sunset when it dropped to 5×10^{10} cm^{-2}, the satellite being near perigee at 220 km. Combining Doppler data for 19 observations with ionosonde data permitted computation of electron densities at satellite altitude and total electron density gradients. The range of the former was $0{\cdot}85\times10^5$ cm^{-3} (600 km altitude) to $6{\cdot}7\times10^5$ cm^{-3} (260 km altitude); that of the latter was from a negative $1{\cdot}4\times10^4$ cm^{-2} km^{-1} to a positive 4×10^4 cm^{-2} km^{-1}. The index for the topside exponential decrease in electron density was computed in six cases where the satellite was more than 100 km above the F-layer maximum; it ranged from 4·5 to 6×10^{-3} km^{-1}. The dimensions of inhomogeneities were computed from density gradient data, the maximum frequencies of occurrence being found for dimensions near 25 km and 135 km. Faraday rotation data were obtained in 75 cases, 28 of which were suitable for analysis. Total electron content was computed by the joint analysis of amplitude and phase records, and from the Faraday rotation data. The results, compared to Doppler frequency difference data, are tabulated below. The values computed by the various methods disagree by no more than 20%. It was concluded that the accuracy of analysis obtainable using Faraday rotation data is somewhat lower than that using Doppler frequency difference records, but that the simplicity of the experimental method was a great advantage.	Mityakov, *et al.* (1963)[54] Mityakova (1965)[56] Yerukhimov *et al.* (1965)[72]

TABLE 9.2 (Continued)

Results of Soviet Rocket and Satellite Studies (Except Ion-Trap) of Ionospheric Electron Densities (1957–1966)

Vehicle/Experiment	Altitude, km	Measurement Technique	Results						References
			Date, 1962	Time	Satellite Altitude, km	Total Electron Content $\times 10^{-13}$ cm^{-2} (1)	(2)	(3)	
			Mar. 23	1015	420	0·95	—	1·1	
			Mar. 22	1017	405	1·21	—	1·2	
			Mar. 22	1157	330	1·1	1·16	1·0	
			Mar. 21	1157	325	0·85	0·843	0·86	
			Apr. 17	1458	670	1·88	2·5	2·3	
			Apr. 18	1448	660	1·85	—	2·2	
			Apr. 16	1510	630	1·7	2·0	1·67	
			Mar. 22	1518	225	0·46	—	0·43	
			Mar. 21	1518	225	0·34	—	0·3	
			Apr. 15	1521	600	1·66	1·9	1·65	
			Apr. 14	1531	600	1·39	1·42	1·48	
			Apr. 13	1541	590	1·42	1·4	1·49	
			Apr. 12	1551	550	1·6	1·63	1·65	
			Apr. 11	1601	520	1·6	1·65	1·91	
			Apr. 10	1610	500	2·0	1·79	2·0	
			Apr. 19	1619	480	0·81	0·9	0·92	
			Apr. 8	1627	490	1·24	1·06	1·14	
			Apr. 7	1636	450	1·08	1·11	1·09	
			Apr. 17	1646	470	1·16	1·46	1·3	
			Apr. 16	1657	450	1·23	1·47	1·31	
			Apr. 15	1708	440	1·46	1·31	1·2	
			Apr. 13	1729	390	0·84	0·87	0·91	
			Apr. 10	1757	360	0·74	0·66	0·72	
			Apr. 9	1806	350	0·525	0·65	0·6	
			Apr. 19	1810	360	0·94	—	0·85	
			Apr. 8	1813	380	0·56	0·76	0·7	
			Apr. 18	1822	340	0·73	—	0·76	
			Apr. 7	1823	320	0·51	0·54	0·6	

(1) Results from Doppler frequency difference measurements
(2) Results from the joint analysis of the amplitude and phase records
(3) Results from Faraday rotation measurements.

Vehicle/Experiment	Altitude, km	Measurement Technique	Results	References
Cosmos 1 and Cosmos 2 (Observations at Moscow and in the Crimea, March 23–April 15 1962)	240–456	Doppler frequency difference (20·005 mc and 90·0225 mc)	Electron densities at satellite altitude, computed from observational data, together with those from ionosonde data for those observations when the satellite was below the F-region maximum, appear in the table below.	Al'pert (1964)[6] Tsedilina, *et al.* (1964)[67]

Date, 1962	Time	Satellite Altitude, km	Electron Densities $\times 10^{-5}$ cm^{-3} Moscow	Crimea	Ionosonde
Mar. 23	1156	336	2·7	—	—
Mar. 23	1157	320	3·2	—	—
Apr. 7	2009	267	—	1·9	1·45
Apr. 7	2009	263	1·1	1·1	—
Apr. 7	2010	257	—	2·3	1·20
Apr. 8	2000	281	2·9	—	3·80
Apr. 8	2000	280	3·5	3·1	—
Apr. 8	2001	270	2·5	3·5	—
Apr. 8	2001	268	2·5	2·8	—
Apr. 8	2001	260	2·7	—	2·6
Apr. 9	1805	379	—	0·5	—
Apr. 9	1805	366	—	1·2	—
Apr. 9	1806	357	2·3	2·6	—
Apr. 9	1951	286	—	2·4	—
Apr. 9	1952	277	3·1	3·2	—
Apr. 9	1953	264	—	1·7	2·9
Apr. 10	1756	393	—	3·8	—
Apr. 10	1756	388	—	3·7	—
Apr. 10	1757	370	4·2	2·6	—
Apr. 10	1757	364	4·1	—	—
Apr. 10	1757	358	4·3	—	—
Apr. 10	1758	346	4·5	—	—
Apr. 11	1747	408	6·6	—	—
Apr. 11	1748	379	5·2	—	—
Apr. 13	1915	308	—	5·0	—
Apr. 13	1915	306	—	5·0	—
Apr. 14	1718	437	1·7	—	—
Apr. 14	1718	425	2·7	—	—
Apr. 14	1719	408	1·4	—	—
Apr. 15	1707	456	—	1·2	—
Apr. 15	1708	441	1·5	—	—
Apr. 15	1708	422	2·1	—	—

TABLE 9.2 (Continued)

Results of Soviet Rocket and Satellite Studies (Except Ion-Trap) of Ionospheric Electron Densities (1957–1966)

Vehicle/Experiment	Altitude, km	Measurement Technique	Results	References
			Al'pert compared these data with the results of computations by another method, and with Faraday rotation data for Moscow. He found fair agreement and concluded that the neglected factors, horizontal gradients, nonstationary effects, and the integral effects of the ionosphere were of little influence. However, other data are presented for observations closely related in time, taking into account gradient and temporal effects. In these instances Al'pert concludes that the gradients can have a strong, and the nonstationary effects a minor influence on the measured densities. Density gradients at satellite altitude were found to be 5×10^2 to 2×10^3 cm^{-3} km^{-1}. This is nearly an order of magnitude larger than Al'pert's earlier results (Al'pert, et al., 1963)[10]. From computed values of the ratio of satellite-height electron densities to those of the F-region max, evidence appears of a second density maximum, having a value of 0·9 to 0·95 of the F-region max density, and from 120–140 km above its level. Al'pert offers no explanation, but points out that only a few observations were made at this altitude. Frequency of occurrence maxima for ionospheric inhomogeneities, considered both by Al'pert and Tsedilina and Kharybina, were found for those having dimensions near 5, 15, 30, and 100 km. Data for Moscow and the Crimea were consistent in this regard. The maximum number of inhomogeneities was observed 50–100 km below the main F-region maximum. These characteristics were apparently independent of the time of day and season. It was further found that the number of inhomogeneities decreased with height and that none of small scale (less than 1 km) occurred above 400 km.	
Cosmos 2 (Observations made in October 1962)	No data reported	Doppler frequency difference (20·005 mc and 90·0225 mc)	Forty-six observations of the total electron content to satellite height permitted determination of its gradient. By night a smooth increase of the gradient was observed as the vehicle moved from west to east into an increasingly illuminated ionosphere; by day the change in gradient was quasi-periodic, indicative of the presence of inhomogeneities.	Misyura, *et al.* (1964)[51]

Vehicle/Experiment	Altitude, km	Measurement Technique	Results	References
Cosmos 11 (Observations at Moscow and at Sverdlovsk October 21–29 1962)	275–506	Doppler frequency difference (20·005 mc and 90·0225 mc)	The studies of inhomogeneities made from data acquired utilizing Cosmos 1 and Cosmos 2, at Moscow and in the Crimea, summarized above (Al'pert, 1964[6]; Tsedilina and Kharybina, 1964)[67], also considered data acquired utilizing Cosmos 11. Six observations were made at Moscow, and six at Sverdlovsk The data were entirely consistent with those obtained utilizing Cosmos 1 and Cosmos 2.	Al'pert (1964)[6] Misyura, *et al.* (1965)[5] Tsedilina, *et al.* (1964)[57]
Electron 1 (Observations at Moscow, Sverdlovsk, Novosibirsk, and Tbilisi, February 1–March 27 1964)	430–1800	Doppler frequency difference (20·005 mc, 30·0075 mc, and 90·0025 mc)	From 52 observations at Moscow, 78 in Sverdlovsk, and 45 in Novosibirsk, about 7000 determinations of inhomogeneity dimensions in the 1–400 km range were determined. Only 41 Moscow observations were utilized in computations. Observational times ranged from 1000–1500 local time. It was found that little change in inhomogeneity dimensions occurred within an altitude range of 100 km, and few inhomogeneities were observed above 1200 km. Data were categorized, therefore, for the 400–800 km and 800–1200 km intervals. Maxima of frequency of occurrence of inhomogeneities in the dimension range 2–4 km appeared at all sites in the lower category, and anomalously at Sverdlovsk in the 800–1200 km interval. This anomaly was found to endure only a few days. Maxima in the 15–30 km range occurred at all sites in both altitude intervals. Considerable variation with longitude in size-frequency of occurrence relationships was evident. Electron densities in inhomogeneities increased most frequently from 5–15%, and up to 35% in small inhomogeneities ($\leqslant 6$ km) in the 800–1200 km level. In 30 out of 1100 cases computed, all dimensions being between 2–10 km in the 400–800 km interval, changes in electron density in the range ± 50–170% occurred. Computed electron densities at satellite altitudes for the three observation sites are available in Reference 18. From these, 10-day (March 17–27) mean profiles of electron density were constructed to about 1100 km. They are quasi-periodic, show five maxima at intervals of 120–160 km, and a deep minimum near 620 km. The authors provide some potential hypotheses with regard to these periodicities. Using data from Tbilisi, a smoothed mean profile to 1800 km is presented. It has a minimum of about $2{\cdot}5 \times 10^5$ cm^{-3} near 600 km, a maximum of 3×10^5 cm^{-3} at 800 km, then decreases apparently nearly exponentially to 2×10^4 cm^{-3} at 1800 km.	Al'pert, *et al.* (1965)[16–18, 20] (1966)[19, 21] Blagonravov (1965)[22]

TABLE 9.2 (Continued)

Results of Soviet Rocket and Satellite Studies (Except Ion-Trap) of Ionospheric Electron Densities (1957–1966)

Vehicle/Experiment	Altitude, km	Measurement Technique	Results	References
Electron 1 (Observations at Gor'kiy, February 12–March 26 1964, and in the Crimean February 10–19 1964)	406–2002	Doppler frequency difference (20·005 mc and 30·0075 mc)	A considerable amount of tabulated data is presented, using 50 of 80 observations made at Gor'kiy between 0800 and 1700 hours, local time, and 14 of 16 made in the Crimea from 1100 to 2200 hours local time. In addition associated ionosonde data is included. Using only data from observations when the satellite was above 600 km, the mean diurnal variation of total electron content to satellite altitude was analyzed. Maxima of 1·2 to $1{\cdot}5\times10^{13}$ cm^{-2} were found to occur near 1300 hours, the magnitudes varying inversely with latitude in the range (51–60° N) of the observations. Total electron content in the Crimea decreased to about $0{\cdot}3\times10^{13}$ cm^{-2} near 2100 hours. Total electron content of the entire atmosphere differed from that to satellite altitude by about 10^{12} cm^{-2}. Based on the assumption of an exponentially decreasing electron density topside, ratios of total electron content above to that below the F-region max were found to be 2–4 from 0900 to 1600, and 5–8 from 1700 to 2100 hours. The mean thickness of the topside ionosphere ranged from a minimum of about 200 km at noon to 300–400 km morning and evening. Magnitudes increased 25–100 km when Wolf numbers exceeded 30. Computer processing of nine Gor'kiy observations has provided values for the exponential index of the topside ionosphere as listed in the table below. The mean value is approximately $6{\cdot}2\times10^{-3}$ km^{-1}.	Mityakov, *et al.* (1966)[55] Yerukhimov, *et al.* (1965)[72]

Date, 1964	Time	Satellite Altitude, km	Total Electron Density $\times10^{-13}$, cm^{-2}	Exponential Index $\times10^{3}$, km^{-1}
Feb. 21	1236	429	0·96	4·3
Feb. 27	1528	786	1·23	7·4
Mar. 2	1524	858	1·1	5·4
Mar. 7	1356	721	1·59	5·6
Mar. 9	1354	753	1·05	5·75
Mar. 10	0932	410	0·765	10·5
Mar. 11	1353	790	1·56	5·05
Mar. 12	0930	413	0·82	5·4
Mar. 13	1353	823	1·29	6·0

In contrast, the Soviets have emphasized indirect radio wave measurement techniques. The principal organizations associated with these activities appear to be the Institute of Terrestrial Magnetism, Ionosphere, and Radio Wave Propagation, and the Radio Engineering Institute, both of the Soviet Academy of Sciences, and both in or near Moscow. Ya. L. Al'pert and K. I. Gringauz, of the respective organizations, apparently are the leading Soviet authorities in this disciplinary area. The former has been concerned with studies utilizing artificial earth satellites, while the latter appears to have been associated principally with work involving geophysical rockets.

With regard to the last-mentioned vehicles, indirect RF measurements have been made primarily by dispersion interferometry, supplemented in a few instances by Faraday rotation studies, and the results have been presented almost entirely in terms of electron density profiles. The Soviets have found no evidence of well-defined layers below the main F-region maximum. In eight of the 14 flights tabulated, however, a minor maximum of electron density was observed near 125 km. In comparing data[26, 63], they found that the F-region maximum was several times less dense and from 50 to 100 km lower during quiet sun years than during periods near the solar activity maximum. They also found that the density above the maximum decreased more rapidly during the quiet periods. During two 1958 flights (February 21 and August 27) the time rate of change of the integral electron content was measured near the trajectory peaks and found to be 5×10^9 cm^{-2} sec^{-1}.[35] This is claimed to be the first determination of a rate of change to vehicle height, as opposed to that referred to the total thickness of the ionosphere. The analysis indicates that this introduces less than a 10% error in density measurements, except for the last five km to the trajectory peak, the data for which, therefore, were not reported. Similar measurements were made during the 1965 flights in which the time rate of change of integral electron content was found to be 10^9 cm^{-2} sec^{-1}.[31]

The Soviets employed their first satellite, Sputnik 1, for electron density studies, utilizing the radio-rise-and-set technique. They thus acquired the first data for the 200–600 km range and presented the first extrapolation to higher levels. With regard to the latter, Al'pert, *et al.* [12–13] point out that above 600 km gas temperatures are higher, and electron lifetimes longer than assumed in the calculations, and the electron densities presented at the higher levels are therefore probably too low. The discussions include consideration of neutral particle concentrations also. Al'pert concluded[2] that electron density declines more slowly above the F-region main max than it increases below, and that little diurnal or spatial variation occurs.

The next Soviet study, made in late 1961 and early 1962, utilized the U.S. satellites Explorer 7 and Discoverer 36, as well as Soviet vehicles Cosmos 1 and Cosmos 2[68, 70, 72]. A semiquantitative Western-originated technique* was employed whereby the signal intensities of satellite transmissions were recorded and their fluctuations analyzed in order to study the incidence of small-scale ionospheric inhomogeneities. These were found to range up to about 2, but to most commonly about 1 km in extent, to occur principally at night in well-delineated regions extending up to hundreds of km, and to be within 70 km above or below the F-region maximum level. The diurnal variation of occurrence of these inhomogeneities was studied utilizing a subjective analytical procedure apparently the same as that described in the paper[43] reporting a later 1962 study utilizing Explorer 7 and made aboard ship in the western Pacific Ocean. The findings of this later experiment were substantially similar.

In a more quantitative early 1962 experiment, utilizing Explorer 7, Faraday rotation measurements were made from which the total electron content to satellite height was computed. The few results selectively reported are included in Table 9.2[56, 72].

Subsequently reported Soviet experiments have utilized the Doppler frequency difference technique almost exclusively, appear to have been conducted with relative sophistication compared to earlier work, and have been extensively reported. There have been two series of experiments. The first occurred during March and April 1962. The observations were made at Gor'kiy, at Moscow, and in the Crimea, and utilized the vehicles Cosmos 1 and Cosmos 2. A few observations were made in October 1962 utilizing Cosmos 11. The results with regard to electron densities include data for densities at orbital altitude, for the total electron content to orbital altitude, for orbital and total electron density gradients, and values for the exponential index above the F-region maximum. With regard to ionospheric inhomogeneities, the results provide data as to their size range, the sizes of maximal frequency of occurrence, and their altitude ranges. Perhaps the most interesting single item of information included was the report by Al'pert[6] of evidence of a second density maximum having a value of 0·9–0·95 of the F-region max density, and existing from 120–140 km above its level. The evidence appeared in computed values of the ratio of electron

* Frihagen, I., and I. Trøim, J. Atmos. and Terr. Phys., 1960, Vol 18, No. 1, p 75. Kent, G. S., and I. R. Koster, Nature, 1961, Vol 191, No. 4793, p 1083.

densities at satellite height to those of the F-region maximum. The experiments of this series are discussed in References 6, 10, 51, 52, 54, 56, 67, and 72.

The second series of experiments were conducted during February and March 1964. Observations were made at Moscow, Sverdlovsk, Novosibirsk and Tibilisi, and at Gor'kiy and in the Crimea. The vehicle utilized was Electron 1, and the categories of information contained in the reported results are basically those outlined for the studies involving Cosmos 1 and Cosmos 2. The most unusual item appears to be a report by Al'pert and Sinel'nikov[16–18, 22] of evidence of an electron density structure that is quasi-periodic above the main F-region maximum. This appeared in 10-day mean profiles to about 1100 km constructed for three locations. The authors comment that if the electron density intensifications are produced from distant sources, the unperturbed profile would probably approximate a curve tangent to the minima. However, if produced by local redistribution of electrons, the profile would probably approximate a secant curve averaging the periodicities. They theorize that these may be due to plasma oscillations between two dissimilar ionospheric regions (atomic oxygen below, helium and hydrogen above about the 1000-km level), the waves being hydrodynamic in nature. They also consider that the periodicity may be the result of limited regions being excited, the released electrons being so channeled by the geomagnetic field as to produce the observed effects. In regard to this topic, Mityakov, *et al.*[55], note that the maxima and minima occurred on days of elevated and reduced solar activity, respectively, and suggest that herein may be the explanation. They also claim that the method employed by Al'pert and Sinel'nikov in determining electron densities was not completely correct. The experiments conducted utilizing Electron 1 are discussed in References 16 through 22, 55, and 72.

Several papers are available in the Soviet literature variously treating the topics, or summarizing the results of experiments, considered in this discussion. A 1962 paper by Yerukhimov[69] is concerned principally with ionospheric inhomogeneities. It is a review of published experimental results for the period 1949–1961, based on 133 references, only 29 of which are from Soviet sources. The author provides extended and detailed summarizing comment. A few salient conclusions he reaches are that inhomogeneities are located essentially in the F layer of the ionosphere, are more frequent in the auroral zones and at the magnetic equator, are in general more frequent at night, tend to be grouped in discrete regions, and are noticeably related to geomagnetic disturbances.

In an early 1964 paper[40], G. S. Ivanov-Kholodnyy, of the Institute of Applied Geophysics, Soviet Academy of Sciences, presents an excellent review of ionospheric D-region space-vehicle investigations. Techniques of measurement, both direct and indirect, are outlined and comparisons with ground measurements are made. Most of the data and much of the discussion is based on non-Soviet work. Of 81 references, only 11 are to Soviet papers. The author's conclusions in essence are that ground and rocket measurements are compatible, that the D region undergoes considerable regular variation produced by solar activity and corpuscular radiation, that the presence of a sporadic E layer is common in the ionosphere at night and it influences D-region variations, which are from one to more than 1·5 orders of magnitude, and that while a broad electron density maximum, associated with the C layer, is sometimes observed from 65–75 km, electron densities around 90 km show a sudden decrease, separating the D and E regions. A 1965 paper[26] briefly reviews the results of ionospheric researches by rockets and satellites in the period 1960—1964. Both positive ion and electron densities are discussed, and Soviet and Western experimental results are presented. The authors conclude with regard to measurement techniques that because of the complexity of the ionosphere, reliable values should be the result of determinations by several methods, that rockets are the most reliable means of obtaining charged-particle density profiles, and that dispersion interferometry is much less suitable for use with satellites than with rockets. Lastly, another 1965 paper by Yerukhimov, *et al.*[72], some detail from which has been presented in Table 9.2, provides a compact summary of some of the results of the two series of studies (1962 and 1964) outlined in the preceding discussion.

RELATED RESEARCH

An interesting comparison of data from three 1958 rocket flights, obtained by dispersion interferometry, with standard ionosonde data appears in a late 1962 paper by Gringauz and Gdalevich[30]. On a rocket-derived profile, layers were so delineated that the electron density was essentially a linear function of height within a layer. At a given frequency the virtual thicknesses of the layers were computed and summed to provide the virtual height of reflection. Height-frequency diagrams were thus constructed for each of the three rocket flights and superimposed on the associated ionogram. Very close correspondence was obtained. The inverse comparison was made by calculating profiles of electron density from the ionograms, using Shim-Kelso coefficients to allow for geomagnetic effects. Again good agreement was found.

A 1965 Soviet paper[42] considers the data from a set of unidentified rocket direct probings of the ionosphere for the altitude range 100–300 km. Electron densities as a function of a parameter characterizing the ambient air mass were studied. Significant seasonal and solar-cyclic variations were found. With regard to the latter during midday near 100 km the values during the solar minimum period were 1·5–2 times lower than those of the intense phase, and above to 200 km the decrease was 20–50% being least near 130 km.

A second 1965 paper[41] considers ionization mechanisms in the lower ionosphere. The discussion is based in part on radiation data obtained by rockets and satellites, both Soviet and Western. In this paper one conclusion reached is that in the height range of 75–90 km, the electron density must vary by 0·5–1 order of magnitude during the solar cycle.

In another 1965 paper[44], the Cherenkov excitation of slow drift waves in the upper ionosphere is examined, and the possibility is discussed of determining electron densities, among other parameters, from measurement of drift-wave frequencies. Drift waves are slow electromagnetic waves in drifting plasma, which in the accompanying system of coordinates are actually ordinary plasma oscillations.

In a number of Soviet papers, authored primarily by only a few writers (e.g., Ya. L. Al'pert, and A. V. Gurevich), the disturbing effects of a moving vehicle upon ambient conditions in the ionosphere are considered, both in terms of conditions immediately adjacent to the vehicle (sheath effects), and of conditions at extended distances (wake and wave effects). An important parameter considered is the electron density. Among these papers are References 7, 8, 14, 15, 37, 38, 39, 48, 57, 58, and 60.

Electron temperature measurements by means of Langmuir probes have been made aboard several Soviet geophysical rockets and at least one satellite, Cosmos 2. Discussions of this subject are contained in References 22, 23, 24, 31 and 32.

References

1. "Recent Research on Radio Wave Propagation Over the Earth's Surface", Gostekhizdat, Moscow-Leningrad, 1945.

2. Ya. L. Al'pert, "Conditions of the Outer Ionosphere", Priroda, 1958, No. 6, pp 86–87.

3. Ya. L. Al'pert, "Method of Investigating the Ionosphere by Means of an Artificial Earth Satellite", Uspekhi Fizicheskikh Nauk, 1958, Vol 64, No. 1, pp 3–14.

4. Ya. L. Al'pert, "Investigation of the Ionosphere and the Interplanetary Gas With the Aid of Artificial Satellites and Cosmic Rockets", Uspekhi Fizicheskikh Nauk, 1960, Vol 71, No. 3, pp 369–409 (Soviet Physics [Uspekhi], 1961, pp 479–502); Iskusstvennyye Sputniki Zemli, 1961, No. 7, pp 125–169 (Planetary and Space Science, Pergamon Press, 1962, Vol 9, pp 391–433).

5. Ya. L. Al'pert, "On the Refraction and Doppler Shift of Radio Waves Radiated by Artificial Earth Satellites in a Three-Dimensional Heterogeneous Ionosphere", Geomagnetizm i Aeronomya, 1963, Vol 3, No. 4, pp 626–634 (Geomagnetism and Aeronomy, pp 505–511).

6. Ya. L. Al'pert, "Results of Ionospheric Investigations by Means of Coherent Radio Waves Emitted From Artificial Earth Satellites", Geomagnetizm i Aeronomya, 1964, Vol 4, No. 3, pp 479–502 (Geomagnetism and Aeronomy, pp 382–398); Space Research V, Proceedings of the Fifth International Space Science Symposium, Florence, Italy, May 12–16 1964, North-Holland Publishing Company, Amsterdam, 1965, pp 652–686.

7. Ya. L. Al'pert, "Interaction of Moving Bodies With a Plasma", Trudy Vsesoyuznoy Konferentsii po Fizike, Kosmicheskogo Prostranstva, Izdatel'stvo "Nauka" Moscow, June 10–16 1965, pp 237–240 (NASA TT F–389, pp 317–321).

8. Ya. L. Al'pert, "Electromagnetic Effects in the Vicinity of a Satellite or a Spacecraft Moving in the Ionosphere of Interplanetary Space", Geomagnetizm i Aeronomiya, 1965, Vol 5, No. 1, pp 3–31 (Geomagnetism and Aeronomy, pp 1–20).

9. Ya. L. Al'pert, "Concerning the Article by K. I. Gringauz, Yu. A. Kravtsov, V. A. Rudakov, and S. M. Rytov: 'Possibility of Determining Local Electron Concentration by the Dispersion Method Using Artificial Earth Satellites and the New Ionization Maximum in the Ionosphere' ", Geomagnetizm i Aeronomiya, 1965, Vol 5, No. 4 pp 766–768 (Geomagnetism and Aeronomy, pp 594–596).

10. Ya. L. Al'pert, V. B. Belyanskiy, and N. A. Mityakov, "Radio Investigations of the Structure of the Ionosphere With the Satellite 'Cosmos' on Coherent Frequencies (Preliminary Results)", Geomagnetizm i Aeronomiya, 1963, Vol 3, No. 1, pp 10–24 (Geomagnetism and Aeronomy, pp 6–17).

11. Ya. L. Al'pert, V. B. Belyanskiy and A. F. Kutyakov, "Unit for Coherent Radio Recording of Doppler-Shift Differences of Waves Radiated From 'Cosmos' Satellites", Geomagnetizm i Aeronomiya, 1963, Vol 3, No. 1, pp 157–170 (Geomagnetism and Aeronomy, pp 127–137).

12. Ya. L. Al'pert, F. F. Dobryakova, Ye. F. Chudesenko, and B. S. Shapiro, "On the Results Obtained by Determining the Electron Concentration of the External Region of the Ionosphere on the Basis of Radio Signals Emitted by the First Satellite", Doklady Akademii Nauk SSSR, 1958, Vol 120, No. 4, pp 743–746.

13. Ya. L. Al'pert, F. F. Dobryakova, Ye. F. Chudesenko, and B. S. Shapiro, "On the Results Obtained by Determining the Electron Concentration of the External Region of the Ionosphere on the Basis of Radio Signals Emitted by the First Satellite", Uspekhi Fizichekskikh Nauk, 1958, Vol 65, No. 2, pp 161–174 (Advances in Physical Sciences, 1964, pp 150–163).

14. Ya. L. Al'pert, A. V. Gurevich, and L. V. Pitayevskiy, "Effects Caused by Artficial Satellites Flying Rapidly Through the Ionosphere or the Interplanetary Medium", Uspekhi Fizicheskikh Nauk, 1963, Vol 79, No. 1, pp 23–80 (Soviet Physics [Uspekhi], 1963, Vol 6, No. 1, pp 13–46).

15. Ya. L. Al'pert, A. V. Gurevich, and L. P. Pitayevskiy, "Artificial Satellites in Rarefied Plasma", Izd-vo "Nauka", Moscow, 1964, 382 pp.

16. Ya. L. Al'pert and V. M. Sinel'nikov, "Altitude-Time Distribution of Electron Concentration in the Outer Ionosphere and Its Stratified-Inhomogeneous Disturbance, I", Geomagnetizm i Aeronomiya, 1965, Vol 5, No. 2, pp 209–219 (Geomagnetism and Aeronomy, pp 159–167).

17. Ya. L. Al'pert and V. M. Sinel'nikov, "Altitudinal-Time Distribution of the Electron Concentration and Nonuniform Formations of the Outer Ionosphere", Trudy Vsesoyuznoy Konferenstii po Fizike, Kosmicheskogo Prostranstva, Izdatel'stvo "Nauka", Moscow, June 10–16 1965, pp 123–137 (NASA TT F-389, pp 166–184).

18. Ya. L. Al'pert and V. M. Sinel'nikov, "On the Altitude-Time Distribution of the Electron Concentration of the Outer Ionosphere and Its Stratum-Inhomogeneous Disturbance. I", Space Research VI, Proceedings of the Sixth International Space Science Symposium, Mar del Plata, Argentina, May 11–19 1965, North-Holland Publishing Company, Amsterdam, 1966, pp 524–541.

19. Ya. L. Al'pert, L. N. Vitshas, and V. M. Sinel'nikov, "On Homogeneous Formations in the Outer Ionosphere. II", Space Research VI, Proceedings of the Sixth International Space Science Symposium, Mar del Plata, Argentina, May 11–19 1965, North-Holland Publishing Company, Amsterdam, 1966, pp 542–557.

20. Ya. L. Al'pert, L. N. Vitshas, and V. M. Sinel'nikov, "Altitude-Time Distribution of Electron Concentration in the Outer Ionosphere and Its Stratified-Inhomogeneous Disturbance. II. Inhomogeneous Formations in the Outer Ionosphere", Geomagnetizm i Aeronomiya, 1965, Vol 5, No. 4, pp 649–657 (Geomagnetism and Aeronomy, pp 502–508).

21. Ya. L. Al'pert, L. N. Vitshas, and V. M. Sinel'nikov, "On Spectra of Inhomogeneities Observed in the Outer Ionosphere", paper presented at the Seventh International Space Science Symposium, Vienna, Austria, May 1966.

22. USSR National Reports to COSPAR, 1960–1967, presented by A. A. Blagonravov (In English).

23. T. K. Breus and G. L. Gdalevich, "Electron and Ion Temperatures in the Ionosphere (Summary)", Trudy Vsesoyuznoy Konferentsii po Fizike, Kosmicheskogo Prostranstva, Izdatel'stvo "Nauka", Moscow, June 10–16 1965, pp 189–192 (NASA TT F-389, pp 254–258).

24. T. K. Breus and G. L. Gdalevich, "Electron and Ion Temperatures in the Ionosphere", Kosmicheskiye Issledovaniya, 1965, Vol 3, No. 6, pp 877–889 (TT-65-1985, pp 108–133).

25. G. G. Getmantsev and N. G. Denisov, "On a Specific Effect Observed in the Ionosphere When Measuring Electron Concentration by the Antenna Probe Method", Geomagnetizm i Aeronomiya, 1962, Vol 2, No. 4, pp 691–693 (Geomagnetism and Aeronomy, pp 575–577).

26. B. N. Gorozhankin and V. A. Rudakov, "Results of Ionosphere Research by Rockets and Satellites in 1960–1964", Trudy Vsesoyuznoy Konferentsii po Fizike, Kosmicheskogo Prostranstva, Izdatel'stvo "Nauka", Moscow, June 10–16 1965, pp 168–177 (NASA TT F-389, pp 225–238).

27. K. I. Gringauz, "Rocket Measurement of Electron Concentration in the Ionosphere by Means of an Ultrashort-Wave Dispersion Interferometer", Iskusstvennyye Sputniki Zemli, 1958, No. 1, pp 62–66 (Artificial Earth Satellites, pp 79–84); Doklady Akademii Nauk SSSR, 1958, Vol 120, No. 6, pp 1234–1237 (Soviet Physics—Doklady, 1958, Vol 3, No. 3, pp 620–623).

28. K. I. Gringauz, "The Structure of the Earth's Ionized Gas Envelope Based on Results of Direct Measurements in the USSR of Charged Particle Local Concentrations", Space Research II, Proceedings of the Second International Space Science Symposium, Florence, Italy, April 10–14 1961, North-Holland Publishing Company, Amsterdam, 1961, pp 574–592.

29. K. I. Gringauz, "The Structure of the Earth's Ionized Gas Envelope According to Data of Direct Measurements of Local Concentrations of Charged Particles Made in the USSR", Iskusstvennyye Sputniki Zemli, 1962, No. 12, pp 105–118 (Artificial Earth Satellites, pp 114–130).

30. K. I. Gringauz and G. L. Gdalevich, "Analysis of the Results of Simultaneous Measurements of the Electron Concentration in the Ionosphere by Means of Iono-

spheric Stations and Rockets", Iskusstvennyye Sputniki Zemli, 1962, No. 13, pp 89–96 (Artificial Earth Satellites, pp 94–101).

31. K. I. Gringauz, G. L. Gdalevich, V. F. Gubsky, I. A. Knorin, V. A. Rudakov, and N. M. Shutte, "Preliminary Results of Experiments Carried Out in the Ionosphere in the Early Morning by Means of Geophysical Rockets Launched in the Autumn of 1965", paper presented at the Seventh International Space Science Symposium, Vienna, Austria, May 1966.

32. K. I. Gringauz, B. N. Gorozhankin, G. L. Gdalevich, V. V. Afonin, R. Ye. Rybchinsky, and N. M. Shyutte, "The Technique and Results of Experiments Conducted on the Cosmos 2 Satellite by Means of Langmuir Probes and Ion Traps of the Honeycomb Type", Space Research V, Proceedings of the Fifth International Space Science Symposium, Florence, Italy, May 12–16 1964, North-Holland Publishing Company, Amsterdam, 1965, pp 733–750.

33. K. I. Gringauz, Yu. A. Kravtsov, V. A. Rudakov, and S. M. Rytov, "Possibility of Determining Local Electron Concentration by the Dispersion Method Using Artificial Earth Satellites and the New Ionization Maximum in the Ionosphere" Geomagnetizm i Aeronomiya, 1965, Vol 5, No. 4, pp 762–766 (Geomagnetism and Aeronomy, pp 591–593).

34. K. I. Gringauz and V. A. Rudakov, "Measurement of Electron Concentration in the Ionosphere According to the Rotation of the Polarization Plane of Radio Wave, Emitted by Rockets", Doklady Akademii Nauk SSSR, 1960, Vol 132, No. 6, pp 1311–1313 (Doklady of the Academy of Sciences of the USSR, Earth Sciences Sections, pp 584–585).

35. K. I. Gringauz and V. A. Rudakov, "Measurements of Electron Concentration in the Ionosphere Up to Altitudes of 420–470 km Conducted During the International Geophysical Year by Means of Radio Waves Emitted From Geophysical Rockets of the Academy of Sciences, USSR", Iskusstvennyye Sputniki Zemli, 1961, No. 6, pp 48–62 (Planetary and Space Science, Pergamon Press, 1961, Vol 8, pp 183–193).

36. K. I. Gringauz, V. A. Rudakov, and A. V. Kaporskiy, "Instruments for Rocket Measurements of Free-Electron Concentrations in the Ionosphere", Iskusstvennyye Sputniki Zemli, 1961, No. 6, pp 33–47 (Planetary and Space Science, Pergamon Press, 1962, Vol 9, pp 247–257).

37. A. V. Gurevich, "Disturbances in the Atmosphere Caused by a Moving Body", Iskusstvennyye Sputniki Zemli, 1961, No. 7, pp 101–124 (Artificial Earth Satellites, pp 101–125).

38. A. V. Gurevich and A. M. Moskalenko, "Braking of Bodies Moving in a Rarefied Plasma", Trudy Vsesoyuznoy Konferentsii po Fizike, Kosmicheskogo Prostranstva, Izdatel'stvo "Nauka", Moscow, June 10–16 1965, pp 241–254 (NASA TT F-389, pp 322–341).

39. A. V. Gurevich and L. P. Pitayevskiy, "Supersonic Motion of a Body in Plasma", Geomagnetizm i Aeronomiya, 1964, Vol 4, No. 5, pp 817–824 (Geomagnetism and Aeronomy, pp 637–642).

40. G. S. Ivanov-Kholodnyy, "Investigation of Electron Concentration in the Lower Part of the Ionosphere (D Region)", Geomagnetizm i Aeronomiya, 1964, Vol 4, No. 3, pp 417–435 (Geomagnetism and Aeronomy, pp 337–350).

41. G. S. Ivanov-Kholodnyy, "Ionization Mechanism of the Lower Ionosphere. I", Geomagnetizm i Aeronomiya, 1965, Vol 5, No. 4, pp 705–720 (Geomagnetism and Aeronomy, pp 544–556).

42. T. V. Kazachevskaya and G. S. Ivanov-Kholodnyy, "Rocket Data on the Behavior of Electron Concentration in the Ionosphere at Altitudes of 100–300 km", Trudy Vsesoyuznoy Konferentsii po Fizike, Kosmicheskogo Prostranstva, Izdatel'stvo "Nauka", Moscow, June 10–16 1965, pp 184–189 (NASA TT F-389, pp 248–254); Geomagnetizm i Aeronomiya, 1965, Vol 5, No. 6, pp 1009–1024 (Geomagnetism and Aeronomy, pp 794–805).

43. Yu. S. Korobkov and V. V. Pisareva, "Study of Electron Concentration Inhomogeneities in the Ionosphere in the Pacific Ocean Region by Means of Artificial Earth Satellites", Geomagnetizm i Aeronomiya, 1965, Vol 5, No. 3, pp 423–428 (Geomagnetism and Aeronomy, pp 327–330).

44. M. S. Kovner and L. M. Obolenskiy, "Possibility of Determining Electron Concentration and the Magnetic Field in Plasma From Measurements of the Frequencies of Drift Waves", Geomagnetizm i Aeronomiya, 1965, Vol 5, No. 5, pp 831–834 (Geomagnetism and Aeronomy, pp 646–649).

45. P. Ye. Krasnushkin and N. L. Kolesnikov, "Investigation of the Lower Ionosphere by Means of Long Radio Waves and Low-Frequency Radiosondes Placed on a Rocket. Detection of a New Ionosphere Layer", Doklady Akademii Nauk SSSR, 1962, Vol 146, No. 3, pp 596–599 (Doklady of the Academy of Sciences USSR, Earth Science Section, pp 10–12).

46. P. Ye. Krasnushkin and N. L. Kolesnikov, "Investigation of the Lower Ionosphere by the Method of the Impedance Low-Frequency Radiosonde", Geomagnetizm i Aeronomiya, 1965, Vol 5, No. 1, pp 55–69 (Geomagnetism and Aeronomy, pp 38–48.)

47. V. I. Krasovskiy, "Soviet Ionospheric Studies Using Rockets and Artificial Earth Satellites", Iskusstvennyye Sputniki Zemli, 1958, No. 2, pp 36–49 (Artificial Earth Satellites, pp 45–62).

48. M. V. Maslennikov and Yu. S. Sigov, "Discrete Model of Matter in the Problem Concerning the Interaction of Rapidly Moving Bodies With a Rarefied Plasma", Trudy Vsesoyuznoy Konferentsii po Fizike, Kosmicheskogo Prostranstva, Izdatel'stvo "Nauka", Moscow, June 10–16 1965, pp 270–271 (NASA TT F–389, pp 365–366).

49. V. A. Misyura, D. D. Osipov, Ye. B. Krokhmal'nikov, and G. K. Solodovnikov, "Certain Possibilities and Results of Ionospheric Measurements From Slant Observation of the Faraday Effect in Geophysical-Rocket Signals", Kosmicheskiye Issledovaniya, 1965, Vol 3, No. 4, pp 604–613 (TT–65–1262, pp 172–189).

50. V. A. Misyura, G. K. Solodovnikov, Ye. B. Krokhmal'nikov, and V. M. Migunov, "Certain Results of Ionosphere Studies Using Artificial Earth Satellites and Geophysical Rockets", Trudy Vsesoyuznoy Konferentsii po Fizike, Kosmicheskogo Prostranstva, Izdatel'stvo "Nauka", Moscow June 10–16 1965, pp 138–147 (NASA TT F–389, pp 185–196).

51. V. A. Misyura, G. K. Solodovnikov, and V. M. Migunov, "The Gradients of the Integral Electron Content in the Ionosphere", Geomagnetizm i Aeronomiya, 1964, Vol 4, No. 6, pp 1124–1125 (Geomagnetism and Aeronomy, pp 872–873).

52. V. A. Misyura, G. K. Solodovnikov, and V. M. Migunov, "Measurement of the Electron Concentration of the Upper Ionosphere With the Cosmos Series Artificial Satellites", Kosmicheskiye Issledovaniya, 1965, Vol 3, No. 4, pp 595–603 (TT–65–1262, pp 157–171).

53. N. A. Mityakov and Ye. Ye. Mityakova, "A Method for Investigation of the Ionosphere by the Ground Reception of Radio Signals From Artificial Earth Satellites", Geomagnetizm i Aeronomiya, 1963, Vol 3, No. 5, pp 858–867 (Geomagnetism and Aeronomy, pp 694–701).

54. N. A. Mityakov, Ye. Ye. Mityakova, and V. A. Cherepovitskiy, "Results of Radio Trackings of the Soviet Satellites Cosmos 1 and Cosmos 2 in the Crimea", Geomagnetizm i Aeronomiya, 1963, Vol 3, No. 5, pp 816–822 (Geomagnetism and Aeronomy, pp 660–665).

55. N. A. Mityakov, Ye. Ye. Mityakova, and V. A. Cherepovitskiy, "Results From an Investigation Into the Distribution of the Electron Concentration in the Ionosphere by the Method of Ground Reception of Radio Signals From the Electron 1 Artificial Earth Satellite", Kosmicheskiye Issledovaniya, 1966, Vol 4, No. 2, pp 249–256 (TT–66–?44, pp 113–127).

56. Ye. Ye. Mityakova, "Measurement of the Electron Concentration in the Ionosphere Using Observations of the Faraday Effect in Radio Signals From Artificial Earth

Satellites", Geomagnetizm i Aeronomiya, 1964, Vol 4, No. 4, pp 668–674 (Geomagnetism and Aeronomy, pp 526–530).

57. A. M. Moskalenko, "Particle Flow Near Rapidly Moving Bodies", Geomagnetizm i Aeronomiya, 1962, Vol 2, No. 3, pp 407–424 (Geomagnetism and Aeronomy, pp 337–350).

58. A. M. Moskalenko, "Perturbed Zone Structure in the Vicinity of a Cylindrical Body in a Plasma", Trudy Vsesoyuznoy Konferentsii po Fizike, Kosmicheskogo Prostranstva, Izdatel'stvo "Nauka", Moscow, June 10–16 1965, pp 264–266 (NASA TT F-389, pp 355–359).

59. S. A. Namazov, "Determination of the Concentration of Electrons in the Ionosphere by Analysis of the Polarization Fading in Signals From a Rocket or an Artificial Earth Satellite", Radiotekhnika i Elektronika, 1962, Vol 7, No. 8, pp 1311–1315 (Radio Engineering and Electronic Physics, pp 1233–1238).

60. Yu. M. Panchenko, "Asymptotic Form of a Trail of a Body Moving in a Rarefied Plasma", Trudy Vsesoyuznoy Konferentsii po Fizike, Kosmicheskogo Prostranstva, Izdatel'stvo "Nauka", Moscow, June 10–16 1965, pp 254–264 (NASA TT F-389, pp 342–354).

61. N. D. Papaleksi, "Reports of the Expedition to Observe the Solar Eclipse of June 19 1936", Izdatel'stvo Akademii Nauk SSSR, 1937, Vol 1, p 115.

62. V. A. Rudakov, "Certain Results of Rocket Measurements of Electron Con-, centration in Ionosphere Up to Altitudes of 200 km (Conducted in 1959–1960)" Iskusstvennyye Sputniki Zemli, 1961, No. 10, pp 102–103 (Artificial Earth Satellites, pp 238–239).

63. V. A. Rudakov, "N(h)-Profiles Obtained With the VHF Dispersion Interferometer Launchings of Academy of Sciences USSR Geophysical Rockets During 1962–1963", Kosmicheskiye Issledovaniya, 1964, Vol 2, No. 6, pp 946–947 (TT–64–1316, pp 217–220); submitted at the Seventh Conference of Geophysicists of the Socialist Countries, Moscow, June 1964.

64. Yu. A. Ryzhov and O. I. Yudin, "On the Effect of Ionospheric Electron-Density Irregularities on Measurement of Some Parameters of the Ionosphere", Izvestiya Vysshikh Uchebnykh Zavedeniy, Radiofizika, 1962, Vol 5, No. 1, pp 13–20 (JPRS: 14 515).

65. Ye. Ya. Shchegolev, K. Ye. Viller, and I. N. Borushko, "Recent Research on Radio Wave Propagation Over the Earth's Surface", Gostekhizdat, Moscow-Leningrad, 1945, p 45.

66. Ye. Ye. Tsedilina, "Space Modulation of the Doppler Frequency Shift of Radio Waves Received From Artificial Earth Satellites", Geomagnetizm i Aeronomiya, 1964, Vol 4, No. 3, pp 584–585 (Geomagnetism and Aeronomy, pp 462–463).

67. Ye. Ye. Tsedilina and A. A. Kharybina, "Investigation of the Inhomogeneous Structure of the Ionosphere From the Results of Radio Observations of Cosmos 1, Cosmos 2, and Cosmos 11 on Coherent Frequencies", Geomagnetizm i Aeronomy, 1964, Vol 4, No. 3, pp 503–508 (Geomagnetism and Aeronomy, pp 399–403).

68. L. M. Yerukhimov, "Preliminary Results of Measurements of the Height of Ionosphere Inhomogeneities by Signals From Artificial Earth Satellites", Geomagnetizm i Aeronomiya, 1962, Vol 2, No. 4, pp 688–690 (Geomagnetism and Aeronomiya, pp 572–574).

69. L. M. Yerukhimov, "Studies of Electron-Density Irregularities in the Ionosphere by Radio Astronomical and Satellite Methods. A Review", Izvestiya Vysshikh Uchebnykh Zavedeniy, Radiofizika, 1962, Vol 5, No. 5, pp 839–865 (JPRS: 18 823).

70. L. M. Yerukhimov, "The Height and Dimensions of the Ionospheric Nonuniformities Responsible for Fluctuations in Signals Received From Artificial Earth Satellites. I. Night Hours", Kosmicheskiye Issledovaniya, 1965, Vol. 3, no. 4, pp 584–594.

71. L. M. Yerukhimov and N. A. Mityakov, "Some Methods of Investigation of

Ionosphere by Means of Space Reception of Signals From Artificial Earth Satellites", Izvestiya Vysshikh Uchebnykh Zavedeniy, Radiofizika, 1964, Vol 7, No. 3, pp 556–559 (JPRS: 27 545).

72. L. M. Yerukhimov, N. A. Mityakov, and Ye. Ye. Mityakova, "Ionosphere Studies by Receiving Artificial Earth Satellite Radio Emission on the Earth", Trudy Vsesoyuznoy Konferentsii po Fizike, Kosmicheskogo Prostranstva, Izdatel'stvo "Nauka", Moscow, June 10–16 1965, pp 147–150 (NASA TT F-389, pp 196–201).

73. Yu. Zaytsev, "Cosmos Ships in Earth Orbits", Aviatsiya i Kosmonavtika, 1966, No. 12, pp 62–67.

74. L. A. Zhekulin, "Distribution of Electron Density as a Function of Height From Data of Experiments With Rockets and Artificial Earth Satellites: Effect on Propagation of Radio Waves", Iskusstvennyye Sputniki Zemli, 1958, No. 1, pp 67–79 (Artificial Earth Satellites, pp 85–101).

Chapter 10

Solar and Cosmic Electromagnetic and Charged-Particle Radiations

*H. H. Guendel**

INTRODUCTION

The investigation of solar and cosmic radiation above much of the earth's atmosphere has been carried out in the Soviet Union since the late 1940's or early 1950's. Following initial experiments with balloons and rockets (up to 100 km), using simple optical and particle radiation sensing equipment, the Soviets developed a variety of sensors and systems for increased use on geophysical rockets (>500 km), artificial satellites, and space probes. As in the U.S., the Soviets have successfully undertaken extensive investigation of the nuclear components of primary cosmic rays, the composition of radiation zones surrounding the earth, and the intensity of ionized gas fluxes in interplanetary space over the past 10 years.

The following discussion presents a brief description of the types and

* Bell Aerosystems Company, Research Department (Consultant, Columbus Laboratories, Battelle Memorial Institute).

results of Soviet solar and cosmic electromagnetic radiation measurements, including ultraviolet, X-ray, and gamma radiation, and charged particle radiation measurements. To best coincide with the nature and characteristics of the sensors involved, the discussion of the latter has been divided into three subject categories, namely, relativistic particle radiation, corpuscular radiation, and ion flux.

Because of the voluminous amount of Soviet publications existing in this area and the rapid change in data interpretations, discussion has necessarily been limited to only the more recent and/or more significant results. Table 10·1 identifies the numerous vehicles involved in solar and cosmic radiation studies since 1957.

INSTRUMENTATION

Several types of radiation sensors have been employed by the Soviets during high altitude tests and space flights. This section contains a description of the earlier filter instruments, gas-discharge counters, scintillation detectors, Cherenkov counters, and ionization chambers, as well as the more sophisticated instruments, like electrostatic spherical analyzers, spectrometers, and ionization calorimeters developed during the past five years. These detectors are used for the mapping of the natural environment (galactic cosmic radiation, solar electromagnetic and particle radiation, Van Allen belts) and radiation monitoring for the protection of astronauts. In the following paragraphs Soviet instruments and instrumentation systems are briefly described in relation to the environment sensed.

Electromagnetic Radiation Measurements

Ultraviolet, Gamma-Ray, and X-Ray Radiation Sensors

Short-Wave Radiation Sensors

For observations of short-wave (soft X-ray and far ultraviolet) radiation aboard high-altitude rockets and earth satellites, the Soviets have depended largely on filter apparatus. The basis of this method is the separation of the different spectral regions of the sun's short-wave radiation by a choice of filters which are moved automatically in front of a receiver. The most suitable receiver is an open-type secondary electron multiplier working under the vacuum condition existing in space. A reduction of photo-emissions in the near ultraviolet and visible regions of the spectrum, a powerful long-wave radiation background, is

TABLE 10.1

Soviet Space Vehicles Associated With Solar and Cosmic Radiation Studies (1957–1966)

Geophysical rockets		Satellites		Planetary Probes	
		Electromagnetic Radiation Studies			
		Ultraviolet, X-Ray, and Gamma-Ray Radiation Measurements			
	(km)				
Aug. 25 1957	212	*Initial Sputnik Series*		Zond 1	Apr. 1 1964
Feb. 21 1958	473	Sputnik 2	Nov. 3 1957	Venus 2	Nov. 12 1965
Aug. 13 1958	212			Venus 3	Nov. 16 1965
Aug. 27 1958	451	*Spaceship-Satellite Series*			
July 2 1959	212	SS 2	Aug. 19 1960		
July 10 1959	212	SS 3	Dec. 1 1960		
July 21 1959	110				
July 21 1959	110	*Electron Series*			
June 15 1960	212	Electron 2	Jan. 30 1964		
Sep. 6 1960	100	Electron 4	July 11 1964		
Sep. 16 1960	212				
Sep. 19 1960	100				
Sep. 21 1960	100	*Proton Series*			
Feb. 15 1961	100	Proton 1	July 16 1965		
June 6 1963	515	Proton 2	Nov. 2 1965		
June 18 1963	510				
Oct. 25 1963	430				
Dec. 24 1963	422				
		Charged Particle Radiation Studies			
		Relativistic Particle Radiation Measurements			
July 2 1958	210	*Initial Sputnik Series*		*Lunar Probes*	
Aug. 2 1958	212	Sputnik 3	May 15 1958	Luna 1	Jan. 2 1959
Aug. 13 1958	212			Luna 2	Sep. 12 1959
July 21 1959	110	*Spaceship-Satellite Series*		Luna 3	Oct. 4 1959
July 21 1959	110	SS 2	Aug. 19 1960		
Sep. 6 1960	100	SS 3	Dec. 1 1960	*Planetary Probes*	
Sep. 16 1960	210			Mars 1	Nov. 1 1962
Sep. 19 1960	100	*Electron Series*			
Sep. 21 1960	100	Electron 2	Jan. 30 1964		
Sep. 22 1960	210	Electron 4	July 11 1964		
Oct. 18 1962	500				
June 6 1963	515	*Cosmos Series*			
June 18 1963	510	Cosmos 6	June 30 1962		
Oct. 25 1963	430	Cosmos 19	Aug. 6 1963		
Dec. 24 1963	422	Cosmos 25	Feb. 27 1964		
Sep. 20 1965	500				
Oct. 1 1965	500	*Proton Series*			
		Proton 1	July 1 1965		
		Proton 2	Nov. 2 1965		

TABLE 10.1 (Continued)

Soviet Space Vehicles Associated with Solar and Cosmic Radiation Studies (1957–1966)

Geophysical Rockets	(km)	Satellites		Planetary Probes	
		Corpuscular Radiation Measurements			
July 2 1958	210	*Initial Sputnik Series*		*Lunar Probes*	
Aug. 2 1958	212	Sputnik 2	Nov. 3 1957	Luna 1	Jan. 2 1959
Aug. 13 1958	212	Sputnik 3	May 15 1958	Luna 2	Sep. 12 1959
Oct. 4 1958	110			Luna 3	Oct. 4 1959
Oct. 10 1958	110	*Spaceship-Satellite Series*		Luna 4	Apr. 2 1963
Dec. 23 1958	110	SS 2	Aug. 19 1960	Luna 9	Jan. 31 1966
Dec. 25 1958	110	SS 3	Dec. 1 1960	Luna 10	Mar. 31 1966
July 21 1959	110				
July 21 1959	110	*Electron Series*			
Sep. 6 1960	100	Electron 1/2	Jan. 30 1964		
Sep. 16 1960	212	Electron 3/4	July 11 1964		
Sep. 19 1960	100				
Sep. 21 1960	100				
Sep. 22 1960	210	*Cosmos Series*		*Planetary Probes*	
Oct. 18 1962	500	Cosmos 3	Apr. 24 1962	Venus 1	Feb. 12 1961
June 6 1963	500	Cosmos 4	Apr. 26 1962	Mars 1	Nov. 1 1962
June 18 1963	510	Cosmos 5	May 28 1962	Zond 1	Apr. 2 1964
Sep. 20 1965	500	Cosmos 7	July 28 1962	Zond 3	July 18 1965
Oct. 1 1965	500	Cosmos 9	Sep. 27 1962	Venus 2	Nov. 12 1965
		Cosmos 10	Oct. 17 1962	Venus 3	Nov. 16 1965
		Cosmos 12	Dec. 22 1962		
		Cosmos 13	Mar. 21 1963		
		Cosmos 15	Apr. 22 1963		
		Cosmos 16	Apr. 28 1963		
		Cosmos 17	May 22 1963		
		Cosmos 41	Aug. 22 1964		
		Ion Flux Measurements			
See Chapter 9 (Table 9.1)		*Initial Sputnik Series*		*Lunar Probes*	
		Sputnik 3	May 15 1958	Luna 1	Jan. 12 1959
		Electron Series		Luna 2	Sep. 12 1959
		Electron 2	Jan. 30 1964	Luna 10	Mar. 31 1966
		Electron 4	July 11 1964	*Planetary Probes*	
		Cosmos Series		Venus 1	Feb. 12 1961
		Cosmos 2	Apr. 6 1962	Mars 1	Nov. 1 1962
		Cosmos 3	Apr. 24 1962	Zond 2	Nov. 30 1964
		Cosmos 5	May 28 1962	Zond 3	July 18 1965
		Cosmos 12	Dec. 22 1962	Venus 2	Nov. 12 1965
		Cosmos 15	Apr. 22 1963	Venus 3	Nov. 16 1965

necessary and is accomplished by the selection of suitable photocathode materials. Materials such as Be0 and SrF_2 have been used by the Soviets. Filters of copper (1·4–3 Å), beryllium (5–10 Å), aluminum (8–21 Å), polystyrene (44–100 Å) lithium and calcium fluoride (Lyman alpha, 1216 Å), and quartz (>1500 Å), mounted in a disk and positioned in front of the electron multiplier, were employed in the early Sputnik and Spaceship-Satellites (see Figures 10.1 and 10.2). The step rotating disk also contained openings with no filter, so that direct (visible) radiation could fall

Fig. 10.1. Soft X-Ray and UV-Radiation Sensor on Sputnik 2[4]

on the photocathode permitting the necessary corrections to be made to account for the angle of incidence of the radiation striking the filters. The output of the electron multiplier consisted of voltage levels, whose amplitudes were proportional to the intensity of the incident radiation. The operation of each multiplier was controlled by two photosensitive resistors which were always irradiated concurrently with their associated multiplier. One resistor controlled the heater circuits (120° view) and the other (60° view) controlled the anode supply[4, 34, 100, 164–166].

End-Window Photon Counters

Electron 2 and Electron 4 carried two groups of geiger photon counters, one mounted outside and the other inside the satellite, to measure solar X-rays in the 2–8 Å (beryllium window) and 8–18 Å (aluminium window) spectral regions (see Figure 10.3). The design of the sensitive counters was similar to that of the Spaceship-Satellites 2 and 3 counters shown schematically in Figure 10.4. The main block of

Fig. 10.2. Soft X-Ray Sensors on Spaceship-Satellite 2[164]

1. Photon counter units
2. Secondary electron multipliers
3. Optical detectors
4. Recording units
5. Input window

Fig. 10.3. Solar X-Ray Sensor on Electron 2[15]

the counters was continuously directed to the sun. The optical axis of the second block, which was rigidly attached to the satellite hull, coincided with the dead zone center of the main block. Silicon phototransducers kept the sun in the field of view of the two blocks of the photon counters[89, 127–128].

High-Energy Gamma-Ray Counter

New prospects for the study of cosmic rays from remote areas of the universe have unfolded due to the recently developed gamma and X-ray astronomy. Gamma-rays with energies of >50 MeV can be generated only by cosmic rays, e.g., decay of the π°-mesons. Therefore, the measurement of the intensity and spectral and spatial distribution of such cosmic rays can provide valuable information about the resulting gamma-rays[33].

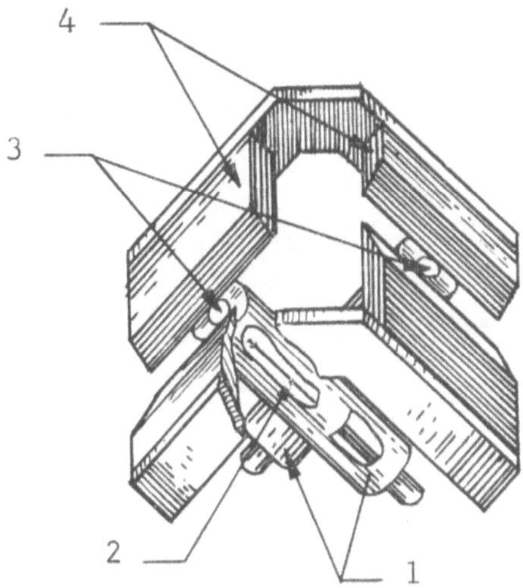

Fig. 10.4. End-Window Photon Geiger Counter on Spaceship-Satellites 2 and 3[98]
1. Photon counters
2. Counter "filament"
3. Beryllium window
4. Magnets

Protons 1 and 2 employed a high-energy gamma-ray counter telescope (GG-1), shown in Figure 10.5, for the integral measurement of the flux of gamma quanta with energy levels between 10^8 eV and 3×10^9 eV. The working principle of this instrument was the measurement of coincidences between a sandwich of fast and slow scintillators and a Cherenkov radiator. The scintillator sandwich was used for the discrimination of gamma-rays from other radiation, and the Cherenkov counter performed the energy measurement. The resolution time of the coincidence circuit was 5×10^{-8} sec. The scintillator sandwich was constructed of nine plastic scintillators having a thickness of 3 mm, which were separated by light CsI(*Tl*) scintillators of a thickness of 2 mm. The overall thickness of the sandwich was about 1·1 shower units. The intensities of the fast components (from the plastic scintillators) of the scintillations in the sandwich were compared to those of the low components (from the CsI-crystal) to discriminate gamma quanta from

Fig. 10.5. Scientific Instrumentation on Protons 1 and 2[2]

1. Low-energy particle charge spectrometer (SEZ–1)
2. Gamma telescope (GG–1)
3. Proportional counter
4. Graphite and polyethylene filter

5/6/8. Ionization calorimeter (SEZ–14)

7. Interaction detector
9. Electron spectrometer (SEZ–12)

other radiations. Whenever a gamma ray was indicated by the sandwich, the energy measurement performed in the Cherenkov counter was stored in the memory of the device. The energy measurement in the Cherenkov counter was performed simply by counting the number of relativistic electrons in the shower produced by the incident gamma ray. This number of electrons being proportional to the incident energy.

The Cherenkov radiator was made of lead glass. Its thickness, 5 cm, corresponded to 2·5 shower units. The discrimination following the Cherenkov detector separated pulses caused by showers containing 1, 3, 10, and 30 or more electrons, which corresponded to incident energies of 10^8, 3×10^8, 10^9, and 3×10^9 eV[41, 43].

Charged-Particle Radiation Measurements

Relativistic Particle Sensors

Cherenkov Counters

Cherenkov counters were employed in several Soviet earth satellites (Sputnik 3, Spaceship-Satellites 2 and 3) and space probes (Lunas 1–3 Mars 1) for the investigation of the charge of nuclear components of the cosmic radiation (see Figures 10.6–10.9). The principal of operation is that a relativistic particle with charge Z passing through a Cherenkov radiator (provisions being made for the path of all registered particles through the radiator to be the same length) will produce a signal proportional to Z^2 [24, 32, 41, 43].

Both integral and differential type Cherenkov counters have been used in Soviet space experiments. The major difference between integral and differential counters is in the electronic processing of the photomultiplier output. In an integral counting system, all particles having a charge higher than a given one are registered. For example, counting systems on space vehicles were designed to count all relativistic particles with charges of $Z \geqslant 2$, $Z \geqslant 5$, or $Z \geqslant 15$. In Protons 1 and 2 the registered radiation was divided into nine Z-groups. In differential counting systems, the amplitudes of all output pulses are measured and recorded individually. Linearity checks and calibration of the Cherenkov counters and of the electronic systems were performed on the ground prior to the flights. The calibration of the whole system was performed by the measurement, on ground, of the μ mesons of the cosmic radiation[41].

1. Integral Counter. The Cherenkov counter radiator used in integral systems of Sputnik 3, the Spaceship-Satellites, lunar and Mars probes was of plexiglass, of cylindrical shape, and had dimensions of $2{\cdot}6 \times 2{\cdot}6$ cm. A single-charged relativistic particle passing through this radiator produced 750 photons of visible light. Particles of higher atomic number produced Z^2 times as many photons. The Cherenkov radiation was registered by FEU-25 photomultipliers. Because of the low efficiency of the photocathode of this photomultiplier, only about 35 to 40 photo-

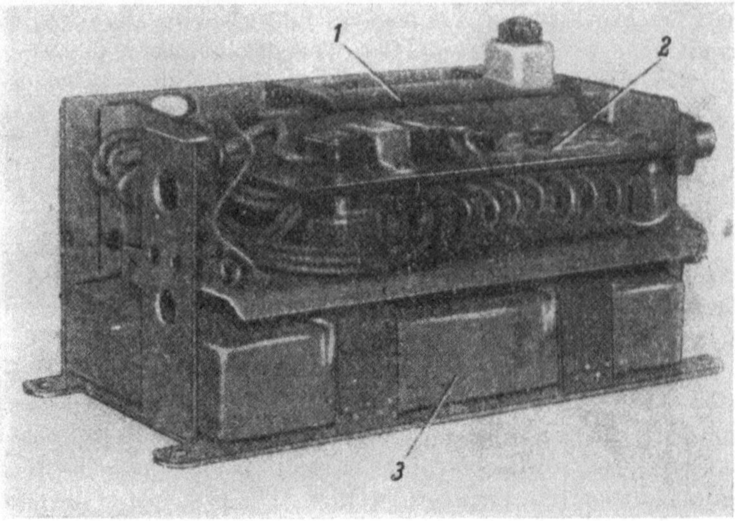

Fig. 10.6. Cherenkov Counter (Integral Type) on Sputnik 3, Lunas 2 and 3, Spaceship Satellites 2 and 3[32]

Fig. 10.7. Cherenkov Counter (Differential Type) on Spaceship-Satellites 2 and 3[32]

electrons were emitted for these 750 Cherenkov photons. This means that, for a gain of about 10^5 and an output capacitance on the order of about 10 $\mu\mu$F, the photomultiplier output pulse had an amplitude of about 50 mV for a single-charged relativistic particle.

The "upper end" surfaces of the Cherenkov radiators were blackened, and the side surfaces were aluminized. Therefore, since Cherenkov

Fig. 10.8. Cherenkov Counter With Gas-Discharge Counter Telescope as Used on Spaceship-Satellites 2 and 3[4]

radiation is emitted under a small angle from the incident particle only light flashes from particles impinging from the direction of the "upper hemisphere" (i.e., having an angle of not more than 90° from the forward direction of the detector), were reflected towards the photomultiplier for detection. The minimum path length through the detector, depending on the density of the produced photons and therefore on Z (the atomic number of the incident particle), defines the "geometric factor" or the aperture of the counter. An effective aperture was calculated by deriving the aperture for each Z and then integrating over all atomic numbers of the incident spectrum of particles.

Because of the large aperture of the integral Cherenkov counter, and because of the inclusion of many atomic numbers in one group of data, the counting rate of this instrument was relatively high. Therefore, the integral Cherenkov counter was conveniently used for the investigation of time variations of the cosmic radiation and for the measurement of nuclei with low flux values (high Z). The output of the photomultipliers was amplified, discriminated (for $Z \geqslant 2$, $Z \geqslant 5$, and $Z \geqslant 15$)

Fig. 10.9. Cherenkov Counter (Integral Type) With Electronic Components on Electron 2[16]

and stored in binary scaling-down circuits (bistable multivibrators). The condition of these binary scaling circuits was read by summing circuits, and transmitted to the radio telemetry system.

The Cherenkov counter on Proton 1 (SEZ-1 device) was made of plexiglass, had a diameter of 15 cm and was 3 cm high and was positioned between a "telescope" of scintillation detectors[41].

2. Differential Counters. Differential Cherenkov counters were used in Spaceship-Satellites 2 and 3. In contrast to the integral counter, which registered all particles above a predetermined atomic number Z, the charge of every particle recorded by the differential Cherenkov counters was determined exactly. The crystals of the differential

counters had a diameter of 7·5 cm and a height of 3·0 cm. To assure uniform path length, the Cherenkov counters were placed between the two layers of a telescope of gas discharge counters, and wired in coincidence with these layers. The output pulses of the photomultiplier were amplified and fed through a pulse height-to-time converter. This converted pulse was then used for the gating of an oscillator, passing a group of pulses for each event. The number of pulses in each group was counted and stored by a series of binary scaling circuits. The conditions of these scaling circuits, indicating the energy of each individual pulse, were read by summing circuits and fed to the radio telemetry system.

3. Summing Circuits. The information obtained by the Cherenkov counters (integral as well as differential) was stored in binary scaling circuits. The use of summing circuits permitted the transmission of a greater amount of information on the condition of these scaling circuits. Basically the summing circuit works on the principle that one resistor is mounted in series with the base supply resistors of several scaling circuit output emitter followers. Usually, the condition of three scaling circuits was transmitted by one summing circuit. When the base resistors of these scaling circuits are chosen to have the ratio 4:2:1, and assuming a constant voltage drop across these resistors, then an activation of these three scaling circuits will cause a voltage drop across the common resistor of the summing circuit in a ratio of 1:2:4. In this case the successive operation of three contiguous scaling circuits would yield a steplike monotonous rise of the voltage across the resistor of the summator. If scaling circuits which are not contiguous are connected to the summing circuit, then the output pattern will be more complex.

Calorimeters

Calorimeters were employed in Protons 1 and 2 for the direct measurement of the energy of relativistic positive particles (device SEZ-14) and electrons (device SEZ-12). A detailed description of this instrumentation is given by Grigorov[41–42]. A photograph of the Proton apparatus is shown in Figures 2.16 and 10.5.

1. SEZ-14. The initials of this setup stand for Spectrometer of Energies and Charges (Z) of particles up to an energy of 10^{14} eV. The basic components of this spectrometer were: energy detectors (calorimeters) for the measurement of the energy of each particle, proportional counters for the measurement of the charge of the primary particle, and filters for the measurement of the cross sections of in-

elastic interactions of protons with carbon and hydrogen. The spectrometers were built in the form of two identical and independent parts, each part representing a complete instrument capable of performing the whole program of measuring E, Z, and σ. A polyethylene and a carbon filter were positioned over the spectrometers. These filters were interchanged and then removed completely from the spectrometers every 12 hours.

The energy detector was an ionization calorimeter made of iron in which the ionization chambers were replaced by plastic scintillators. The principle of the design and ground operation of an ionization calorimeter were described by Grigorov[41]. The calorimeters of Protons 1 and 2 were constructed of nine steel plates 5·5 cm thick separated by 5-cm thick layers of bars of plastic scintillators. All scintillators were viewed from both sides of the stack, by two large scintillation detectors (diameter of photocathode: 15 cm). Shower particles of showers produced in the iron plates of the calorimeter were detected by the interspersed plastic scintillators and the two photomultipliers. The added outputs of the two photomultipliers were divided (by integral discriminators) into nine (logarithmically spaced) energy groups between 10^{10} and 10^{14} eV. The aperture of the device was determined by a "telescope" of plastic scintillators (one layer positioned above and the other one below the spectrometer), wired in coincidence with each other and with the calorimeter. Proportional counters, positioned over the calorimeters, were used for the determination of the charge of the primary particles. These counters were mounted in a single hermetically sealed aluminum body.

2. SEZ-12. This spectrometer was designed to measure the electron component of the cosmic radiation in the energy range of 3×10^8 to 10^{10} eV. Since the galactic electron current is only about 1% of the proton current, special provisions were made to exclude the measurement of positive particles by the SEZ-12 spectrometer. This was achieved by positioning a gaseous Cherenkov counter over the energy measuring section of the spectrometer, wired in coincidence with the calorimeter. This Cherenkov counter had a height of 53 cm and a diameter of 20 cm. It was filled with Freon 13 at a pressure of 11 atmospheres. The design of this Cherenkov counter was such that only protons with energies $E_p \geqslant 10^{10}$ eV would trigger it. Such protons, however, are likely to generate showers of pions. An anticoincidence shower detector surrounded the gas Cherenkov radiator and, whenever triggered, excluded the energy measurement from registration. This shower detector was a plastic scintillating cylinder completely surround-

ing the gas Cherenkov radiator and viewed by four photomultiplier tubes. The energy detector was of the ionization calorimeter type and consisted of four lead plates and four scintillator plates, each one centimeter thick. The scintillators were viewed by two photomultipliers. The combined outputs of the photomultipliers were subjected to integral discrimination at levels corresponding to 1, 3·3, 33, 100, 330, and 1000 times the output generated by cosmic μ mesons [41–42].

Primary Cosmic Particle Sensors

Ionization chambers and gas-discharge counters were used in geophysical rockets and in satellites of the Cosmos series for the determination of height—intensity relations of the primary cosmic radiation and for the measurement of the mean specific ionization of these particles. A photograph of the commonly used, spherical ionization chamber is shown in Figure 10.10. Basically, the chamber was an argon-filled sphere pressurized to $\sim$100 psig and had an outside diameter of 24 cm. The diameter of the internal electrode was 2·8 cm[118]. In addition to ionization chambers, gas-discharge counters of the STS-6 type were used in geophysical rockets and of the STS-5 type in Cosmos 6 and 19 satellites to investigate the intensity and direction of incidence of the primary cosmic radiation outside the earth's radiation zones.

The first experiments in space were performed in a rather crude manner, measurements including the albedo radiation and particles from showers originating in the structure of the rocket. In later experiments (since 1962), the rockets carrying the instruments were stabilized with an accuracy of $\pm 2°$ in space, and since the lower hemisphere was shielded by the rocket the albedo radiation was not included in the measurements. Also, the associated electronic equipment was removed sufficiently from the sensors and did not contribute shower particles in significant numbers. This was substantiated by the fact that shielded and unshielded gas-discharge counters showed the same counting rate. In these experiments, the vertical and 60° inclined components of the cosmic radiation were measured, the east-west asymmetry was determined, and the magnitude of the albedo radiation was established[116–117].

Nuclear Emulsions

Nuclear emulsions provide a means for permanently registering the effects which an ionizing agent (ray or particle) produces as it passes through the emulsion. The ionization tracks which are recorded by the emulsion furnish information about the nature of the ionizing particle or ray. Such information may include energy, direction of propogation, point of origin and charge. The Soviets indicated that Spaceship-

Fig. 10.10. Equipment for Recording Cosmic Rays on Geophysical Rockets[118]

1. Horizontal counter
2. Vertical counter (in a thin-walled duraluminum container)
3. Ionization chamber

Satellite 2 contained emulsion blocks which were chemically treated inside the spacecraft and developed on the ground after recovery[131].

Soviet literature also contains a description of a photon shower detector located below an emulsion block for the detection of particles with an energy of 10^{12} eV and higher[34, 106]. In this case, emulsions were placed between seven 5-mm thick lead plates and scintillator plates. Automatic film (emulsion) scanning methods and systems have also been developed by the Soviets, especially in connection with primary cosmic radiation particle investigations[6].

Corpuscular Radiation Sensors

Scintillation and Gas-Discharge Counters

Scintillation counters using single crystals (sodium iodide or cesium iodide) are one of the most versatile types of radiation sensors. Depending on the crystal and electronic processing of the output pulses, the counter can be used for the detection of X-rays, gamma-rays, neutrons, electrons, and protons up to an energy of about 100 MeV/nucleon. The pulses arising at the output of the photomultiplier are amplified by a semiconductor amplifier and fed to a binary counting system. The threshold of the counter circuit corresponds to an energy liberation of several tens of keV in the crystal. Simultaneous measurements of the anode and one or more of the dynode currents permits not only measurement of the total ionization taking place in the crystal (dynode current) but also an estimate of the average composition of the incident radiation. By the use of discriminators and coincidence circuitry, it is possible to register separately and selectively the number of X-ray quanta (energy release 40–500 keV), of gamma-rays (energy release of 500–5000 keV), and of high-energy particles ($>$5 MeV)[133].

Another common radiation sensor is the gas-discharge counter which usually consists of a metal tube with a filament stretched along its axis. The tube is filled with a mixture of gases of the halogen group and a potential difference of several hundred volts is applied between the casing and the filament. When a charged particle passes through the counter, ions are formed along the path of the particle and are set in motion by the electric field. These moving ions collide with the gas molecules in the counter creating further ions until an avalanche is formed. This avalanche creates an electric pulse which is registered by the counter circuitry. The total number of charged particles passing through the counter is registered only, since the sensitivity to γ-rays (Bremsstrahlung) is low (1%). This characteristic makes it possible, whenever a gas-discharge counter is used in combination with a scintillation counter, to separate the total counts of a scintillation counter into charged particles and Bremsstrahlung. Since the Relative Biological Effect (RBE) is different for charged particles and γ-rays, the knowledge of the composition of the total flux is necessary for the determination of the absorbed dosage in rem (Roentgen Equivalent Man).

In the past 10 years the Soviets have employed various combinations of the gas-discharge and scintillation counter for corpuscular radiation studies aboard rockets, satellites, and space probes. Some of this instrumentation is illustrated in Figures 10.11–10.19. Such equipment has been employed in the initial Sputnik series, in Spaceship-Satellites 2 and 3, Electron satellites, Vostok satellites, and numerous satellites of the Cosmos series, largely for dosimetric purposes[63]. A more sophisticated radiometric instrumentation system, including proton detectors, was, however, used on Cosmos 41. In several cases gas-discharge counter telescopes were installed (Spaceship-Satellites 2 and 3) with multiple coincidences (geophysical rockets, Cosmos 6 and 19) or in a group surrounding one counter for intensity, coincidence and anticoincidence measurements (Zond 1). For soft electrons the Soviets have used fluorescent screens with attached photomultipliers. Changing the screen potential also changed the penetrating ability of the electrons and thereby determined the sensitivity of the equipment[9, 35].

Semiconductor Proton Counters

Solid-state (sometimes called semiconductor or silicon) detectors, for the measurement of protons directly, have been employed by the Soviets in both earth satellites and space probes. This detector has certain advantages over the scintillation counter: it is compact, has linear response toward charged particles, and does not respond to gamma, beta, and X-ray radiation backgrounds.

Cosmos 41 used two proton detectors covering energy ranges from 400 keV to 7 MeV and 3 to 8 MeV. One detector was shielded with an aluminum foil 70 μ thick[113–114]. Interplanetary space probes Venus 2 and 3 measured proton flux increases with a solid-state (n-p) detector covering a range from 1 to 5 MeV[139]. Satellites of the Electron series employed silicon detectors to sense protons in the energy range from 0·9 to 5·5 MeV. Also the Soviets have described a combined proton spectrometer consisting of a silicon detector and a scintillation detector which reportedly permitted registration of four ranges of proton energies: 20–200 MeV, 40–100 MeV, 60–100 MeV, and 3–20 MeV[110].

Fig. 10.11. Cosmic-Ray Counter on Sputnik 2[5]

Fig. 10.12. Cosmic-Ray Counter on Sputnik 3[5]

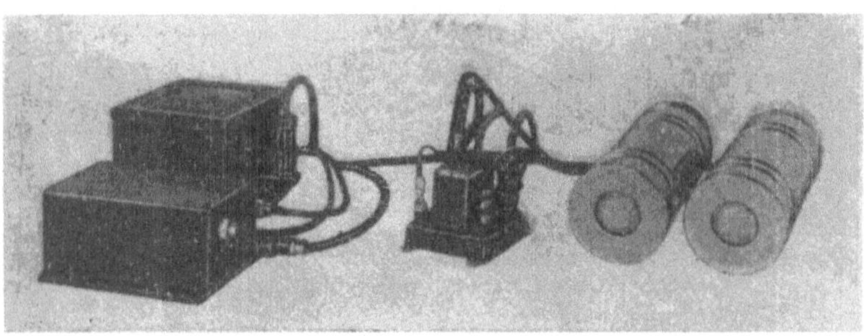

Fig. 10.13. Soft-Electron Detector on Sputnik 3[5]

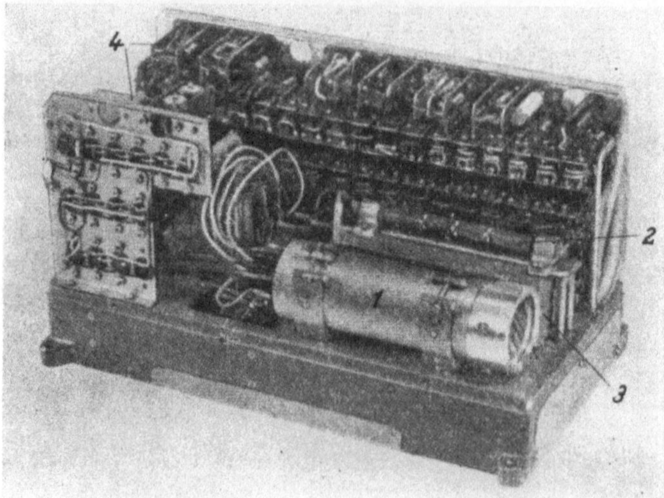

Fig. 10.14. Radiometric Instrumentation Internal Unit Scintillation and Gas-Discharge Counters on Spaceship-Satellite 2[104]

1. Scintillation counter
2. Gas-discharge counter C_{I}
3. Gas-discharge counter C_{II}
4. Electronic circuit

Fig. 10.15. Radiometric Instrumentation External Unit Scintillation Counter on Spaceship-Satellite 2[104]

1. Scintillation counter
2. Discharge-tube pulse shaping circuit
3. Supply battery for photomultiplier

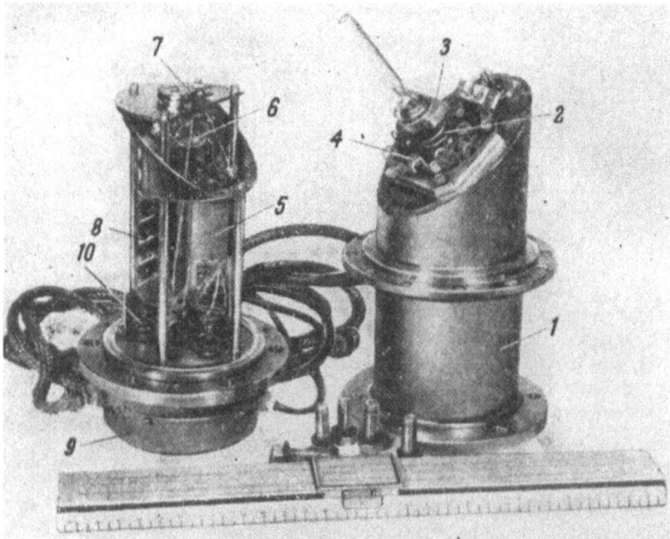

Fig. 10.16. Scintillation Counter With Filters and Open-Type Secondary Electron Photomultiplier Installed in Rockets[10]

1. Receiver housing
2. Glass entry fitting
3. Breaker half-rings
4. Puncturing mechanism
5. Electromagnet
6. Filter disk
7. Guard rings
8. VEU secondary electron multiplier
9. Mount with pulse amplifier
10. Hermetic glass beads

Fig. 10.17. Corpuscular Radiation Counters Aboard Electron 1[15]

Fig. 10.18. Corpuscular Radiation Counter Aboard Electron 2[15]

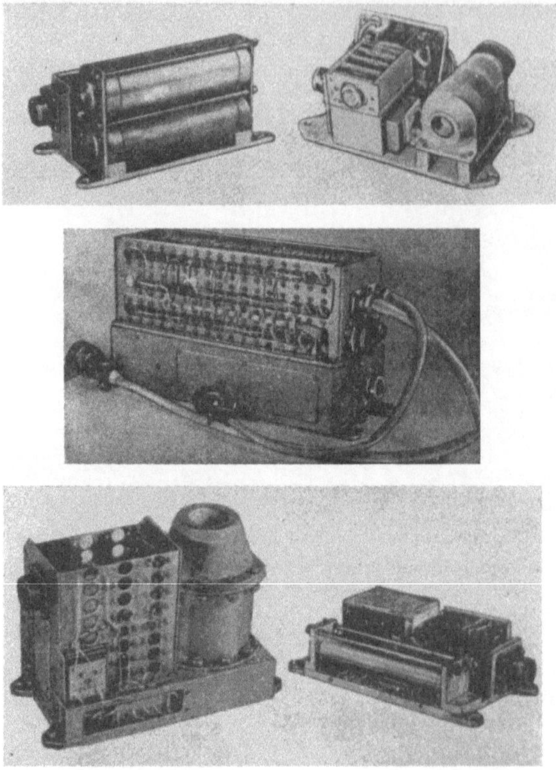

Fig. 10.19. General View of Radiation-Belt Measuring Apparatus on Electrons 1 and 2[15]

Ion Sensors

Ion Traps

The measurement of ion densities in the atmosphere and in space were made by ion traps of different designs employed in Soviet geophysical rockets, earth satellites, and space probes. Two-electrode ion traps, consisting of an outer spherical wire mesh electrode, and an inner spherical collector were installed in Sputnik 3 and Cosmos 2. Three-electrode traps of semispherical form and multielectrode (modulation) traps with plane grids were employed on several of the Electron and Cosmos satellites as well as on many of the Soviet space probes. Both trap types are shown in Figures 10.20–10.22. Also a honeycomb type of ion trap shown in Figure 10.23 was used for the first time on Cosmos 2 for the determination of positive ion temperatures[3, 5, 15].

Fig. 10.20. Sputnik 3 Spherical Ion Traps[5]

Fig. 10.21. Close-up View of Spherical Ion Trap[167]

Fig. 10.22. Semispherical Ion Trap on Electron 2[15]

Fig. 10.23. Honeycomb-Type Ion Trap[3]

Fig. 10.24. Electrostatic Analyzer Used on Electron 2[15]

Spherical Electrostatic Analyzers

The Cosmos 12 and 15 and Electron 2 and 4 satellites had a spherical electrostatic analyzer on board to measure low-energy charged-particles (see Figure 10.24). Particles which passed through the spherical plates of the analyzer were detected by a Faraday cup which was connected to a storage capacitor. The collector inlet window was covered with two grids. One grid was kept at − 20 volt potential relative to the instrument frame to suppress secondary electrons from the cup. The spherical capacitor plates were supplied with symmetrical voltages which were changed during flight in accordance with a preset program. The inlet of the analyzer of Cosmos 15 was covered with a grid having a + 12 volt potential relative to the satellite frame, to prevent ionospheric thermal ions from entering the analyzer. Electrons 2 and 4 had two identical analyzers which were adjusted to particle energies of 0·1, 0·2, 0·4, 1·0, 2·5, 5·0, and 10 keV. The analyzer of Cosmos 12 registered electrons with energies of 0·5 and 1·0 keV or 1·0 keV ions and the Cosmos 15 analyzer was adjusted to pass 1·0 keV electrons or ions only[143, 145].

SUMMARY OF RESULTS

Electromagnetic Radiation Measuiements

Ultraviolet and X-Ray Radiations

Measurements of the ultraviolet and the soft X-ray spectrum of the sun have been made by the Soviets using spectrographs, camerae obscurae, secondary electron multipliers with various passband filters, and end-window gas-discharge counters, on both geophysical rockets and earth satellites. Published results of the activity are briefly noted in Table 10.2.

According to available Soviet reports, the ultraviolet spectrum of the sun was initially photographed in the USSR on May 31 1956 by a spectrograph, employed in a geophysical rocket, in the region 2471 to 2635 Å at altitudes up to 100 km. The description of the spectrometer (with a 600 lines/mm concave diffraction grating) as well as results obtained were presented in a paper by Kachalov and associates[64]. In June 1963 tests were carried out with geophysical rockets reaching an altitude of 500 km. In this experiment 12 camerae obscurae with aluminum filters of various thicknesses (0·08 to 0·2 μ) were used, which covered parts of the spectral region from 10 to 400 Å[168]. Also in 1963, the Soviets successfully measured the energy distribution in the ultraviolet and X-ray region of the sun using a thermoluminescent phosphor, $CaSO_4$ (Mn), in the spectrum below 1300 Å. When irradiated by short-wave-length ultraviolet radiation the phosphor stored up energy, and after reheating reradiated the energy in the visible region with a maximum in the green part of the spectrum (5000 Å)[71]. More recently, in 1964 and 1965, rocket measurements up to 500 km of the scattered ultraviolet radiation (1216 Å and 1300 Å) were made with photon counters and lithium-fluoride and calcium-fluoride end-windows[65, 70]. Sputnik 3 and Spaceship-Satellites 2 and 3 were the first Soviet satellites to successfully measure the solar ultraviolet. Their payloads employed three ultraviolet radiation sensors consisting of a rotating disk with various filters and an open-type secondary electron multiplier as a detector. The filters were films of beryllium, aluminum and polyethylene of various thicknesses. A lithium-fluoride and a calcium-fluoride filter were used also to record the Lyman-alpha radiation. Subsequent studies were undertaken aboard the Electron satellites and Venus and Zond probes. Major results of direct measurements reported by the Soviets to date include the following[4, 93, 97–99, 164–165]:

1. Measurements in the 44 to 110 Å range were fairly constant at $1{\cdot}5 \times 10^4 \pm 8\%$ photons/cm² sec.
2. In the 8 to 21 Å range the flux was constant at $6{\cdot}2 \times 10^4$ photons/cm² sec except for two periods of 9 and 4 minutes of high solar activity, during which the flux increased by a factor of 3·2 and 63%, respectively.
3. In the range shorter than 8 Å low-activity solar radiation was very weak and frequently could not be distinguished from the background radiation of nonsolar origin.
4. In the 5 to 10 Å range the flux increased by a factor of 11 during high solar activity as compared to the background radiation previously recorded.
5. In the 1·4 to 3 Å range only the background radiation from non-solar origin was recorded.
6. The maximum intensity of the green coronal line Fe XIV (5303 Å) occurred during high solar activity.
7. X-ray radiation of solar origin is the main "hard" radiation found at altitudes of 200–300 km in the latitude range of approximately 35° N to 35° S.
8. Radiation caused by radiation zone particles is superimposed on the "hard" radiation in higher latitudes.

9. Results of Soviet rocket and satellite radiation flux measurements during 1959–1960 below 10 Å have varied between $2{\cdot}5 \times 10^{-4}$ erg/cm^2 sec and 8×10^{-4} erg/cm^2 sec.

10. On February 2 1964 the Electron 2 counters registered solar X-ray emission fluxes of $3{\cdot}3 \times 10^{-4}$ erg/cm^2 sec in the 2–8 Å region, and $2{\cdot}4 \times 10^{-3}$ erg/cm^2 sec in the 8–15 Å region.

11. The brightness of the Lyman-alpha radiation at large distances from the earth was recorded by Venus 2 instrumentation and amounted to 10^{-4} erg/cm^2 sec ster. The same value was obtained by Zond 1 instrumentation in 1964.

Gamma Radiations

The Soviet Proton 1 and 2 satellites employed an instrument to measure the flux and to investigate the energy spectrum of galactic gamma quanta in an energy range from 20 MeV to 1·5 GeV. Results reported to date indicated the existence of a flux of gamma quanta with energies exceeding 5 MeV of $\sim 2 \times 10^{-3}$ cm^2 sec ster[37].

Charged-Particle Radiation Measurements

The intensity and composition of solar and primary cosmic radiation have been investigated by the Soviets during numerous experiments with instrumented payloads of high-altitude geophysical rockets, artificial earth satellites and interplanetary probes. More recently emphasis has been on determining the composition, geometry, and particle intensities of the inner and outer radiation zones of the earth. A fairly detailed summary of available results is presented in Table 10.2. The following paragraphs review briefly experimental results and hypotheses concerning charged-particle radiations of solar and cosmic origin.

Relativistic Particle Radiations

In early rocket flights, geiger counters and ionization chambers were used for the measurement of the intensity of the cosmic radiation. Later, these detectors (as well as scintillation detectors) were used only for supplementary measurements since single instruments did not give information on the energy or charge of the particles. The first satellite measurements showed that at heights of 200-300 km the radiation flux was several times higher than the expected flux of primary cosmic radiation. This was confirmed by measurements with single gas-discharge counters as well as with counter telescopes on Spaceship-Satellites 2 and

3. With both methods, counting rates were measured which were about three times as high as the flux of primary cosmic rays ($\approx$2.3 particles/ cm^2 sec)[46, 118].

Stacks of 489 NIKFI "R" photoemulsion layers were carried on Spaceship-Satellites 2 and 3 to investigate the nature of this "excess" radiation. The photoemulsions were evaluated by measuring the grain density of the recorded tracks and by counting the stars (with three or more tracks) produced per unit volume. The Soviets found that 45% of the "excess" particles consisted of relativistic particles with tracks of low grain density. These particles did not react with the nuclei of the emulsion and most probably were electrons. The remaining 55% of the "excess" tracks had high grain density and produced stars. They were produced by fission products and not by protons of the radiation belt. All of the "excess" particles (electrons, as well as fission products) were genetically associated with the primary cosmic radiation at the point of observation[46].

Summaries of early results obtained with Cherenkov counters aboard Sputnik 3 were given by Ginzburg and Kurnosova and associates[32, 88]. Cherenkov counter results from Spaceship-Satellites 2 and 3, Electron 2 and 4 satellites, as well as Luna 2 and 3 and Mars 1 probes are summarized in several Soviet papers[28, 31, 85, 86]. The amplitude of the output pulse of Cherenkov counters is proportional to the square of the charge of a relativistic particle, Z^2. Therefore, these counters have been conveniently used for the measurement of the mass spectrum of the cosmic radiation. Z^2 measurements were performed by the Soviets with differential as well as integral counters. Kurnosova and associates compared the results of the integral measurements performed with counters on Electron 2, Luna 2, Spaceship-Satellite 3, and Mars 1, which indicated that for particles with $Z \geqslant 2$, the corresponding results were 343/151/128/333 particles/m^2 sec, respectively. The values for $Z \geqslant 5$ were 18/11/10/– particles/m^2 sec and for $Z \geqslant 15$ the numbers were 0·7/0·4/ – / – particles/ m^2 sec for the above-mentioned vehicles[85].

The mass spectrum was measured up to values of $Z = 30$–40 (or $Z = 35$) on Sputnik 3. Comparison of the abundance of elements in the cosmic radiation (mass spectrum) to the abundance of elements in nature showed that the factor K (the ratio of cosmic to natural abundance for a given element) increased with increasing atomic number in approximate proportion to Z. In other words, if K would be standardized to be $K=1$ for $Z=1$, a value of $K=35$ could be found for $Z=35$. This regular increase of K with increasing Z indicated that the acceleration of heavy nuclei in the cosmic ray sources played a major role[32].

Layers of photographic emulsions can also be used for the determination of the mass spectrum of the cosmic radiation. The disadvantage of this method is, that particles measured through the whole mission are accumulated, and that a differentiation of the data with respect to time and latitude is not possible. On Spaceship-Satellite 2, however, an attempt was made to create an installation providing a limited exposure of the emulsions[131].

In contrast to the emulsion technique, results obtained by Cherenkov counters can be differentiated in time and latitude of the measurements. Integral measurements performed on Spaceship-Satellites 2 and 3 were evaluated in terms of intensity versus latitude for $Z \geqslant 2$, $Z \geqslant 5$, and $Z \geqslant 12$. It was found that the intensity decreased with decreasing latitude, and that this decrease was (relatively) the same for all atomic numbers (Z) considered. From these data the Soviets concluded that the mass spectrum of the cosmic radiation does not change with latitude[32].

The decrease of the measured intensity with the decreasing latitude was explained in Ginzburg's report by the interaction of the earth's magnetic field with the "low-energy" cosmic radiation. Particles up to 2×10^{10} eV/nucleon for nuclei and up to 4×10^{10} eV for protons were effected by the earth's magnetic field, which allowed a certain geomagnetic latitude to be reached from a definite direction only by those particles whose "rigidity" exceeded a definite value ("cutoff rigidity"). This was and is the cause of the latitudinal dependence of the intensity of the cosmic radiation. Calculation of the cutoff rigidity for the different latitudes and measurement of the intensity of the cosmic radiation at different latitudes permits the determination of the energy spectrum of the cosmic radiation. This can be done only up to energies that are sensitive to the "cutoff" by the earth's magnetic field (2×10^{10} eV/nucleon). Such measurements were carried out with Spaceship-Satellites 2 and 3. With these satellites, values of $n = 0{\cdot}83$ and $n = 1{\cdot}41$ for the exponent of the commonly used integral energy spectrum $N(>E) = AE^{-n}$ were measured[32].

The Soviets obtained interesting results with Cherenkov detectors on board Luna 2. The apparatus consisted of two independently acting integral Cherenkov counters, one of which registered nuclei with $Z \geqslant 2$, and the other nuclei with $Z \geqslant 5$ and $Z \geqslant 15$. A most noticeable case of short-duration increase in intensity registered by these Cherenkov counters took place on September 12 1959 at 11:27 world time. At this time the number of counts increased by 1·3 for $Z \geqslant 2$, by 1·5 for $Z \geqslant 5$, and by 11·8 for $Z \geqslant 15$. Thus an appreciable variation was observed in the heavy nuclei with $Z \geqslant 15$. No variations in the proton flux were observed during the corresponding time interval, and the chromospheric flare associated with this event was relatively weak. Usually, solar particle emissions lead to increased proton flux values particularly in cases involving large chromospheric flares. The Soviets concluded therefore that two different mechanisms of particle generation must have been effective on the sun[32, 88].

Corpuscular Radiations

The detectors used to sense particles in the earth's radiation zones were similar to those used to detect nonrelativistic cosmic radiation particles. Gas-discharge counters and scintillation counters with various shieldings were employed in the initial Sputnik satellites and lunar probes. Two gas-discharge counters were installed in Luna 1 and six counters in Luna 2. Scintillation counters with various threshold levels were mounted inside and outside of the various vehicles to detect electrons and protons of various energy levels. Sputnik 3 contained two soft-electron detectors (about 10 keV) consisting of fluorescent screens covered with an aluminum foil and mounted in front of a photo-multiplier tube[82–83].

Various combinations of the different types of instruments were employed on later Soviet satellites, such as the Spaceship-Satellites, and the Cosmos and Electron satellite series, as well as on a majority of their space probes. A rather complete summary of Soviet corpuscular radiation studies is contained in Table 10.2.

Although many geophysical rockets had been launched in the USSR since 1957 to investigate primary cosmic radiation particles, studies made with gas-discharge and scintillation counters on Sputniks 2 and 3, Spaceship-Satellites 2 and 3 and space probes Luna 1 to 3 soon eclipsed the earlier rocket results. These satellite studies made it possible for the Soviets to recognize and partly map the inner and outer earth radiation zones initially predicted and investigated by Van Allen in the U.S.

These early measurements indicated the existence of a flux of protons of $10^3/cm^2$ sec for energies of 10^8 eV. At the edge of the inner zone, at the geometric latitudes 35° to 40°, particles of low energy were found which were assumed to be electrons. A flux of electrons with an energy of more than 0·5 to 1·0 MeV was found in the outer zone which amounted to $10^5/cm^2$ sec steradian. A second, weaker, flux of less than $0{\cdot}1/cm^2$ sec ster with an electron energy more than 5 MeV was also found in the outer zone. Sputnik 3 passed over the Antarctic at an altitude of 1800 km where a larger flux intensity was observed than in the Northern Hemisphere at a maximum flight altitude of 300 to 500 km. The average increase in ionization in the scintillation crystal was 5×10^8 eV/sec in the Northern Hemisphere. However, the ionization in the crystal amounted to 2×10^{10} eV/sec in the Antarctic area, indicating a rise in intensity of 40-fold. Luna 1 instrumentation registered the maximum intensity within the limits of the outer zone. At a distance of 16 000 km from the center of the earth the ionization was equal to 3×10^{11} eV/sec which was 15 times greater than at an altitude of 1800 km. This fact demonstrated that the intensity increased appreciably along the lines of force and that charged particles, in this case electrons, were held in a magnetic trap formed around the earth[132].

Spaceship-Satellite 2 counters registered an enhanced radiation intensity during passages above the regions of the South Atlantic magnetic anomaly as well as in a number of other regions on earth (Northern Hemisphere). The Soviets concluded the following from the numerous data obtained from the instruments of Spaceship-Satellite 2[84, 155]:

1. The enhanced radiation intensity found at an altitude of 320 km above the Brazilian magnetic anomaly is due to the less intense magnetic field in that region. The inner belt does not manifest itself at comparable heights to the north of the geomagnetic equator because an equal value of the magnetic field strength and consequently the mirror points are situated at greater heights than in the anomalous region.

2. At low geomagnetic latitudes in the anomalous region the proton component of the inner radiation belt is the predominant form of radiation. The intensity of the X-rays (Bremsstrahlung) produced by slowing down of electrons by the hull of the spaceship increases with increasing latitude, while the intensity of the proton component decreases.

3. The outer radiation belt manifests itself at geomagnetic latitudes above 40°.

4. There is a transition region between the inner and the outer radiation zone, detected with Sputnik 3 instrumentation in the Northern Hemisphere, which is practically absent in the Southern anomaly.

With the beginning of the launching of the Cosmos scientific satellites in March 1962, the energy flux of electrons was studied with soft-electron detectors. Blagonravov reported in 1964 that the energy flux recorded at sunrise at satellite altitudes from 200 to 1500 km reached an average of 5×10^8 electrons/cm^2 sec ster with energies from 40 eV to 5 keV and was isotropic. An electron flux with energies of $\geqslant$30–35 keV, amounting to 2×10^5/cm^2 sec ster, was evaluated with the use of absorbers. A soft boundary of electron fluxes below 8 keV (possibly 1 keV or softer) was observed which gave an estimated energy flux at daytime of $\leqslant 0{\cdot}6$–$0{\cdot}1$ erg/cm^2 sec ster[8, 17].

In 1963, Krasovskiy and his associates presented results of investigations carried out with Cosmos 3 and 5[77]. The results indicated that protons (820/cm^2 sec) with energies $\geqslant 50$ MeV were not predominant at low altitudes ($\sim$1520 km, 19° S, 10° E). Values for other corpuscular particle fluxes were:

Electrons 100 keV: 7×10^7/cm^2 sec with an energy flux of 10 erg/cm^2 sec, and

Electrons <5 keV: 2×10^8/cm^2 sec with an energy flux of $1{\cdot}6$ erg/cm^2 sec.

In discussing the results of the first 15 Cosmos satellites, Vernov, *et al.*, noted that a sharp distinction must be drawn between the electron components having different energies when studying these zones. The term "outer radiation zone", he suggested, applies to electrons with an energy exceeding 100 keV in the region between $L=3$ to 8. The term "inner radiation zone" applies to protons with energies of tens of hundreds of MeV which are trapped by the geomagnetic field within $L<2$. The spatial region $2<L<3$ is called the gap[147].

A similar classification was made by Tverskoy, who distinguished the following four groups of particles, which differ in terms of composition, energy spectrum, spatial distribution, and variation in time:

1. Soft electrons with energies from less than 50 keV up to 100 keV within the magnetosphere.
2. Heavy electrons in the outer zone with energies of >200 keV which form a sharp peak at $4<L<6$ and are located between $L<3$ and >7.
3. Protons of the inner zone with energies greater than 40–50 MeV at $L \lesssim 2{\cdot}5$.
4. Protons of the outer zone with energies of 100 keV—1 MeV, but also including those with energies up to 20 MeV and partially up to 40 MeV[123].

The Soviets have also reported the following interesting findings from Electron 1 and 2 experiments [135]:

1. A zone of artificially injected electrons had a maximum intensity at $L=1\cdot35$ in February 1964. The flux of electrons having an energy of >2 MeV at the maximum was 1×10^7/cm^2 sec ster.
2. At the maximum of the inner zone, the mean directional flux of protons having an energy of 45–70 MeV is $\sim1\cdot5\times10^3$/cm^2 sec ster for $L=1\cdot45$. For $L=2\cdot2$ there is a change in the intregal spectrum for energies >50 MeV. The spectrum becomes harder in this energy region and can apparently be explained by the theory of albedo neutrons.
3. The mean directional flux of protons having an energy >2 MeV is $4\cdot5\times10^5$/cm^2 sec ster in the upper equatorial plane for $L=2\cdot8$.
4. A zone of high-energy electrons was discovered at $L=2\cdot75$. The mean directional flux of electrons, having an energy of >6 MeV, is $\sim10^2$/cm^2 sec ster.
5. A minimum was observed in the distribution of electrons having an energy of >150 keV in the $3<L<4$ region.
6. The maximum of the outer zone, both on the nocturnal and on the morning side, was located on the average at $L=4\cdot8$. The mean directional flux of electrons, having an energy of >70 keV at the outer zone maximum, was $\sim5\times10^6$/cm^2 sec ster.

Ion Fluxes

In addition to the study of electron densities employing various radiowave measurement techniques aboard rockets and satellites, which are discussed in Chapter 9, the Soviets have almost routinely employed ion traps and, to a lesser extent, Langmuir probes to investigate the positive ion or proton concentration in the upper atmosphere and interplanetary space. These same traps have also been successfully employed by the Soviets in various satellite and space probe schemes to provide data on solar and cosmic radiation conditions in deep space. A complete description of various types of traps as well as individual results reported to date are noted in Table 10.2.

Major findings associated with Soviet ion-trap measurements include the following:

1. Negative ions in the F region of the ionosphere were present only in insignificant quantities. Measured electron densities were approximately equal to the positive ion concentrations[48].

2. The ionization of the neutral air particles by the satellite's motion does not significantly influence the results measured by the ion trap[48].
3. Luna 2 ion traps indicated the existence of a region of constant density (approximately 800 ions or electrons per cm^3) from an altitude of 2000 to 12 000 km[48].
4. Ion-trap measurements made near maximum solar activity, indicated that the earth was surrounded by an ionized gas envelope with a thickness up to four earth radii, with ion concentrations of the order of $10^3/cm^3$, which considerably exceeded the background concentration in the interplanetary medium[48, 61].
5. At distances from 55 000 to 75 000 km from the earth only negative ion currents were recorded by all four traps of Luna 2, which demonstrated the existence of electrons with sufficient energies (>200 eV) to overcome the inner grid retardation field of the trap and reach the collector. Since corpuscular particle sensing instruments with a threshold of about 20 keV were also installed in the lunar probe but did not register any particles, it was concluded by Gringauz that the upper limit of these electrons was at 20 keV[47, 61].
6. The evaluation of the flux density recorded at altitudes from 50 000 to 75 000 km showed $1{\cdot}5$ to 4×10^8 electrons/cm^2 sec. These measurements gave Gringauz reason to postulate the existence of an outermost zone of charged particles surrounding the earth beyond the well known Van Allen zones[47].
7. A correlation has been established between solar corpuscular stream intensity variations and geomagnetic field variations during magnetic storms[51].
8. Electron 2 results provided evidence of the existence of a soft component of electron fluxes ($<1{\cdot}0$ MeV) in the outer radiation zone whose variability in time is much greater than the variability of high-energy particle fluxes[54].
9. A possibly permanent flux of (1 keV electrons) $\sim 2 \times 10^8$ particles/cm^2 sec keV was recorded during the flights of Cosmos 12 and 15 in the vicinity of Southern New Zealand. This enhanced flux occurred in the region of the maximum southern isochasm. Later measurements indicated en-

hanced fluxes of up to $\sim 10^7$ particles/cm^2 sec keV, and ions were observed in the equatorial regions of the Pacific Ocean. An association of these fluxes with solar flares were mentioned as possible explanation for the enhancement[143].

10. Regions of increased intensity of up to 10^9 electrons/cm^2 sec keV with energies from 0·1 to 10 keV were detected by the Electron 2 satellite at distances of 8 to 4 earth radii at southern latitudes $>40°$. An appearance of sporadic fluxes of electrons with an energy of 1 keV at distances of more than 9 earth radii were also recorded. Ion fluxes of energies from 0·1 to 10 keV inside the magnetosphere did not exceed 5×10^7 ions/cm^2 sec which also was the sensitivity threshold of the instrument. This result substantiates data from ion measurements carried out by Explorer 10, 12, and 18 satellites inside the magnetosphere[144–145].

RELATED RESEARCH

Numerous studies of the influence of radiation on the operation of satellite instrumentation systems and evaluations of the lifetime of satellite equipment, as well as investigations about the radiation hazards to astronauts during earth orbital flights, have been made by the Soviets since 1960. The analysis of initial data obtained with the instruments on board Spaceship-Satellite 2 enabled Soviet investigators to delineate the position of the radiation zones and to estimate the amount of radiation hazard present[108–109].

Savenko and associates reported that data of a scintillation counter (mounted outside the satellite) indicated that the radiation in the zones was anisotropic and that the energy flux under a layer of matter of 2 mg/cm^2 was 10^{10} eV/cm^2 sec. The mean energy released per quantum in the internal scintillation detector was computed to be 2×10^5 eV, and the flux of electrons in the outer zone to be 5×10^4 particles/cm^2 sec, with an average energy of 200 keV. The radiation dosage absorbed inside Spaceship-Satellite 2 was determined by dividing the amount of energy liberated in the sodium iodide crystal of the scintillation counter by the weight of the crystal. A dosage value of 7 mrad/day was obtained, which in terms of the RBE (relative biological effect) of the cosmic charged particles amounted to 5 mrad/day. This value was considered relatively safe along trajectories of Spaceship-Satellite 2 (approximately in a circular orbit at an altitude of 320 km) when the sun was quiet.

TABLE 10.2

Results of Solar and Cosmic Radiation Studies Aboard Soviet Rockets, Satellites, and Space Probes (1957–1966)

Vehicle/Launch Date	Height Perigee-Apogee, km	Instrument Employed	Type and Range of Measurement	Major Results Reported	References
		Electromagnetic Radiation Measurements (Ultraviolet, X-Ray and Gamma-Ray Radiation Intensity)			
Geophysical Rockets					
July 21 1959	105	End-window photon geiger counters type SBT-9 (two groups of two counters) Mica window: 4 mm diameter, 1·6 mg/cm^2 thick Two counters directed towards the sun (one counter had magnetic safeguard effective for particle energies up to 15–20 keV) Aluminum 2 μ thick was sprayed on mica to suppress low sensitivity to UV radiation One counter of each group was a control counter sensing X-ray photons and charged particles	Soft X-rays Up to 10Å	During the morning flight the maximum count of 80/sec was obtained at an altitude of 105 km. The maximum count was 120/sec at 105 km during the evening flight. One control counter, turned 15° from the direction of the sun, registered the level of the cosmic background: 3 counts/sec.	Mandel'shtam, *et al.* (1961)[97]
1959–1960 February 15 1961	Up to 100	Thermoluminescent phosphor sensor ($CaSO_4$[Mn]) with FEU-35 photomultiplier Phosphor stored energy which is reradiated (max at 5000 Å) when heated Total phosphor area: 0·8 cm^2 FEU-35 photomultiplier with antimony-cesium photo-cathode was calibrated during flight with ZnS(Cu) phosphor activated by a C^{14} isotope	UV- and X-ray radiation of the sun Lyman alpha radiation 1–1300 Å	Measurements characterize the intensity of the ionizing radiation in the atmosphere. The results indicated too high measurement data which were explained as being caused by electron fluxes with an intensity of 0·06–0·07 erg/cm^2 sec at an altitude of 95 km. Measurements were made during eclipse.	Kazachevskaya, *et al.* (1963)[71–72]
June 6 and 18 1963	Up to 500	Ionization chamber type KFL-2 with lithium-fluoride window Chamber filled with nitric oxide Mouted on selforienting container	UV radiation 1100–1345 Å Lyman alpha radiation	The absorption of Lyman alpha radiation was investigated during the descent of the instrument container.	Katyushina (1965)[69]

Vehicle/Launch Date	Height Perigee-Apogee, km	Instrument Employed	Type and Range of Measurement	Major Results Reported	References
October and December 1963	Up to 500 <200	End-window photon geiger counter type SFM-1 (two) Lithium-fluoride window: 1·5 mm diameter Calcium fluoride window Geiger counter filled with nitric oxide	UV radiation 1050–1340 Å Lyman alpha radiation	Intensities of scattered radiation versus altitude for different angular distances from the sun were obtained. For some altitude intervals (below 200 km) the intensities of the triplet OI (1302, 1304, and 1308 Å) were obtained for different zenith distances with an altitude of the sun of 3°.	Kaplan, *et al.* (1964)[65], (1965)[66], Kasatkin (1964)[67] Katyushina (1965)[68] Katyushina and Kurt (1965)[70]
Artificial Earth Satellites					
Sputnik 2/ November 3 1957		Short-wave radiation sensor (three) Secondary emission photo-multiplier insensitive to radiation above 2400 Å (beryllium bronze photo-cathode) Step-rotating disk with open aperture and filters: beryllium foil 200 μ thick aluminum foil 5 μ thick (Beryllium foil transmission subtracted) Polyethylene foil 3 μ thick (Subtracting transmission radiation from 4 μ beryllium foil) Lithium fluoride filters The three radiation sensors were mounted 120° apart Photosensitive resistor (two) 120° view and 60° view Control for anode and heater of photomultiplier	Ultraviolet and soft X-ray radiation (visible light Lyman alpha radiation) 1–10 Å 10–22 Å 44–120 Å 1216 Å (Lyman alpha)	Solar electromagnetic radiation in the soft X-ray and ultraviolet region was measured (1 to 120 and 1216 Å).	Aleksandrov and Fedorov (1959)[4] Mandel'shtam and Yefremov (1957)[100]
Spaceship-Satellite 2/ August 19 1960		End-window photon geiger counter (six) Beryllium window: 7 mm diameter, 0·1 mm thick Magnetic shielding of window against low-energy electrons	Soft X-rays 2–10 Å	On the sunny side of the orbit the counters registered counting rates of the order of several thousand pulses per second On the shadow side of the orbit the counting rate decreased to several tens of pulses per second characterizing the cosmic background. The X-ray flux was $7{\cdot}6 \times 10^{-4}$ ergs/cm^2 in the 2–10 Å region.	Mandel'shtam, *et al.* (1961)[98], (1962)[99] Vasil'yev, *et al.* (1961)[128], (1963)[129]

TABLE 10.2 (Continued)

Results of Solar and Cosmic Radiation Studies Aboard Soviet Rockets, Satellites, and Space Probes (1957-1966)

Vehicle/Launch Date	Height Perigee-Apogee, km	Instrument Employed	Type and Range of Measurement	Major Results Reported	References
		Short-wave radiation sensors (three sets of two sensors)	UV and soft X-ray radiation	The flux of radiation in the region 44–110 Å [$(CH)_x$-filter] was constant within an accuracy of ±8% and corresponded to $1{\cdot}5 \times 10^7$ impulses/cm² sec. The flux in the region 8–21 Å (aluminum filter) was constant ($6{\cdot}2 \times 10^4$ impulses/cm² sec) and increased 3·2 times or by 63% at times. In the region less than 8 Å (beryllium filter) the radiation of the quiet sun was very small and often did not stand out above the background of radiation of nonsolar origin. In times of increased activity of the sun the stream in the region 5–10 Å (beryllium filter) increased eleven-fold in comparison with the background registered up to this time. In the region 1·4–3 Å (Cu filter) only the background from radiation of nonsolar origin was registered. The flux of radiation in the chromospheric line of hydrogen L_α was 5–6 erg/cm² sec.	Goltsoff (1960)[34], Yefremov, *et al.* (1961)[164–165] (1963)[166]
		Two secondary emission photomultipliers (BeO and SrF_2 photocathode) with filters:	Lyman alpha radiation		
		Copper 16 mg/cm², 0·20 cm²	1·4–3 Å		
		Beryllium 10 mg/cm², 0·20 cm²	1–10 Å		
		Aluminum 1·4 mg/cm², 0·11 cm²	8–21 Å		
		Polystyrene 0·5 mg/cm², $0{\cdot}18 \times 10^{-3}$ cm²	44–100 Å		
		Lithium-fluoride 0·5 mm thick, 2×10^{-5} cm²	1216 Å (Lyman alpha)		
		Calcium-fluoride 0·5 mm thick, 2×10^{-5} cm²	Attenuates Lyman alpha >1500 Å		
		Quartz 0·5 mm thick, $1{\cdot}8 \times 10^{-4}$ cm²			
		Outside container			
		Optical switch directed at the sun			
		Off-on control of short-wave radiation sensor			
Spaceship-Satellite 3 December 1 1960		End-window photon geiger counters (three sets of two counters)	Soft X-rays and UV-radiation	The maximum counting rate of the counters with beryllium windows was $2{\cdot}4 \times 10^{-4}$ erg/cm² sec in the 2 to 10 Å region. On December 1 1960 the maximum counting rate of each counter set was 280, 175, and 1900 pulses/sec.	Mandel'shtam, et al. (1961)[98], (1962)[99] Vasil'yev, *et al.* (1961)[128], (1963)[129]
		Mica window: 4 mm diameter 1·6 mg/cm², covered by two layers of aluminum foil 5 μ thick for two counters (counters were always directed at the sun)	2–10 Å		
		Two similar counters directed perpendicular to the sun direction. Tantalum plate mounted 45° with respect to the direction of sunlight in front of counters.	X-rays generated by electrons striking a tantalum plate		

Vehicle/Launch Date	Height Perigee-Apogee, km	Instrument Employed	Type and Range of Measurement	Major Results Reported	References
		Two counters with beryllium window 1 mm thick Magnetic shielding of counters	2–10 Å		
	200–210	Concave diffraction grating spectrometer (two-channel type) Concave grating: 1200 lines/mm Radius of curvature: 500 mm Scanning slot (0·2 × 7 mm) moved at 0·05 mm/sec, operation cycle: 177 sec Dispersion: 16 Å/mm Open-type secondary emission photomultiplier	1700–2000 Å Far ultraviolet spectrum near 303·8 Å	The amplitude average of the obtained maxima were 1·6 pulses/sec for 303·8 Å. The energy flux of the 303·8 Å wavelength at altitudes from 200 to 210 km for the solar zenith angle of 50° was calculated to be 0·5 erg/cm² sec. Satellite was not oriented to the sun, however, tracking system for sun was employed.	Bruns and Prokof'yev (1961)[20–21] Vasil'yev and Shapov (1961)[130]
Electron 2/ January 30 1964 and Electron4/ July 11 1964		End-window photon geiger counter block (two blocks) Each block consisted of: Counter with 25 mg/cm² beryllium window Counter with 2·7 mg/cm² aluminum window Counter with aluminum window and gold foil Counter with aluminum window and silver foil Silicon phototransducer for position control relative to sun direction	Soft X-ray and UV-radiation 2–8 Å 8–18 Å Protons Electrons	The counters registered solar X-ray emission fluxes of $3{\cdot}3 \times 10^{-4}$ erg/cm² sec in the 2–8 Å region, and $2{\cdot}4 \times 10^{-3}$ erg/cm² sec in the 8–18 Å region. The sensitivity threshold was 1×10^{-5} erg/cm² sec for the Be-window counter and $1{\cdot}5 \times 10^{-4}$ erg/cm² sec for the aluminum-window counter. The pulse counting rate was registered with an accuracy of ±15 to 17%.	Kurnosova, *et al.* (1966)[89] Vasil'yev (1966)[127]
Proton 1/ July 16 1965 and Proton 2/ November 2 1965		High-energy gamma ray counter type GG-1 consisting of: Fast scintillator: nine plastic plates 3 mm thick and eight CsI(Tl) plates 2 mm thick with photomultiplier Cherenkov counter with lead glass radiator Scintillator hood with photomultipliers	Gamma quanta $10^8 - 3 \times 10^9$ eV	A flux of gamma quanta with energies exceeding 5 MeV of approximately 2×10^{-3}/cm² sec ster was registered.	Grigorov, *et al.* (1965)[41–42] (1966)[37]

TABLE 10.2 (Continued)

Results of Solar and Cosmic Radiation Studies Aboard Soviet Rockets, Satellites, and Space Probes (1957–1966)

Vehicle/Launch Date	Height Perigee-Apogee, km	Instrument Employed	Type and Range of Measurement	Major Results Reported	References
Lunar and Interplanetary Probes					
Zond 1/April 1 1964	37 000 to 47 000	End-window photon geiger counter (two) Lithium-fluoride window Window Area coverage: 0·7 g/cm² Field of vision: $1{\cdot}15 \times 10^{-2}$ ster Lead shielding: 3·5 g/cm² Calcium-fluoride window in front of second counter Geiger counter filled with nitric oxide	Scattered UV radiation 1225–1350 Å Recording rate range 3 to 10^4 pulses/sec	At a distance of 37 000 to 47 000 km, four complete photometer cross sections of the hydrogen corona of earth were obtained. No conclusions were possible yet regarding the presence of an extended hydrogen "tail" at the earth.	Kurt (1965)[92]
Venus 2/November 12 1965	Up to 20 R_E	End-window photon geiger counter	Scattered UV radiation 1225–1350 Å	A neutral hydrogen distribution has been obtained up to 20 earth radii. Data have been registered for three-month measurements of Lyman-alpha radiation in the interplanetary medium. Measurements have demonstrated that brightness is approximately constant and amounts to 10^{-4} erg/cm² sec ster. Variations of the diurnal values do not exceed 20 per cent.	Blagonravov (1966)[15] Kurt (1966)[93]
Venus 3/November 16 1965		End-window photon geiger counter	Scattered UV radiation 1225–1350 Å		
		Charged-Particle Radiation Measurements (Relativistic Particle Radiation Intensity)			
Geophysical Rockets					
July 2 1958	Up to 210 23	Ionization chamber Stainless steel sphere with wall thickness: 0·4 mm diameter: 240 mm Hollow thin-walled sphere electrode, inside chamber Electrode diameter: 28 mm Chamber filled with argon Pressurized to ~100 psig Halogen gas-discharge counter Type STS-6 (two)	Ionization level and height-intensity relation of primary cosmic radiation	Count rates registered were 20 to 30% higher than obtained by other experimenters. High rates were explained by showers produced in structure of rocket. Ionization maximum was measured at 23 km altitude which coincided with data related to global intensity of particle flux as a function of altitude. Flux intensities measured at 100, 150, and 200 km altitude were (after albedo corrections): 0·126, 0·120, and 0·123 particles/cm² sec ster, respectively.	Shafer and Yarygin (1960)[118]

Vehicle/Launch Date	Height Perigee-Apogee, km	Instrument Employed	Type and Range of Measurement	Major Results Reported	References
		Counter dimensions: 18 mm diameter, 150 mm long Stainless steel wall: 1 mm thick Operating voltage: 380–400 V Effective area: 15 cm^2 One counter placed vertically above ionization chamber Eight per cent of upper portion shielded with >50 g/cm^2 material Other counter, unshielded, placed horizontally at the side of the chamber Lower side of instrumentation was shielded by the body of the rocket by a solid angle of $0{\cdot}16\pi$ ster.			
1962	Up to 500	Ionization chambers as above	East-west asymmetry, height-intensity relation of primary cosmic radiation	Flux intensities measured at 100, 200, 300, 400, and 500 km altitude were: 0·202, 0·180, 0·207, 0·210, and 0·204 particles/cm^2 sec ster, respectively, in the western plane.	Shafer, *et al.* (1964)[116–117]
1963	Up to 500	Group of vertical gas-discharge counter telescope Group of 60° inclined gas-discharge counter telescope Gas-discharge counter telescope with double coincidences and small separation Unshielded solitary gas-discharge counter Solitary gas-discharge counter shielded with lead 10 mm thick Total of 11 gas-discharge counters type STS-6 Counter dimension: 18 mm diameter, 150 mm long (effective area) Overall instrumentation shielding: no shielding in the upper hemisphere except instrument housing sides shielded by duraluminum 2 mm thick Shielding of the lower hemisphere: $\sim$300 g/cm^2 over solid angle of π ster	Albedo particle intensity determination	0·148, 0·148, 0·142, 0·166, and 0·164 particles/cm^2 sec ster, respectively, in the eastern plane, 0·120, 0·121, 0·133, 0·140, and 0·142 particles/cm^2 sec ster, respectively, in the zenith direction. The albedo particle intensity was at 100, 200, 300, 400, and 500 km altitude 0·056, 0·047, 0·047, 0·053, and 0·046, particles/cm^2 sec ster, respectively. Albedo particle contributed to the east-west measurements but not to the vertical telescope readings.	

TABLE 10.2 (Continued)

Results of Solar and Cosmic Radiation Studies Aboard Soviet Rockets, Satellites, and Space Probes (1957–1966)

Vehicle/Launch Date	Height Perigee-Apogee, km	Instrument Employed	Type and Range of Measurement	Major Results Reported	References
Artificial Earth Satellites					
Sputnik 3/ May 15 1958		Cherenkov counter (integral type) Plexiglass detector with photo-multiplier Detector dimensions: 26 mm diameter, 26 mm high Shielding of detector: Front: aluminum 2 mm thick Sides: aluminum 3 mm thick (0·8 g/cm²) Inside container with aluminum shielding of 1 g/cm²	Nuclear component of cosmic radiation	The counter registered nuclei with charges of $Z \geqslant 15$ and $Z \geqslant 30$. Only one nucleus with $Z \sim 30$–40 was recorded in nine days.	Pravda (1958)[1] Ginzburg, *et al.* (1962)[31] Vakulov, *et al.* (1965)[124]
Spaceship-Satellite 2/ August 19 1960		Cherenkov counter (integral type) Plexiglass detector with photo-multiplier Detector dimensions: 26 mm diameter, 26 mm high Shielding of detector: Front: aluminum 2 mm thick Sides: aluminum 3 mm thick (0·8 g/cm²) Inside container with aluminum shielding of 2 g/cm²	Nuclear component of cosmic radiation	The counter registered nuclei with charges of $Z \geqslant 5$, $Z \geqslant 15$, and $Z \geqslant 34$.	Ginzburg, *et al.* (1961)[29], (1962)[30–31], (1964)[32] Kurnosova, *et al.* (1962)[87, 91]
		Cherenkov counter (differential type) Plexiglass detector with photo-multiplier (two counters) Detector dimensions: 75 mm diameter, 30 mm high A halogen gas-discharge counter telescope opened photomultiplier channel only when incoming nuclei particles approach through a specifically defined solid angle Inside container with aluminum shielding of 1 g/cm²	Nuclei of atomic rumber 2 to 8	The counters registered nuclei with charges of $Z \geqslant 2$. Nuclei of lithium, beryllium, and boron were detected by the Cherenkov counters. Data of flux intensity were not published.	Ginzburg, et al. (1962)[30]

Vehicle/Launch Date	Height Perigee-Apogee, km	Instrument Employed	Type and Range of Measurement	Major Results Reported	References
		Nuclear photoemulsion stack 20 unpacked pellicles 300 μ thick each, 90 mm in diameter (Processing of emulsions exposed to relativistic particles was carried out aboard)	Energy, charge and mass of primary cosmic radiation	Final processing of emulsions was made on the ground after recovery of satellite. Tracks of relativistic nuclei of cosmic radiation were obtained and cases of interactions could be seen in microphotographs of the nuclear emulsions.	Veprik, *et al.* (1961)[131]
		Nuclear photoemulsion stack 489 layers of Br-type emulsion, 10×10 cm^2, 400 μ thick each (emulsion volume: 656 cm^3)	Nuclei of atomic number 2 to 8	Primary-alpha particles were practically unobserved. Particles with charges of $Z \geqslant 2$ were registered.	Grigorov, *et al.* (1961)[45], (1963)[46]
		Nuclear photoemulsion stack Ten layers of 20 bundles of NIKFI-Br type emulsions, 100 μ thick, 10×10 cm^2	Nuclei of atomic number 2 to 8	Nuclei with a charge of $Z = 3$ (Li), 4 (Be), 5 (B), 6 (C), 7 (N), and 8 (O) were registered in photoemulsions during microscopic evaluations.	Alekseyeva, *et al.* (1962)[6]
Spaceship-Satellite 3/ December 1 1960		Cherenkov counter (integral type) Plexiglass detector with photomultiplier Detector dimensions: 26 mm diameter, 26 mm high Shielding of detector: Front: aluminum 2 mm thick Sides: aluminum 3 mm thick (0·8 g/cm^2) Inside container with aluminum shielding of 2 g/cm^2	Nuclear component of cosmic radiation	The counter registered nuclei with charges of $Z \geqslant 5$, $Z \geqslant 12$–14, $Z \geqslant 31$–34, and $Z \geqslant 34$. A flux of nuclei with $Z \geqslant 5$ of 2·3 particles/m^2 sec ster and 9·7 particles/m^2 sec ster was measured at geomagnetic latitudes 0–10° and 60–70°, respectively One nucleus per day with $Z \geqslant 31$–34 was recorded during time of flight. Nuclei with $Z \geqslant 34$ were not noted.	Ginzburg, *et al.* (1961)[29], (1962)[30–31], (1964)[32] Kurnosova, *et al.* (1962)[87, 91]
		Cherenkov counter (differential type) Plexiglass detector with photomultiplier (two counters) Detector dimensions: 75 mm diameter, 30 mm high A halogen gas-discharge counter telescope opened photomultiplier channel only when incoming nuclei particles approached through a specifically defined solid angle. Inside container with aluminum shielding of 1 g/cm^2	Nuclei of atomic number 2 to 8	The counter registered nuclei with charges of $Z \geqslant 2$. A flux of nuclei with $Z \geqslant 2$ of 26 particles/m^2 sec ster and 155 particles/m^2 sec ster was recorded at geomagnetic latitudes 0–10° and 60–70°, respectively. A maximum flux of 22·7 particles/m^2 sec ster was recorded of nuclei with $Z \geqslant 4$.	Kurnosova, *et al.* (1962)[97]

TABLE 10.2 (Continued)

Results of Solar and Cosmic Radiation Studies Aboard Soviet Rockets, Satellites, and Space Probes (1957–1966)

Vehicle/Launch Date	Height Perigee-Apogee, km	Instrument Employed	Type and Range of Measurement	Major Results Reported	References
Cosmos 6/June 30 1962 Cosmos 19/August 6 1964	270–355 270–518	Ionization chambers Stainless steel sphere with wall thickness: 0·4 mm diameter: 240 mm Hollow thin-walled sphere electrode inside chamber Electrode diameter: 28 mm Chamber filled with argon pressurized to ~100 psig Unshielded solitary gas-discharge counters Gas-discharge counters shielded with lead 15 mm thick Circular telescope of six gas-discharge counters with triple coincidence All gas-discharge counters of Type STS-5 Counter dimensions: 10 mm diameter, 53 mm long (effective area)	Ionization level and height—intensity relation of primary cosmic radiation Albedo particle intensity determination	Data about the energy of the albedo particles were obtained from an analysis of measurement results provided by the ionization chamber. The specific ionization of the measured radiation did not indicate variations due to altitude or time The specific ionization amounted to 200–220 ion pairs/cm at 760 mm Hg in argon. Results about intensities of cosmic radiation registered by the spherical ionization chamber, shielded and unshielded counters and counter telescope have not been published yet for Cosmos 6 and 19 experiments.	Shafer, *et al.* (1964)[117]
Electron 2/ January 30 1964		Cherenkov counter (integral type) Plexiglass detector with FEU-35 photomultiplier Detector dimensions: 26 mm diameter, 26 mm high Inside container with aluminum shielding of 1·5 g/cm^2	Nuclear component of cosmic radiation Cosmic ray nuclei $E_{kin} \geqslant 600$ MeV/nucleon	The counter registered nuclei with charges of $Z \geqslant 2$, $Z \geqslant 5$, and $Z \geqslant 15$. A flux of nuclei with $Z \geqslant 5$ of 18·5 particles/m^2 sec ster, and a flux of nuclei with $Z \geqslant 15$ of 0·69 particles/m^2 sec ster were measured	Blokh, *et al.* (1965)[16] Kurnosova, *et al*, (1965)[95]
		Cherenkov counter (integral type) Plexiglass detector with FEU-35 photomultiplier Detector dimensions: 26 mm diameter, 26 mm high	Nuclear component of cosmic radiation Cosmic ray nuclei $E_{kin} \geqslant 600$ MeV/nucleon	The counter registered nuclei with charges of $Z \geqslant 2$ and $Z \geqslant 30$. A flux of nuclei with $Z \geqslant 2$ of 343 particles/m^2 sec ster was measured	Blokh, *et al.* (1965)[16] Kurnosova, *et al.* (1965)[95]

Vehicle/Launch Date	Height Perigee-Apogee, km	Instrument Employed	Type and Range of Measurement	Major Results Reported	References
		A scintillation counter, connected to the detector of the Cherenkov counter, formed a wide-angle telescope with the counter Outside container with aluminum shielding of $\sim 0{\cdot}6$ g/cm^2			
Cosmos 25/ February 27 1964		Ionization chamber Gas-discharge counters Shielded and unshielded Scintillation counter array Shielded	Threshold energy 4 BeV	The ionization chamber readings characterized the ionizing power of particles more than it did their intensity. Chamber reading increased when particle intensity, recorded by gas-discharge counter, decreased. The authors assumed that an additional radiation counting at low latitudes by the shielded counter could be attributed to either the recording of electron-positron pairs developed by gamma quanta, the effect of X-ray on the shield of the counter or solar X-ray radiation during atmospheric flares. The information obtained from the scintillation counter array and gas-discharge counters indicated a 30-day lag in the variations in cosmic ray intensity in respect to the variations in solar activity.	Kuzhevskiy, *et al.* (1966)[94]
Proton 1/July 16 1965 Proton 2/ November 2 1965		Ionization spectrometer type SEZ-14 consisting of: Ionization calorimeter with carbon and polyethylene filter, 12 photomultipliers, plastic scintillators, two proportional counters, two interaction detectors and lead and steel plate absorbers.	Energy spectrum of primary particles, nuclear mass of cosmic radiation in the energy range from 10^{10} to 10^{14} eV, and effective cross-section determinations	A spectrum of cosmic ray primary particles was measured irrespective of their charge in the energy range 10^{10}–10^{14} eV. The primary proton energy spectrum was measured in the energy range 10^{10}–10^{12} eV. Measurements were made of the effective cross-section of inelastic interaction with carbon nuclei in the energy range $\sim 5\times10^{9}$–10^{12} eV. Values from 200 to 400 millibarns were obtained from measurements from Proton 1.	Blagonravov (1966)[15] Grigorov, *et al.* (1965)[41–42, 44], (1966)[38, 43] Pomanskiy (1966)[105]
		High-energy spectrometer type SEZ-12 consisting of: Two-scintillator telescope, gas Cherenkov counter, six photomultipliers, two anti-shower scintillators, scintillator for energy detector, lead and steel absorbers.	Electrons 0·3–10 BeV	A large flux of electrons with energies $\geqslant 300$ MeV in the region of the geomagnetic equator has been detected ($2{\cdot}5\times10^{-2}$/cm^2 sec ster). This flux exceeded the primary proton flux in the same region by a factor of 2.	

TABLE 10.2 (Continued)

Results of Solar and Cosmic Radiation Studies Aboard Soviet Rockets, Satellites, and Space Probes, (1957–1966)

Vehicle/Launch Date	Height Perigee-Apogee, km	Instrument Employed	Type and Range of Measurement	Major Results Reported	References
	500	Nuclear charge spectrometer type SEZ-1 consisting of: Cherenkov counter with plexiglass radiator and FEU-49 photomultiplier Radiator dimensions: 165 mm Diameter: 30 mm high Two plastic scintillators with FEU-13 photomultipliers Thickness of plastic scintillator: 15 mm Geometric factor: 133 ± 6 cm^2 ster	Chemical composition and energy spectrum of solar and cosmic rays in the energy range from 0·4 to 30 GeV/nucleon Proton 1: Protons <100 MeV Proton 2: Protons >100 MeV, Electrons >20 MeV	Energy spectra of different groups of nuclei were measured by geomagnetic effects in an energy range up to 20 GeV/nucleon. The intensity of nuclei of C, N, O, and heavier nuclei measured by Proton 1 instruments differed from earlier measurements with other satellites. Proton 1 and 2 satellites repeatedly passed over the Brazilian anomaly at an altitude of 500 km. A ratio of concentrations of protons with energies >100 MeV and protons >400 MeV varied from $18 \pm 0·5$ to $8·5 \pm 0·2$. Particles with $Z=1$ (protons) and $Z=2$ (helium nuclei) were registered in the equatorial region	Volodichev, *et al.* (1966)[162–163]
Lunar and Interplanetary Probes					
Luna 1/ January 2 1959	27 000	Cherenkov counter Plexiglass detector with photomultiplier Inside container	Nuclear component of cosmic radiation	The counter registered nuclei with charges of $Z \geqslant 2$ and $Z \geqslant 15$. A maximum of 130 counts/sec was obtained at a distance of 27 000 km from the center of the earth.	Dragun, *et al.* (1961)[24] Ginzburg, *et al.* (1962)[30] Kurnosova, *et al.* (1960)[96] Vakulov, *et al.* (1965)[124]
Luna 2/ September 12 1959	17 000 80 000 to 160 000	Cherenkov counter (integral type) Plexiglass detector with photomultiplier Detector dimensions: 26 mm diameter, 26 mm high Shielding of detector: Front: aluminum 2 mm thick Sides: aluminum 3 mm thick (0·8 g/cm^2) Inside container with aluminum shielding of 1 g/cm^2	Nuclear component of cosmic radiation	The counter registered nuclei with charges of $Z \geqslant 5$ and $Z \geqslant 15$. The number of nuclei recorded during entire time of flight was 3000 particles with $Z \geqslant 5$ and 100 particles with $Z \geqslant 15$. A maximum of 780 counts/sec was obtained at a distance of 17 000 km from the center of the earth. Mean values of ~ 2 particles/min for $Z \geqslant 5$ and $\sim 0·1$ particle/min for $Z \geqslant 15$ were recorded at distances from 80 000 to 160 000 km from the center of the earth.	Ginzburg, *et al.* (1962)[31], (1964)[32] Kurnosova, *et al.* (1960)[96] Vakulov, *et al.* (1965)[124]

Vehicle/Launch Date	Height Perigee-Apogee, km	Instrument Employed	Type and Range of Measurement	Major Results Reported	References
	10 000 to 30 000	Cherenkov counter (integral type) Plexiglass detector with photo-multiplier Detector dimensions: 26 mm diameter, 26 mm high Shielding of detector: Front: aluminum 2 mm thick Sides: aluminum 3 mm thick (0·8 g/cm^2) Outside container	Nuclei with atomic number 2 to 8	The counter registered nuclei with charges of $Z \geqslant 2$. The number of nuclei recorded during entire time of flight was 30 000. At 10 000 to 30 000 km from the center of the earth a large increase in the counting rate was recorded. The particle flux recorded during the time of flight was 150 particles/m^2 sec ster.	Ginzburg, *et al.* (1962)[31], (1964)[32] Kurnosova, *et al.* (1960)[86] Vakulov, *et al.* (1965)[124]
Luna 3/ October 4 1959		Cherenkov counter (integral type) Plexiglass detector with photo-multiplier Detector dimensions: 26 mm diameter, 26 mm high Shielding of detector: Front: aluminum 2 mm thick Sides: aluminum 3 mm thick (0·8 g/cm^2) Inside container with aluminum shielding of 1 g/cm^2	Nuclear component of cosmic radiation	The counter registered nuclei with charges of $Z \geqslant 2$ $Z, \geqslant 15$, and $Z \geqslant 28$. The particle flux recorded during the time of flight was 150 particles/m^2 sec ster for $Z \geqslant 2$, and 0·4–0·5 particles/m^2 sec ster for $Z \geqslant 15$ which was comparable with data obtained with Luna 2 counters. A flux of 0·5 particles/m^2 sec ster was calculated for $Z \geqslant 28$.	Ginzburg, *et al.* (1962)[31], (1964)[32] Kurnosova, *et al.* (1960)[86], (1961)[90] Vakulov, *et al.* (1965)[124]
Mars 1/ November 30 1962		Cherenkov counter (integral type) Plexiglass detector with photo-multiplier	Nuclei with atomic number 2 to 8	The counter registered nuclei with charges of $Z \geqslant 2$. The particle flux recorded during the time of flight was 333 particles/m^2 sec.	Kurnosova, *et al.* (1965)[95]
		Charged-Particle Radiation Measurements (*Corpuscular Radiation Intensity*)			
Geophysical Rockets					
July 21 1959	Up to 105	Scintillation counter w th fluorescent screen and photomultiplier Screen material: ZnS(Ag) 0·18 mg/cm^2 Aluminum foil: 2 mg/cm^2	Electron <40 keV	The average electron energy estimated was equal to 10^{-2} erg/cm^2 sec ster at altitudes from 80 to 105 km.	Antonova (1961)[7]

TABLE 10.2 (Continued)

Results of Solar and Cosmic Radiation Studies Aboard Soviet Rockets, Satellites, and Space Probes (1957–1966)

Vehicle/Launch Date	Height Perigee-Apogee, km	Instrument Employed	Type and Range of Measurement	Major Results Reported	References
October 18 1962	180–500	Scintillation counter with filters and open-type secondary photomultiplier	Electrons	The total intensity of the electron stream at altitudes from 200 to 500 km was found to be 5×10^7 electrons/cm^2 sec ster including the softest electrons (<10 keV). A most frequent count was 400 counts/sec which was accurate to 30%.	Antonova (1965)[8] Antonova, *et al.* (1965)[10]
		(SF-type instrument)		Almost all filters gave comparatively high and uniform counts which indicated a recording of high background in addition to useful signals. Background signal level diminished systematically with altitude.	
		Rotating disk with filters:	Energy limit:	The reduced signal registered with filters was below 15–20 counts/sec or 3–5% of the background level. The electron energy value evaluated was 0·1–0·6 erg/cm^2 sec ster.	
		Aluminum foil $1{\cdot}6\times10^{-1}$ cm^2, 6×10^{-3} g/cm^2	60 keV		
		Teflon film $7{\cdot}8\times10^{-3}$ cm^2, $2{\cdot}2\times10^{-3}$ g/cm^2	35 keV		
		Copper foil $1{\cdot}6\times10^{-1}$ cm^2, $2{\cdot}2\times10^{-2}$ g/cm^2	130 keV		
		Aluminum foil $7{\cdot}8\times10^{-3}$ cm^2, $2{\cdot}0\times10^{-3}$ g/cm^2	30 keV		
		Terylene film $1{\cdot}8\times10^{-6}$ cm^2, $1{\cdot}6\times10^{-4}$ g/cm^2	8 keV		
		Beryllium foil $1{\cdot}6\times10^{-1}$ cm^2, $3{\cdot}7\times10^{-2}$ g/cm^2	200 keV		
		Terylene film $1{\cdot}1\times10^{-4}$ cm^2, $6{\cdot}0\times10^{-4}$ g/cm^2	14 keV		
		Aluminum $1{\cdot}9\times10^{-1}$ cm^2, $1{\cdot}3\times10^{-1}$ g/cm^2	500 keV		
		Open windows 2×10^{-7} and $7{\cdot}9\times10^{-5}$ cm^2 (area), 5 μ and 0·1 mm (diameter)			
		Two standard filters (material unknown)			
		Guard rings in front of instrument entry port, with ±12 V applied to rings, prevented entry of charged particles of the ionosphere			
		Instrument was installed on stabilized and orientation monitored container which rotated once in 12 sec (initially)			

Vehicle/Launch Date	Height Perigee-Apogee, km	Instrument Employed	Type and Range of Measurement	Major Results Reported	References
Artificial Earth Satellites					
Sputnik 2/ November 3 1957		Gas-discharge counter (two) Dimensions: 18 mm diameter, 100 mm long Inside satellite with shielding of 10 g/cm^2 of matter		The counters registered short-period variations in cosmic radiation intensity. A high intensity (50% increase) of 73 counts/sec was recorded on November 7 1957 at the 64° N latitude.	Aleksandrov, *et al.* (1961)[5] Vakulov, *et al.* (1965)[124] Vernov (1958)[142] Vernov, *et al.* (1960)[132]
Sputnik 3/ May 15 1958		Scintillation counter with NaI(Tl)-crystal and photomultiplier Crystal Dimensions: 40 mm diameter, 40 mm high Inside container with shielding of 1 g/cm^2 of matter	Threshold energy: 35 keV	A sharp increase in radiation intensity was measured at approximately 60° N latitude, as measured by Sputnik 2 instrumentation. The energy of photons responsible for this increased counting rate was evaluated to be 100 keV. The origination of these photons was explained by Bremsstrahlung caused by the bombardment of the satellite's shell by electrons with an energy of 100 keV and a flux of 10^3–10^4/cm^2 sec ster. The existence of the outer radiation belt was indicated by these measurements.	Chudakov (1958)[22] Vernov, *et al.* (1960)[132] Vernov, *et al.* (1958)[159]
	Up to 1900	Soft-electron detector (two) ZnS fluorescent screen activated by 2 mg/cm^2 Ag (one screen was covered with a foil 0·4 mg/cm^2, the other with foil 0·8 mg/cm^2) photomultiplier, restricting diaphragms in front of screen to limit registration to a solid angle of 1/4 steradian.	Electrons >10 keV	Electron energies between 10 and 40 keV were recorded. The energy of the electrons exceeded 100 erg/cm^2 sec at an altitude of 1900 km.	Aleksandrov, *et al.* (1961)[5] Krasovskiy (1960)[76] Krasovskiy, *et al.* (1959)[80–81], (1960)[82], (1961)[83]
Spaceship-Satellite 2/ August 19 1960		Halogen gas-discharge counter telescope Two groups of counters with areas of 120 and 125 cm^2 Five STS-5 counters Eight STS-2 counters Geometric factor of telescope: $2{\cdot}5 \times 10^{-4}$ m^2 ster. Distance between the two groups of counters, 35·8 cm, was filled with plexiglass (4 g/cm^2)	Protons >60 MeV Electrons >8 MeV	Flux intensities registered in the southern and South Atlantic region (anomalies) were 70 counts/sec. In other geomagnetic latitudes the counting rates measured were 0·83, 1·3, and 2·6 counts/sec (or 1·36, 2·0, and 3·3 particles/cm^2 sec) at latitudes 0, 30, and 60°, respectively.	Ginzburg, *et al.* (1961)[27, 29] Kurnosova, *et al.* (1961)[84]

TABLE 10.2 (Continued)

Results of Solar and Cosmic Radiation Studies Aboard Soviet Rockets, Satellites, and Space Probes (1957–1966)

Vehicle/Launch Date	Height Perigee-Apogee, km	Instrument Employed	Type and Range of Measurement	Major Results Reported	References
	320	Radiometric instrumentation consisting of following three parts: Scintillation counter with NaI(Tl) crystal and FEU-16 photomultiplier Crystal dimensions: 30 mm diameter, 14 mm high	Threshold energy: 25 keV	The counting rate gradually increased from 3–5 pulses/cm² sec to 10–12 pulses/cm² sec during the satellite's flight from the equator to latitude of ±40–50°. In most cases the counting rate increased sharply to 20–600 pulses/cm² sec in the region of geomagnetic latitudes of 50–60° (South Atlantic anomaly).	Papkov, *et al.* (1961)[104] Savenko, *et al.* (1961)[107] Vernov, *et al.* (1961)[111, 146, 149]
	320	Halogen gas-discharge counter type STS-5 (two) One counter was shielded with brass 1 mm thick and steel 1 mm thick The scintillation counter and both gas-discharge counters were mounted inside the container with shielding of matter between 8 and 15 g/cm² within 50% of solid angle, between 15 and 150 g/cm² within 35% of solid angle and 5–8 g/cm² in all other directions.		The radiometric instruments mounted inside the container were employed to perform dosimetric measurements. During a flight from the equatorial region to higher latitudes the radiation intensity sensed by the gas-discharge counters increased from 0·8 to 3·2 particles/cm² sec and to 3·5 to 13 pulses/cm² sec for the scintillation counter. This increase was caused by electrons impinging on the satellite's hull when the satellite passed through the lower region of the outer belt. The resulting X-rays inside the container were registered by the scintillation counter 100%, but only <1% by the gas-discharge counter (low efficiency to X-rays). The radiation dose changed from 0·35 to 0·70 mrad during one revolution around the earth. The absorbed dose was equal to 7 mrad/day.	Savenko, *et al.* (1961)[108, 111] Vernov, *et al.* (1963)[154]
	320	Scintillation counter with CsI(Tl)-crystal and FEU-15 photomultiplier Crystal dimensions: 30 mm diameter, 2·2 mm thick Shielded within a solid angle of 0·15 steradian by 2 mg/cm² of aluminum and from 1 to 150 g/cm² in all other directions Outside container.	Protons >1 MeV Electrons >30 keV X-ray Quanta	The counter registered maximum energy libration in the crystal ($>50\times10^{10}$ eV) in regions over southeast Asia, the Central Pacific Ocean and South America.	Savenko, *et al.* (1962)[113] Vernov, *et al.* (1961)[146]

Vehicle/Launch Date	Height Perigee-Apogee, km	Instrument Employed	Type and Range of Measurement	Major Results Reported	References
Spaceship-Satellite 3/ December 1 1960		Radiometer instrumentation consisting of: Scintillation counter with NaI(Tl)-crystal and FEU-15 photomultiplier Crystal dimensions: 30 mm diameter, 14 mm high Shielding of counter with material 3 g/cm² thick Inside container	Protons >50 MeV Electrons >6 MeV Threshold energy: 25 keV	Maximum counting rates were obtained south of 40° of the southern magnetic latitude and amounted to 530 pulses/cm² sec for the scintillation counter, to 130 MeV/cm² sec for the energy release in the NaI-crystal and to 5·1 impulses/cm² sec as the maximum counting rate for the gas-discharge counter. As a result of the investigations carried out during Spaceship-Satellites 2 and 3, the existence of the earth's outer (electron) radiation zone was distinctly revealed as limited to the region of high geomagnetic latitudes. The high counting rate of the scintillation counter represents gamma-radiation (Bremsstrahlung) with an average energy of 100–300 keV. An electron flux of 2×10^5 particles/cm² sec with an average of 3×10^5 eV corresponds to a counting rate of 400 pulses/cm² sec for the scintillation counter. The total number of impulses recorded by the STS 5 counter was $1{\cdot}8\times10^5$/cm² day. The power of the absorbed dosage was 7 mrad/day and the calculated tissue dose rate amounted to 8 mrad/day.	Savenko, *et al.* (1961)[107], (1962)[109] Vernov, *et al.* (1961)[146], (1962)[150], (1963)[154], (1965)[147, 153]
		Halogen gas-discharge counter Type STS-5 Inside container shielded with material from 5 to 150 g/cm²			
		Scintillation counter with CsI(Tl)-crystal and FEU-16 photomultiplier Crystal dimensions: 30 mm diameter, 2·2 mm thick Shielded with aluminum foil 2 mg/cm² Outside container	Electrons ⩾100 keV		
		Gas-discharge counter telescope consisting of two groups of counters Five STS-5 and eight STS-2 counters Counter areas 120 and 125 cm² Material between counter areas (mainly plexiglass): 4 g/cm² Distance between counter groups: 35·8 cm	Protons ⩾60 MeV Electron ⩾8 MeV	Particle fluxes registered by the counter telescope were 0·72, 1·1, and 2·2 particles/cm² sec at 0, 30, and 60 degrees northern latitude, respectively. In the region of the southern anomaly the maximum counting rate was 400 particles/cm² sec.	Ginzburg, *et al.* (1961)[29] Kurnosova, *et al.* (1961)[34]
Cosmos 3/April 24 1962	650	Scintillation counter with fluorescent screen and photo-multiplier (three) Screen material: $Sr_3(PO_4)_2$(Eu) 1·4 mg/cm² Aluminum foils: 0·4, 0·6, and 1·1 mg/cm² Accelerating voltages applied to aluminum foil: 0·1, 1·6, 2·3, and 4·3 kV	Protons Electrons 40 eV to ⩾50 keV	Contours of electron flux intensities with energies in excess of 50 keV over the South Atlantic magnetic anomaly were measured at an altitude of 650 km. Electrons of energies less than 5 keV were observed mostly in the sunlit atmospheric regions. Electrons with energies of 7 to 50 keV appeared at high geomagnetic latitudes and at high altitudes. Their fluxes amounted to $\sim10^6$ electrons/cm² sec ster.	Gal'perin and Krasovskiy (1963)[25] Krasovskiy, *et al.* (1963)[79] Mulyarchik (1966)[102] Mulyarchik, *et al.* (1964)[103]

TABLE 10.2 (Continued)

Results of Solar and Cosmic Radiation Studies Aboard Soviet Rockets, Satellites, and Space Probes (1957–1966)

Vehicle/Launch Date	Height Perigee-Apogee, km	Instrument Employed	Type and Range of Measurement	Major Results Reported	References
		Metal grid voltage: –40 V Outside container		Hard electrons with energies of hundreds of keV with a flux of 5×10^7 electrons/cm² sec ster were recorded at an altitude of about 1000 km.	
	1000	Halogen gas-discharge counter type STS-5 Shielded with 3·4 g/cm² lead plus structural shielding of 0·8 g/cm² aluminum over a solid angle of 2π ster and $\sim$25 g/cm² in other directions Inside container	Protons $\geqslant$ 50 MeV Electrons $\geqslant$ 10 MeV X-ray Bremsstrahlung from low-energy electrons (>400 keV)		Gal'perin, *et al.* (1963)[25] Krasovskiy, *et al.* (1963)[78–79] Temnyy (1963)[120–122]
Cosmos 4/April 26 1962		Radiometer instrumentation consisting of: Scintillation counter with CsI(Tl)-crystal and FEU-16 photomultiplier Crystal dimensions: 30-mm diameter, 30 mm high Crystal shielded with material 3 g/cm² thick Inside container	Threshold energy 60 keV	The total number of impulses recorded by the STS-5 counter was $2{\cdot}5 \times 10^5$/cm² day. The power of absorbed dosage was 13 mrad/day and the calculated tissue dose rate amounted to 16 mrad/day. The results obtained indicated a shift of intensity maxima in the outer radiation zone on magnetically quiet days onto magnetic shells with greater L-values.	Savenko, *et al.* (1963)[112] Vernov, *et al.* (1963)[154], (1964)[152], (1965)[147]
		Gas-discharge counter type STS-5 (two) Inside container	Protons $\geqslant$ 40 MeV Electrons $\geqslant$ 6 MeV	The maximum counting rate registered was 30 counts/cm² sec.	Savenko, *et al.* (1963)[112]
		Scintillation counter with CsI(Tl)-crystal and FEU-16 photomultiplier Crystal dimensions: 30-mm diameter, 20 mm high Shielded with aluminum foil 2 mg/cm² within 30% of complete solid angle Outside container	Threshold energy 60 keV Electrons $\geqslant$ 100 keV		

Vehicle/Launch Date	Height Perigee-Apogee, km	Instrument Employed	Type and Range of Measurement	Major Results Reported	References
Cosmos 5/May 28 1962	1600	Scintillation counter with fluorescent screen and photomultiplier (three) Screen material: $Sr_3(PO_4)_2(Eu)$ 1·4 mg/cm² (two counters) Aluminum foils: 0·4 and 0·6 mg/cm² Screen material: ZnS(Ag), 5 mg/cm² (one counter) Aluminum foil: 1·1 mg/cm² Accelerating voltages applied to aluminum foil: 0·15, 3·0, 5·5, and 1·1 kV Metal grid voltage: –40 V Outside container	Protons Electrons 40 eV to >50 keV	Electrons with an energy of 40 eV to 5 keV were registered mostly on the sunlit portion of the orbit. Their maximum intensity reached 5×10^9 electrons/cm² sec ster. The average intensity was 5×10^8 electrons/cm² sec ster. The data obtained from Cosmos 5 whose apogee reached heights of up to 1600 km, indicated considerable variations in the spectrum of particles.	Gal'perin, *et al.* (1963)[25] Krasovskiy, *et al.* (1963)[79] Mulyarchik (1966)[102] Mulyarchik, *et al.* (1964)[103]
		Halogen gas-discharge counter type STS-5 Shielded with 3·4 g/cm² lead plus structural shielding of 0·8 g/cm² aluminum over a solid angle of 2π ster and $\sim$25 g/cm² in other directions Inside container	Protons $\geqslant 50$ MeV Electrons $\geqslant 2$ MeV X-ray Bremsstrahlung from low-energy electron (>400 keV)		Gal'perin, *et al.* (1963)[25] Krasovskiy, *et al.* (1963)[78–79] Temnyy (1963)[120–122]
Cosmos 7/April 26 1962		Radiometer instrumentation consisting of:		Detailed instrumentation descriptions are presented in Table 10.3.	Papkov, *et al.* (1961)[104]
Cosmos 9/September 27 1962 Cosmos 10/October 17 1962 Cosmos 12/December 22 1962 Cosmos 13/March 21 1963 Cosmos 15/April 22 1963 Cosmos 16/April 28 1963 Cosmos 18/May 24 1963		Scintillation counter with NaI(Tl)- or CsI(Tl)-crystal and photomultiplier One or two gas-discharge counters type STS-5 Both counters mounted inside satellite container Scintillation counter with CsI(Tl)-crystal and photomultiplier Shielding of material 2 mg/cm² or 180 mg/cm² thick, or	Threshold energies: 30 to 160 keV Threshold energies: 45 to 75 keV	The total number of pulses recorded by the gas-discharge (or geiger) counter, the absorbed dosage and calculated tissue dosage rate are presented in Table 10.4. The information presented in Tables 10.3 and 10.4 were taken from literature referenced. The following additional results were obtained with gas-discharge and scintillation counters, functioning also as dosimetric equipment: In the region of the Brazil anomaly on magnetic shells with $L=1\cdot2$ to $1\cdot4$, a counting rate longitude dependence differed from that of electron intensity on greater L shells.	Savenko, *et al.* (1963)[112] Vernov, *et al*, (1963)[154], (1964)[153], (1965)[147] Vernov, *et al.* (1964)[151–153], (1965)[148]

TABLE 10.2 (Continued)

Results of Solar and Cosmic Radiation Studies Aboard Soviet Rockets, Satellites, and Space Probes (1957–1966)

Vehicle/Launch Date	Height Perigee-Apogee, km	Instrument Employed	Type and Range of Measurement	Major Results Reported	References
	250	Spherical electrostatic analyzer, or Telescopes with gas-discharge counters of type STS-5 Counters mounted outside satellite container		The spatial distribution of the artificial radiation belt electrons has been obtained at minimum altitude of 250 km. The artificial belt electron lifetime in the region of the Brazil anomaly at L = 1·3 was calculated to be 170 days and decreasing to 70 days at greater L values.	
Cosmos 17/May 22 1963		Scintillation counter with NaI(Tl)-crystal and photo-multiplier Crystal dimensions: 20-mm diameter, 20 mm high Inside container with aluminum shielding 1 g/cm² for solid angle of about 2π ster.	Threshold energies: 50 keV 3·3 MeV Protons >30 MeV Electrons >2·5 MeV Gamma quanta >20 keV (Minimum energies indicated)	Data were obtained on the intensity and spatial distribution of particles of the radiation zones and cosmic rays. Data were also obtained concerning the intensity of electrons at the forbidden region of space, i.e., at those drift trajectories of electrons which approach the low altitude at the region of the Atlantic anomaly. The various threshold levels of the scintillation counters permitted to register protons or electrons of certain energies. The thresholds of 45 keV and 160 keV permitted to record minimum scattering electrons and also Bremsstrahlung. A flux of 3·5 particles/cm² sec was registered by the employed instrumentation which was caused by cosmic rays and albedo particles. A flux of 0·9 particles/cm² sec was estimated to be accounted for by albedo at an altitude of 800 km and 45° latitude. At 800 km in the equatorial region the flux measured was 0·5 particles/cm² sec, of which 0·05 particles/cm² sec were accounted for by the true cosmic radiation, 30% were made up by secondary particles from the shielding material and 0·3 particles/cm² sec by albedo particles.	Vernov, *et al.* (1964)[134]
		Gas-discharge counter type STS-5 Effective area: 4·3 cm² Minimum shielding of 2 g/cm² of aluminum in solid angle of about π ster.	Protons >30 MeV Electrons >3 MeV Gamma quanta >30 keV		
	800	Inside container Scintillation counter with CsI(Tl)-crystal and photo-multiplier Crystal dimensions: 6-mm diameter, 3 mm thick Crystal covered by aluminum foil 2 mg/cm² thick	Threshold energies: 45 keV 160 keV 5·4 MeV 8·5 MeV		

Vehicle/Launch Date	Height Perigee-Apogee, km	Instrument Employed	Type and Range of Measurement	Major Results Reported	References
		Crystal surrounded with 5 mm of lead and 11 mm of aluminum, except aperture angle of 40° Outside container	Protons >500 keV Electrons >50 keV Gamma quanta >5 keV (minimum energies indicated)		
		Instrument I Wide-range radiation intensity-meter consisting of: Four gas-discharge counters type STS-5 Mounted in single plane and connected in parallel Effective area: 17·2 cm^2 Shielding minimum: 0·95 g/cm^2	Protons >27 MeV Electrons >2 MeV Counting rate range: 1·6–8 × 10^4 pulses/sec	According to the composition of the charged particles penetrating a shielding of 1 g/cm^2, the inner earth radiation zone was divided into two regions (based on data obtained by Cosmos 17 instrumentation during period from May 22 through 30 1963 at altitudes from 260 to 780 km): With L from 1·15 to 1·6 where main portion of radiation consisted of electrons injected during high-altitude nuclear explosion. With L from 1·6 to 2·5 where main portion of radiation consisted of protons. The flow of protons with energies higher than 30 MeV increased in comparison with 1958 observations which was explained by decreased solar activity during the 11-year solar cycle. A narrow region between the inner and outer radiation zone was detected where electrons with energies of 0·1 ⩽ E_e ⩽ 1·5 MeV were recorded. Beneath the lower boundary of the inner zone, stable flows of soft corpuscles (electrons) with energies of 50–100 keV and fluxes amounting to ∼10^5–10^6 particles/cm^2 sec were estimated to exist. This information was based on measurement made with the SBT-9 counter and instruments of Cosmos 3 and 5.	Kirdina, *et al.* (1965)[73], (1966)[74]
		Gas-discharge counter type SI-ZBG Effective area: 0·11 cm^2 Shielding minimum: 1·35 g/cm^2	Protons >33 MeV Electrons >2·8 MeV Counting rate range: 1·6–8 × 10^4 pulses/sec		
		Instrument II Radiation intensity-meter Four gas-discharge counters type STS-5 Effective area: 4·3 cm^2 Shielding minimum: 0·75 g/cm^2 Counters mounted inside satellite at corners of instrumentation panel	Protons >24 MeV Electrons >1·5 MeV Counting rate range: < 5 × 10^3 pulses/sec		
		Instrument III Radiation intensity-meter Four scintillation counters with NaI(Tl)-crystal with photomultiplier Crystal dimensions: 30-mm diameter, 20 mm long Effective area: 7·0 cm^2 Shielding minimum: 1·4 g/cm^2	Threshold energy approx 100 keV Protons >32 MeV Electrons >3 MeV Gamma quanta >0·1 MeV Counting rate range: < 5 × 10^3 pulses/sec		

TABLE 10.2 (Continued)

Results of Solar and Cosmic Radiation Studies Aboard Soviet Rockets, Satellites, and Space Probes (1957–1966)

Vehicle/Launch Date	Height Perigee-Apogee, km	Instrument Employed	Type and Range of Measurement	Major Results Reported	References
		Instrument IV Electromagnetic radiation intensity-meter Four end-window gas-discharge counters type SBT-9 Mica window effective area: 0·26 cm^2, window thickness: 1 mg/cm^2 Three counters mounted outside satellite on one side of hull directed along line of maximum sun-sensor sensitivity The fourth counter mounted on opposite side of hull Shielding minimum: 1 mg/cm^2	Protons $>0{\cdot}5$ MeV Electrons >40 keV X-rays >3 keV Counting rate range $<3\times10^3$ pulses/sec		
Electrons 1/2/ January 30 1964 Electrons 3/4/ July 11 1964	5000–7000	Soft-electron detector with photomultiplier (two) Fluorescent screen of $Sr_3(PO_4)_2$ [Eu]: 1·4 mg/cm^2 thick (One screen was covered with aluminum foil 0·6 mg/cm^2 thick. The other screen was covered with aluminum foil 0·4 mg/cm^2 thick) Modulation voltages of one counter: 10, 6, 3, and 0 volts Sweeping magnet on other counter to exclude electrons with energies $>1{\cdot}1$ MeV Shielding and viewing angle of both counters: 1/30 ster	Protons >150 keV Electrons >20 keV	At low altitudes near the equator both indicators registered signals due mainly to electrons with energies $>1{\cdot}1$ MeV. Preliminary data indicated relativistic electrons of the artificial belt with intensities from 0·5 to $12\times10^8/cm^2$ sec with $E<1{\cdot}1$ MeV, intensities from 0·4–6/cm^3 with $E>1{\cdot}1$ MeV, and soft protons with intensities from 1·0–10 $\times 10^7/cm^2$ sec. The mean intensity of the electron flux in the inner zone at $L=1{\cdot}5$ reached $2\times10^9/cm^2$ sec for energies 20 keV to 1·1 MeV. In the middle latitudes ($L\gg2$) a high intensity of trapped soft protons with energies >150 keV were registered. During magnetically quiet conditions the electron intensity was $5\times10^6/cm^2$ sec ster at $L\sim5$ to 6 for energies $\geqslant 20$ keV. Especially strong electron intensity fluctuations of $1{\cdot}6\times10^9/cm^2$ sec and $8\times10^8/cm^2$ sec ($E\geqslant20$ keV) at great heights (5000–7000 km) were observed during magnetic storms of January 31 and March 5 1964, respectively.	Bolyunova, *et al.* (1965)[19]

Vehicle/Launch Date	Height Perigee-Apogee, km	Instrument Employed	Type and Range of Measurement	Major Results Reported	References
		Scintillation counter with NaI(Tl)-crystal and photo-multiplier Crystal dimensions: 20 mm diameter, 20 mm high Counter covered with ~7·0 g/cm² matter Shielding: aluminum 0·135 g/cm² thick Geometric factor: 4·7 cm² ster Inside container with aluminum shielding 0·8 g/cm² thick	Range expansions by electronic means: 1. Protons >25 MeV Electrons >1·7 MeV 2. Protons >30 MeV Electrons >5·5 MeV	On the basis of the results obtained it was concluded that apparently two mechanisms lead to the formation of high-energy protons below $L=2$: one of them produces protons having an energy mainly up to ~50 MeV. The second mechanism begins to operate from $L=2$ towards smaller L, and is very probably due to protons from albedo neutron decay. For the region of intermediate L between 2·5 and 4, it is characteristic to have a maximum in the distribution of low-energy protons of <10 MeV and a minimum distribution of electrons with energies of several hundred keV. A zone of electrons with energies up to 6 MeV was also found, but its origin has not been clarified. A flux of protons of $4{\cdot}5\times10^5$/cm² sec ster with energies >2 MeV was found in the equatorial plane and the mean directional flux of electrons having the energy ~6 MeV at the equator at $L=2{\cdot}75$ was 10^2/cm² sec ster. In the space between the radiation zones, $3<L<4$, a minimum in intensity distribution of electrons with energies of several hundred keV was registered.	Kuznetsov (1966)[95] Tverskoy (1965)[123] Vernov, *et al.* (1965)[135, 136]
		Gas-discharge counter type STS-5 (three) 1. Shielding: aluminum 1·35 g/cm² thick 2. Shielding: aluminum 4·0 g/cm² thick 3. Shielding: lead 17 g/cm² thick	Protons >45 MeV Electrons >4·3 MeV Protons >70 MeV Electrons >9 MeV Protons >110 MeV		
		Coincidence counter 2 and 3 Geometric factor of all counters: 4·3 cm² ster Inside container with aluminum shielding 0·8 g/cm² thick	Protons >150 MeV		
		Scintillation counter with CsI(Tl)-crystal and photo-multiplier Crystal dimensions: 6 mm diameter, 3 mm thick Counter covered with 1·35 g/cm² lead Shielding: aluminum 2 mg/cm² thick in a 40° angle Geometric factor: 10^{-1} cm² ster Outside container	Range expansion by electronic means: 1. Protons >650 keV Electrons >40 keV 2. Protons >700 keV Electrons >60 keV 3. Protons >750 keV Electrons >180 keV 4. Protons >850 keV Electrons >400 keV 5. Protons >2·2 MeV Electrons >1·6 MeV 6. Protons >1·0 MeV Electrons >600 keV 7. Protons >2·7 MeV 8. Protons >1·5 MeV Electrons >1·2 MeV 9. Protons 6 to 75 MeV 10. Protons 4·5–120 MeV	Low-energy protons with energies from 1 to 9 MeV were recorded with scintillation counters and semiconductor counter at parameters $L=1{\cdot}3$, 1·6 and 2 with Electron 3 and at $L=3$ with Electrons 1 and 2.	Sosnovets (1966)[119]

TABLE 10.2 (Continued)

Results of Solar and Cosmic Radiation Studies Aboard Soviet Rockets, Satellites, and Space Probes (1957–1966)

Vehicle/Launch Date	Height Perigee-Apogee, km	Instrument Employed	Type and Range of Measurement	Major Results Reported	References
		Scintillation counter with CsI(Tl)-crystal and photo-multiplier Crystal dimensions: 6 mm diameter, 0·15 mm thick Counter covered with 67·5 mg/cm^2 lead Shielding: aluminum 2 mg/cm^2 thick in a 40° angle Geometric factor: 10^{-1} cm^2 ster Outside container	Range expansion by electronic means: 1. Protons >650 keV Electrons >40 keV 2. Protons 2–10 MeV	The electron spectrum in the outer radiation zone, $4<L<11$, undergoes considerable variations. The spectrum becomes softer on the average during years of a quiet sun. However, high-energy electrons may appear after large magnetic perturbations. The maximum of the outer zone, both on the nocturnal and morning side, was located at $L=4{\cdot}8$. The mean directional flux of electrons with energies of >70 keV was $5\times10^6/cm^2$ sec ster at the outer zone maximum.	Tverskoy (1965)[123] Vernov, *et al.* (1965)[135]
		n-p solid-stated type detector (silicon, 35 μ sensitive layer) Counter covered with 8·5 mg/cm^2 aluminum Shielding: aluminum 2 mg/cm^2 thick in a 40° angle Geometric factor: 10^{-1} cm^2 ster Outside container	Protons 0·9–5·5 MeV		
Electrons 1–4		In terms of radiation sensor composition, the equipment of Electrons 1 and 2 was completely identical, the only difference being the magnitude of the individual energy thresholds. Each of the four Electron satellites carried identical sets of instruments, therefore results obrained in various parts in space at various times could be compared.			Sosnovets (1966)[119] Vernov, *et al.* (1965)[135]

Vehicle/Launch Date	Height Perigee-Apogee, km	Instrument Employed	Type and Range of Measurement	Major Results Reported	References
Cosmos 41/Aug. 22 1964		Radiometric instrumentation consisting of: n-p solid-state type conductor (silicon layer 8 mg/cm²) (two)		From the data obtained it was concluded that the external part of the protonosphere starting at $L \geqslant 5{\cdot}5$, underwent substantial variations. The largest time variations in electron intensity for a fixed L occur in the region $L = 5$–7 and the count rate variations were associated with a change in the position of the belt in space and with the change in the intensity of electrons with an energy exceeding the detector threshold. Proton flux intensities of 10^4 and 3×10^4/cm² sec ster with energies of 0·4 to 3 MeV were registered at an altitude of 6600 and 26 000 km, at geomagnetic latitude 49° N and 0, at $B = 5{\cdot}4 \times 10^{-2}$ and $0{\cdot}25 \times 10^{-2}$ gauss and with an angle of n-p counter axis to magnetic force line of 60° and 30°, respectively.	Savenko, *et al.* (1965)[114] Vernov, *et al.* (1965)[156–157]
		1. Shielding: silver 200 mg/cm² thick and aluminum Geometric factor: 0·06 cm² ster	Protons 0·4–7 MeV		
	6600 26 000	2. Shielding: aluminum 20 mg/cm² thick Geometric factor: 0·07 cm² ster Field of view: 1 steradian	3–8 MeV		
		Scintillation counter with NaI(Tl)-crystal with FEU-16 photomultiplier Crystal dimensions: 30 mm diameter, 14 mm high Minimum shielding: aluminum 0·18 g/cm² thick Geometric factor: 6·8 cm² (isotropic)	For direct passage: protons >10 MeV electrons >600 keV Threshold energy: 90 keV		
		Gas-discharge counter type STS-5 Minimum shielding: aluminum 0·15 g/cm² and steel Geometric factor: 4·3 cm² (isotropic)	Protons >9 MeV Electrons >500 keV		
		Gas-discharge counter type Si-3BG Minimum shielding: aluminum 0·84 g/cm² and steel Geometric factor: 0·1 cm² (isotropic)	Protons >25 MeV Electrons >2 MeV		
		End-window gas-discharge counter type SBT-9 (three) 1. Mica window 1 mg/cm² thick Geometric factor: 0·1 cm² ster	Protons >0·5 MeV Electrons >25 keV		
		2. Mica window 1 mg/cm² thick Magnetic shielding Geometric factor: 0·1 cm² ster	Protons >0·5 MeV Electrons >120 keV		

TABLE 10.2 (Continued)

Results of Solar and Cosmic Radiation Studies Aboard Soviet Rockets, Satellites, and Space Probes (1957–1966)

Vehicle/Launch Date	Height Perigee-Apogee, km	Instrument Employed	Type and Range of Measurement	Major Results Reported	References
		3. Aluminum window: 20 mg/cm² Mica window: 1 mg/cm² Magnetic shielding Geometric factor: 0·1 cm² ster	Protons >3 MeV Electrons >120 keV		
		All instruments outside container			
		Gas-discharge counter type Si-3BG	Protons >45 MeV		
		Minimum shielding: aluminum 3 g/cm² thick Geometric factor: 0·1 cm² (isotropic) Inside container	Electrons >7 MeV		
Lunar and Interplanetary Probes					
Luna 1/ January 2 1959	50 000 35 000 25 000	Scintillation counter with NaI(Tl)-crystal and FEU-25 photomultiplier Inside container with aluminum shielding of 1 g/cm²	Threshold energy: 1. 45 keV 2. 450 keV 3. 4·5 MeV	The counting rate with threshold 45 keV was 1000 particles/cm² sec at a distance of 50 000 km from the center of the earth. With threshold 450 keV the counting rate was 2500 particles/cm² sec at a distance of 35 000 km; and with threshold 4·5 MeV the counting rate was 25 particles/cm² sec at a distance of 25 000 km (maximum readings).	Ginzburg, *et al.* (1962)[31] Gringauz (1961)[49] Gringauz, *et al.* (1960)[61] Vernov, *et al.* (1960)[132–133]
		Scintillation counter with CsI(Tl)-crystal and photomultiplier Thickness of crystal: 0·3 g/cm² Outside container with aluminum shielding of 1·9 mg/cm²	Electrons: 50–1000 keV	The first scintillation counter measured the energy flux of X-rays produced by electrons of energies over 200 keV. (The threshold is determined by the absorption of X-rays in the aluminum shielding.)	
	25 000			The second scintillation counter registered the energy flux of all electrons of energy greater than 50 keV. The maximum total energy release in the counter crystal in one second (maximum ionization) was 300 BeV/sec at a distance 25 000 km from the center of the earth.	

Vehicle/Launch Date	Height Perigee-Apogee, km	Instrument Employed	Type and Range of Measurement	Major Results Reported	References
	35 000	Gas-discharge counter (two) Cross-sectional areas: 4 and 15 cm^2		Measurement of the total number of charged particles. The counting rate was 2000 particles/cm^2 sec at a distance of 35 000 km from the center of the earth. Electrons of energy >1 MeV contributed to the ionization and readings of the gas-discharge counters.	
	>70 000			At distances exceeding 70 000 km from the center of the earth the radiation intensity measured by all instruments carried by Luna 1 remained constant (20 particles/cm^2 sec).	
Luna 2/ September 12 1959	17 000	Scintillation counter with NaI(Tl)-crystal and photo-multiplier Crystal dimensions: 39·5 mm diameter, 40 mm high Inside container with aluminum shielding of 1 g/cm^2	Threshold energy: 1. 60 keV 2. 600 keV 3. 3·5 MeV	The counting rate with threshold 3·5 MeV was 2·2 particles/cm^2 sec at a distance of 17 000 km from the center of the earth. The counting rate of the two gas-discharge counters located inside the container was 100 and 120/cm^2 sec at the same distance (maximum readings).	Chudakov, *et al.* (1960)[23]
		Gas-discharge counter, 10 mm diameter, 50 mm long Shielding of copper 1·5 mm thick Inside container with aluminum shielding of 1 g/cm^2		It was assumed that the discharge counters registered X-ray radiation (photons with energies >400 keV) caused by an electron flux of energy ⩾1 MeV or proton flux of energy ∼10 MeV impinging on the container hull.	
		Gas-discharge counter, 10 mm diameter, 50 mm long Shielding of lead 3·0 mm thick and aluminum 1·0 mm thick Inside container with aluminum shielding 1 g/cm^2			
		Gas-discharge counter (unshielded) Inside container with aluminum shielding 1 g/cm^2		The unshielded counter was used for program switching of telemetry channels and was tuned to a counting rate of about 500 counts/sec.	Chudakov, *et al.* (1960)[23]
	>70 000	Scintillation counter with NaI(Tl)-crystal and photo-multiplier Crystal dimensions: 39·0 mm diameter, 40 mm high Outside container shielded with 1 g/cm^2 of aluminum	Threshold energy: 1. 45 keV 2. 450 keV	The radiation intensity registered with threshold 450 keV at a distance of >70 000 km was 2·0 pulses/sec and constant to the lunar region.	Chudakov, *et al.* (1960)[23]

TABLE 10.2 (Continued)

Results of Solar and Cosmic Radiation Studies Aboard Soviet Rockets, Satellites, and Space Probes (1957–1966)

Vehicle/Launch Date	Height Perigee-Apogee, km	Instrument Employed	Type and Range of Measurement	Major Results Reported	References
		Scintillation counter with CsI(Tl)-crystal and photo-multiplier Crystal dimensions: 30 mm diameter, 3·0 mm thick Outside container shielded with 1·2 mg/cm^2 aluminum foil	Recording of total ionization produced in crystal		
		Gas-discharge counter Window area: 0·28 cm^2 Shielding of lead 3·0 mm thick and aluminum 1·0 mm thick Outside container			
		Gas-discharge counter Window area: 1·6 cm^2 Shielding of copper foil 0·2 mm thick Outside container		Counter was energized to operate in the zone of high particle intensities only.	
		Gas-discharge counter Window area: 1·6 cm^2 Shielding of copper foil 0·5 mm thick Outside container		Counter was energized to operate in the zone of high particle intensities only.	
		The walls of all gas-discharge counters were of stainless steel with a thickness of 50 mg/cm^2			
Luna 3/ October 4 1959		Scintillation counter with NaI(Tl)-crystal and photo-multiplier Crystal dimensions: 39·5 mm diameter, 40 mm high Inside container with aluminum shielding of 1 g/cm^2	Threshold energy: 1. 45 keV 2. 3·6 MeV	The average counting rate recorded was equal to 6·79 pulses/cm^2 sec for Threshold 1 and 2·15 pulses/cm^2 sec for Threshold 2.	Vernov, *et al.* (1961)[138]

Vehicle/Launch Date	Height Perigee-Apogee, km	Instrument Employed	Type and Range of Measurement	Major Results Reported	References
		Gas-discharge counter 10 mm diameter, 50 mm long Shielding of lead 3·0 mm thick and aluminum 1·0 mm thick Inside container with aluminum shielding 1 g/cm^2		The average intensity recorded with two gas-discharge counters for the flight time from October 4–18 1959 was 2·12 particles/cm^2 sec.	
		Gas-discharge counter 10 mm diameter, 50 mm long Shielding of aluminum 2·55 mm thick Inside container with aluminum shielding 1 g/cm^2			
		Gas-discharge counter 10 mm diameter, 50 mm long Shielding of lead 3·0 mm thick and aluminum 1·0 mm thick with window 0·28 cm^2 covered with aluminum 0·2 mm thick Outside container			
Venus 1/ February 12 1961		Scintillation counter Cross section: 4·7 cm^2 Shielded	Protons $\geqslant$32 MeV Electrons $\geqslant$3 MeV Bremsstrahlung Photons $\geqslant$30 keV	The data indicated that the mean energy liberation in the crystal per pulse of the scintillation counter, equal to 130 keV, remained constant in the range from 32 000–40 000 km, and that the mean energy of the bremsstrahlung photons produced in the shell of the probe, did not change with distance.	Vakulov, *et al.* (1962)[126]
	32 000–40 000	Gas-discharge counter Shielded			
	32 000			The counting rate of the scintillation counter was 5×10^4 particles/cm^2 sec and of the gas-discharge counter $1{\cdot}5 \times 10^2$ particles/cm^2 sec at a distance of 32 000 km from the center of the earth.	

TABLE 10.2 (Continued)

Results of Solar and Cosmic Radiation Studies Aboard Soviet Rockets, Satellites, and Space Probes (1957–1966)

Vehicle/Launch Date	Height Perigee-Apogee, km	Instrument Employed	Type and Range of Measurement	Major Results Reported	References
Mars 1/ November 1 1962		Scintillation counter with NaI(Tl)-crystal and photomultiplier Crystal dimensions: 20 mm diameter, 20 mm high Inside container with aluminum shielding 1 g/cm^2	Threshold energy: 1. 30 keV 2. 2·5 MeV	No considerable cosmic ray intensity variations were recorded by the probe's instruments during November–December 1962 and December 1963 at increasing distances from the sun. Within the error limitations of the measurements (2–3%) the intensity of the primary cosmic radiation remained constant during flight distances from the sun approaching 1·24 astronomical units. The cosmic ray intensity measured by the Cherenkov counter (threshold >30 keV) was 64 impulses/sec and 19 impulses/sec measured by the geiger counter at distances from the sun from 1·0 to 1·2 astronomical units.	Vakulov, *et al.* (1963)[125] Vernov (1964)[137]
	1·24 astronomical units	Gas-discharge counter (two) 10 mm diameter, 50 mm long Heavily shielded Inside container with aluminum shielding 1 g/cm^2			
		Scintillation counter with CsI(Tl)-crystal and photomultiplier Crystal dimensions: 20 mm diameter, 3 mm thick Crystal covered by $\sim$2·2 mg/cm^2 aluminum foil Solid angle: $\sim$3 steradians Outside container shielded with aluminum 1 g/cm^2 thick	Protons $\geqslant$500 keV Electrons $\geqslant$70 keV		
Luna 4/ April 2 1963	100 000–700 000	Gas-discharge counter 10 mm diameter, 50 mm long Inside container with aluminum shielding 1 g/cm^2 Remaining side of counter (insensitive area) was shielded with material >10 g/cm^2 thick		The cosmic ray flux recorded by the counter was 4·45 particles/cm^2 sec at a distance of 100 000–700 000 km from the center of the earth.	Vakulov, *et al.* (1963)[125] Vernov (1964)[137]

Vehicle/Launch Date	Height Perigee-Apogee, km	Instrument Employed	Type and Range of Measurement	Major Results Reported	References
Zond 1/ April 2 1964	0·97 astronomical units	Eight STS-5 gas-discharge counters (one central counter shielded by the others) Shielded with material 10 g/cm^2 thick, and maximum shielding on one side with material 150 g/cm^2 thick	Protons Electrons Gamma quanta	The cosmic ray flux recorded by the ionizing radiation counter was 3·37 particles/cm^2 sec at a distance of 0·97 astronomical unit from the center of the earth. Data indicated that gamma quanta with energies <1 MeV caused large portions of anti-coincidence countings. The gamma quanta were apparently formed by cosmic primary particles interacting with the probe material.	Avdyushin, *et al.* (1965)[11]
Zond 3/ July 18 1965	36 000 000	STS-5 gas-discharge counter Inside container with aluminum shielding 1 g/cm^2		The counting rate increased with time and distance of the probe from its earth-parking orbit. The cosmic ray intensity measured at a distance of 36 000 000 km was 26·3 particles/sec. A peak measured in the early part of October 1965 was associated with a solar flare and consequently with an increase in corpuscular radiation.	Vernov, *et al.* (1966)[161]
		n-p solid state type detector Semiconductor sensing area: 0·2 cm^2	Low-energy protons 1 to 5 MeV	The intensity of protons increased in moving away from the sun.	Vernov, *et al.* (1966)[139–140]
Venus 2/ November 12 1965		Gas-discharge counter type STS-5 Counter dimensions: 10 mm diameter, 50 mm long Inside container with shielding 1 g/cm^2	Protons >30 MeV Electrons >2 MeV	The variations in cosmic radiation intensity were irregular during the flight toward the sun which was attributed to changes in the character and magnitude of the Forbush effect.	Vernov, *et al.* (1966)[139–140]
		n-p solid state detector Effective area: 0·2 cm^2 Aluminum shielding in solid angle of $\sim$1 ster: 1 mg/cm^2 Shielding in other directions: 1 g/cm^2 Outside container	Low-energy protons 1–5 MeV	Considerable increases in counting rates were recorded during the interplanetary flight.	
Venus 3/ November 16 1965		n-p solid state detector Effective area: 0·2 cm^2 Aluminum shielding in solid angle of $\sim$1 ster: 1 mg/cm^2 Shielding in other directions: 1 g/cm^2 Outside container	Low-energy protons 1–5 MeV	Considerable increases of counting rates were recorded during the interplanetary flight.	Vernov, *et al.* (1966)[139]

TABLE 10.2 (Continued)

Results of Solar and Cosmic Radiation Studies Aboard Soviet Rockets, Satellites, and Space Probes (1957–1966)

Vehicle/Launch Date	Height Perigee-Apogee, km	Instrument Employed	Type and Range of Measurement	Major Results Reported	References
Luna 9/ January 31 1966		Gas-discharge counter: 6 mm diameter, 10 mm long		The average counting rate registered by the instrument was 3·27 particles/sec during the flight from the earth to the moon and 2·06 particles/sec near and on the moon. The rather high absolute flux of cosmic ray particles of 5·35/cm² sec detected by the gas-discharge counter can be contributed to the uncertainty (~10%) in regard to the operational dimensions of the counter.	Vernov, *et al.* (1966)[160]
Luna 10/ March 31 1966	350–1017 178·3 minutes 71·9° (Lunar orbital parameters)	SBT-9 end-window gas-discharge counter Shielded with copper 2·5 g/cm² thick Outside satellite	Hard cosmic and solar radiation Protons >50 MeV Electrons >5 MeV	Corresponding data concerning solar and cosmic radiation near the moon, as well as radiation reflected from the surface of the moon are being analyzed and will be published later.	Grigorov, *et al.* (1966)[39–40]
		SBT-9 end-window gas-discharge counter End-window of mica, area 0·2 cm² and 1·2 mg/cm² thick, was covered with gold 0·3 mg/cm² thick to obtain sensitivity for soft X-ray radiation Outside satellite	Protons >500 keV Electrons >40 keV X-rays <10Å		
		SF gas-discharge counter Self-quenching type filled with a neon-oxygen mixture (three) Aluminum shielded Aperture materials: aluminum foil: 2·7 mg/cm² thick; organic material: 1·1 mg/cm² thick Aperture area: 0·5 cm² Field of view: 1 steradian Outside satellite	Protons >800 keV Electrons >50 keV X-rays <14Å		Grigorov, *et al.* (1966)[39] Mandel'shtam, *et al.* (1966)[96]

<table>
<tr><th>Vehicle/Launch Date</th><th>Height Perigee-Apogee, km</th><th>Instrument Employed</th><th>Type and Range of Measurement</th><th>Major Results Reported</th><th>References</th></tr>
<tr><td colspan="6">Charged–Particle Radiation Measurements (Ion Density)</td></tr>
<tr><td>Artificial Earth Satellites</td><td></td><td></td><td></td><td></td><td></td></tr>
<tr><td>Sputnik 3/
May 15 1958</td><td>226–1881</td><td>Spherical two-electrode ion trap mounted on long rod
(two)
Outer electrode (wire-mesh) connected to satellite hull through 10^3 ohm resistor
Inner sphere–collector
Voltage: –150 volts
Sawtooth voltage pulses ± 20 volts applied to outer electrode every two seconds of 0·2-second duration.</td><td>Positive ions</td><td>Ion density measured during different passes of the satellite at altitudes from 500–795 km were
<table>
<tr><th>Altitude, km</th><th>Number of Passes</th><th>Ion Density at an Altitude cm^{-3}</th></tr>
<tr><td>From 500–600</td><td>68</td><td>$5{\cdot}9 \times 10^5$</td></tr>
<tr><td></td><td>216</td><td>$4{\cdot}0 \times 10^5$</td></tr>
<tr><td></td><td>256</td><td>$2{\cdot}0 \times 10^5$</td></tr>
<tr><td>From 600–700</td><td>5</td><td>$2{\cdot}5 \times 10^5$</td></tr>
<tr><td></td><td>42</td><td>$2{\cdot}6 \times 10^5$</td></tr>
<tr><td></td><td>56</td><td>$4{\cdot}2 \times 10^5$</td></tr>
<tr><td>795</td><td>–</td><td>$1{\cdot}8 \times 10^5$</td></tr>
</table></td><td>Gringauz (1962)[48]
Gringauz, et al. (1957)[62]
Krasovskiy (1959)[75]</td></tr>
<tr><td>Cosmos 2/
April 6 1962</td><td>212–1546</td><td>Three-electrode type ion trap with flat grids
(eight)
External grid connected to satellite hull</td><td>Positive ions</td><td>Steep ion density changes (from 1 to 16×10^5 ions/cm^3) were observed at altitudes of $\sim$600 km, a transitional region, where changes from the predominance of heavy ions at lower altitudes to the predominance of lighter ions at higher altitudes occur.</td><td>Afonin, et al. (1965)[3]</td></tr>
<tr><td></td><td></td><td>Spherical three-electrode ion trap mounted on rod 65 cm long
(two)
Sawtooth voltage pulses applied to outer grid
Negative voltage on inner grid relative to collector</td><td></td><td>An analysis of the data from spherical ion traps made it possible to estimate ion density at altitudes from 520–650 km.</td><td>Gringauz, et al. (1963)[59]</td></tr>
<tr><td></td><td></td><td>Langmuir probe
(three)
Cylindrical probe, 1 cm diameter, 20 cm long
Sawtooth voltage pulses applied to probe</td><td>Electron density
Electron temperature</td><td>Electron density of $2{\cdot}5 \times 10^5/cm^3$ was measured at 285 km altitude and $0{\cdot}5 \times 10^5/cm^3$ at 500 km.</td><td></td></tr>
</table>

TABLE 10.2 (Continued)

Results of Solar and Cosmic Radiation Studies Aboard Soviet Rockets, Satellites, and Space Probes (1957–1966)

Vehicle/Launch Date	Height Perigee-Apogee, km	Instrument Employed	Type and Range of Measurement	Major Results Reported	References
		Honeycomb-type three-electrode ion trap with collector and antiphoto-electron grid (two) The external honeycomb adapter consisted of hexagonal tubes Traps mounted on rod 20 cm long Outer grid connected to satellite hull Inner grid voltage: −100 volts Collector voltage: −60 volts	Positive ion temperature	The trap had directional properties. Peak current occurred when the velocity vector of the incoming ion stream coincided with the normal to the collector.	Gringauz, *et al.* (1965)[56]
		Three-electrode type semi-spherical ion trap (three) Collector functioned as photo-electron emitter Outer grid voltage: +36 volts Inner grid voltage: −63 to −35 volts Emitter (collector) voltage: −41 volts	Photoemissions	The effect of solar radiation on the photo-electron emission from metal surfaces (collector of ion trap) was studied.	Gringauz, *et al.* (1963)[58]
Cosmos 3/ April 24 1962		Four-electrode ion trap with flat grids and ring electrode (two) Outer and fourth grid connected to housing Second grid voltage: −40 V Third grid supplied with variable voltage: 0·15, 3·0, 6·0, and 11 kV Ring electrode collector placed in permanent magnetic field	Positive ions >0·15, 3·0, 6·0, and 11·0 keV Electrons >5 keV	The negative potential of the second grid to the vehicle body prevented the penetration of thermal electrons of the ionosphere into the trap. The magnetic field surrounding the ring collector reduced the sensitivity of the trap to electrons of energies below 5 keV. The range of the trap current amplifier permitted registration of positive isotropic ion fluxes inside a solid angle of about 1 steradian between 3×10^5 and 5×10^6 ions/cm^2 sec ster and anisotropic fluxes between 3×10^6 and 5×10^7 ions/cm^2 sec ster.	Gal'perin, *et al.* (1963)[25–26] Krasovskiy, *et al.* (1963)[79]

Vehicle/Launch Date	Height Perigee-Apogee, km	Instrument Employed	Type and Range of Measurement	Major Results Reported	References
Cosmos 5/ May 28 1962		Four-electrode ion trap with flat grids and ring electrode (two) Outer, third and fourth grid connected to housing Second grid voltage: +24 V Ring electrode collector placed in permanent magnetic field	Ions >24 eV	The range of the trap current amplifier permitted registration of positive isotropic ion fluxes between 3×10^5 and 3×10^8 ions/cm² sec ster and anisotropic fluxes between 3×10^6 and 3×10^9 ions/cm² sec ster.	Gal'perin, *et al.* (1963)[25–26] Krasovskiy, *et al.* (1963)[79]
Cosmos 12/ December 22 1962		Spherical electrostatic analyzer Collector inlet grid voltage: −20 V Grid close to collector connected to housing	Positive ions >1·0 keV Electrons >0·5 keV >1·0 keV	The fluxes of 1 keV electrons and ions on the daylight side of the earth were not more than a few times 10^7/cm² sec keV.	Vernov, *et al.* (1965)[143–145]
Cosmos 15/ April 22 1963		Spherical electrostatic analyzer Entrance grid voltage: +12 V Collector inlet grid voltage: −20 V Grid close to collector connected to housing	Positive ions >1·0 keV Electrons >1·0 keV	An apparently permanent flux of 1 keV electrons with an intensity of 2×10^8/cm² sec keV was recorded during Cosmos 12 and Cosmos 15 flights in the vicinity of southern New Zealand (longitude 180±30° E, latitude 60–65° S).	Vernov, *et al.* (1965)[143–145]
Electron 2/ January 30 1964	67 000	Spherical electrostatic analyzer (two) Collector inlet grid voltage: −20 V Grid close to collector connected to housing	Positive ions Electrons 1. 0·1, 0·2, 0·5, 1·0 keV 2. 1·0, 2·5, 5·0, 10 keV	Fluxes of positive ions with energies from 0·1 to 10 keV did not exceed $\sim 5 \times 10^7$/cm² sec. Inside the magnetosphere, an extensive spatial region of increased electron intensity with energies from 0·1 to 10 keV has been measured outside the trapped electron zone with energies ⩾150 keV. The electron fluxes reached values of $\sim 10^9$/cm² sec keV with an energy of 0·2 keV and $\sim 5 \times 10^7$/cm² sec keV for 10 keV. The region extended farther for electrons with energies of ⩽1 keV than for those of ∼10 keV.	Vernov, *et al.* (1965)[144–145]
	26 000 20 000 2000	Three-electrode type ion trap with flat grids Outer grid potential equal to satellite body potential Inner grid voltage: −100 Volts	Positive ions Electrons ⩾100 eV	The maximum electron flux of 3×10^8/cm² sec was measured on February 2 1964 at an altitude of 26 000 km during a measuring period from January 30 to February 17 1964. The positive ion density was 100/cm³ at 20 000 km and 5000/cm³ at 2000 km on February 2 1964.	Bezrukikh, *et al.* (1965)[14] Gringauz, *et al.* (1965)[54, 60]

TABLE 10.2 (Continued)

Results of Solar and Cosmic Radiation Studies Aboard Soviet Rockets, Satellites, and Space Probes (1957–1966)

Vehicle/Launch Date	Height Perigee-Apogee, km	Instrument Employed	Type and Range of Measurement	Major Results Reported	References
Lunar and Interplanetary Probes					
Luna 1/ January 2 1959	50 000 to 82 000	Three-electrode type semi-spherical ion traps with semispherical inner anti-photoelectron grid (four) Radius of thin metallic outer grid: 30 mm Radius of thin metallic inner grid: 22·5 mm Radius of collector: 20 mm Outer grid voltages: Trap 1 −10 volts Trap 2 0 volts Trap 3 +15 volts Trap 4 +15 volts (The voltage potential is negative, equal, or positive in relation to the probe body) Inner-grid voltage: −200 volts Collector voltage: 90 volts	Protons >25 eV Electrons >200 eV	The statistically most valuable data were obtained from the probe Luna 2 although the motion of the probe along its trajectory, accompanied by complex rotational movements, caused oscillations in the collector currents as the orientation of each trap relative to the velocity vector and to the direction to the sun changed continuously. The electron flux in the outer radiation zone did not exceed an upper value of 2 to $3 \times 10^7/cm^2$ sec. An electron flux of 1·5 to $4 \times 10^8/cm^2$ sec has been recorded at distances from 56 400 to 81 400 km from the center of the earth. These measurements, confirmed by the data of Luna 1, gave reason to postulate the existence of an outermost zone of charged particles which surrounds the earth and is located beyond the radiation zones (van Allen belts). The earth is surrounded by a plasma with an ion density of about $1000/cm^3$ extending up to a height of 22 000 km above the earth's surface. The density of this plasma decreases regularly with height. The density of the interplanetary gas in the vicinity of the earth is less than $100/cm^3$ and in all probability less than $30/cm^3$. The energy spectrum of electrons in the region of the maximum density of the outer radiation zone is substantially harder than beyond its outer boundary.	Gringauz (1961)[47–48] Gringauz, *et al.* (1960)[55,61] (1964)[50]
Luna 2/ September 12 1959		Three-electrode type semi-spherical ion trap with flat inner grid (four) Radius of outer nickel grid: 30 mm Outer grid voltages: Trap 1 −10 volts Trap 2 − 5 volts Trap 3 0 volts Trap 4 +15 volts Inner flat tungsten grid voltage: −200 volts Flat nickel collector voltages: −60 to −40 volts	Protons >25 eV Electrons >200 eV		

Vehicle/Launch Date	Height Perigee-Apogee, km	Instrument Employed	Type and Range of Measurement	Major Results Reported	References
Luna 3/ October 4 1959		Three-electrode type semi-spherical ion trap with flat inner grid (four) Radius of outer nickel grid: 30 mm Outer grid voltages: Trap 1 +25 volts Trap 2 −10 volts Traps 3 and 4 +9 to −19 volt sawtooth pulses of 18 sec periods Inner flat grid voltage: −200 volts Flat collector voltage: −90 volts	Protons >25 eV Electrons >200 eV		
Venus 1/ February 12 1961	30 000 to 45 000 1 900 000	Three-electrode type semi-spherical ion trap with flat inner grid (two) Radius of outer nickel grid: 30 mm Outer Grid voltages: Trap 1 0 volt Trap 2 +50 volts Inner flat tungsten grid voltage: −200 volts Flat nickel collector voltages: −60 to −40 volts	Positive ions >50 eV Electrons >200 eV	During the measurements the traps retained definite orientation with respect to the direction to the sun and to the velocity vector. During the first reception of signals the current of both traps oscillated near the zero value. The probe was at a distance of 30 000 to 45 000 km from the center of the earth, i.e., in the outer part of the outer radiation zone. During the third radio contact (1 900 000 km) the current of the trap 1 (0 volt) was $3{\cdot}3 \times 10^{-9}$ amp. and of trap 2 (+50 volts) was 2×10^{-9} amp. This value apparently determined the flux of the positive corpuscular stream ($\sim 10^{-9}/cm^2$ sec) which took place during the radio contact. During the third radio contact the commencement of a magnetic storm was observed on earth. The solar corpuscular streams recorded with the Venus probe were similar to those recorded by Luna 3, however, more intensive than those obtained during the flight of Luna 2.	Gringauz (1961)[47,49] Gringauz, *et al.* (1961)[51]

TABLE 10.2 (Continued)

Results of Solar and Cosmic Radiation Studies Aboard Soviet Rockets, Satellites, and Space Probes (1957–1966)

Vehicle/Launch Date	Height Perigee-Apogee, km	Instrument Employed	Type and Range of Measurement	Major Results Reported	References
Mars 1/ November 1 1962		Three-electrode type semi-spherical ion traps with flat inner grid (two) Radius of outer nickel grid: 30 mm Outer grid voltages: Trap 1 0 volt; Trap 2 +50 volts Inner flat tungsten grid voltage: −80 volts Flat nickel collector voltage: 0 volt	Positive ions >50 eV Electrons >80 eV	Measurements of charged particles in the outermost zone were made at much lower distances from the earth than first observed with Luna 2 because the Mars probe passed through regions of considerably higher geomagnetic latitude. The measured maximum electron flux in the outermost zone was approximately $4 \times 10^8/cm^2$ sec.	Bezrukikh, *et al.* (1965)[12–13] Gringauz, *et al.* (1963)[53]
		Multielectrode (modulation) ion trap with flat electrodes (one trap) Similar in design to modulation trap of Zond 2			
Zond 2/ November 30 1964		Integral trap with four plane electrodes (several traps) Outer grid was connected to the probe container Second grid voltage: ±50 volts Third grid voltage: −70 volts	Positive ions >50 eV Electrons >70 eV	Measurements of solar wind proton streams and their energy spectra were made with integral and modulation ion traps during radio communications with the probe. The modulation trap faced the sun during the measurements. A positive ion flux intensity of 3 to $7 \times 10^7/cm^2$ sec in the energy range up to 900 eV was recorded at a distance of $1{\cdot}7 \times 10^6$ km from the earth. Solar plasma fluxes of $\sim 10^9/cm^2$ sec were measured during strong geomagnetic disturbances on November 22 and 30 1964.	Bezrukikh, *et al.* (1965)[12–13] Gringauz, *et al.* (1965)[60]
	1 700 000	Modulation trap with flat modulation and screen grids The modulation voltage was the sum of a constant voltage of eight values from ∼230 to ∼3200 volts and an alternating voltage of square shape, amplitude ∼450 volts and modulation frequency of ∼1000 Hz	Protons <3·6 keV Electrons >70 eV		

Vehicle/Launch Date	Height Perigee-Apogee, km	Instrument Employed	Type and Range of Measurement	Major Results Reported	References
		A system of screen grids separated the modulation grid from the collector Voltage of electron suppression grid: −70 volts			
Zond 3/ July 18 1965	120 000	Modulation traps with flat modulation grid and screen grids	Positive ions keV Electrons eV	Continuous measurements were made by Venus 2 up to distances of 120 000 km from the earth. A cold plasma envelope was recorded up to a height of about 20 000 km. The recording of solar plasma streams (which corresponded to the space vehicle's passage through the collisionless shock wave front) began at a geocentric distance of about 90 000 km. The maximum registered energy was 3·2 keV. Considerable streams of solar wind particles (about $10^9/cm^2$ sec) of low energy corresponded with observations of small geomagnetic disturbances.	Blagonravov (1966)[15]
Venus 2/ November 12 1965	20 000	Modulation traps with flat modulation grid and screen grids	Positive ions keV Electrons eV		
Venus 3/ November 16 1965	90 000	Modulation traps with flat modulation grid and screen grids	Positive ions keV Electrons eV		
Luna 10/ March 31 1966	350–1017 178·3 minutes 71·9° (lunar orbital elements)	Four-electrode type ion trap with plane electrodes (two) Outer grid was connected to the lunar satellite container Second grid voltage: change in two-minute intervals from 0 to +50 volts	Positive ions >50 eV Electrons >70 eV	Ion-density measurements were made with the moon (and Luna 10 satellite) in the tail of the earth's magnetosphere and outside the earth's magnetosphere.	Gringauz, *et al.* (1966)[52]
		Modulation trap	Low-energy positive ions		
		Ion trap	Positive ions Electrons		

Owing to latitude and longitude effects and to the specific distribution of radiation at the altitude of Spaceship-Satellite 2, the dosage per orbit ranged from 0·35 to 0·70 mrad[108–109].

One scintillation counter with a sodium iodide crystal and two gas-discharge counters were installed in Spaceship-Satellite 2, and one scintillation counter with a cesium iodide crystal was mounted outside the satellite. This group of counters functioned as a radiometric instrumentation system, which was employed also in other Soviet satellites, with slight variations, to obtain data on the absorbed dosage on-board. Table 10.3 presents the various instrumentation combinations used in Spaceship-Satellite 3 and several satellites of the Cosmos series[154]. Detailed description of this dosimetric instrumentation has been published by Papkov, *et al.*[104].

The outside scintillation counter was sometimes replaced by a gas-discharge counter telescope which allowed the counting of particles arriving from a well defined direction only. In two other satellites, Cosmos 12 and 15, the Soviets used a spherical electrostatic analyzer which was sensitive to soft radiation particles, electrons, and ions, with energies $\sim$1 keV. Table 10.4 presents the counting rates obtained from the STS–5 gas-discharge counter aboard unmanned satellites which had been used for the determination of internal dose rates[112, 154]. Total radiation dosages for Soviet biosatellites through Voskhod 2 are shown in Table 14.1.

More sophisticated radiometric equipment was installed on the Cosmos 41 satellite. This instrumentation consisted of a scintillation counter with a sodium iodide crystal (threshold 90 keV) and a FEU–16 photomultiplier, three type SBT–9, two type Si–3BG, and one type STS–5 gas-discharge counters, and two solid state *n-p* counters (proton sensors). This group of instruments was mounted outside the satellite. One additional Si–3BG counter was located inside Cosmos 41. Experimental data in regard to absorbed dosage or tissue dose rates have not been published. However, the magnitude of the proton intensity, registered by the solid state detector amounted to 2×10^5 protons/cm^2 sec ster with energies from 0·4 to 3 MeV and to $\leqslant 3$ protons/cm^2 sec ster with energies from 3 to 8 MeV at $L = 3{\cdot}5$ and $B = 5{\cdot}4 \times 10^{-2}$ gauss. The maximum intensity of protons with energies of 0·4 to 3 MeV at the geomagnetic equator ($L = 3{\cdot}5$) was calculated to be 6×10^6 protons/cm^2 sec ster[114–115, 156].

The various measurement techniques and results associated with radiation studies aboard Soviet manned satellites Vostoks 1–6 and Voskhods 1 and 2 are discussed in Chapter 14.

TABLE 10.3

Radiometric Instrumentation on Board Soviet Satellites (1957–1966)

Satellite	Instrumentation	
	Inside Satellite	Outside Satellite
Spaceship-Satellite 2	Scintillation counter with NaI(Tl), d = 30 mm, h = 14 mm crystal. Recording was performed with threshold of 25 keV anode current. Two STS–5 gas-discharge counters.	Scintillation counter with CsI(Tl), d = 30 mm, h = 2·2 mm crystal. Anode current was recorded. Minimum screening of 2 mg/cm^2 aluminum.
Spaceship-Satellite 3	Scintillation counter with NaI(Tl), d = 30 mm, h = 14 mm crystal. Recording was done with threshold of 25 keV and anode current. STS–5 gas-discharge counter.	Scintillation counter with CsI(Tl), d = 30 mm, h = 2·2 mm crystal. Anode current was recorded. Minimum screening of 2 mg/cm^2 aluminum.
Cosmos 4	Scintillation counter with CsI(Tl), d = 30 mm, h = 20 mm crystal. Recording was done with threshold of 60 keV and anode current. Two STS–5 counters.	Scintillation counter with CsI(Tl), d = 30 mm, h = 20 mm crystal. Recording was done with threshold of 60 keV and anode current. Minimum screening of 2 mg/cm^2 aluminum.
Cosmos 7	Scintillation counter with CsI(Tl), d = 30 mm, h = 20 mm crystal. Recording was done with threshold of 30 KeV and anode current. Two STS–5 counters.	Scintillation counter with CsI(Tl), d = 30 mm, h = 20 mm crystal. Recording was done with threshold of 45 keV and anode current. Minimum screening of 180 mg/cm^2.
Cosmos 9	Scintillation counter with NaI(Tl), d = 30 mm, h = 20 mm crystal. Recording was done with threshold of 15 keV and anode current. STS–5 counter.	Scintillation counter with CsI(Tl), d = 30 mm, h = 20 mm crystal. Recording was done with threshold of 60 keV and anode current. Minimum screening of 180 mg/cm^2.
Cosmos 10	Scintillation counter with CsI(Tl), d = 30 mm, h = 20 mm crystal. Recording was done with threshold of 160 keV and anode current. Two STS–5 counters.	Scintillation counter with CsI(Tl), d = 26 mm, h = 4 mm crystal. Recording was done with threshold of 75 keV. Minimum screening of 2 mg/cm^2 aluminum.
Cosmos 12	Scintillation counter with CsI(Tl), d = 30 mm, h = 20 mm crystal. Recording was done w th a threshold of 30 keV and anode current. Two STS–5 counters.	Spherical electrostatic analyzer.
Cosmos 13	Gas-discharge counter STS–5.	Telescope of STS–5 counters.
Cosmos 15	Scintillation counter with CsI(Tl), d = 30 mm, h = 20 mm crystal. Recording was done with threshold of 100 keV and anode current. Two STS–5 counters.	Spherical electrostatic analyzer.
Cosmos 16	Gas-discharge counter STS–5.	Gas-discharge counters STS–5.
Cosmos 18	Gas-discharge counter STS–5.	Telescope of STS–5 counters.

TABLE 10.4

Tissue Dose Rates for Unmanned Soviet Earth Satellites[112, 154]

Satellite	Launch Date	Perigee and Apogee, km	Total Number of Impulses Recorded by Geiger Counter for 24 hours, imp/cm² day	Power of Absorbed Dosage, mrad/day	Calculated Tissue Dose Rate, mrad/day
Spaceship-Satellite 2	Aug. 19 1960	306–339	$1{\cdot}8 \times 10^5$	7*	8
Spaceship-Satellite 3	Dec. 1 1960	187–265	$1{\cdot}8 \times 10^5$	7*	8
Cosmos 4	Apr. 26 1962	298–330	$2{\cdot}5 \times 10^5$	13	16
Cosmos 7	July 28 1962	210–369	$5{\cdot}8 \times 10^5$	45	55
Cosmos 9	Sep. 27 1962	301–353	$4{\cdot}3 \times 10^5$	30	35
Cosmos 10	Oct. 17 1962	210–380	$4{\cdot}4 \times 10^5$	30	35
Cosmos 12	Dec. 22 1962	211–405	$4{\cdot}6 \times 10^5$	35	40
Cosmos 13	Mar. 21 1963	205–337	$3{\cdot}0 \times 10^5$	18	20
Cosmos 15	Apr. 22 1963	173–371	$2{\cdot}7 \times 10^5$	15	18
Cosmos 16	Apr. 28 1963	207–401	$4{\cdot}3 \times 10^5$	30	35
Cosmos 18	May 24 1963	209–301	$2{\cdot}4 \times 10^5$	12	14

* Value of absorbed dosage was measured directly with respect to energy release in the crystal of the scintillation counter.

References

1. "Details of Sputnik III", Pravda, May 18 1958.

2. "Cosmic Scientific Station—Proton 1", Pravda, August 7 1965.

3. V. V. Afonin, T. K. Breus, G. L. Gdalevich, B. N. Gorozhankin, K. I. Gringauz, and R. Ye. Rybchinskiy, "Brief Survey of the Results of Physical Experiments on the Satellite 'Cosmos 2' in the Ionosphere", Trudy Vsesoyuznoy Konferentsii po Fizike Kosmicheskogo Prostranstva, Izdatel'stvo "Nauka", Moscow, June 10–16 1965, pp. 151–167 (NASA TT F-389, pp. 202–224).

4. S. G. Alexsandrov and R. Ye. Fedorov, "Soviet Satellites and Cosmic Rockets", Izdatel'stvo Akademii Nauk SSSR, Moscow, 1959, First edition (MCL–418).

5. S. G. Aleksandrov and R. Ye. Fedorov, "Soviet Satellites and Spaceships", Izdatel'stvo Akademii Nauk SSSR, Moscow, 1961, Second edition (TT–62–121 and JPRS: 15 805).

6. K. I. Alekseyeva, L. L. Gabuniya, G. B. Zhdanov, Ye. A. Zamchalova, M. N. Shcherbakov, and M. I. Tretyakova, "Investigation of the Primary Cosmic Ray Composition at an Altitude of 320 km", Izkusstvennyye Sputniki Zemli, 1962, No. 12, pp. 6–15 (Artificial Earth Satellites, pp. 7–17).

7. L. A. Antonova, "Measurement of Corpuscular Radiation at Altitudes up to 100 km", Izvestiya Akademii Nauk SSSR, Seriya Geofizicheskaya, 1961, No. 9, pp. 1437–1438 (Byulleten [Izvestiya], Academy of Sciences USSR, Geophysics Series, pp. 935–936).

8. L. A. Antonova, "Measurement of Intensity and Spectrum of Soft Electron Flux in the Ionosphere at Altitudes of 200–500 km", Kosmicheskiye Issledovaniya, 1965, Vol. 3, No. 1, pp. 89–101 (TT–65–170, pp. 137–160).

9. L. A. Antonova and G. S. Ivanov-Kholodnyy, "Ionization in the Night Ionosphere (Corpuscular Hypothesis)", Space Research II, Proceedings of the Second International Space Science Symposium, Florence, Italy, April 10–14 1961, North-Holland Publishing Company, Amsterdam, 1962, pp. 981–992.

10. L. A. Antonova, G. S. Ivanov-Kholodnyy, N. D. Masanova, and V. S. Medvedev, "Measurement of Soft-Electron Fluxes in the Upper Atmosphere Using a Secondary-Electron Multiplier", Kosmicheskiye Issledovaniya, 1965, Vol. 3, No. 1, pp. 82–88 (TT–65–170, pp. 125–136).

11. S. I. Avdyushin, N. K. Pereyaslova, and I. Ye. Petrenko, "Ionizing Radiation Intensity Based on Measurements on Zond–1", Trudy Vsesoyuznoy Konferentsii po Fizike Kosmicheskogo Prostranstva, Izdatel'stvo "Nauka", Moscow, June 10–16 1965, pp. 511–512 (NASA TT F–389, pp. 689–691).

12. V. V. Bezrukikh, K. I. Gringauz, M. Z. Khokhlov, L. S. Musatov, and R. Ye. Rybchinskiy, "Preliminary Results of Measurements Carried Out by Means of Charged Particle Traps on the Interplanetary Station Zond 2", Space Research VI, Proceedings of the Sixth International Space Science Symposium, Mar Del Plata, Argentina, May 11–19 1965, North-Holland Publishing Company, Amsterdam, 1966, pp. 862–869.

13. V. V. Bezrukikh, K. I. Gringauz, L. S. Musatov, R. Ye. Rybchinskiy, and M. Z. Khokhlov, "Investigation of Solar Plasma Streams by the Interplanetary Station Zond 2", Doklady Akademii Nauk SSSR, 1965, Vol. 163, No. 4, pp. 873–876 (Doklady of the Academy of Sciences of the USSR, Earth Science Sections, pp. 5–7).

14. V. V. Bezrukikh, K. I. Gringauz, L. S. Musatov, and Ye. K. Solomatina, "Possible Existence of a Soft Electron Component in the Outer Radiation Zone and Its Variations", Trudy Vsesoyuznoy Konferentsii po Fizike Kosmicheskogo Prostranstva, Izdatel'stvo "Nauka", Moscow, June 10–16 1965, pp. 418–419 (NASA TT F–389, pp. 566–568).

15. USSR National Reports to COSPAR, 1960–1966, presented by A. A. Blagonravov (In English).

16. Ya. L. Blokh, L. I. Dorman, L. V. Kurnosova, V. I. Logachev, G. F. Platonov, L. A. Razorenov, V. G. Sinitsina, A. A. Suslov, and M. I. Fradkin, "Results Derived From Studying the Cosmic Ray Nuclear Component on the Satellite 'Electron 2' ", Trudy Vsesoyuznoy Konferentsii po Fizike Kosmicheskogo Prostranstva, Izdatel'stvo "Nauka", Moscow, June 10–16 1965. pp. 514–528 (NASA TT F–389, pp. 693–710).

17. A. D. Bolyunova, "Radioactivity of the Cosmos 3 Satellite After the Explosion of 9 July 1962", Kosmicheskiye Issledovaniya, 1966, Vol. 4, No. 1, pp. 167–169 (TT–66–76, pp. 273–278).

18. A. D. Bolyunova, O. L. Vaysberg, Yu. I. Gal'perin, B. P. Potapov, V. V. Temnyy, and F. K. Shuyskaya, "Preliminary Results of the Studies of Corpuscles on the Electron 1 Satellite", Space Research VI, Proceedings of the Sixth International Space Science Symposium, Mar Del Plata, Argentina, May 11–19 1965, North-Holland Publishing Company, Amsterdam, 1966, pp. 649–661.

19. A. D. Bolyunova, O. L. Vaysberg, Yu. I. Gal'perin, B. P. Potapov, V. V. Temnyy, and F. K. Shuyskaya, "Preliminary Results Derived for Corpuscle Investigations on the Satellite 'Electron 1' ", Trudy Vsesoyuznoy Konferentsii po Fizike Kosmicheskogo Prostranstva, Izdatel'stvo "Nauka", Moscow, June 10–16 1965, pp. 406–417 (NASA TT F–389, pp. 551–566).

20. A. V. Bruns and V. K. Prokof'yev, "Measurement of the Solar Far-Ultraviolet Radiation of Helium", Iskusstvennyye Sputniki Zemli, 1961, No. 11, pp. 15–22 (Artificial Earth Satellites, pp. 17–25).

21. A. V. Bruns and V. K. Prokof'yev, "Spectrometer for Measuring Solar Radiation in the Far Ultraviolet", Iskusstvennyye Sputniki Zemli, 1961, No. 11, pp. 23–29 (Artificial Earth Satellites, pp. 26–32).

22. A. Ye. Chudakov, "Investigation of Photons With the Aid of the Third Man-Made Satellite", Priroda, 1958, No. 12, pp. 88–90 (FTS–9974, pp. 77–82).

23. A. Ye. Chudakov, S. N. Vernov, P. V. Vakulov, Yu. I. Logachev, and A. G. Nikolayev, "Radiation Measurements During the Flight of the Second Soviet Space Rocket", Space Research, Proceedings of the First International Space Science Symposium, Nice, France, January 11–16 1960, North-Holland Publishing Company, Amsterdam, 1960, pp. 845–851.

24. G. S. Dragun, L. V. Kurnosova, V. I. Logachev, L. A. Razorenov, I. A. Sirotkin, and M. I. Fradkin, "Equipment for the Investigation of the Nuclear Component of Cosmic Rays on Cosmic Rockets and Artificial Earth Satellites", Iskusstvennyye Sputniki Zemli, 1961, No. 9, pp. 86–110 (Artificial Earth Satellites, pp. 115–138).

25. Yu. I. Gal'perin and V. I. Krasovskiy, "Investigations of the Upper Atmosphere Using the Artificial Earth Satellites Cosmos 3 and Cosmos 5", Space Research IV, Proceedings of the Fourth International Space Science Symposium, Warsaw, Poland, June 4–10 1963, North-Holland Publishing Company, Amsterdam, 1964, pp. 563–571.

26. Yu. I. Gal'perin and V. I. Krasovskiy, "Investigations of the Upper Atmosphere With Satellites Cosmos 3 and Cosmos 5. 1. Instrumentation", Kosmicheskiye Issledovaniya, 1963, Vol. 1, No. 1, pp. 126–131 (TT-63-968, pp. 223–234).

27. V. L. Ginzburg, L. V. Kurnosova, V. I. Logachev, L. A. Razorenov, and M. I. Fradkin, "Observation of the Radiation Anomalies at the Altitudes of 200–300 km", presented at the International Conference on Cosmic Rays and the Earth Storm, Kyoto, Japan, September 4–15 1961.

28. V. L. Ginzburg, L. V. Kurnosova, V. I. Logachev, L. A. Razorenov, and M. I. Fradkin, "Short-Lived Increments in the Intensity of the Nuclear Component of Cosmic Radiation, Associated With Solar Activity, and Investigation of the Radiation Intensity at 200–300 km Altitude", Izvestiya Akademii Nauk SSSR, Seriya Fizicheskaya, 1962, Vol. 26, No. 6, pp. 782–798 (Bulletin [Izvestiya] Academy of Sciences, USSR, Physics Series, pp. 784–799).

29. V. L. Ginzburg, L. V. Kurnosova, V. I. Logachev, L. A. Razorenov, I. A. Sirotkin, and M. I. Fradkin, "Investigating the Intensity of Charged Particles During Flight of the Second and Third Satellite Ships", Iskusstvennyye Sputniki Zemli, 1961, No. 10, pp. 22–33 (Artificial Earth Satellites, pp. 157–170).

30. V. L. Ginzburg, L. V. Kurnosova, L. A. Razorenov, and M. I. Fradkin, "Some Space Research Utilizing Satellites and Rockets", Aviatsiya i Kosmonavtika, 1962, No. 6, pp. 11–22.

31. V. L. Ginzburg, L. V. Kurnosova, L. A. Razorenov, and M. I. Fradkin, "Some Investigations of the Cosmic Ray Nuclear Component and of the Radiation Belts of the Earth on Soviet Satellites and Rockets. Review", Geomagnetizm i Aeronomiya, 1962, Vol. 2, No. 2, pp. 193–232 (Geomagnetism and Aeronomy, pp. 165–195).

32. V. L. Ginzburg, L. V. Kurnosova, L. A. Razorenov, and M. I. Fradkin, "Investigations of the Nuclear Component of Cosmic Rays Made on Soviet Satellites and Rockets", Uspekhi Fizicheskikh Nauk, 1964, Vol. 82, No. 4, pp. 585–647 (JPRS: 25 029).

33. V. L. Ginzburg, L. V. Kurnosova, L. A. Razorenov, S. I. Syrovatskiy, and M. I. Fradkin, "Problems of and Prospects for Primary Cosmic Ray Research", Trudy Vsesoyuznoy Konferentsii po Fizike Kosmicheskogo Prostranstva, Izdatel'stvo "Nauka", Moscow, June 10–16 1965, pp. 486–501 (NASA TT F-389, pp. 657–677).

34. G. Goltsoff, "The Second Soviet Cosmic Ship", Izvestiya, September 7 1960, No. 213.

35. Ye. V. Gorchakov, "The Spatial Orientation of the Outer Radiation Belt of the Earth and the Auroral Zones", Iskusstvennyye Sputniki Zemli, 1961, No. 8, pp. 81–83 (Artificial Earth Satellites, pp. 242–244).

36. Ye. V. Gorchakov, "The Outer Radiation Belt and Aurorae", Iskusstvennyye Sputniki Zemli, 1961, No. 9, pp. 66–70 (Artificial Earth Satellites, pp. 96–100).

37. N. L. Grigorov, L. F. Kalinkin, A. S. Melioranskiy, V. Ye. Nesterov, Ye. A. Pryakhin, I. A. Savenko, and I. V. Estulin, "A Study of High-Energy Gamma Quanta

at the Upper Limits of the Atmosphere", Izvestiya Akademii Nauk SSSR, Moscow, Seriya Fizicheskaya, 1966, Vol. 30, No. 11, pp. 1765–1767 (JPRS: 40 863).

38. N. L. Grigorov, Yu. S. Klitsov, V. Ye. Nesterov, I. D. Rapoport, I. A. Savenko, and B. M. Yakovlev, "Study of High-Energy Electrons Using the Proton 1 and 2 Artificial Earth Satellites", Izvestiya Akademii Nauk SSSR, Seriya Fizicheskaya, 1966, Vol. 30, No. 11, pp. 1773–1775 (JPRS: 40 863).

39. N. L. Grigorov, V. L. Maduyev, S. L. Mandel'shtam, N. F. Pisarenko, I. A. Savenko, and I. P. Tindo, "Study of Soft Corpuscular Radiation by the Lunar Satellite 'Luna 10' ", Doklady Akademii Nauk SSSR, 1966, Vol. 170, No. 3, pp. 567–569 (JPRS: 38 795).

40. N. L. Grigorov, V. L. Maduyev, N. F. Pisarenko, and I. A. Savenko, "Investigation of Cosmic Radiation by the Artificial Satellite 'Luna 10' ", Doklady Akademii Nauk SSSR, 1966, Vol. 170, No. 3, pp. 565–566 (JPRS: 38 794).

41. N. L. Grigorov, V. Ye. Nesterov, I. D. Rapoport, I. A. Savenko, and G. A. Skuridin, "Investigation Into High-Energy and Ultra High-Energy Particles and the Electron-Photon Component of Primary Cosmic Rays From Proton–1 Satellite", Paper presented to the International Cosmic Ray Conference, London, September 1965.

42. N. L. Grigorov, V. Ye. Nesterov, I. D. Rapoport, I. A. Savenko, and G. A. Skuridin, "Measurements of High Energy Cosmic Rays by Proton–1", 16th International Astronautical Congress, Athens, Greece, September 13–18 1965.

43. N. L. Grigorov, I. A. Savenko, and G. A. Skuridin, "Proton Space Stations and First Results Obtained by Them", COSPAR Information Bulletin, April 1966, No. 31, pp. 25–31.

44. N. L. Grigorov, I. A. Savenko, G. A. Skuridin, "The Proton Space Stations", Pravda, November 16 1965.

45. N. L. Grigorov, D. A. Zhuravlev, M. A. Kondrat'yeva, I. D. Rapoport, and I. A. Savenko, "Search for Antimatter in Cosmic Rays and in the Cosmic Space", Iskusstvennyye Sputniki Zemli, 1961, No. 10. pp. 96–97 (Artificial Earth Satellites, pp. 232–233).

46. N. L. Grigorov, D. A. Zhuravlev, M. A. Kondrat'yeva, I. D. Rapoport, and I. A. Savenko, "Investigations of Cosmic Radiation Beyond the Atmosphere", Kosmicheskiye Issledovaniya, 1963, Vol. 1, No. 3, pp. 436–442 (TT–64–90, pp. 163–176).

47. K. I. Gringauz, "Some Results of Experiments in Interplanetary Space by Means of Charged Particle Traps on Soviet Space Probes", Space Research II, Proceedings of the Second International Space Science Symposium, Florence, Italy, April 10–14 1961, North-Holland Publishing Company, Amsterdam, 1962, pp. 539–553.

48. K. I. Gringauz, "The Structure of the Earth's Ionized Gas Envelope Based on Local Charged Particle Concentrations Measured in the USSR", Space Research II, Proceedings of the Second International Space Science Symposium, Florence, Italy, April 10–14 1961, North-Holland Publishing Company, Amsterdam, 1962, pp. 574–592.

49. K. I. Gringauz, "Investigation of Interplanetary Plasma and Planetary Ionospheres by Means of Charged Particle Traps on Space Rockets", 12th International Astronautical Congress, Washington, D.C., October 1–7 1961, Academic Press, Inc., New York and London, 1963, Vol. 2, pp. 702–713.

50. K. I. Gringauz, S. M. Balandina, G. A. Bordovskii, and N. M. Shutte, "On the Results of Experiments With Three-Electrode Charged-Particle Counters in the Second Radiation Belt and the Outermost Charged-Particle Belt", Iskusstvennyye Sputniki Zemli, 1964, No. 15, pp. 92–97 (Artificial Earth Satellites, pp. 94–99).

51. K. I. Gringauz, V. V. Bezrukikh, S. M. Balandina, V. D. Ozerov, and R. Ye. Rybchinskiy, "Direct Observations of Solar Plasma Streams at a Distance of 1 900 000 km From the Earth on February 17 1961, and Simultaneous Observations of the

Geomagnetic Field", Space Research III, Proceedings of the Third International Space Science Symposium, Washington, D.C., May 2–8 1962, North-Holland Publishing Company, Amsterdam, 1963, pp. 602–607.

52. K. I. Gringauz, V. V. Bezrukikh, M. Z. Khokhlov, G. N. Zastenker, A. P. Remizov, and L. S. Musatov, "Experimental Results From Observations of the Lunar Ionosphere Performed by the First Artificial Lunar Satellite", Doklady Akademii Nauk SSSR, 1966, Vol. 170, No. 6, pp. 1306–1309 (JPRS: 39 455).

53. K. I. Gringauz, V. V. Bezrukikh, L. S. Musatov, R. Ye. Rybchinskiy, and S. M. Sheronova, "Measurements Made in the Earth's Magnetosphere by Means of Charged Particle Traps Aboard the Mars 1 Probe", Space Research IV, Proceedings of the Fourth International Space Science Symposium, Warsaw, Poland, June 4–10 1963, North-Holland Publishing Company, Amsterdam, 1964, pp. 621–626.

54. K. I. Gringauz, V. V. Bezrukikh, L. S. Musatov, R. Ye. Rybchinskiy, Ye. K. Solomatina, "Some Results of Measurements Carried Out by Means of Charged Particle Trap on the Electron–2 Satellite", Space Research VI, Proceedings of the Sixth International Space Science Symposium, Mar Del Plata, Argentina, May 11–19 1965, North-Holland Publishing Company, 1966, pp. 850–861.

55. K. I. Gringauz, V. V. Bezrukikh, V. D. Ozerov, and R. Ye. Rybchinskiy, "Study of the Interplanetary Ionized Gas, the Electrons and the Corpuscular Radiation of the Sun by Means of Three-Electrode Traps for Charged Particles Fitted in the Second Soviet Space Rocket", Doklady Akademii Nauk SSSR, 1960, Vol. 131, No. 6, pp. 1301–1304 (Soviet Physics Doklady, 1960, Vol. 5, No. 2, pp. 361–364).

56. K. I. Gringauz, B. N. Gorozhankin, G. L. Gdalevich, V. V. Afonin, R. Ye. Rybchinskiy, and N. M. Shutte, "The Technique and Results of Experiments Conducted on the Cosmos–2 Satellite by Means of Langmuir Probes and Ion Traps of the Honeycomb Type", Space Research V, Proceedings of the Fifth International Space Science Symposium, Florence, Italy, May 12–16 1964, North-Holland Publishing Company, Amsterdam, 1965, pp. 733–750.

57. K. I. Gringauz, B. N. Gorozhankin, N. M. Shutte, and G. L. Gdalevich, "Changes of the Distribution of Charged Particle Density With Height and of the Ionic Composition of the Outer Ionosphere Since the Solar Activity Maximum According to Data Collected by Ion Traps on the Cosmos–2 Satellite", Space Research IV, Proceedings of the Fourth International Space Science Symposium, Warsaw, Poland, June 4–10 1963, North-Holland Publishing Company, Amsterdam, 1964, pp. 473–479.

58. K. I. Gringauz, B. N. Gorozhankin, N. M. Shutte, and G. L. Gdalevich, "Some Experiments Carried Out Aboard the Satellite Cosmos 2", Astronautyka, 1963, No. 4, pp. 2–4; 14th International Astronautical Congress, Paris, France, September 1963.

59. K. I. Gringauz, B. N. Gorozhankin, N. M. Shutte, and G. L. Gdalevich, "Altitude Distribution of Charged Particles in the Ionosphere and the Transitional Region Between the Oxygen and Helium Ionic Layers From Cosmos 2 Ion-Trap Experiments", Doklady Akademii Nauk SSSR, 1963, Vol. 151, No. 3, pp. 560–563.

60. K. I. Gringauz and M. Z. Khokhlov, "The Outermost Zone of Charged Particles", Trudy Vsesoyuznoy Konferentsii po Fizike Kosmicheskogo Prostranstva, Izdatel'stvo "Nauka", Moscow, June 10–16 1965, pp. 467–482 (NASA TT F–389, pp. 633–652).

61. K. I. Gringauz, V. G. Kurt, V. I. Moroz, and I. S. Shklovskiy, "The Results of Observations Obtained by Means of Charged Particle Traps on Soviet Cosmic Rockets up to R = 100 000 km", Astronomicheskiy Zhurnal, 1960, Vol. 37, No. 4, pp. 716–735 (Soviet Astronomy AJ, 1961, Vol. 4, No. 4, pp. 680–695).

62. K. N. Gringauz and M. Kh. Zeligman, "Measurement of the Concentrations of Positive Ions Along the Orbit of an Artificial Satellite", Uspekhi Fizicheskikh Nauk, 1957, Vol. 63, No. 1, pp. 239–252 (Advances in Physical Sciences, pp. 321–339).

63. V. I. Ivanov, I. B. Keirim-Markus, and Ye. Ye. Kovalev, "Doses of Cosmic Radiation", Iskusstvennyye Sputniki Zemli, 1961, No. 12, pp. 35–47 (Artificial Earth Satellites, pp. 40–51).

64. V. P. Kachalov, N. A. Pavlenko, and A. V. Yakovleva, "Ultraviolet Spectrum of the Sun in the Region 2471–2635Å", Izvestiya Akademii Nauk SSSR, Seriya Geofizicheskaya, 1958, No. 9, pp. 1099–1104 (Bulletin [Izvestiya] Academy of Sciences, USSR, Geophysics Series, pp. 636–640).

65. S. A. Kaplan, V. V. Katyushina, and V. G. Kurt, "Intensity Measurements of Scattered Ultraviolet Radiation (1216 Å and 1300 Å) in the Upper Atmosphere", Space Research V, Proceedings of the Fifth International Space Science Symposium, Florence, Italy, May 12–16 1964, North-Holland Publishing Company, Amsterdam, 1965, pp. 595–611.

66. S. A. Kaplan and V. G. Kurt, "Investigation of Scattered Ultraviolet Radiation in the Upper Atmosphere of the Earth. 5. Interpretation of Observations of the OI Triplet (λ 1300 Å) in the Upper Atmosphere", Kosmicheskiye Issledovaniya, 1965, Vol. 3, No. 2, pp. 256–261 (TT–65–414, pp. 124–134).

67. A. M. Kasatkin, "High-Altitude Optical Station for the Investigation of the Atmosphere", Iskusstvennyye Sputniki Zemli, 1963, No. 15, pp. 3–21 (Artificial Earth Satellites, pp. 3–21).

68. V. V. Katyushina, "Investigation of the Scattered Ultraviolet Radiation in the Upper Atmosphere of the Earth. 3. Measurement of the Intensity of the Upper-Atmosphere Glow in the Lines of the OI Triplet ($\lambda \sim 1300$ Å) at Altitudes of 100–500 km", Kosmicheskiye Issledovaniya, 1965, Vol. 3, No. 2, pp. 248–251 (TT–65–414, pp. 110–116).

69. V. V. Katyushina, "Measurements of Solar Lα Radiation Absorption in the Earth's Upper Atmosphere", Kosmicheskiye Issledovaniya, 1965, Vol. 3, No. 3, pp. 496–498 (TT–65–828, pp. 256–263).

70. V. V. Katyushina and V. G. Kurt, "Investigation of Scattered Ultraviolet Radiation in the Upper Atmosphere of the Earth. 2. Measurements of Scattered Lα-Radiation in the Upper Atmosphere at Altitudes up to 500 km", Kosmicheskiye Issledovaniya, 1965, Vol. 3, No. 2, pp. 243–247 (TT–65–414, pp. 102–110).

71. T. V. Kazachevskaya, V. A. Arkhangel'skaya, G. S. Ivanov-Kholodnyy, V. S. Medvedev, T. K. Razumova, and A. V. Chudakin, "Measurement of X-Rays and Ultraviolet Radiation With a Thermoluminescent Phosphor, $CaSO_4$(Mn)", Iskusstvennyye Sputniki Zemli, 1963, No. 15, pp. 71–80 (Artificial Earth Satellites, pp. 73–82).

72. T. V. Kazachevskaya and G. S. Ivanov-Kholodnyy, "Interpretation of Upper-Atmosphere Rocket Measurements Obtained With a Thermoluminescent Phosphor", Iskusstvennyye Sputniki Zemli, 1963, No. 15, pp. 81–84 (Artificial Earth Satellites, pp. 83–86).

73. G. A. Kirdina, Yu. M. Kulagin, A. B. Malyshev, M. N. Nazarova, P. M. Svidskiy, and I. S. Yudkevich, "Study of Radiation Intensity in the Radiation Zones of the Earth on the 'Cosmos 17' AES", Trudy Vsesoyuznoy Konferentsii po Fizike Kosmicheskogo, Prostranstva, Izdatel'stvo "Nauka", Moscow, June 10–16 1965, pp. 464–465 (NASA TT F-389, pp. 629–630).

74. G. A. Kirdina, Yu. M. Kulagin, A. B. Malyshev, M. N. Nazarova, P. M. Svidskiy, and I. S. Yudkevich, "Investigation of Emission Intensity in the Radiation Belts of the Earth with the 'Cosmos 17' Artificial Earth Satellite", Kosmicheskiye Issledovaniya, 1966, Vol. 4, No. 2, pp. 257–267 (HT–66–244, pp. 128–143).

75. V. I. Krasovskiy, "Exploration of the Upper Atmosphere With the Help of the Third Soviet Sputnik", Proceedings of the Institute of Radio Engineers, February 1959, Vol. 47, No. 2, pp. 289–297.

76. V. I. Krasovskiy, "Corpuscular Heating of the Outer Atmosphere", Space Research, Proceedings of the First International Space Science Symposium, Nice,

France, January 11–16 1960, North-Holland Publishing Company, Amsterdam, 1961, pp. 90–94.

77. V. I. Krasovskiy, Yu. I. Gal'perin, N. V. Dzhordzhio, T. M. Mulyarchik, and A. D. Bolyunova, "Investigations of the Upper Atmosphere by Cosmos 3 and Cosmos 5 Satellites", Space Research IV, Proceedings of the Fourth International Space Science Symposium, Warsaw, Poland, June 4–10 1963, North-Holland Publishing Company, Amsterdam, 1964, pp. 572–581.

78. V. I. Krasovskiy, Yu. I. Gal'perin, V. V. Temnyy, T. M. Mulyarchik, N. V. Dzhordzhio, M. Ya. Marov, and A. D. Bolyunova, "Some New Results of Geophysical Investigations by Means of the 'Cosmos 3' and 'Cosmos 5' Satellites", Geomagnetizm i Aeronomiya, 1963, Vol. 3, No. 3, pp. 408–416 (Geomagnetism and Aeronomy, pp. 338–344).

79. V. I. Krasovskiy, Yu. I. Gal'perin, V. V. Temnyy, T. M. Mulyarchik, N. V. Dzhordzhio, M. Ya. Marov, A. D. Bolyunova, O. L. Vaysberg, B. P. Potapov, and M. L. Bragin, "Some Characteristics of Geoactive Particles", Geomagnetizm i Aeronomiya, 1963, Vol. 3, No. 3, pp. 401–407 (Geomagnetism and Aeronomy, pp. 333–337).

80. V. I. Krasovskiy, I. S. Shklovskiy, Yu. I. Gal'perin, and Ye. M. Svetlitsky, "On Fast Corpuscles of the Upper Atmosphere", 10th International Astronautical Congress, London, England, August–September 5 1959.

81. V. I. Krasovskiy, I. S. Shklovskiy, Yu. I. Gal'perin, and Ye. M. Svetlitskiy, "The Detection of 10-keV Electrons in the Upper Atmosphere by the Third Russian Satellite", Izvestiya Akademii Nauk SSSR, Seriya Geofizicheskaya, 1959, No. 8, pp. 1157–1163 (Bulletin [Izvestiya] Academy of Sciences, USSR, Geophysics Series, pp. 829–832).

82. V. I. Krasovskiy, I. S. Shklovskiy, Yu. I. Gal'perin, and Ye. M. Svetlitskiy, "Energetic Particles of the Upper Atmosphere", Proceedings of the International Conference on Cosmic Rays, Moscow, 1960, Vol. 3, pp. 69–74 (JPRS: 10 983).

83. V. I. Krasovskiy, I. S. Shklovskiy, Yu. I. Gal'perin, Ye. M. Svetlitskiy, Yu. M. Kushnir, and G. A. Bordovskii, "Discovery of Approximately 10-keV Electrons in the Upper Atmosphere", Iskusstvennyye Sputniki Zemli, 1961, No. 6, pp. 113–126 (Artificial Earth Satellites, pp. 137–155).

84. L. V. Kurnosova, T. N. Kolobyanina, V. I. Logachev, L. A. Razorenov, I. A. Sirotkin, and M. I. Fradkin, "Discovery of Radiation Anomalies Above the South Atlantic at Heights of 310–340 km", Iskusstvennyye Sputniki Zemli, 1961, No. 8, pp. 90–93 (Planetary and Space Science, Pergamon Press, 1962, Vol. 9, pp. 513–516).

85. L. V. Kurnosova, V. I. Logachev, G. F. Platonov, L. A. Razorenov, V. G. Sinitsina, A. A. Suslov, and M. I. Fradkin, "Preliminary Results of Investigation of the Nuclear Component of Cosmic Rays by Equipment Carried by Electron 2". Izvestiya Akademii Nauk SSSR, Seriya Fizicheskaya, 1965, Vol. 29, No. 10, pp. 1853–1858 (Bulletin [Izvestiya] Academy of Sciences USSR, Physics Series, pp. 1686–1690); also presented at the All-Union Conference on Cosmic Ray Physics held at Apatity, August 24–31 1964.

86. L. V. Kurnosova, V. I. Logachev, L. A. Razorenov, and M. I. Fradkin, "Cosmic Ray Investigation by the Second Cosmic Rocket Landed on the Moon", Space Research, Proceedings of the First International Space Science Symposium, Nice, France, January 11–16 1960, North-Holland Publishing Company, Amsterdam, 1960, pp. 852–862.

87. L. V. Kurnosova, V. I. Logachev, L. A. Razorenov, and M. I. Fradkin, "Energy Spectra of Various Groups of Cosmic Ray Nuclei Which Were Obtained in Measurements by Means of Cherenkov Counters on Satellite-Spaceships", Iskusstvennyye Sputniki Zemli, 1962, No. 12, pp. 16–30 (Artificial Earth Satellites, pp. 18–35).

88. L. V. Kurnosova, V. I. Logachev, L. A. Razorenov, and M. I. Fradkin, "Some Results of Investigation of Cosmic Rays on Soviet Artificial Earth Satellites and Space Rockets", Trudy Fizicheskogo Instituta imeni P. N. Lebedeva, 1964, Vol 26, pp. 3–16.

89. L. V. Kurnosova, S. L. Mandel'shtam, L. A. Razorenov, I. P. Tindo, and M. I. Fradkin, "Cases of Transitory Increase in Heavy-Nucleus Flux Accompanied by Outbursts of X-Ray Radiation", Kosmicheskiye Issledovaniya, 1966, Vol. 4, No. 1, pp. 170–172 (TT–66–76, pp. 279–283).

90. L. V. Kurnosova, L. A. Razorenov, and M. I. Fradkin, "The Observation of the Nuclear Component of Cosmic Rays on the Third Cosmic Rocket", Iskusstvennyye Sputniki Zemli, 1961, No. 8, pp. 87–89 (Artificial Earth Satellites, pp. 248–249).

91. L. V. Kurnosova, L. A. Razorenov, and M. I. Fradkin, "A Case of a Short-Term Rise in the Intensity of Heavy Nuclei During the Flight of Spaceship-Satellite III", Iskusstvennyye Sputniki Zemli, 1962, No. 12, pp. 31–34 (Artificial Earth Satellites, pp. 36–39).

92. V. G. Kurt, "Measurement of Scattered Lα-Radiation in the Vicinity of the Earth and in Interplanetary Space", Trudy Vsesoyuznoy Konferentsii po Fizike Kosmicheskogo Prostranstva, Izdatel'stvo "Nauka", Moscow, June 10–16 1965, pp. 576–581 (NASA TT F-389, pp. 769–775).

93. V. G. Kurt, "The Investigation of the Distribution of the Neutral Hydrogen in the Geocorona to the Distance 20 R_e", Paper presented at the Seventh International Space Science Symposium, Vienna, Austria, May 10–19 1966.

94. B. M. Kuzhevskiy, R. B. Salimsibarov, N. G. Skryabin, and Yu. G. Shafer, "Some Preliminary Results of a Study of Cosmic Ray Intensity Variations Made Using the Cosmos 25 Satellite", Izvestiya Akademii Nauk SSSR, Seriya Fizicheskaya, 1966, Vol. 30, No. 11, pp. 1776–1777.

95. S. N. Kuznetsov, "The Behavior of the Outer Radiation Belt According to Data From Electron 1 and Electron 2 Satellites", Izvestiya Akademii Nauk SSSR, Seriya Fizicheskaya, 1966, Vol. 30, No. 11, pp. 1827–1829 (JPRS: 40 863).

96. S. L. Mandel'shtam, I. P. Tindo, and V. I. Karev, "Investigation of Lunar X-Ray Radiation With the Aid of the Luna–10 Lunar Satellite", Kosmicheskiye Issledovaniya, 1966, Vol. 4, No. 6, pp. 827–837.

97. S. L. Mandel'shtam, I. P. Tindo, Yu. K. Voron'ko, A. I. Shurygin, and B. N. Vasil'yev, "Investigation of the Sun's X-Radiation. 1. Measurements With the Aid of Geophysical Rockets", Iskusstvennyye Sputniki Zemli, 1961, No. 10, pp. 12–21 (Planetary and Space Science, Pergamon Press, 1962, Vol. 9, pp. 977–986).

98. S. L. Mandel'shtam, I. P. Tindo, Yu. K. Voron'ko, B. N. Vasil'yev, and A. I. Shurygin, "Investigations of Solar X-Ray Radiation. II. Measurements by Means of Spaceships", Iskusstvennyye Sputniki Zemli, 1961, No. 11, pp. 3–14 (Artificial Earth Satellites, pp. 1–16).

99. S. L. Mandel'shtam, B. N. Vasil'yev, Yu. Voron'ko, I. P. Tindo, and A. I. Shurygin. "Measurements of Solar X-Ray Radiation", Space Research III, Proceedings of the Third International Space Science Symposium, Washington, D.C., May 2–8 1962, North-Holland Publishing Company, Amsterdam, 1963, pp. 822–835.

100. S. L. Mandel'shtam and A. I. Yefremov, "Research on Shortwave Solar Ultra-Violet Radiation", Uspekhi Fizicheskikh Nauk, Moscow, 1957, Vol. 63, No. 1, pp. 163–180 (Advances in Physical Sciences, pp. 218–242).

101. V. V. Mel'nikov, I. A. Savenko, B. I. Savin, and P. I. Shavrin, "Experiment in the Use of an Electrostatic Analyzer on the Cosmos–12 Satellite", Geomagnetizm i Aeronomiya, 1965, Vol 5, No. 1, pp. 148–154 (Geomagnetism and Aeronomy, pp. 107–112).

102. T. M. Mulyarchik, "Investigation of the Softest Corpuscles With the Cosmos 3 and Cosmos 5 Satellites", Kosmicheskiye Issledovaniya, 1966, Vol. 4, No. 1, pp. 95–104 (TT–66–76, pp. 152–167).

103. T. M. Mulyarchik and O. L. Vaysberg, "Extremely Low Energy Electrons Measured by 'Cosmos 3' and 'Cosmos 5' Satellites", Space Research V, Proceedings of the Fifth International Space Science Symposium, Florence, Italy, May 12–16 1964, North-Holland Publishing Company, Amsterdam, 1965, pp. 500–510.

104. S. F. Papkov, N. F. Pisarenko, I. A. Savenko, A. F. Tupkin, and P. I. Shavrin,

'Radiometric Equipment of the Second Soviet Spacecraft-Satellite", Iskusstvennyye Sputniki Zemli, 1961, No. 9, pp. 78–85 (Artificial Earth Satellites, pp. 108–114).

105. A. A. Pomanskiy, "Physics of Cosmic Rays", Vestnik Akademii Nauk SSSR, 1966, No. 3, pp. 163–166.

106. I. D. Rapoport, "Photographic Method of Detection of Dense Showeɪs of Charged Particles", Zhurnal Eksperimental'noy i Teoreticheskoy Fiziki, 1958, Vol. 34, No. 4, pp. 998–1000 (Soviet Physics JETP, October 1958, Vol. 34, No. 4, pp 689–690).

107. I. A. Savenko, V. Ye. Nesterov, P. I. Shavrin, and N. F. Pisarenko, "The Cosmic Ray Equator as Determined From the Data of the Third Soviet Satellite-Ship", Iskusstvennyye Sputniki Zemli, 1961, No. 11, pp. 30–44 (Artificial Earth Satellites, pp. 33–37).

108. I. A. Savenko, N. F. Pisarenko, and P. I. Shavrin, "Dosimetric Measurements on the Second Spaceship Satellite", Iskusstvennyye Sputniki Zemli, 1961, No. 9, pp. 71–77 (Artificial Earth Satellites, pp. 101–107).

109. I. A. Savenko, N. F. Pisarenko, and P. I. Shavrin, "Space Flights and the Radiation Hazard", Priroda, 1962, No. 2, pp. 40–48 (TT–62–450).

110. I. A. Savenko, O. I. Savun, P. I. Shavrin, and B. M. Yakovlev, "Combined Proton Spectrometer for Space Studies", Geomagnetizm i Aeronomiya, 1965, Vol. 5, No. 3, pp. 546–549 (Geomagnetism and Aeronomy, pp. 420–422).

111. I. A. Savenko, P. I. Shavrin, V. Ye. Nesterov, and N. F. Pisarenko, "Equator of Cosmic Rays According to Data of the Second Soviet Cosmic Ship", Iskusstvennyye Sputniki Zemli, 1961, No. 10, pp. 45–47 (Artificial Earth Satellites, pp. 182–184).

112. I. A. Savenko, P. I. Shavrin, V. Ye. Nesterov, N. F. Pisarenko, and M. V. Tel'tsov, "The Radiation Situation in Space on the Eve of the Flight of Vostok 3 and Vostok 4", Kosmicheskiye Issledovaniya, 1963, Vol. 1, No. 1, pp. 172–175 (TT–63–968, pp. 308–315).

113. I. A. Savenko, P. I. Shavrin, and N. F. Pisarenko, "Soft Corpuscular Radiation at an Altitude of 320 km in the Latitudes Near the Equator", Iskusstvennyye Sputniki Zemli, 1962, No. 13, pp. 75–80 (Artificial Earth Satellites, pp. 80–85).

114. I. A. Savenko, M. V. Tel'tsov, V. L. Maduyev, O. I. Savun, and A. V. Yurovskiy, "Radiometric Equipment of the Cosmos 41 Satellite", Geomagnetizm i Aeronomiya, 1965, Vol. 5, No. 6, pp. 1129–1132 (Geomagnetism and Aeronomy, pp. 889–891)

115. I. A. Savenko, M. V. Tel'tsov, and P. I. Shavrin, "Variations in the Intensity of Outer Radiation Belt Protons and Electrons", Geomagnetizm i Aeronomiya, 1966, Vol. 6, No. 2, pp. 377–379 (Geomagnetism and Aeronomy, pp. 298–300).

116. Yu. G. Shafer, V. D. Sokolov, N. G. Skryabin, S. K. Dergeym, and R. B. Salimsibarov, "Some Results Obtained in the Measurement of East–West Asymmetry in the Intensity of Primary Cosmic Radiation", Kosmicheskiye Issledovaniya, 1964, Vol. 2, No. 6, pp. 933–935 (TT–64–1316, pp. 194–198).

117. Yu. G. Shafer, V. D. Sokolov, N. G. Skryabin, V. F. Lyutenko, A. V. Yarygin and R. B. Salimsibarov, "The Distribution of Cosmic-Ray Intensity in the Atmosphere at an Altitude of 500 km", Kosmicheskiye Issledovaniya, 1964, Vol. 2, No. 6, pp. 928–932 (TT–64–1316, pp. 185–193).

118. Yu. G. Shafer and A. V. Yarygin, "Measurements of Cosmic Rays on Geophysical Rockets", Iskusstvennyye Sputniki Zemli, 1960, No. 4, pp. 184–194 (Planetary and Space Science, Pergamon Press, 1961, pp. 165–172).

119. Ye. N. Sosnovets, "Recording of Low-Energy Protons on Satellites of the Electron Series", Izvestiya Akademii Nauk SSSR, Seriya Fizicheskaya, 1966, Vol. 30, No. 11, pp. 1820–1823 (JPRS: 40 863).

120. V. V. Temnyy, "Investigations of the Upper Atmosphere by the Cosmos 3 and Cosmos 5 Satellites", Space Research IV, Proceedings of the Fourth International Space Science Symposium, Warsaw, Poland, June 4–10 1963, North-Holland Publishing Company, Amsterdam, 1964, pp. 582–587.

121. V. V. Temnyy, "Investigations of the Upper Atmosphere With Satellites Cosmos 3 and Cosmos 5. 3. High-Energy Particles", Kosmicheskiye Issledovaniya, 1963, Vol. 1, No. 1, pp. 139–143 (TT-63-968, pp. 248–255).

122. V. V. Temnyy, "Atlas of the Intensity Distributions of Trapped Corpuscles Measured by the Cosmos 3 and Cosmos 5 Satellites", Space Research V, Proceedings of the Fifth International Space Science Symposium, Florence, Italy, May 12–16 1964, North-Holland Publishing Company, Amsterdam, 1965, pp. 489–497.

123. B. A. Tverskoy, "Anomalous Diffusion of Charged Particles in the Radiation Zones of the Earth", Trudy Vsesoyuznoy Konferentsii po Fizike Kosmicheskogo Prostranstva, Izdatel'stvo "Nauka", Moscow, June 10–16 1965, pp. 314–325 (NASA TT F-389, pp. 426–442).

124. P. V. Vakulov, Ye. V. Gorchakov, and Yu. I. Logachev, "Results of Researches on Soviet Satellites and Rockets in 1957–1959", Kosmicheskiye Luchi, VII Razdel Programmy MGG, 1965, No. 6.

125. P. V. Vakulov, S. N. Vernov, Ye. V. Gorchakov, Yu. I. Logachev, A. N. Charakhchyan, T. N. Charakhchyan, and A. Ye. Chudakov, "Investigation of Cosmic Rays", Space Research IV, Proceedings of the Fourth International Space Science Symposium, Warsaw, Poland, June 4–10 1963, North-Holland Publishing Company, Amsterdam, 1964, pp. 26–59.

126. P. V. Vakulov, S. N. Vernov, Ye. V. Gorchakov, Yu. I. Logachev, V. Ye. Nesterov, A. G. Nikolayev, N. F. Pisarenko, I. A. Savenko, A. Ye. Chudakov, and P. I. Shavrin, "Investigation of Radiation in Flights of Satellites, Cosmic Vehicles, and Rockets", Izvestiya Akademii Nauk SSSR, Seriya Fizicheskaya, 1962, Vol. 26, No. 6, pp. 758–781 (Bulletin [Izvestiya] Academy of Sciences USSR, Physics Series, pp. 760–783).

127. B. N. Vasil'yev, "Measuring Solar X-Ray Emission With Satellite-Borne Instruments", Kosmicheskiye Issledovaniya, 1966, Vol. 4, No. 5, pp. 748–754.

128. B. N. Vasil'yev, S. L. Mandel'shtam, I. P. Tindo, and A. I. Shurygin, "Preliminary Results of Investigation of Sun's X-Ray Radiation With the Help of Rockets and Spaceships", Doklady Akademii Nauk SSSR, 1961, Vol. 140, No. 5, pp. 1058–1061.

129. B. N. Vasil'yev, A. I. Shurygin, I. P. Tindo, and Yu. K. Voron'ko, "Solar X-Ray Research. III. Electronic Apparatus", Iskusstvennyye Sputniki Zemli, 1963, No. 15, pp. 85–91 (Artificial Earth Satellites, pp. 87–93).

130. I. G. Vasil'yev and A. I. Shapov, "Solar Tracking Head", Iskusstvennyye Sputniki Zemli, 1961, No. 11, pp. 87–93 (Artificial Earth Satellites, pp. 91–98).

131. Ya. M. Veprik, L. V. Kurnosova, L. A. Razorenov, K. D. Tolstov, M. I. Fradkin, and V. S. Chukin, "Processing of Nuclear Emulsions on Board the Second Spacecraft-Satellite", Iskusstvennyye Sputniki Zemli, 1961, No. 11, pp. 35–41 (Artificial Earth Satellites, pp. 38–44).

132. S. N. Vernov and A. Ye. Chudakov, "Terrestrial Corpuscular Radiation and Cosmic Rays", Space Research, Proceedings of the First International Space Science Symposium, Nice, France, January 11–16 1960, North-Holland Publishing Company, Amsterdam, 1960, pp. 751–796.

133. S. N. Vernov and A. Ye. Chudakov, "Investigations of Cosmic Radiation and of the Terrestrial Corpuscular Radiation by Means of Rockets and Satellites", Uspekhi Fizicheskikh Nauk, 1960, Vol. 70, No. 4, pp. 585–619 (Soviet Physics Progress, 1960, Vol. 3, No. 2, pp. 230–249).

134. S. N. Vernov and A. Ye. Chudakov, "Radiation Carried Out on Board the Cosmos 17 Satellite", Space Research V, Proceedings of the Fifth International Space Science Symposium, Florence, Italy, May 12–16 1964, North-Holland Publishing Company, Amsterdam, 1965, pp. 404–422.

135. S. N. Vernov, A. Ye. Chudakov, P. V. Vakulov, Ye. V. Gorchakov, S. N. Kuznetsov, Yu. I. Logachev, A. G. Nikolayev, Ye. N. Sosnovets, I. A. Rubinshteyn,

V. G. Stolpovskiy, and V. A. El'tekov, "Results Derived From Studying the Geometric Position and Particle Composition of Radiation Zones of the Earth Based on Data From the Satellites 'Electron–1' and 'Electron–2' ", Trudy Vsesoyuznoy Konferentsii po Fizike Kosmicheskogo Prostranstva, Izdatel'stvo "Nauka", Moscow, June 10–16 1965, pp. 394–406 (NASA TT F–389, pp. 535–550).

136. S. N. Vernov, A. Ye. Chudakov, P. V. Vakulov, Ye. V. Gorchakov, S. N. Kuznetsov, Yu. I. Logachev, A. G. Nikolayev, Ye. N. Sosnovets, and V. G. Stolpovskiy, "Results of Investigations of Geometrical Position and of Composition of the Earth's Radiation Belt Particles According to Electron 1 and Electron 2 Data", Space Research VI, Proceedings of the Sixth International Space Science Symposium, Mar Del Plata, Argentina, May 11–19 1965, North-Holland Publishing Company, Amsterdam, 1966, pp. 829–846.

137. S. N. Vernov, A. Ye. Chudakov, P. V. Vakulov, Ye. V. Gorchakov, Yu. I. Logachev, G. P. Lyubimov, and A. G. Nikolayev, "Investigation of Radiation With the Flights of the 'Mars 1' and 'Luna 4' Interplanetary Automatic Stations", Kosmicheskiye Issledovaniya, 1964, Vol, 2, No. 4, pp. 633–640 (TT–64–892, pp. 202–216).

138. S. N. Vernov, A. Ye. Chudakov, P. V. Vakulov, Ye. V. Gorchakov, Yu. I. Logachev, and A. G. Nikolayev, "Measurements of Radiation During the Flight of the Third Cosmic Rocket", Doklady Akademii Nauk SSSR, 1961, Vol. 136, No. 2, pp. 322–324 (Soviet Physics, Doklady, 1961, Vol. 6, No. 1, pp. 43–45).

139. S. N. Vernov, A. Ye. Chudakov, P. V. Vakulov, Yu. I. Logachev, G. P. Lyubimov, A. G. Nikolayev, and N. V. Pereslegina, "Measurements of Protons of Solar Origin With Energies of 1–5 MeV During Flight of 'Venus–2', 'Venus–3', and 'Zond–3' ", Doklady Akademii Nauk SSSR, 1966, Vol. 171, No. 4, pp. 847–850.

140. S. N. Vernov, A. Ye. Chudakov, P. V. Vakulov, Yu. I. Logachev, G. P. Lyubimov, and N. V. Pereslegina, "Cosmic Ray Variations According to Data From Zond–3 and Venera–2", Doklady Akademii Nauk SSSR, 1966, Vol. 171, No. 3, pp. 583–586.

141. S. N. Vernov, Ye. V. Gorchakov, Yu. I. Logachev, V. Ye. Nesterov, N. F. Pisarenko, I. A. Savenko, A. Ye. Chudakov, and P. I. Shavrin, "Investigations of Radiation During Flights of Satellites, Space Vehicles, and Rockets", presented at the International Conference on Cosmic Rays and the Earth Storm, Kyoto, Japan, September 4–15 1961.

142. S. N. Vernov, N. L. Grigorov, Yu. I. Logachev, and A. Ye. Chudakov, "Measurements of Cosmic Radiation by Means of an Artificial Earth Satellite", Doklady Akademii Nauk SSSR, 1958, Vol. 120, No. 6, pp. 1231–1233.

143. S. N. Vernov, V. V. Mel'nikov, I. A. Savenko, and B. I. Savin, "Investigation of Low-Energy Charged Particles With the Cosmos 12, Cosmos 15, and Electron 2 Satellites", Izvestiya Akademii Nauk SSSR, Seriya Fizicheskaya, 1965, Vol. 29, No. 10, pp. 1794–1799 (Bulletin [Izvestiya] Academy of Sciences USSR, pp. 1627–1631); presented at the All-Union Conference on Cosmic Ray Physics held at Apatity, August 24–31 1964.

144. S. N. Vernov, V. V. Mel'nikov, I. A. Savenko, and B. I. Savin, "Measurements of Low Energy Particle Fluxes From the 'Cosmos' and 'Electron' Satellites", Space Research VI, Proceedings of the Sixth International Space Science Symposium, Mar Del Plata, Argentina, May 11–19 1965, North-Holland Publishing Company, Amsterdam, 1966, pp. 746–756.

145. S. N. Vernov, V. V. Mel'nikov, I. A. Savenko, B. I. Savin, and T. I. Pervaya, "Recording Charged Particles With an Energy of 0·1–10 keV With a Spherical Electrostatic Analyzer", Trudy Vsesoyuznoy Konferentsii po Fizike Kosmicheskogo Prostrantstva, Izdatel'stvo "Nauka", Moscow, June 10–16 1965, pp. 381–387 (NASA TT F–389, pp. 518–525).

146. S. N. Vernov, V. Ye. Nesterov, N. F. Pisarenko, I. A. Savenko, and P. I. Shavrin, "Investigation of Cosmic Radiation on Space Vehicles", 12th International

Astronautical Congress, Washington, D.C., October 4 1961, Academic Press, New York and London, 1963, Vol. 2, pp. 855–878.

147. S. N. Vernov, V. Ye. Nesterov, N. F. Pisarenko, I. A. Savenko, L. V. Tverskaya, and P. I. Shavrin, "Study of the Earth's Outer Radiation Zone at Low Altitudes During Spacecraft-Satellite Flights and the 'Cosmos' AES Flights Between 1960 and 1963", Trudy Vsesoyuznoy Konferentsii po Fizike Kosmicheskogo Prostranstva, Izdatel'stvo "Nauka", Moscow, June 10–16 1965, pp. 434–447 (NASA TT F-389, pp. 588–606).

148. S. N. Vernov, V. Ye. Nesterov, I. A. Savenko, P. I. Shavrin, and K. N. Sharvina, "Discovery and Investigations of the Brazil Anomaly by Spaceships and the Cosmos Series Satellites", Space Research VI, Proceedings of the Sixth International Space Science Symposium, Mar Del Plata, Argentina, May 11–19 1965, North-Holland Publishing Company, Amsterdam, 1966, pp. 165–175.

149. S. N. Vernov, I. A. Savenko, P. I. Shavrin, V. Ye. Nesterov, and N. F. Pisarenko, "The Outer Circumterrestrial Radiation Belt at a Height of 320 km", Iskusstvennyye Sputniki Zemli, 1961, No. 10, pp. 34–39 (Artificial Earth Satellites, pp. 171–176).

150. S. N. Vernov, I. A. Savenko, P. I. Shavrin, V. Ye. Nesterov, and N. F. Pisarenko, "Radiation Belts of the Earth at Heights of From 180 to 250 km", Geomagnetizm i Aeronomiya, 1962, Vol. 2, No. 1, pp. 41–47 (Geomagnetism and Aeronomy, pp. 31–36).

151. S. N. Vernov, I. A. Savenko, P. I. Shavrin, V. Ye. Nesterov, N. F. Pisarenko, and R. N. Basilova, "Investigation of Cosmic Rays at High Altitudes", Izvestiya Akademii Nauk SSSR, Seriya Fizicheskaya, 1964, Vol. 28, No. 12, pp. 2045–2048 (Bulletin [Izvestiya] Academy of Sciences USSR, Physics Series, pp. 1936–1939).

152. S. N. Vernov, I. A. Savenko, P. I. Shavrin, V. Ye. Nesterov, N. F. Pisarenko, and K. N. Sharvina, "Investigation of the Earth's Radiation Belts at the Altitudes of 200–400 km", Space Research V, Proceedings of the Fifth International Space Science Symposium, Florence, Italy, May 12–16 1964, North-Holland Publishing Company, Amsterdam, 1966, pp. 343–359.

153. S. N. Vernov, I. A. Savenko, P. I. Shavrin, V. Ye. Nesterov, N. F. Pisarenko, and K. N. Sharvina, "Some Data on the Earth's Radiation Belts, Obtained During Flights of Cosmos Satellites at Altitudes of 200 to 400 km", Izvestiya Akademii Nauk SSSR, Seriya Fizicheskaya, 1964, Vol. 28, No. 12, pp. 2049–2057 (Bulletin [Izvestiya] Academy of Sciences USSR, Physics Series, pp. 1940–1947.)

154. S. N. Vernov, I. A. Savenko, P. I. Shavrin, V. Ye. Nesterov, N. F. Pisarenko, M. V. Tel'tsov, T. I. Pervaya, and V. N. Yerofeyeva, "Some Results of Radiometric Measurements at Altitudes of 200–400 km From 1960–1963", Kosmicheskiye Issledovaniya, 1963, Vol. 2, No. 1, pp. 136–146 (TT-64-499, pp. 216–235).

155. S. N. Vernov, I. A. Savenko, P. I. Shavrin, and N. F. Pisarenko, "Detection of the Inner Radiation Belt at the Altitude of 320 km in the Region of the Southern Atlantic Magnetic Anomaly", Iskusstvennyye Sputniki Zemli, 1961, No. 10, pp. 40–44 (Artificial Earth Satellites, pp. 177–181).

156. S. N. Vernov, I. A. Savenko, M. V. Tel'tsov, and P. I. Shavrin, "Measurements of Low-Energy Protons From the Cosmos 41 Satellite", Space Research VI, Proceedings of the Sixth International Space Science Symposium, Mar Del Plata, May 11–19 1965, North-Holland Publishing Company, Amsterdam, 1966, pp. 734–745.

157. S. N. Vernov, I. A. Savenko, M. V. Tel'tsov, and P. I. Shavrin, "Measurement Results of the Satellite 'Cosmos 41' in the Outer Radiation Zone", Trudy Vsesoyuznoy Konferentsii po Fizike Kosmicheskogo Prostranstva, Izdatel'stvo "Nauka", Moscow, June 10–16 1965, pp. 460–464 (NASA TT F-389, pp. 623–629).

158. S. N. Vernov, I. A. Savenko, L. V. Tverskaya, B. A. Tverskoy, and P. I. Shavrin, "The Electron Intensity of the Radiation Belts at Altitudes of 180 to 330 km in Regions Conjugate With Negative Geomagnetic Anomalies", Kosmicheskiye Issledovaniya, 1965, Vol. 3, No. 1, pp. 128–134 (TT-65-170, pp. 203–215).

159. S. N. Vernov, P. V. Vakulov, Ye. V. Gorchakov, Yu, I. Logachev, and A. Ye. Chudakov, "Studying the Mild Component of Cosmic Rays Beyond the Atmosphere", Iskusstvennyye Sputniki Zemli, 1958, No. 2, pp. 61–69 (Artificial Earth Satellites, pp. 78–89).

160. S. N. Vernov, P. V. Vakulov, Ye. V. Gorchakov, Yu. V. Logachev, G. P. Lyubimov, A. G. Nikolayev, and N. V. Pereslegina, "Measurement of Intensity of Penetrating Radiation on the Moon's Surface", Paper presented at the Seventh International Space Science Symposium, Vienna, Austria, May 1966.

161. S. N. Vernov, P. V. Vakulov, S. N. Kuznetsov, Yu. I. Logachev, G. P. Lyubimov, A. G. Nikolayev, N. V. Pereslegina, and A. Ye. Chudakov, "Measurements of the Cosmic-Ray Intensity During the Flight of the Zond–3 Automatic Space Probe", Seventh International Space Science Symposium, Vienna, Austria, May 1966.

162. N. N. Volodichev, N. L. Grigorov, V. Ye. Nesterov, I. D. Rapoport, I. A. Savenko, and B. M. Yakovlev, "A Study Made Using the Proton–1 Satellite of the Chemical Composition of Primary Cosmic Rays in the Moderate Energy Region", Izvestiya Akademii Nauk SSSR, Seriya Fizicheskaya, 1966, Vol. 30, No. 11, pp. 1763–1764 (JPRS: 40 863).

163. N. N. Volodichev, V. Ye. Nesterov, I. A. Savenko, and K. N. Sharvina, "Study of the Proton Component of the Inner Radiation Belt in the Brazilian Anomaly by the Artificial Earth Satellites Proton 1 and Proton 2", Izvestiya Akademii Nauk SSSR, Seriya Fizicheskaya, 1966, Vol. 30, No. 11, pp. 1768–1770 (JPRS: 40 863).

164. A. I. Yefremov, A. L. Podmoshenskiy, M. A. Ivanov, V. N. Nikiforov, and O. N. Yefimov, "Filter Apparatus for Studying the Sun's Short-Wave Emission", Iskusstvennyye Sputniki Zemli, 1961, No. 10, pp. 48–54 (Planetary and Space Science, Pergamon Press, 1962, Vol. 9, pp. 987–992).

165. A. I. Yefremov, A. L. Podmoshenskiy, O. N. Yefimov, and A. A. Lebedev, "The Investigation of the Sun's Short-Wave Radiation", Iskusstvennyye Sputniki Zemli, 1961, No. 10, pp. 3–11 (Planetary and Space Science, Pergamon Press, 1962, Vol 9, pp. 969–976).

166. A. I. Yefremov, A. L. Podmoshenskiy, O. N. Yefimov, A. A. Lebedev, "Investigations of Solar X-Rays and Lyman Alpha Radiation on August 19–20, 1960", Space Research III, Proceedings of the Third International Space Science Symposium, Washington, D.C., May 2–8 1962, North-Holland Publishing Company, Amsterdam, 1963, pp. 843–854.

167. Yu. Zaitsev, " 'Cosmos' Ships in Earth Orbit. 2. Stories Told by Satellites", Aviatsiya i Kosmonavtika, 1966, No. 12, pp. 62–66.

168. I. A. Zhitnik, V. V. Krutov, L. P. Malyavkin, and S. L. Mandel'shtam, "The Image of the Sun in the Distant Ultraviolet Region of the Spectrum", Kosmicheskiye Issledovaniya, 1964, Vol. 2, No. 6, pp. 920–927 (TT–64–1316, pp. 170–184).

Chapter 11

Astronomical Investigations above the Terrestrial Atmosphere

*N. T. Bobrovnikoff**

INTRODUCTION

In the U.S., space astronomy has been developing at a rapid pace since early efforts to obtain balloon photographs of the solar spectrum. This development has involved attempts to utilize all available platforming techniques including aircraft, balloons, rockets, satellites, and space probes for astronomical observing. Although initially limited to solar studies, these platforms have now been successfully utilized for stellar, lunar, and planetary investigations. Published Soviet space astronomy accomplishments and plans do not appear to be nearly as comprehensive or extensive, particularly with regard to the use of balloons, rockets, and satellites, as those in the U.S. However, Soviet efforts to use space platforming capabilities for lunar and planetary exploration have been equally ambitious although considerably less successful than U.S.

* Consultant, Columbus Laboratories, Battelle Memorial Institute; Professor Emeritus, The Ohio State University.

efforts. The specific Soviet rockets, satellites, and space probes associated with astronomical studies to date are noted in Table 11.1.

TABLE 11.1

Soviet Space Vehicles Associated With Astronomical Studies (1957–1966)

Solar Studies	Stellar Surveys	Planetary Studies	Lunar Studies
Since 1957 several geophysical rockets and satellites have undertaken solar UV and X-ray studies. These studies are discussed in Chapter 10.	Cosmos 51 and 92	Radio astronomical studies: Electron series Mars 1 Venera 2 and 3 Zond 2 and 3 Photographic studies: Mars 1 Venera 2 Zond probes? Surface studies: Venera 3	Radio astronomical studies: Luna 11, 12 Photographic studies: Luna 3, 9, 12, 13 Zond 3 Geophysical studies: Magnetic field, meteorics, radiation, etc., discussed in other chapters Surface studies: Direct— Luna 9 and 13 Indirect— Luna 10, 11, 12 Zond 3

SUMMARY OF STUDIES AND RESULTS

The Sun

Numerous rocket and satellite measurements of solar ultraviolet and X-ray radiation have been undertaken in the USSR. The types of studies undertaken, techniques employed, and results obtained are summarized in Chapter 10, relating to cosmic and solar radiation.

In general, Soviet study of the sun by means of rockets and artificial satellites has a distinctly utilitarian connotation. It is pursued with a view of establishing reliable criteria for the forecast of solar flares which might be dangerous for cosmonauts, and which affect all sorts of processes in the atmosphere of the earth. This theme is also emphasized in numerous proposals for a solar observatory on the moon.

It has always been a puzzle why the Soviets have not developed balloon techniques for solar studies, as they are much less expensive and less complex than are satellites and rockets in such investigations. The first indications that the Soviets have begun such studies are two items[2, 11] discussing an "automatic astronomical station" lifted to a height of more than 20 km on November 1 1966. Reportedly the platform was equipped with a heliograph and a spectrograph for solar work and was controlled from the earth by radio. The total weight of the station was 7·5 tons. The main mirror has a diameter of one meter, but for first flights only a 50-cm mirror was used. A photograph of the balloon and apparatus is shown in Figure 11.1.

To date, no results of these studies have been announced. From the statement by V. Krat, the new director of the Pulkovo Observatory, it appears that this apparatus is to be used principally in the investigation of solar flares which begin in small areas of about $0''\!.1$ diameter, difficult to see from the surface of the earth. This new technique is hailed as a great achievement of Soviet science and technology. There is no mention made of similar apparatus used in the U.S. by M. Schwarzschild of Princeton more than 10 years ago.

Stellar Systems

At a conference held at the Shternberg Astronomical Institute (GAISh) on January 27–28 1964 the following resolutions were adopted, which characterized the needs associated with stellar observations outside the terrestrial atmosphere[24]:

1. Formation of stellar catalogues in the ultraviolet and infrared parts of the spectrum
2. Study of distribution of energy in stellar spectra
3. Investigation of the law of interstellar absorption of light and of the properties of the absorbing medium
4. Study of interstellar lines in star spectra
5. Observations outside the atmosphere of the earth in which the brightness of the background is substantially lowered. This will allow the recording of fainter objects, such as diffuse nebulae of low surface brightness and outside regions of the Milky Way.

An analysis of results obtained to date was presented, but it is based wholly on Western (mostly U.S.) data. Much of the quoted material was taken from the Liege Coloquium, July 11–14 1960, "Les Spectres des Astres dans l'Ultraviolet Lointain".

There is very little published Soviet material in support of the resolutions. The closest paper is that by N. A. Dinov and A. B. Severnyy[7] containing a brief review of photometric studies aboard Cosmos 51.

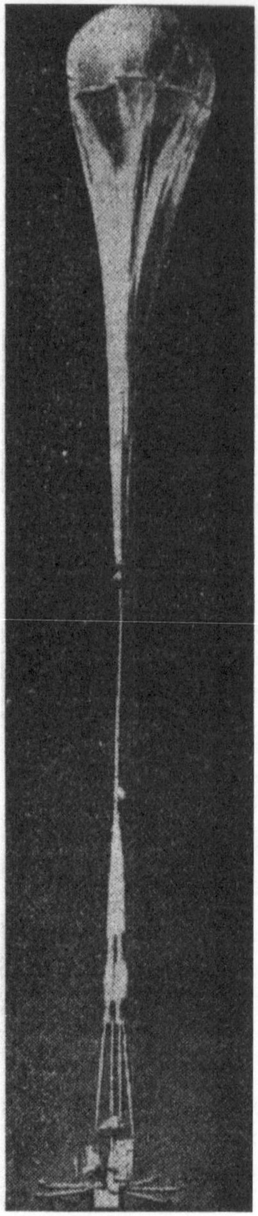

Fig. 11.1. Balloon Solar Observatory[4]

These studies involved the determination of background luminosity of the sky in three wavelengths: 3700–5500 Å, 2200–3000 Å, and 1300–1760 Å. Reportedly, photoelectric spectrophotometer readings showed luminosity of the sky in the visible region to be about 30 per cent of that observed on the earth's surface. Nothing was noted, however, concerning the ultraviolet intensity. Another communication by A. I. Lebedinskiy[12] notes that Cosmos 92 also contained an ultraviolet spectrophotometer for stellar observations but reports no results.

The Planets

In spite of their numerous attempts to undertake measurements in the vicinity of Mars and Venus (see Table 11.1) the Soviets have had a remarkable lack of success to date. The availability of the recent paper comparing U.S. and Soviet efforts to explore Mars, by Murray and Davies,* makes further discussion of Martian exploration superfluous. As for the exploration of Venus, the Soviets pinned high hopes on the recent flights of Venera 2 and Venera 3. Both these vehicles functioned satisfactorily until radio contact was lost as they approached the vicinity of the planet. Venera 2 (launched on November 12 1965), which was to make physical measurements and photograph the planet, reportedly passed within 24 000 km from the surface of Venus. Venera 3 (launched four days after Venera 2) was the first spacecraft to impact on a planet (March 1 1966). It reportedly included a soft-landing instrument capsule.

Most of the published results of Soviet planetary exploration efforts relate to radio astronomical studies. Benediktov, Getmantsev, and Ginzburg showed in 1961 that radio astronomical studies from satellites and space rockets were possible and that such studies contained high potential for obtaining valuable data on solar and planetary radio emissions[4]. Initial results of such studies aboard Electron 2 (launched on January 30 1964) were reported by Benediktov, *et al.*, to the May 1965 COSPAR symposium[5]. The authors found a sharp increase in intensity of radio emission at 725 kc/s, in the nature of separate outbursts. At that frequency the intensity of sporadic emission was found at times to exceed the general cosmic background by two orders of magnitude and to be much higher than at 1525 kc/s. The authors stated that these outbursts are very similar to the sporadic radio emission of the sun and Jupiter at other wavelengths.

More recently, an unexpected result, potentially of fundamental importance in planetary physics, was obtained from low frequency (20 to

* B. C. Murray and M. E. Davies, "A Comparison of U. S. and Soviet Efforts to Explore Mars", Science, 1966, Vol. 151, pp. 945–954.

2200 kc/s) studies aboard Zond 2, Zond 3, and Venera 2. According to V. I. Slysh, the record of Zond 2 showed that at frequencies of 210 kc/s the intensity of radio emissions was about 100 times that of the general galactic background at frequencies 985, 2000, 2200 kc/s. This result was confirmed by measurements on board Zond 3 and Venera 2. It was found that at least 60 to 80 per cent of the longer wavelength radiation can be ascribed to Jupiter since it varied in accordance with the distance from that planet. Also the sharp drop in intensity of the radiation (recorded on Venera 2, November 12 1965) could be identified with the occultation of Jupiter by the moon as projected from Venera 2. The author stated that the previous experiments on rockets and satellites failed to show any radio emission from Jupiter, except for some indefinite radio noise on Electron 2 which now appears to be also connected with Jupiter. Therefore the Jupiter radiation at low frequency may be variable, as the results described above refer only to the time interval, November 30 1964 to November 12 1965[25].

Thus there exists a disagreement between the Electron results as reported by Benediktov and the space-probe data of Slysh. The latter found the intensity of radio emission in space at 985 kc/s to be quite comparable with that of the galactic background, and the substantial increase was noticeable only at the 20–200 kc/s frequencies. It may be that the situation is, as usual in such phenomena, much more complicated than it initially appears. Subsequent reports on this subject may clarify the discrepancy.

The Moon

Some of the most significant contributions of Soviet space-science research to date have been associated with their efforts to investigate the moon using the Luna and Zond series of space probes. A complete description of all lunar vehicles is contained in Part I, Chapter 3.

Initial Lunar Studies

Exploitation of the earlier results obtained by Luna 3 in October 1959 still continues in Soviet literature. In December 1960 an international conference (IAU Symposium No. 14) on the moon was called in Pulkovo and the Soviets contributed four papers based on the Luna 3 photographic data[6, 16, 19, 21]. The general result of all these studies was announced by A. A. Mikhaylov[21], Member of the Academy of Sciences, USSR, former Director of the Pulkovo Observatory, and the usual spokesman on astronomical matters in Soviet space astronomy. He considered all sorts of suggestions advanced so far to explain the striking difference in the aspect of the two sides of the moon, such as the tidal force of the earth, the screening or focusing of meteorites by the earth's gravitational field, etc., and concluded that no external agent can account for the difference. It should be ascribed to the internal activity of the moon itself. A more recent study of the same material is made in an article by A. V. Markov[20].

Similar, but more definite, results by Zond 3 (July 18 1965) have been only partially studied and reported by the Soviets[17]. The pictures by Zond 3, which passed within 11 600 km of the moon, were by all accounts very impressive when displayed by Mikhaylov at the 16th Astronautical Congress in Athens on September 15 1965. Crater chains much longer than those on the visible side, and extending up to 1500 km, were well marked. Another feature was the thalassoids, depressions of some 500–km diameter but not as dark as the maria on the visible side. Also, results of Zond 3 ultraviolet spectrophotometric measurements of the moon in the 1900–3500 Å range were reviewed at the 1966 Vienna COSPAR meeting[10]. Data from Luna 3 and Zond 3 have been significant and bear directly on the problem of the origin of the moon, and indirectly on the origin of the solar system. Also significant results can be expected from the more recent orbiter and soft-landing series (Lunas 9–13).

Lunar Orbiter Studies

Following the first successful lunar satellite launching (Luna 10), Mikhaylov noted the following data that can be obtained from an analysis of the satellite motion around the moon[22]:

1. The mass of the moon, or rather, the ratio of the mass of the earth to the mass of the moon. At present this ratio is assumed to be $81{\cdot}53 \pm 0{\cdot}05$, with no possibility of further refinement by the observations made on the surface of the earth. For this problem, as well as for other similar problems, the fact that the lunar satellite moves in an orbit highly inclined to the plane of the lunar equator is very important, as it means a thorough investigation of practically every point on the surface of the moon.
2. Distribution of mass within the moon. This will help us to elucidate some observational facts, such as the lack of coincidence between the optical and gravitational centers of the moon.

3. The exact shape of the moon. The bulge toward the earth indicates that the moon was formed much nearer the earth, at a distance of say, 150 000 km, rather than 384 000 km as it is now. This argument is based on the fact that the lunar axis directed toward the earth is much longer than would follow from the law of gravitation at the distance of the moon.
4. General field of gravity on the moon and its local irregularities should be determined more easily by means of the satellite than is possible from the earth.
5. The internal structure of the moon can be established from the gravity observations. This would be of great cosmological significance.

A similar study of the lunar gravity field was included in the programs for Lunas 11 and 12 but the results are not yet available. So far as Luna 10 is concerned, some information has, however, been published[3]. The Luna 10 satellite functioned from April 3 to May 30 1966 during which time it performed over 460 revolutions around the moon. From the perturbations of the node and periselene of its orbit, the existence of a non-central field of the moon has been confirmed. In other words the shape of the moon markedly differs from an ideal sphere. The result is given in Figure 11.2 in which the cross-section of the moon in planes *XY*, *XZ*, and *YZ* are indicated. The axis *X* is directed toward the earth and *Z* is perpendicular to the plane *XY*. The largest departure of the lunar figure is in *X* and *Z* directions. Although the numerical values are not given in the article, the diagrams show a maximum departure on the order of 4 km, with the greatest bulge unexpectedly occurring on the back side of the moon. This gravitational result corresponds to Item (3) in Mikhaylov's statement above.

The γ-ray measurements of the lunar surface were carried out by A. P. Vinogradov (Institute of Geochemistry and Analytical Chemistry) and his collaborators[26]. Radioactivity of the moon's surface was found to be approximately the same as that of the terrestrial crust. It is inferred that the process of formation of the crust of the earth and of the moon must have been similar, and also that no great quantity of iron meteorites can be present on the surface of the moon.

The X-ray radiation from the lunar surface has been studied by S. L. Mandel'shtam and his collaborators at the Physical Institute of the Academy of Sciences[18]. The most likely source of the X-rays on the moon was assumed to be due to solar X-rays which would cause the lunar surface to fluoresce and might reveal the relative abundance of

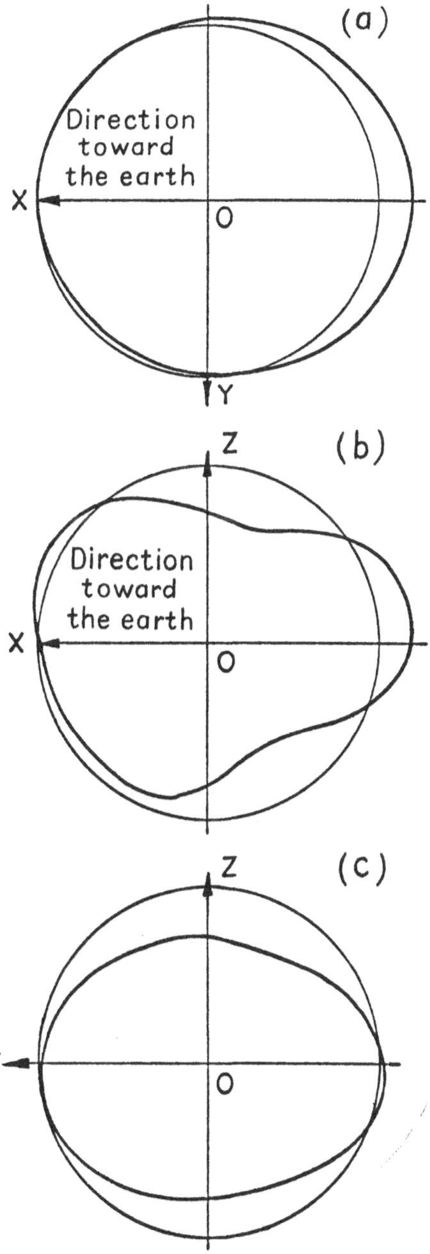

Fig. 11.2. Cross-Section of the Equipotential Surface of the Moon by
(a) XY-plane (equatorial)
(b) XZ-plane (meridional $\lambda = 0°$)
(c) YZ-plane (meridional $\lambda = 90°$)[3]
Departures from the spherical shape are exaggerated.

Si, Al, and Mg on the lunar surface. The interpretation of records, however, is not easy, and it is admitted by the authors that no conclusive results have been obtained. Micrometeorites seem to be rather abundant in the vicinity of the moon, but these results are discussed in Chapter 4.

The study of the lunar radiation in the infrared and visible interval was undertaken by A. I. Lebedinskiy, *et al.* To date published information contains only the description of the apparatus, with results promised for the future[14]. The same can be said of the study of the lunar ionosphere by K. I. Gringauz, *et al.*[8]. However, preliminary results indicate that the concentration of ions around the moon is very feeble and does not exceed $100/cm^3$.

Nothing has yet been published in connection with Luna 11, except for the general statement that the program resembled that of Luna 10. There is only slightly more information about the performance of Luna 12. The same program of investigations as for Lunas 10 and 11 has been announced and, in addition, some photographs of the lunar surface were made at periselene at 100 km from the surface of the moon. However, the quality of photographs does not appear very high. Twenty-one photographs were transmitted but the phototelevision device evidently did not function properly. The area of crater Aristarchus, of 25 km^2, was reproduced. The smallest objects visible on it are 15 to 20 meters across. The Soviet scientists found the photographs good enough to show that the bright rays from that crater are due to accumulations of craterlets along the axes of rays. It is inevitable that this information will be supplemented and extended by the future lunar satellites, both from the USSR and U.S., as it is a very important link in the preparation for a manned landing on the moon.

Lunar Soft-Landing Studies

To date successful Soviet soft-landings on the moon have been accomplished by Luna 9 and Luna 13. Luna 9, launched on January 31 1966, accomplished the first soft landing on the moon, on February 3 1966. The capsule with instruments weighed about 100 kg and was ejected from the spacecraft (1583 kg) just before landing. It intermittently transmitted panoramic pictures of the moon until February 6 1966, when its radio went dead. Spectacular as this feat appeared to be, it was to be eclipsed within four months by the U.S. Surveyor 1 which landed on the moon on June 1 1966, and continued to send pictures until June 15. For this reason the achievements of Luna 9 do not appear as impressive as those of Luna 10, although it was undoubtedly a first rate technological accomplishment. A complete collection of pictures obtained from the lunar surface by Luna 9 has been published by the Academy of Sciences, USSR[1].

Luna 9 landed at a point with lunar coordinates 7°08′N, 64°22′W in the region known as the Oceanus Procellarum (the Ocean of Storms), to the west of crater Reiner and to the east of Kepler. The television camera of Luna 9 when on the moon was only 60 cm above the lunar surface, so that its visible horizon was not more than 1·5 km distant. Whatever else the panoramic pictures of the moon in this restricted area show, one result seems to be clear. There is no dust layer, postulated by some theoreticians, in at least that particular locality of the moon. This result was confirmed by Surveyor 1, June 2 1966, for another lunar locality. The absence of any appreciable dust layer is of the greatest importance for any further landings on the moon. Other important results

Fig. 11.3. Radiation Densitometer Soft-Landed on the Moon by Luna 13*

of Luna 9 were the presence of erosion, and of a system of white streaks on the lunar surface.

Luna 13 landed in Oceanus Procellarum between crater Krafft and Seleucus, in the same general locality as Luna 9. The only significant feature of Luna 13 was that it contained both a penetrometer and radiation densitometer for determining the average density of the lunar surface. The radiation densitometer (see Figure 11.3) consisted of a gamma-ray source, three blocks of gamma-quanta counters, and a shield protecting the counters from direct gamma-ray penetration of the sensor. Upon contact with the lunar surface the latter was irradiated by gamma quanta. By recording the flow intensity it was possible to estimate the density of the lunar material which was found by this method to be about 0·8 g/cm³.*

* USSR National Report to COSPAR, 10th Plenary Meeting, London, England, July, 1967.

The soil penetrometer (Figure 11.4) had a conic prodding device 35 mm in diameter made of titanium. The penetrometer employed a solid-fuel device which provided a thrust on the moon equal to 6·7 kg/cm² for approximately 0·8 sec. The depth of penetration in the case of Luna 13 turned out to be 45 mm, which like the preceding method, indicated the existence of a weakly cemented granular mineral soil on the moon having a density less than 1·0 g/cm³, considerably less dense than the superficial layers of the earth. These data were also reportedly confirmed by dynamic-deceleration data during soft-landing.*

Fig. 11.4. Soil Penetrometer Soft-Landed on Moon by Luna 13*

According to Professor A. I. Lebedinskiy[13], radiation was also measured by Luna 9 near and on the surface of the moon. He expressed his conviction that a further study of the results thus obtained may establish the chemical composition of the superficial layer of the moon. The same idea is expounded by G. A. Leykin[15].

Much vague and indefinite talk on the desirability of establishing an observatory on the moon can be found in Soviet literature. Usually the advantage of such an observatory for the study of the sun is emphasized. For example, an optical telescope for the moon was briefly considered

* USSR National Report to COSPAR, 10th Plenary Meeting, London, England, July, 1967.

by V. M. Mozhzherin, *et al.*, at the Athens IAF meeting (1965)[23]. However, more recently, Kondrat'yev, *et al.*, have considered the value of a lunar-based telescope for terrestrial meteorological observing[9]. The Soviets have also participated in discussions relating to the establishment of a Lunar International Laboratory (LIL) at the 1964 and 1965 meetings of the International Astrnoautical Congress. These technical sessions, and earlier official political discussions concerned with international cooperation in lunar exploration, have been equally ineffective.

References

1. "The First Panoramic Views of the Lunar Surface", Izdatel'stvo Nauka, Moscow, 1966, 129 pp. (NASA TT F–393).

2. "Observatory in the Stratosphere", Pravda, November 10 1966, p. 4.

3. E. L. Akim, "Determining the Gravitational Field of the Moon", Doklady Akademii Nauk, 1966, Vol. 170, No. 4, pp. 799–802; Kosmicheskiye Issledovaniya, 1966, Vol. 4, No. 6, pp. 823–826.

4. Ye. A. Benediktov, G. G. Getmantsev, and V. L. Ginzburg, "Radio Astronomical Explorations Made in the USSR Using Artificial Earth Satellites and Space Rockets", Iskusstvennyye Sputniki Zemli, 1961, Vol. 7, pp. 3–23 (NASA TT F–8175).

5. Ye. A. Benediktov, G. G. Getmantsev, N. A. Mityakov, V. O. Rapoport, Ya. A. Sazonov, and A. F. Tarasov, "Intensity Measurements of Radiation at Frequencies 725 and 1525 kc by Means of the Receiver on the Satellite 'Electron 2' ", Space Research VI, Proceedings of the Sixth International Space Science Symposium, Mar Del Plata, Argentina, May 11–19 1965, North-Holland Publishing Company, Amsterdam, 1966, pp. 110–113.

6. I. I. Breydo and D. Ye. Shchegolev, "Schematic Chart of the Far Side of the Moon", "The Moon", Symposium No. 14 of the International Astronomical Union held at Pulkovo Observatory near Leningrad December 1960, Academic Press, London and New York, 1962, pp. 25–38.

7. N. A. Dinov and A. B. Severnyy, presentation at the 16th International Astronautical Congress, Athens, Greece, September 1965 (In English).

8. K. I. Gringauz, V. V. Bezrukikh, M. Z. Khokhlov, G. N. Zastenker, A. P. Remizov, and I. S. Mysatov, "Experimental Results From Observations of the Lunar Ionosphere Carried Out in the First Artificial Satellite of the Moon", Doklady Akademii Nauk, 1966, Vol. 170, No. 6, pp. 1306–1309 (JPRS: 39 455).

9. K. Ya. Kondrat'yev, V. L. Gayevskiy, V. N. Konashenok, and A. L. Reshednikov, "Lunar Meteorological Observatory", Zemlya i Vselennaya, 1966, No. 1, pp. 27–32; Kosmicheskiye Issledovaniya, 1966, Vol. 4, No. 3, pp. 427–438.

10. V. A. Krasnopol'skiy, A. A. Krysko, A. I. Lebedinskiy, "Spectrophotometric Measurements of the Moon in the Region 1900–7750 Å From Zond 3", Paper presented at the Seventh International Space Science Symposium, Vienna, Austria, May 1966.

11. V. Krat, "Rendezvous With the Sun", Pravda, February 5 1967, No. 36 (17718), p. 3.

12. A. Lebedinskiy, "Scientific Experiments on the Artificial Earth Satellite Cosmos 92", Krasnaya Zvezda, October 17 1965.

13. A. I. Lebedinskiy, "Exploring the Moon. First Automatic Station on the Moon", Soviet Life, May 1966, pp. 22–25 (In English).

14. A. I. Lebedinskiy, A. A. Lozhnikov, and V. I. Tulupov, "Measurements of Lunar Radiation Flux in the Infrared and Visible Region of the Spectrum by Means of Luna 10 Satellite", Kosmicheskiye Issledovaniya, 1966, Vol. 4, No. 6, pp. 838–841.

15. G. A. Leykin, "The Moon From the Point of View of a Cosmonaut", Zemlya Vselennaya, 1966, No. 2, pp. 36–43.

16. Yu. N. Lipskiy, "A Study of the Photographs of the Far Side", "The Moon", Symposium No. 14 of the International Astronomical Union held at Pulkovo Observatory near Leningrad December 1960, Academic Press, London and New York, 1962, pp. 7–23.

17. Yu. N. Lipskiy, "Zond 3 Photographs of the Moon's Far Side", Sky and Telescope, December 1965, pp. 338–341 (In English).

18. S. L. Mandel'shtam, I. P. Tindo, and V. I. Karev, "Investigation of the X-Ray Radiation of the Moon With the Aid of Luna 10 Satellite", Kosmicheskiye Issledovaniya, 1966, Vol. 4, No. 6, pp. 827–837.

19. A. V. Markov, "Probable Structure and Nature of the Formations on the Reverse Side of the Moon According to Photometric Measurements of Lunar Photographs", "The Moon", Symposium No. 14 of the International Astronomical Union held at Pulkovo Observatory near Leningrad December 1960, Academic Press, London and New York, 1962, pp. 39–44.

20. A. V. Markov, "Results of Investigations of the Moon and Structure of the Upper Layers of the Crust", Izvestiya Pulkovo Observatorii, 1965, Vol. 24, No. 178, pp. 162–167.

21. A. A. Mikhaylov, "The Reverse Side of the Moon", "The Moon", Symposium No. 14 of the International Astronomical Union held at Pulkovo Observatory near Leningrad December 1960, Academic Press, London and New York, 1962, pp. 3–6.

22. A. A. Mikhaylov, "Station Luna 10: Results of the Flight", Izvestiya, June 3 1966, p. 1.

23. V. M. Mozhzherin, V. B. Nikonov, V. K. Prokof'yev, and N. S. Chernykh, "The Optical Telescope for the Moon", Paper presented at the 16th International Astronautical Congress, Athens, Greece, September 1965 (In English).

24. A. S. Sharov, Byulleten' Abastumani Astrofizicheskoy Observatorii, 1965, No. 33.

25. V. I. Slysh, "Measurement of Kilometer Cosmic Radioemission in Interplanetary Space", Kosmicheskiye Issledovaniya, 1966, Vol. 4, No. 6, pp. 923–931.

26. A. P. Vinogradov, Yu. A. Surkov, G. M. Chernov, F. F. Kirnozov, and G. B. Nazarkina, "Preliminary Results of Gamma Measurements of the Lunar Surface on Cosmic Station Luna 10", Kosmicheskiye Issledovaniya, 1966, Vol. 4, No. 6, pp. 871–879; Geokhimiya, 1966, No. 8, pp. 891–899; "Studies of the Intensity and Spectral Composition of Lunar γ Radiation by Means of the Automatic Station Luna 10", Doklady Akademi Nauk SSSR, 1966, Vol. 170, No. 3, pp. 561–564.

Chapter 12

Technical and Scientific Studies Aboard Manned Satellites

R. A. Duffee and G. E. Wukelic

In the period from April 12 1961 through March 18 1965 the Soviets launched eight manned satellites. All flights had orbital inclinations of 65°, while the apogees varied between 233 and 495 km. Flight duration extended from the single orbit flight of Vostok 1 to the five-day mission of Vostok 5. Popular press releases coinciding with these launchings alluded in general terms to manned physical, geophysical, and astronomical studies as part of the mission objectives. Only for the Voskhod 1 event, which included the scientist K. P. Feoktistov as a crew member, did the announced program objectives refer directly to visual, photometric, and photographic observations of the earth, its atmosphere and horizon, polar aurorae, luminescent particles, and the stellar background. Accordingly, at least the Voskhod 1 mission had experimental objectives similar to those scheduled for the U.S. Gemini program, which included geoastronomical observations, dim-light photography experiments, synoptic terrain photography, and synoptic weather photography.*

* Gemini Mid-Program Conference, Part II Experiments, February 23–25 1966, Manned Spacecraft Center, Houston, Texas.

Based on the avowed objectives of the manned space flights, the published results can only be described as disappointing. Only six technical publications related to these studies have been noted, and their technical content is meager at best. One 52-page publication of a semipopular nature is essentially an album of photographs of the earth's surface, and particularly, various cloud formations taken during the flights of Vostoks 2 through 6[1]. Photos in this album are considerably inferior to those obtained by Gordon Cooper during the U.S. MA–9 flight, and by Borman and Lovell during their 14-day Gemini 7 mission. While the album also contains discussions of other applications of manned satellites for meteorological or aeronomical purposes, no results of such studies are given.

Results of geoastronomical studies during the Vostok series are presented in only two articles, one coauthored by the female cosmonaut V. V. Nikolayeva-Tereshkova[2, 8]. A discussion of the stratospheric aerosol layers, based on analysis of photographs of the earth's twilight horizon taken during the Vostok 6 flight, is presented. The authors conclude that two aerosol layers are found at the heights of $11{\cdot}5 \pm 1$ km and $19{\cdot}5 \pm 1$ km. Estimates of the optical thickness of the layers and the scattering radius of the particles are also presented.

The Voskhod 1 results are also available in only two papers. The senior author of the first is the scientist-cosmonaut K. P. Feoktistov[3]. An analysis of the horizon aureole as it appears in daylight is presented. The altitude of the apparent visual edge of the aureole is concluded to be a function of meteorological (tropospheric) conditions and the spectral region considered. According to the authors this accounts for the observed gradual change in color from white near the earth's horizon to blue at higher altitudes. The observable height of the aureole is estimated to be approximately 30 km. In the other article by V. Komarov, the estimated height of the brightness layer is given as 60–100 km[4].

For the structure of the aureole on the twilight side of the earth, a series of vertical photometric outlines, based on Rayleigh's law, is also contained in these articles. The authors conclude that the Voskhod 1 findings substantiate the Vostok results, viz., that two aerosol layers are generally observed at altitudes of 11 km and 19 km. The mean particle size of the aerosols is concluded to be slightly in excess of the wavelength of the observed brilliance (Rayleigh scattering). A weak maximum of luminosity $2^{\circ}\!.5$ to 3° from the edge of the nocturnal horizon of the earth, discovered by John Glenn and observed by other astronauts, was also observed during the Voskhod 1 event. The possibility that the luminosity is not a true emission of the atmosphere, but a diffusion of lunar light through an atmospheric aerosol layer, as indicated by other investigators, is suggested. During the Voskhod 1 flight, the layer was clearly observed in the Southern Hemisphere against a polar dawn background. The layer thinned out just before sunrise and was invisible at and subsequent to sunrise.

The luminous particles ("fireflies") first observed by John Glenn were also noted during the Voskhod 1 flight. Feoktistov and his coauthors conclude that the particles: (1) are related to the spacecraft, (2) diffuse solar light rather than being self-luminous; and (3) are approximately 10 microns in diameter, assuming an albedo of 0·2. They are ascribed as a possible cause of automatic astronavigational equipment malfunctions. In another article[5] the suggestion is made that the particles are earth-originated dust particles which remain attached to the craft after its insertion into orbit. The possibility that they evolve from the exterior surface of the craft because of the action of radiation, aerodynamic current, or micrometeorite erosion is also suggested.

The final section of the article by Feoktistov, *et al.*[3], presents a statistical analysis of the results of photographic observations of clouds at the periphery of both cyclones and anticyclones and the cloud cover of the earth obtained during the Voskhod 1 flight. Data are given for the dispersion of the brilliance, asymmetry of the brilliance, and average duration of transference in terms of the brilliance at various levels. Results are given in relative units (associated with logarithms of the brightness) and indicate that most of the statistical characteristics depend on cloud abundance.

Two articles by G. V. Rozenberg[6, 7], a coauthor with both Nikolayeva-Tereshkova and Feoktistov, treat the use of brightness profile of both the twilight and daytime horizons for determining stratospheric scattering coefficients. The latter article[7] presents data obtained during the Vostok 6 and Voskhod 1 flights.

Physical studies concerned with the behavior of liquids during weightlessness were conducted during the Vostok 4 and Voskhod 1 flights, but no technical publications have been noted discussing details and results of these experiments. During the press conference held after his flight on Vostok 4, Pavel Popovich reported briefly on an experiment with a half-filled flask of water. He stated that in a weightless condition the air formed a sphere in the middle of the flask with water below and above the air.

References

1. "Our Planet as it Looks From Outer Space", Hydrometeorological Publishing House, Leningrad, 1964, 57 pp.

2. A. Ya. Driving, I. M. Mikhaylin, G. V. Rosenberg, A. B. Sandomirskiy, and G. I. Trifonova, "Photometric Analysis of the Twilight Aureole Photographs Taken From the Vostok 6 Spaceship", Fizika Atmosfery i Okeana, 1966, Vol. 2, No. 10, pp. 1046–1054 (Atmospheric and Oceanic Physics, pp. 630–634).

3. K. Feoktistov, G. Rozenberg, *et al.*, "A Few Observational Results Made During the Space Flight of the Voskhod", Quelques Resultats de Observations Effectuees au Cours du Vol Cosmique du "Voskhod", 1964, 15 pp.; Trudy Vsesoyuznoy Konferentsii po Fizike Kosmicheskogo Prostrantsva, Izdatel'stvo "Nauka", Moscow, June 10–16, 1965, pp. 62–64 (NASA TT F–389, pp. 82–85).

4. V. Komarov, "Scientific Expeditions in Space", Aviatsiya i Kosmonavtika, 1966, No. 9, pp. 40–43.

5. A. I. Lazarev, "Assumptions on the Nature of Luminous Particles Observed by Astronauts", Doklady Akademii Nauk SSSR, 1964, Vol. 156, No. 2, pp. 306–307 (Doklady of the Academy of Sciences USSR, Earth Sciences Sections, pp. 3–4).

6. G. V. Rozenberg, "On Twilight Studies of Planetary Atmospheres From Spaceships", Fizika Atmosfery i Okeana, 1965, Vol. 1, No. 4, pp. 377–385 (Atmospheric and Oceanic Physics, pp. 223–227).

7. G. V. Rozenberg, "The Investigation of the Atmosphere by Observing the Brightness Profile of the Planet Edge From Spaceships", 17th International Astronautical Congress, Madrid, Spain, October 1966 (abstract only).

8. G. V. Rozenberg and V. V. Nikolayeva-Tereshkova, "Stratospheric Aerosol Measured From Spaceship", Fizika Atmosfery i Okeana, 1965, Vol. 1, No. 4, pp. 386–394 (Atmospheric and Oceanic Physics, pp. 228–232); Trudy Vsesoyuznoy Konferentsii po Fizike Kosmicheskogo Prostrantsva, Izdatel'stvo "Nauka", Moscow, June 10–16, 1965, p. 61 (NASA TT F–389, p. 81).

Part III

Soviet Biomedical Space Research

Summaries of the subjects investigated, methods employed, and results of direct Soviet biomedical space experimentation are presented in Chapters 13 and 14 which comprise Part III of this handbook. Chapter 13 deals with studies of a physiological nature on canine or human subjects. More general biological studies, especially of a genetic nature, are summarized in Chapter 14. For convenience, this latter chapter includes a discussion of some experiments on small laboratory animals, specifically mice, rats, and guinea pigs, that could be considered within the province of physiology.

The material in each chapter is organized into four sections. The first contains a description of the nature and history of the pertinent Soviet biomedical space research, including identification of the specific Soviet space vehicles associated with the experiments. The second section comprises a discussion of the subjects studied and the methods of investigation. The third is devoted to a summary of the major results and conclusions of the studies, as published by the Soviet scientists.

Finally, a list of the most pertinent Soviet publications is included. These bibliographies have been selected partially because they provide the most detailed discussions of the specific experiments of interest, and partially because they are the most readily obtainable. A few articles of a general review nature have been included within each reference list, primarily for the nonspecialist.

Since the primary purpose of this handbook is to summarize the results of direct Soviet space experimentation and to provide a guide to pertinent Soviet literature, many topics integral to biomedical space research either are not discussed or, at most, are only briefly mentioned. Similarly, comparisons with the methods and results of the U.S. manned space-flight programs have been avoided.

Among the topics omitted are Soviet cosmonaut selection and training procedures, and the technological details of their biotelemetry systems. Life-support techniques employed by the Soviets are only briefly mentioned, principally because of the dearth of technical details contained in the Soviet literature. In addition, much of the reported life-support research has been conducted in laboratory experiments. The results of this research indicate possible variants in Soviet life-support techniques for future manned space flights, but do not describe the presently utilized systems. The reader interested in these related topics is referred to the included bibliographies. Many of the referenced literature sources also contain articles directly pertaining to these topics

Chapter 13

Physiological Methods and Results

R. A. Wright and R. A. Duffee

The Soviet biomedical vertical rocket flight program involved physiological experimentation on a total of 35 dogs, several of which made multiple flights[20–22, 47]. Several rabbits, mice, and rats were also used to evaluate weightlessness effects[51]. Various methods of physiological monitoring and biotelemetry were developed and tested during this series, as were life support and recovery techniques which were later adapted for orbital animal and, finally, manned flights. As indicated in Table 13.1 emphasis on in-flight physiological monitoring was placed on the cardiovascular and respiratory systems. As described in Chapter 1, postflight evaluations were directed toward assessment of flight effects on the hemopoietic and skeletomuscular systems.

The first Soviet orbital biomedical experiment was, of course, the flight of the mongrel dog "Laika" aboard the 508·3-kg Sputnik 2 launched on November 3 1957, only one month after the world's first successful satellite launching[51]. Telemetry data of cardiovascular responses, respiration frequency, body temperature, and motor activity,

TABLE 13.1

Physiological Measurements Aboard Soviet Bioprobes and Biosatellites (1951–1966)

Vehicle/Launch Date	Period of Weight-lessness	Experimental Subjects	On-Board Measurements																	
			ECG MX Lead	ECG DS Lead	Seismo-cardiogram	Kineto-cardiogram	Phono-cardiogram	Pulse Rate	Arterial Pressure	Sphygmogram	Pneumogram (Respiration)	Body Temp.	EEG	EOG	GSR	Actogram	Dynamogram	Behavior (Neuro-muscular) Reflex	Movement Coordination	Electro-myogram
Vertical Rockets																				
100 km; 6 flights; 1951–1952	Partial ~3 min.	6 mongrel dogs (2 per flight)	−	−	−	−	−	+	−	−	−	+	−	−	−	−	−	+	−	−
100–110 km; 9 flights; 1952–1956	Partial ~3 min.	12 mongrel dogs (2 per flight)	−	−	−	−	−	+	+	−	+	+	−	−	−	−	−	+	−	−
212–473 km; 10 flights; 1957–1960	~6 or 12 min.	14 mongrel dogs (2 per flight) 2 rabbits, mice, rats	+	+	−	−	−	+	+	−	+	+	−	−	−	−	−	+	−	−
Sputnik 2/Nov. 4 1957	7 days	Female mongrel dog (Laika)	+	+	−	−	−	+	+	−	+	+	−	−	−	+	−	+	−	−
Spaceship-Satellite Series																				
Spaceship-Satellite 2/ Aug. 1960	24 hours	2 mongrel dogs (Belka & Strelka)	+	+	−	−	+[a]	+	+[b]	−	+	+	−	−	−	+	−	+	−	−
Spaceship-Satellite 3/ Dec. 1960	24 hours	2 mongrel dogs (Pchelka & Mushka)	+	+	+[c]	−	+[d]	+	−	−	+	+	−	−	−	+	−	+	−	+
Spaceship-Satellite 4/ Mar. 1961	1·5 hours	1 mongrel dog (Chernushka)	+	+	−	−	−	+	−	+	+	+	−	−	−	−	−	+	−	−
Spaceship-Satellite 5/ Mar. 1961	1·5 hours	1 mongrel dog (Zvezdochka)	+	+	−	−	−	+	−	+	+	+	−	−	−	−	−	+	−	−
Vostok Series (Manned)																				
Vostok 1/Apr. 12 1961	1·4 hours	Yu. A. Gagarin	+	+	−	−	−	+	+	−	+	−	−	−	−	−	−	+	−	−
Vostok 2/Aug. 6 1961	25 hours	G. Titov	−	+	−	+	−	+	+	−	+	−	−	−	−	−	−	+	−	−
Vostok 3/Aug. 11 1962	95 hours	A. Nikolayev	−	+	−	−	−	+	+	−	+	−	+	+	+	−	−	+	−	+
Vostok 4/Aug. 12 1962	71 hours	P. R. Popovich	−	+	−	−	−	+	+	−	+	−	+	+	+	−	−	+	−	−
Vostok 5/Jun. 14 1963	119 hours	V. Bykovskiy	−	+	+	−	−	+	+	−	+	−	+	+	+	−	−	+	−	+
Vostok 6/Jun. 16 1963	71 hours	V. Tereshkova	−	+	+	−	−	+	+	−	+	−	+	+	+	−	−	+	−	−

Vehicle/Launch Date	Period of Weight-lessness	Experimental Subjects	On-Board Measurements																	
			ECG MX Lead	ECG DS Lead	Seismo-cardiogram	Kineto-cardiogram	Phono-cardiogram	Pulse Rate	Arterial Pressure	Sphygmogram	Pneumogram (Respiration)	Body Temp.	EEG	EOG	GSR	Actogram	Dynamogram	Behavior (Neuro-muscular) Reflex	Movement Coordination	Electro-myogram
Voskhod Series (Manned)																				
Voskhod 1/Oct. 12 1964	24 hours	V. Komarov B. B. Yegorov K. P. Feoktistov	−	+	+	−	−	+[e]	+	−	+	−	+	+	−	−	+	+	+	−
Voskhod 2/Mar. 18 1965	26 hours	P. Belyayev A. Leonov	−	+	+	−	−	+[e]	+	−	+	+	+	+	−	−	−	+	+	−
Cosmos Series																				
Cosmos 110/Feb. 22 1966	22 days	2 mongrel dogs Ugolek Veterok	+	+	+	−	−	+	−	+	+	−	−	−	−	−	−	+	−	−

(a) Belka only
(b) Strelka only
(c) Pchelka only
(d) Mushka only
(e) Pulse and respiration combined

were received over a seven-day period when the transmitters became silent[39, 51].

Final preparations for manned space flights were completed during the Spaceship-Satellite flights which began on May 15 1960 and culminated in the single-orbit flight of Spaceship-Satellite 5 on March 25 1961. The first flight in this series was essentially a systems test and did not involve direct biomedical experimentation. During the other four flights, mongrel female dogs once again served as the primary experimental subjects. Four of the dogs made 24-hour (17-orbit) flights: "Belka" and "Strelka" on Spaceship-Satellite 2; and "Pchelka" and "Mushka" on Spaceship-Satellite 3. The latter vehicle was destroyed during re-entry into the atmosphere thereby preventing any postflight analyses on the dogs. Spaceship-Satellites 4 and 5 were single-orbit flights with one dog per vehicle: "Chernushka" on Spaceship-Satellite 4 and "Zvezdochka" on Spaceship-Satellite 5. Rats, mice, and guinea pigs were also carried on board during these flights. Results of the studies on rodents are presented in Chapter 14.

Physiological monitoring and telemetry techniques were considerably expanded during this series with primary emphasis again placed on cardiovascular and respiratory responses. Motor coordination, behavior, and neuro-muscular reflexes were also studied. As shown in Table 13.1, cardiovascular studies included not only ECG measurements with two leads but also the first attempts at seismocardiography, a variant of ballistocardiography, and phonocardiography. For the first time direct television observation of the experimental subjects was incorporated within the medical monitoring system. The recording of the televised images of the dogs was synchronized with the telemetry data making it possible to compare observed behavior with physiological response data.

Both variants of landing techniques used in the manned Vostok series were evaluated during the Spaceship-Satellite flights; landing within the cabin or catapulting from the spacecraft at an altitude of 7 km with parachute descent[32, 51]. Cabin pressure was maintained at approximately 760–780 mm Hg equivalent to one atmosphere with an oxygen content of 21 to 24 per cent. As was the case during the vertical rocket program the diluent gas was nitrogen[32, 51].

On April 12 1961 man first emerged into space in the person of Yuriy Gagarin on board Vostok 1, launched only three weeks after the successful mission of Spaceship-Satellite 5. Physiological monitoring and telemetry techniques including television observation used in this flight were essentially identical to those employed during the flights of Spaceship-Satellites 4 and 5. Gagarin landed in the spacecraft cabin, apparently the only Vostok cosmonaut to use this landing option.

Gagarin's single orbit flight was followed quickly by the 17-orbit flight of Gherman Titov on Vostok 2 launched August 6 1961. The only change in the physiological monitoring system used during this flight was the inclusion of a kinetocardiogram, another variant of ballistocardiography. Titov developed acute symptoms of motion sickness during his flight, causing considerable consternation among the Soviet bioastronautics specialists as to the effectiveness of their cosmonaut selection and training programs. Accordingly, the Soviets modified their cosmonaut training to include exercises aimed at "conditioning" the vestibular and other orientation sensors of the body.

Their concern with central nervous system functions and psychological reactions to stress was evinced in the biomedical monitoring program used during the Vostok 3 and Vostok 4 missions in August 1962. In addition to the usual battery of respiratory and cardiovascular sensors, cosmonauts Andrian Nikolayev and Pavel Popovich were suitably instrumented to record electroencephalograms (EEG), electrooculograms (EOG), and galvanic skin response (GSR). The same parameters were recorded during the group flight of Valeriy Bykovskiy on Vostok 5 and the first woman in space, Valentina Tereshkova, on Vostok 6, in June 1963. Methodology for evaluating cardiovascular system responses to space flight was expanded in these latter missions to include seismocardiography[46].

The physiological studies conducted during the multi-crew Voskhod flights in October 1964 and March 1965 were basically the same as those evaluated during the Vostok 5 and Vostok 6 flights. However, the inclusion of the physician Boris Yegorov in the crew of Voskhod 1 allowed for a considerable expansion of the program. Dr. Yegorov had a system of detachable biosensors available to him to supplement the telemetry data. He directly obtained EEG's and measures of psychomotor performance and visual acuity from the other crew members. Blood samples were also taken during flight for subsequent evaluation of blood morphology. The Voskhod 1 flight is of particular interest as the crew did not wear space suits at any time during the mission, and a soft landing was successfully accomplished.

An important study of the effects of prolonged weightlessness and radiation akin to lunar mission exposures was conducted on the Cosmos 110 flight. Two mongrel dogs, "Veterok" (experimental) and "Ugolek" (control), were launched into space on February 22 1966 and recovered on March 16 1966 after 22 days in orbit. With an apogee of 904 km the spacecraft penetrated the inner radiation belt. The two dogs were carried in separate pressurized cabins, similar to those used in the Spaceship-Satellites, and provided with individual life-support systems. Their restraint harness allowed some freedom of movement; however, due to feeding hose and electrical connections it appears that movement was limited to standing, sitting, or lying on the sternum. Feeding was accomplished through a gastric fistula as opposed to *ad lib.* feeding employed on previous orbital dog studies. As indicated in Table 13.1 physiological measurements were more extensive than on any previous mission. However, little detailed information as to the results of this mission have been published to date.

METHODS

Physiological Techniques

The Soviets realized early in their space program that an understanding of the effects of space flight upon humans was most important for the success of their program. Collection of various physiological data was thus begun on the early animal flights. As has been previously mentioned, extensive research using dogs and other species began in the early 1950's[5, 19, 22, 31–33].

There are many difficulties associated with gathering physiological data from an animal or human in space. Radiotelemetry is the main method for data transfer; however, the methods used to record, amplify, and transmit data to earth must be rugged enough to withstand the launch stresses and small enough to be integrated into the spacecraft with the other instrumentation[19, 43]. The sensors which attach to the body must be reliable for long periods of time. If the resistance of an electrode builds up over a period of time, the signal may become too weak for amplification and recording.

The Soviets used various forms of physiological measurements on board their spacecraft; particular measurements depended upon the flight (see Table 13.1).

Descriptions of the measurements used to monitor and evaluate the well being of the animal or human follows.

1. Electrocardiography. In the studies on animals, implanted electrodes in the form of a tantalum alloy net were used. The electrocardiograms were recorded from standard and chest leads. The recordings were made in the 0·5–40 cps frequency band. On the Vostok ships the cosmonauts were electrocardiographed with bipolar chest leads DS and MX. In the DS leads the electrodes were placed in the fifth

intercostal space along the middle axillary line from left to right, and the MX leads, in the xiphoid process region and the upper portion of the sternum. In these leads interference from muscular bioelectrical potential and skin shifting is minimal. A system of long-term electrode fixation by a chest belt ensured reception of good recordings during vigorous activity over many days. For continuous monitoring of the pulse rate during flights of many days, an electrocardiophone was used. This is an instrument which transforms the bioelectrical potentials of the heart into square waves with a period of about 0·15 sec.

2. Phonocardiography. Heart tones from the dogs Belka and Mushka were recorded on a Type TG–7 telephone used as a microphone. The frequency response of the microphone together with the amplification channel were within the limits of 50–500 cps. To use a narrow-band telemetric channel to transmit the phonocardiograms, a detector with an integrator, which ensured reception of the "envelope" of sound vibrations, was mounted at the amplifier output. This phonocardiogram was called "integrated". It is convenient for determining the time and amplitude interrelationships in the phonocardiographic curve but does not allow analyses of the frequency composition of heart tones. However, the "integrated phonocardiography" method is fully applicable to space flight conditions in which it is difficult to expect the appearance of heart noises caused by organic injury to the cardiac valve apparatus.

3. Sphygmography. Sphygmograms of the carotid artery in the dogs Chernushka and Zvezdochka were recorded during the brief orbital flights of Spaceship-Satellites 4 and 5. The pickup for their recording consisted of a sleeve worn on the carotid artery exposed in a skin flap. A piezoelectric device which picked up the mechanical vibrations of the vessel wall was mounted in the sleeve. The signal was amplified by a standard electrocardiographic channel.

4. Arterial oscillography. Arterial pressure was measured by the oscillator method on the carotid artery which was exposed in a skin flap. Arterial oscillograms were recorded in the dog Strelka on board Spaceship-Satellite 2. The magnitude of arterial pressure was determined by simultaneously recording the oscillator and pressure curves.

5. Kinetocardiography. A kinetocardiogram is a recording of the local vibration of the chest wall. A miniature electromagnetic microphone placed in the region of the apex beat recorded G. S. Titov's kinetocardiogram. The frequency response of the microphone and amplifier were within 10–40 cps limits. Kinetocardiograms allowed an evaluation of the phases of the heart cycle.

6. Seismocardiography. The seismocardiographic method is a modification of ballistocardiography suitable for cosmic research. The first seismocardiogram was recorded during the flight of Spaceship-Satellite 3. A pickup consisting of two induction coils and a seismic mass-magnet which was firmly connected to the body of the pickup by springs was placed on the spine of the dog Pchelka. The counter resonance frequency was about 20 cps, the damping period of a single vibration was 0·1 sec. During each heart contraction two vibration cycles arose: systolic and diastolic. Their amplitude is a derivative of the forces generated by the heart during systole and diastole, but the period is connected to the degree of coordination of heart contractions. Thus, the cycle amplitude and period of a seismocardiogram indirectly reflect the state of the contracting function of the myocardium. A similar but even smaller pickup, having a uniform frequency response within a 10–100 cps range (due to critical damping of the seismic mass), was used during the flights of V. F. Bykovskiy and V. V. Nikolayeva-Tereshkova. The pickup was placed in the region of the base of the sternum.

7. Electromyography. Motor coordination and muscle tonus were measured using muscle biopotentials.

8. Actography. Overall motor activity for both man and animals was recorded on an actograph. Potentiometer pickups were used providing a record of displacement in three axes. Quantitative criteria for the assessment of actograms have yet to be developed.

8. Thermometry. Body temperature measurements of the cosmonauts were important to assess the functioning of heat regulation. Body temperature will indicate metabolic rate and in conjunction with oxygen consumption can give a complete picture of metabolic processes during space flight activities.

9. Pneumography. A resistance pneumograph around the chest gave respiration information. These data were used to monitor the cosmonauts' condition and were telemetered to the earth.

To investigate higher nervous activity and motor coordination, special transducers were used. Conditioned responses to painful stimuli were used on animals to evaluate effects under abnormal conditions while the human central nervous system was studied with the use of electroencephalography, galvanic skin resistance, and psychological testing. The cosmonaut's condition was monitored by listening to radio conversations and analyzing flight logs and diaries[28].

Table 13.1 lists the physiological measurements made on each flight.

RESULTS

Vertical Rockets

The cabin of the geophysical rockets was pressurized to about 760 mm Hg with O_2 and N_2; however, if the cabin pressure dropped to 460 mm Hg, the composition of gases was adjusted so that the oxygen partial pressure never fell lower than 100–110 mm Hg. Thus, oxygen content in the cabin was maintained at about 37%. Soda lime was used to absorb CO_2 and silica gel to absorb water vapors[22].

A typical flight was as follows: lift off with a maximum of 6 g acceleration; maximum velocity during ascent, 1·72 km per second; deceleration occurred after rocket burnout and the nose cone separated at apogee and began a descent of 1·75 km per second. A deceleration chute opened at 4 km and at 2 km the main chute opened. Total flight time was from 600–660 seconds. Maximum period of dynamic weightlessness was about 360–370 seconds[22]. Angular accelerations did occur upon reentry. Short-term decelerations were observed upon chute opening and landing. Due to the cabin insulation, the inside air temperature remained relatively stable around 25° C. Cabin pressures held steady throughout the flight; variations were not more than ± 5 mm Hg[22].

The dogs were trained exceptionally well before flight and were not nervous or excited before launch. Upon engine ignition, an orientating reflex was expressed. No abnormal physiological functions were noted. X-ray, blood, and urinalysis were normal after flight. Physiological functions were not measured throughout the flight. The failure of the recording systems was caused partly by the weightless state and varying accelerative forces.

Although a difference was noted between dogs drugged and dogs not drugged, certain general rules can be formulated. Respiration increased during an increase in g-load and decreased during weightlessness. Heart rates increased in all dogs during active flight and decreased to normal values during weightlessness and after landing.

In the drugged animals, breathing rates and heart rates changed less than the non-narcotized. Heart rates changed little and were normal by the weightless portion of the flight. Thus, it appears as though the change in condition recognition may play a role in physiological response.

Although changes were noticed in blood pressure and respiration during periods of high g-loading, no departure from normal was witnessed in the weightless phase of the flight. Therefore, no radical difference in behavior was expected or observed. Startle reaction was noticed during engine ignition, acceleration, and engine shutoff. Hexanal decreased the response of flight to near-normal conditions. The only changes of significance after a flight was a leukocytosis. This was assumed to be a normal reaction to the acceleration forces imposed upon the animals since it has been observed previously in animals undergoing vibration experiments. Internal hemorrhage may theoretically be the cause for the leukocytosis.

In conclusion, rocket flights to altitudes of 100–473 km do not result in sudden physiological changes. Trained animals withstand the flights without disturbance and are quite normal in their behavior during and after the flights[5, 19, 22, 32–33].

Orbital Investigations

Cardiovascular System

Most of the physiological data collected from Soviet spacecraft were concerned with the reaction of the cardiovascular system to the weightless state. Due to the large amount of data obtained during the flight experiments, only that material used by the Soviets to evaluate the effects of weightlessness on physiological functions is presented.

In both the animal and human flights, pulse rates increased during the launch and early orbital phases[4, 6, 16, 30–31, 49–50]. After a period of what might be called acclimation to weightlessness, the pulse rate began to return to normal values. In most instances the period for the heart rate to normalize was slower than was indicated during centrifuge training; however, actual space flight conditions present exposure to more variables and anxiety. Differences in the pulse normalization times were attributed to variations in emotional stress of the cosmonauts. This variation was not as great in the animal experiments.

A daily periodicity in pulse rate dynamics was witnessed in all flights with a higher rate in the mornings and decreasing toward the end of the day. During sleep, however, the pulse rate was lower than previously predicted from centrifuge training runs. Variability of pulse rate was given special attention. It is known that the dog has an inherent arrythmia during rest and periods of emotional stability which disappears when demands on the cardiovascular system are increased. The periods of arrythmia were noticed to disappear during the launch phase of the flights and acceleration. After a short period under weightless conditions, the arrythmias increased in the dog and often exceeded the number witnessed before launch[33].

People do not normally show arrythmias and a cardiac cycle of a duration greater than 0·20–0·25 sec may be considered abnormal. Considerably greater cardiac cycle durations were noticed in the cosmonauts during weightless flight which caused concern[6]. This led to greater research effort to study cardiovascular effects of space flight, and it was later shown that the distribution time values of the cardiac cycle changed substantially. The variation curves were widened and flattened. The Soviets suggest that the effects of a lack of a hydrostatic force reflects itself in the cardiac sequence. The changes in time sequences were noticed when prelaunch ECG values were compared to in-flight values for the dogs, Belka and Pchelka[31]. During the first half of the flight the atrioventricular conduction time was somewhat longer than prelaunch values while during the last half the values slowed to less than prelaunch conduction times. For most of the cosmonauts, the atrioventricular conduction times were somewhat greater than those taken on earth; however, there was a daily periodicity. This varied among individuals. Intraventricular conduction times were not significantly different for animals or humans in a weightless state.

The Q-T interval in animals shortens during launch and powered flight, but then lowers to near prelaunch values after a period of weightless flight. For human cosmonauts the Q-T interval varied; it did shorten during launch, but showed much daily variation between cosmonauts.

Arterial blood pressure was measured on several orbital flights; however, it was studied most thoroughly in the dog, Strelka, aboard Spaceship-Satellite 2. In the weightless state a biphasic reaction occurred. Systolic pressure dropped 50–80 mm Hg compared to the powered portion of the flight and was 20–30 mm Hg lower than the prelaunch values[6]. After this drop in maximum systolic pressure, an increase in pressure was witnessed which did not reach prelaunch values. A simultaneous drop in diastolic pressures was observed with a rise in pulse pressure. Beginning with the second orbit, the mean arterial pressure dropped to quite low figures and averaged 40–50 mm Hg lower than prelaunch values. After approximately 10–12 hours of weightlessness, the systolic pressures began to rise and held steady at 10–30 mm Hg below the prelaunch values throughout the remainder of the flight. Diastolic pressures also remained below prelaunch values.

Phono- and kinetocardiograms indicated the length of mechanical systole. For Titov the length of mechanical systole before launch was 0·35 sec and while in orbit was 0·41–0·47 sec. The time for mechanical systole in dogs decreased at the beginning of the flight, then increased, and finally decreased toward the end of the day to below prelaunch levels. From the data it is clear that there is a definite variation between the mechanical and electrical systole[6]. Coefficient K, which is the relationship of the lengths of the mechanical and electrical systoles, is relatively shorter at the end of an orbital flight than at the beginning. Data indicate that during Titov's flight the electromechanical retardation (the time of the ECG-Q wave until the beginning mechanical systole) clearly increased from 0·033 seconds before launch to 0·046–0·048 seconds during orbital flight[6].

Soviet attempts to measure cardiac contraction was by the use of the seismocardiogram[6, 16, 50]. The first cycle amplitude of the seismocardiogram appears to be proportional to the rate of blood ejection. Data from animal and human experiments indicate an increase in the first cycle of the seismocardiogram. The second phase of the seismocardiogram indicates the force of the reverse hydraulic stroke of the blood, and after 2–3 days of weightless flight the difference in the amplitude between the two cycles slowly decreases. This may be due to an absolute increase in the amplitude of the second cycle which is thought to depend upon the elastic properties of the great vessels and the level of the arterial pressure. Data obtained during the flight of Tereshkova revealed a third cycle of the seismocardiogram[6]. This third cycle was thought to be due to a change in intracardiac hemodynamics since the ECG showed no change in excitability or conductivity. Another theory for the appearance of a third cycle is an increase in ventricular filling rate during diastole in the weightless state.

Comparisons were made between data from the seismocardiograph and phonocardiograph to determine coordination of the right and left heart. The activity of the right and left heart is known to be strictly determined in both magnitude and time. Any small change in the existing relationships changes the period of the seismocardiographic cycle and the rate of the phonocardiographic tone[6, 50]. Data from the dog, Belka, revealed an increase in phonocardiographic tone during weightless flight and a persistent widening of the first seismocardiographic cycle[6, 31]. Thus, there are objective data that under weightlessness a rearrangement of heart activity occurs together with corresponding changes in time and amplitude relationships between the forces generated by the right and left heart.

The previously mentioned data allowed the Soviets to separate into three phases the reaction of the cardiovascular system to the effect of weightlessness[38]. The first phase is a relative decrease in pulse rate and an increase in pulse arrhythmia in animals, and a similar variability in humans. This phase is thought to be due to the after-effect of acceleration during launch. The second phase is incomplete adaptation to weightlessness lasting for 10–12 hours. Pulse rate normalizes, atrioventricular condition period lengthens, and the Q–T interval increases somewhat; mechanical systole shortened then lengthened and arterial pressure fell and began to rise near the end of a certain period; and the first seismocardiographic cycle shortened. The third phase indicates an adaptation to the weightless condition. The pulse rate settles at a rate lower than prelaunch levels, A–V conduction lengthens, mechanical systole decreases, and the electro-mechanical coefficient drops. There is a stable widening of the first and second seismocardiographic cycles. From the above data it has been possible for the Soviets to separate a number of specific changes in the functional state of the circulatory system which we assume are due to the weightless state[19, 30, 34–35, 41].

Respiratory System

The respiration rate of the Soviet cosmonauts was quite variable, but within normal limits. For example, Bykovskiy's respiration rate was 13–20 prelaunch, 24–27 in late countdown stages, and 15–27 in propelled flight. Respiration rate slowed after a period of powered flight in spite of quite high g and quickened again only during orbit insertion. According to the Soviets, this temporary increase in the respiration rate is obviously connected with periods of great emotional stress caused by such things as separation of the various rocket stages and viewing of stellar space. Since the influence of g-forces, weightlessness, and emotional excitation caused much variation in the respiration rate among the various cosmonauts, it is undoubtedly a reflection of the individual peculiarities of the cosmonauts and must be considered when analyzing the data. No distinct pathological changes were noted due to changes in respiration[6, 16, 18, 24, 35].

Central Nervous System

Bioelectric activity of the brain was measured on several orbital flights[6, 15–16, 31, 33–34, 49]. Difficulties with electrode stability prevented accurate data on some flights. Using a record of Bykovskiy's electroencephalogram (EEG) as an example, it indicated an increase in alpha rhythms in the first two orbital rotations from 35–58 per cent. The alpha rhythms decreased over the next five orbits to 29 per cent.

From the seventh orbit until the 55th orbit the alpha rhythms increased nearly 50%. The increase in alpha rhythm possibly reflects the drowsy state of the cosmonaut during the daylight hours and from the 55th orbit until the end of the flight reflects a state of fatigue. The index showed a lowering of alpha rhythm only immediately before descent[6].

Index of beta rhythms for Bykovskiy was 22% at the beginning of the flight. This increased, however, as the flight progressed and no doubt reflected the excitation of the cosmonaut. A drop in the beta index after sleep indicates a relatively rested state of the cosmonaut. Beta rhythm index for Bykovskiy tended generally to decrease indicating that fatigue was not sharply defined.

Galvanic skin reactions were recorded on some of the Soviet spacecraft. Discussion of results are limited to those from V. F. Bykovskiy on board Vostok 5[17]. During the propelled portion of the flight the galvanic skin reactions were almost twice the number recorded during a five-minute period before launch. After the first orbit the number of galvanic reactions was decreased to near before-launch levels and dropped below prelaunch levels during subsequent orbits. Decreases in pulse rate and respiratory rate were observed to follow the decrease in galvanic skin resistance (GSR). The amplitude of the positive phase of the GSR tended to increase during the flight which was thought to reflect a drop in emotional tension in connection with the proper orbit and normal flight. The negative phase of the GSR was less expressed and showed less amplitude at the same time the positive phase increased in amplitude. Near the end of the flight, both the positive and negative phases increased in amplitude. We can generalize from the GSR data to say that these reactions are intimately connected to the emotional state, activity, and the pulse and respiration rate.

Electrooculogram (EOG) activity increased in the beginning of the flights in which it was used[6]. Again a correlation of eye movements could be made with pulse and respiration rate. Nystagmoid movements were recorded on cosmonautess Tereshkova; however, these movements had been recorded before launch after rotation in a Barani chair. The EOG may be linked with the general well being of the individual and follows pulse rate and respiration rate. Before reentry was made by Bykovskiy, EOG recordings increased as did other measurements. EOG measurements can also determine rest, sleep, drowsiness, and emotional states. No asymmetry of oculomotor reactions was noticed which are characteristic for disturbance of the tonus of the eye muscles in connection with changes in the state of the vestibular apparatus.

Vestibular Function

One of the first indications of the state of the vestibular apparatus and its interactions with other systems during transition into the weightless condition is the presence or absence of illusions of body position in space[6]. Nikolayev noted a sensation of his body trunk sloping forward and a sense of lightness. Popovich had a sense of hovering and his head and body leaning forward and down. Bykovskiy and Tereshkova had no illusions in connection with weightlessness. Special vestibular tests accomplished during flight were: correctness of drawing geometric figures (spirals, stars, parallel lines), equilibrium of outstretched hands with eyes closed, the finger-to-nose test, and motor coordination of the eyes. Assessment of the fulfillment of normal working operations such as manipulating toggle switches, sending morse code, and taking orders from ground stations gave good indications of normal coordination. On the basis of the tests performed one may assume that the changes were not all caused by weightlessness, but to a certain degree by cosmonaut fatigue. Also there were no changes in function of the vestibular and visual analyzers.

Biochemistry

Blood samples were taken before and immediately after each flight. The usual measurements were hemoglobin, RBC's, WBC's, sedimentation rate, differential white cell count, thrombocyte count and the number of reticulocytes[3, 6]. Few changes were found from analysis of the blood. A slight leukocytosis was observed immediately following some of the flights. However, by the second day the entire blood picture had returned to normal.

Analysis of urine after Bykovskiy's flight showed occasional leukocytes, hyaline coats and a few RBC's. Chlorides tended to decrease for two days after the flight. Urinalysis returned to normal after approximately one week.

In general there were no pathological findings from the animal or manned Soviet space flights. Titov, in Vostok 2, developed motion sickness during his fourth earth orbit. These unpleasant sensations continued throughout the flight, although their degree of expression decreased after sleep. This problem has not been noted among American astronauts and may depend upon selection and training of personnel.*

SUMMARY

The Soviet biomedical space effort has been based upon sound planning. They began with very extensive research to determine biological effects from acceleration, weightlessness, radiation and long-term confinement. Many different biological species were used to investigate

* A. D. Calterson, *et al.*, Project Mercury Summary Results, NASA, May 15–16 1963.

these possible effects. Dogs were used extensively in vertical and orbital flights during which the various physiological measurements were made to observe abnormalities. The cardiovascular system effects were investigated in detail, and several modifications of existing tests were devised. Motor activity was investigated with EEG activity and muscle coordination. Insects were carried on board many of the early flights to indicate any effects of radiation which might be encountered.

The Soviet space capsule was similar in design to the tried and proven balloon gondola used in the 1930's for high-altitude flights. Their booster development was advanced sufficiently to provide enough power for propelling a sizeable load into orbit. This advantage reduced the necessity to develop miniaturized components to be used in the spacecraft. This meant they could use proven instrumentation to measure physiological parameters. The findings from these measurements have been reviewed and reported previously in this chapter. Proper selection and training of personnel was found to be very important to maintain orientation in space. Few physiological changes due to weightless space flight were noted and acceleration and confinement were tolerated well by most of the Soviet cosmonauts.

References

1. Problemy Kosmicheskoy Biologii, Izdatel'stvo Akademii Nauk SSSR, 1962, Vol. 1 (NASA TT–F–174); 1962, Vol. 2 (JPRS: 18 395); 1964, Vol. 3 (JPRS: 25 287); and 1965, Vol. 4 (NASA TT–F–368).

2. "The Famous Vostok First Manned Space Ship", Le Celebre Vostok, Editions Del'Agence De Presse (APN) Novosti, 32 page (TT–65–763).

3. "Aviation and Space Medicine", Akademiya Meditsinkikh Nauk SSSR, Moscow, 1963 (NASA TT–F–228).

4. "Biological Studies Under Conditions of Space Flight and Weightlessness", Izvestiya Akademii Nauk SSSR, Seriya Biologicheskaya, 1964, No. 3, pp. 327–387 (JPRS: 25 844).

5. Proceedings of the First International Symposium on Basic Environmental Problems of Man in Space, Paris, October 29–November 2 1962, Springer-Verlag, New York, 1965, 440 pp.

6. "Second Group Space Flight, and Certain Results of Flights of Soviet Cosmonauts on the Vostok Ships", Izdatel'stvo "Nauka", Moscow, 1965, 228 pages (MT–65–256).

7. Soviet Bioastronautics and Manned Spaceflight, ATD Report P–65–14, March 18 1965.

8. Scanback of the Soviet Manned Spaceflight Program, ATD Report P–65–19, April 20 1965.

9. Voskhod 1 and Voskhod 2 Flights, Review articles, ATD Report P–65–46, July 8 1965.

10. Abstracts, XV Congressius Internationalis Medicinai Aviaticae et Cosmicae, Prague, 1966.
11. "A Human Embarked Into Outer Space", Izdatel'stvo Znaniye, Moscow, 1966 (JPRS: 34 576).
12. Problems of Space Medicine, ATD Report 66–116, September 13 1966.
13. Abstracts, XVII Congress of the International Astronautical Federation, Madrid, Spain, October 1966.
14. Materials From a Conference on Space Biology and Medicine, November 10–14 1964, Moscow, 1966 (JPRS: 38 596).
15. I. T. Akulinichev, *et al.*, "Results of Some Electrophysiological Studies on the Voskhod Spacecraft", 16th International Astronautical Congress, Athens, Greece, September 13–18 1965 (JPRS: 32 808).
16. I. T. Akulinchev, *et al.*, "Medical Monitoring of the Condition of the Cosmonauts Belyayev and Leonov During Training and Orbital Flight", Kosmicheskiye Issledovaniya, 1966, Vol. 4, No. 2, pp. 311–319 (HT–66–244, pp. 224–240).
17. R. M. Bayevskiy, "Medical Services and Facilities on Board Spacecraft", Sluzhba Zdorov'ya v Kosmose, Izdatel'stvo "Znaniye", Moscow, 1966, No. 15 (JPRS: 40 383).
18. R. M. Bayevskiy, "Physiological Measurements During Space Travel", Priroda, 1967, No. 1, pp. 68–72 (JPRS: 40 381).
19. R. M. Bayevskiy and O. G. Gazenko, "The Reaction of the Cardiovascular System of Man and Animals Under Weightless Conditions", Kosmicheskiye Issledovaniya, 1964, Vol. 2, No. 2, pp. 307–319 (TT–64–547, pp. 212–232).
20. A. A. Blagonravov, "Investigation of the Upper Layers of the Atmosphere With the Aid of High-Altitude Rockets", Vestnik Akademii Nauk SSSR, 1957, No. 6, pp. 25–32 (NRC–C2456).
21. B. G. Bugrov, *et al.*, "Investigations of the Vital Activity of Animals When Flying in Hermetically Sealed Cabins of Rockets up to a Height of 110 km". Preliminary Results of Scientific Researches on the First Soviet Artificial Earth Satellites and Rockets, 1958, No. 1.
22. A. M. Galkin, *et al.*, "Investigations of the Vital Activity of Animals When Flying in Hermetically Sealed Cabins of Rockets up to a Height of 212 km", Preliminary Results of Scientific Researches on the First Soviet Artificial Earth Satellites and Rockets, 1958, No. 1.
23. S. Gavrilov, "Report on the Cosmos 110 Space Flight", Krasnaya Zvezda, March 19 1966, p. 5.
24. O. G. Gazenko, "Some Results of Physiological Reactions to Space Flight Conditions", 12th International Astronautical Congress, Washington, D.C., October 4 1961.
25. O. G. Gazenko, "Certain Problems of Space Biology", Vestnik Akademii Nauk SSSR, 1962, No. 1, pp. 30–34 (TT–62–412).
26. O. G. Gazenko, "Medical Research on Space Vehicles Vostok and Voskhod", Monograph, Moscow, 1964 (JPRS: 27 925).
27. O. G. Gazenko, "Results of Certain Medical Investigations on the Voskhod Ships", "A Human Embarked Into Outer Space", Izdatel'stvo "Znaniye", Moscow, 1966, pp. 17–22 (JPRS: 34 576).
28. O. G. Gazenko and R. M. Bayevskiy, "Physiological Methods in Space Medicine", Iskusstvennyye Sputniki Zemli, 1961, No. 11, pp. 68–77 (Artificial Earth Satellites, pp. 72–81).
29. O. G. Gazenko, R. A. Grigor'yan, L. A. Kitayev-Smyk, and A. M. Klochkov, "The Increase in Extensor Tonus in Totally or Partially Cerebellectomized Cats During Weightlessness", Byulleten' Eksperimental'noy Biologiy i Meditsiny, 1965, Vol. 60, No. 7, pp. 7–12 (Bulletin of Experimental Biology and Medicine, pp. 727–731).
30. O. G. Gazenko and A. A. Gyurdzhian, "Physiological Effects of Gravitation",

presented at the Sixth International Space Science Symposium, Mar del Plata, Argentina, May 11–19 1965 (JPRS: 30 893).

31. O. G. Gazenko, *et al.*, "Physiological Reactions of Animals During Flights on Spaceship-Satellites 3, 4, and 5", Izvestiya Akademii Nauk SSSR, Seriya Biologicheskaya, 1964, No. 4, pp. 497–510 (JPRS: 26 994).

32. A. M. Genin, *et al.*, "Man in Outer Space", Medgiz, Moscow, 1963, 160 pages (JPRS: 25 825).

33. N. N. Gurovskiy and M. A. Gerd, "In the Spaceflight Laboratory", Nauka i Zhizn', 1961, No. 10, pp. 21–28 (TT–62–652).

34. I. I. Kas'yan, *et al.*, "Motor Reactions Under Conditions of Weightlessness", Izvestiya Akademii Nauk SSSR, Seriya Biologicheskaya, 1964, No. 5, pp. 677–689 (JPRS: 27 591).

35. I. I. Kas'yan, *et al.*, "Certain Physiological Reactions of Man Under Conditions of Brief Weightlessness", Izvestiya Akademii Nauk SSSR, Seriya Biologicheskaya, 1965, No. 5, pp. 633–646 (JPRS: 33 115).

36. I. I. Kas'yan, *et al.*, "Physiological Responses of Cosmonauts in Support-Free Space", Izvestiya Akademii Nauk SSSR, Seriya Biologicheskaya, 1966, No. 1, pp. 3–13 (JPRS: 35 278).

37. I. I. Kas'yan and V. I. Kopanev, "The State of Weightlessness and Artificial Gravity", Izvestiya Akademii Nauk SSSR, Seriya Biologicheskaya, 1963, No. 6, pp. 880–891 (TT–64–140).

38. I. I. Kas'yan and V. I. Kopanev, "On the Physiological Mechanism of the Effect of Weightlessness on the Organism of Man", Izvestiya Akademii Nauk SSSR, Seriya Biologicheskaya, 1965, No. 1, pp. 10–17 (JPRS: 29 433).

39. I. I. Kas'yan, V. I. Kopanev, and V. I. Yazdovskiy, "Blood Circulation Under Conditions of Weightlessness", Izvestiya Akademii Nauk SSSR, Seriya Biologicheskaya, 1964, No. 3, pp. 352–368 (JPRS: 25 844).

40. L. S. Khachatur'yants, "Physiology of Work in Space", Meditsinskaya Sestra, Moscow, 1967, No. 1, pp. 340–347 (JPRS: 40 399).

41. A. A. Leonov, *et al.*, "Orientation of Man in Space", Kosmicheskiye Issledovaniya, 1965, Vol. 3, No. 6, pp. 940–945 (TT–65–1985, pp. 220–231).

42. Yu. V. Natochin, M. M. Sokolova, V. F. Vasil'yeva, and I. S. Balakhovskiy, "Investigation of Kidney Function in the Crew of a Voskhod Spacecraft", Kosmicheskiye Issledovaniya, 1965, Vol. 3, No. 6, pp. 935–939 (TT–65–1985, pp. 212–219).

43. V. V. Parin and R. M. Bayevskiy, "Certain Problems of Contemporary Biological Telemetry", Fiziologicheskiy Zhurnal SSSR imeni I. M. Sechenova, 1964, Vol. 50, No. 8, pp. 924–933 (MT–64–416).

44. V. V. Parin and O. G. Gazenko, "Soviet Experiments Aimed at Investigating the Influence of Space Flight Factors on the Physiology of Animals and Man", Life Sciences and Space Research, Third COSPAR Symposium, North-Holland Publishing Company, 1963, pp. 113–127.

45. V. V. Parin, O. G. Gazenko, and V. I. Yazdovskiy, "The Possibility of Protective Adaptation of the Organism and Limits of Adaptation Under Conditions of Maximum Overstrain and Weightlessness", Vestnik Akademii Meditsynskikh Nauk SSSR, 1962, No. 4, pp. 76–81 (JPRS: 15 187).

46. V. V. Parin, Yu. M. Volynkin and P. V. Vasil'yev, "Manned Space Flight—Some Scientific Results", Life Sciences and Space Research, Fifth COSPAR Symposium, North-Holland Publishing Co., 1965, pp. 1–21.

47. A. V. Pokrovskiy, "Study of the Vital Activity of Animals During Rocket Flights in the Upper Layers of the Atmosphere", International Congress on Guided Missiles and Rockets, Paris, France, December 1956.

48. V. A. Popov, *et al.*, "Analysis of Intonational Characteristics of Speech as a Criterion of the Emotional State of Man Under Conditions of Space Flight", Zhurnal Vyshey Nervnoy Deyatel'nosti, 1966, Vol. 16, No. 6, pp. 974–983 (JPRS: 39 906).

49. Yu. M. Volynkin, V. I. Yazdovskiy, *et al.*, "The First Manned Space Flights", Mediko-Biologicheskiye Issledovaniya, 1962 (TT–62–1619).

50. A. D. Voskresenskiy and M. D. Venttsel, "Application of Correlation Analysis Methods to Study of the Reactions of the Human Cardiovascular System in Space Flight Aboard a Voskhod Vehicle", Kosmicheskiye Issledovaniya, 1965, Vol. 3, No. 6, pp. 927–934 (TT–65–1985, pp. 198–211).

51. V. I. Yazdovskiy, "Fundamental Problems of Space Biology and Medicine", "Medical and Biological Problems of Space Flights. Space Biology and Medicine", Izdatel'stvo "Nauka", Moscow, 1966, 462 pp., Chapter 4, pp. 68–104 (JFRS: 38 935).

Chapter 14

Biological Experimentation: Methods and Results

R. A. Duffee and H. T. Kemp

Initial Soviet biological experiments have been concerned primarily with the genetic effects of cosmic radiation. The effects of other environmental parameters encountered in space flight, e.g., acceleration, vibration, and weightlessness, however, were also evaluated particularly in ground experiments in the laboratory, as well as in flight experiments. From the outset of their biological experimentation on orbital spacecraft, Soviet scientists have used a variety of biological specimens ranging from subcellular materials through tissue and microorganisms to highly organized intact plants and animals. Methodological techniques employed for evaluating effects of radiation and stress factors included physiological, biochemical, morphological, cytological, histological, and microbiological analyses. In addition, numerous investigations of the composition and energy of cosmic radiation and the dynamics of its temporal and spatial variations were conducted (see Chapter 10). Induced radioactivity of some of the biological specimens and their containers was also measured, but proved negative[44]. For comparison purposes the period of weightlessness and total adsorbed radiation dose for each of the 11 biosatellites are shown in Table 14.1.

The launching of the stratosphere balloon USSR–1 to a height of 16 km in 1935, which carried fruit flies on board, may be considered the first Soviet biological flight experimentation. True space experimentation began in 1951 with the launching of dogs, rabbits, rats, and mice to altitudes of 100 km and subsequently in the 1956–1960 time period to altitudes of 212 and 473 km. Analyses of the morphological and biochemical properties of the peripheral blood of dogs were made to

TABLE 14.1[60]

Weightlessness and Radiation Characteristics of Soviet Biosatellites

Vehicle	Launch Date	Duration of Weightlessness, hours	Total Radiation Dosage, mrads*
Spaceship-Satellite 2	8/19/60	24	11
Spaceship-Satellite 4	3/9/61	1·1	–0·4
Spaceship-Satellite 5	3/25/61	1·1	2·1
Vostok 1	4/12/61	1·1	2·2
Vostok 2	8/6/61	25	13·1
Vostok 3	8/11/62	95	64±1
Vostok 4	8/12/62	71	42±1
Vostok 5	6/14/63	119	75±2
Vostok 6	6/16/63	71	48±1
Voskhod 1	10/12/64	24	30±5†
Voskhod 2	3/18/65	26	70±5†

* The ILK dosimeter was a luminescent type with a range of 2×10^{-3} to 1×10^{3} rad[36] and with 3·2 mm aluminum shielding.
† Dosimeter type not reported.

evaluate radiation and other effects[35]. However, the short duration of these flights precluded reliable assignation of adverse effects on these animals. The experiment with the dog "Laika" on Sputnik 2 was of limited value for determining the effects of cosmic radiation since the animal was not recovered[12]. Similarly, the experiments on the third Spaceship-Satellite (December 1 1960) were unsuccessful from the radiobiological viewpoint due to the destruction of the vehicle during re-entry into the atmosphere[30]. The Cosmos 110 experiment, launched on February 22 1966 and extended 22 days in orbital flight passing through part of the radiation belt with successful recovery, remains the most significant study to date of cosmic radiation effects. In addition to the two dogs, "Ugolek" and "Veterok", the vehicle contained 300 dosimeters and 10 nuclear emulsions along with biological indicators. Some of the biological indicators identified by the Soviets in "Izvestiya" (March 1 1966) were lysogenic bacteria (*Escherichia coli* K–12 [λ]), strains of yeast and Chlorella, samples of blood serum and various albumins. The basic purpose of the experiment, according to Soviet scientists, was to test the effect of prolonged weightlessness on the neuroreflex regulation of the cardiovascular system. At the 17th Congress of the International Astronautical Federation held in Madrid October 9–15 1966, Professor V. V. Parin of the Soviet Academy of Medical Sciences stated that the radiation dosage received by the dogs was negligible and that no antiradiation drugs were used*. To date, however, no details have been published by the Soviets relating to adverse effects during the Cosmos 110 flight on the dogs or the other biological specimens aboard.

Because of the aforementioned brevity of the vertical rocket flights, the nonrecovery of Sputnik 2 and Spaceship-Satellite 3, and the lack of published data regarding Cosmos 110, discussions of Soviet space biological studies in this chapter are limited to 11 orbital vehicles that were recovered successfully. These are Spaceship-Satellites 2, 4, and 5, Vostoks 1 through 6, and the two Voskhod flights. As shown in Chapter 2 all of these flights had similar orbital characteristics. The maximal distance from the earth's surface (apogee) was realized in the Voskhod flights, 495 km. All flights had an orbital inclination of 65° to the equator and, with the exception of Voskhod 1, were launched in the spring-summer period.

METHODS

The long-range goal of Soviet space biology is to determine genetic, cytological, and histological changes, particularly of a pathological nature, on the elements of ecological systems that may be employed in future long-duration manned flights. These elements, according to Soviet philosophy, will include microorganisms, higher and lower plants, and other representatives of the animal and plant kingdoms. To this intent, the Soviets have investigated the effects of cosmic radiation and other factors integral to space flight, both individually and in combination, on a number of biological specimens. These specimens can be

* B. Mandrovsky, Foreign Science Bulletin, ATD Library of Congress, December 1966, Vol. 2, No. 12, pp. 24–26.

considered as indicator organisms or biological systems that have well-defined responses to various environmental stresses in the laboratory.

The biological materials studied on each vehicle are shown in Table 14.2. References to the Soviet literature that best describe the methodology and experimental results for each biological object are indicated in this table. Because of the reticence of the Soviets to publish specific details of experimental procedures and results, some inaccuracies may have inadvertently crept in. This is particularly true with regard to the so-called "bioelements", the AMN-1, in which the bacterium *Clostridium butyricum* was used; and the "BIOS" which involved use of frog ova and sperm. Although initial descriptions of these elements have been given, Soviet scientists have not clearly indicated which vehicles carried the devices. It was assumed by the authors that these devices were included as part of the payload on all of the Soviet biosatellites. For comparison purposes, the reader is referred to a similar table prepared by J. Tyson*. The same biological materials are identified, but assignment of the various experimental objects to specific vehicles in his paper differs from that shown in Table 14.2.

TABLE 14.2

Biological Materials Used in Soviet Space Flights

	Vehicles											
	Spaceship-Satellites			Vostoks						Voskhods		
Biological Specimens	2	4	5	1	2	3	4	5	6	1	2	References
Animals												
Dogs, mongrel	+	+	+	–	–	–	–	–	–	–	–	2, 4–6, 30, 32, 33, 58
Mice, C57/B1 and white, mixed breed	+	+	+	–	–	–	–	–	–	–	–	4–6, 9–11, 19, 29–33, 50, 58
Rats	+	–	–	–	–	–	–	–	–	–	–	4–6, 19, 29–33, 41–43, 50
Guinea Pigs	–	+	+	+	+	–	–	–	–	–	–	4–6, 19, 29, 30, 50, 58
Animal Tissue or Cells												
HeLa Cancer Cells (human lung epithelium)	+	+	+	+	+	+	+	–	+	+	+	4–7, 19, 29–31, 57, 60–62, 65, 66

* J. Tyson, "Russian Life in Space", George Washington University Magazine, Winter 1965, p. 9.

TABLE 14.2 (Continued)
Biological Materials Used in Soviet Space Flights

	Vehicles											
	Spaceship-Satellites			Vostoks						Voskhods		
Biological Specimens	2	4	5	1	2	3	4	5	6	1	2	References
				Animal Tissue or Cells (Continued)								
Amnion (embryonic) and Fibroblast Cells (connective, human)	−	+	+	+	+	+	−	+	+	+	?	6, 7, 19, 29–31, 60–62, 65, 66
Bone Marrow (rabbit)	−	+	+	+	+	−	−	−	−	−	−	6, 19, 29, 30
Fertilized Loach Roe	−	−	−	−	+	+	+	−	−	−	−	5, 19, 57–58
Skin (human)	+	−	−	−	−	−	−	−	−	−	−	5, 6, 19, 29, 31, 60–62, 65
Skin (rabbit)	+	−	−	−	−	−	−	−	−	−	−	5, 6, 19, 29, 31, 60–62, 65
Eggs of Ascaris (roundworm)	−	−	−	−	−	+	+	−	−	−	−	7, 57–58
Frog Ova and Sperm	−	+	+	+	+	+	+	+	+	+	+	4, 6, 30
Calf Thymus DNA	+	−	−	−	−	−	−	−	−	−	−	5, 6, 60
				Insects								
Drosophila melanogaster (D–32, D–18)	+	+	+	+	+	+	+	+	+	+	+	1, 4–6, 8, 18, 20, 21, 23, 25, 26, 29, 30, 45, 46–48, 56, 59
				Plants								
Tradescantia paludosa	+	+	−	−	−	+	+	+	+	+	+	4–6, 8, 13–16, 29–30, 56, 59
Seeds (air dried)												4–6, 13, 17, 22, 24, 27–30, 34, 37, 38, 54–56, 67
Pea	+	+	+	+	+	+	+	−	−	−	−	
Corn	+	−	−	−	−	−	+	−	−	−	−	
Winter Wheat (PPG–186)	+	+	+	+	+	+	+	+	+	+	+	
Spring Onion	+	−	−	−	−	−	−	−	−	−	−	
Nutmeg	+	−	−	−	−	−	−	−	−	−	−	
Beans, broad	−	+	+	−	−	+	+	+	+	−	−	
Rice	−	−	−	−	−	+	+	−	−	−	−	
Onion, annual	−	−	−	−	−	+	+	−	−	−	−	
Watercress	−	−	−	−	−	+	+	−	−	−	−	
Mustard	−	−	−	−	−	+	+	+	+	−	−	
Pine	−	−	−	−	−	+	+	+	+	+	+	
Cabbage	−	−	−	−	−	+	+	−	−	−	−	
Radish	−	−	−	−	−	+	+	−	−	−	−	
Tomato	−	−	−	−	−	+	+	+	+	−	−	
Carrot	−	−	−	−	−	+	+	+	+	−	−	
Lettuce	−	−	−	−	−	+	+	+	+	−	−	
Sugar beet	−	−	−	−	−	+	+	−	−	−	−	
Cucumber	−	−	−	−	−	−	−	+	+	−	−	
Prickwood	−	−	−	−	−	+	−	+	−	−	−	

TABLE 14.2 (Continued)
Biological Materials Used in Soviet Space Flights

	Vehicles											
	Spaceship-Satellites			Vostoks						Voskhods		
Biological Specimens	2	4	5	1	2	3	4	5	6	1	2	References
Plants (Continued)												
Seed (sprouting)												
Spring onion	–	+	+	–	–	–	–	–	–	–	–	5, 37
Nutmeg	–	+	+	–	–	–	–	–	–	–	–	5, 37
Microorganisms												
Algae												
Chlorella	+	+	+	–	–	–	–	+	+	+	+	4, 5, 19, 51–52, 56
Fungi												4, 6, 22–24, 29
Actinomycete 2577	+	–	–	–	–	–	–	–	–	–	–	
Actinomycete 8594	+	–	–	–	–	–	–	–	–	–	–	
Actinomycete LS3	+	–	–	–	–	–	–	–	–	–	–	
Actinomycete LSB-2201	–	+	+	–	–	–	–	–	–	–	–	
Bacteria												4, 6, 29–31,
E. coli K–12 (λ)	+	+	+	+	+	+	+	+	+	+	+	60–67
E. coli B	+	+	+	–	–	–	–	–	–	–	–	
Aerobacter a 1321	+	+	+	–	–	–	–	–	–	–	–	
Staphylococcus aureus	+	+	+	–	–	–	–	–	–	–	–	
Clostridium butyricum	+	+	+	+	+	+	+	+	+	+	+	
Bacteriophages												
T–2	+	+	+	–	–	–	–	–	–	–	–	60
1321	+	+	+	–	–	–	–	–	–	–	–	60
Yeast	–	–	–	–	+	–	–	–	–	–	–	39, 40
Others												
Enzymes: pepsin, trypsin, alkaline phosphatase, catalase, peroxidase, and ribonuclease	–	+	+	–	–	–	–	–	–	–	–	4–6, 29, 30
Viruses: tobacco mosaic and influenza	–	+	+	+	–	–	–	–	–	–	–	4–6, 29, 30
Ferments: wheat cell nuclei, wheat cell cytoplasm, wheat cell buds	–	+	+	–	–	–	–	–	–	–	–	4–6, 29, 30

Animals

Evaluation of the combined effects of space radiation and other flight factors in mammals was attempted through studies of:

1. The biochemical changes in the blood and urine of dogs, rats, and mice flown on the Spaceship-Satellites 2, 4, and 5. Specifically the protein composition and nonspecific cholinesterase activity of blood serum, and the corticosteroid and nucleic acid metabolite content of the urine, i.e., deoxycytidine and Dische-positive substances.
2. Cytogenetic and histologic analyses of the hematopoietic organs (marrow and spleen) of mice.
3. Determination of the serotonin (5-hydroxytryptamine) concentration of the blood of dogs and mice.
4. Pathological changes in the internal organs of mice and guinea pigs.
5. Immunological reactions (microflora and bactericidal properties of the skin, and phagocytic and bactericidal properties of the blood) in dogs.
6. Pigmentation changes in the hair of C 57/B1 strain mice.
7. Higher nervous activity of white rats by means of conditioned-reflex activity to sound and light stimuli.

Tissue or Cell Cultures

In addition to studies on intact animals, viability of human tissue cultures (HeLa cancer cells, lung epithelium tissue), amnion cells (embryonic tissue) and fibroblasts (connective tissue) were also investigated. Excised pieces of human and rabbit skin were exposed to space flight conditions on Spaceship-Satellite 2, and reimplanted into donors after return to earth. A number of the human tissue cultures, particularly the HeLa cells, were re-exposed to space flight conditions in multiple flights. Cultures of rabbit bone marrow cells and calf thymus desoxyribonucleic acid (DNA) were also investigated.

Although Soviet scientists report experiments on influenza and tobacco viruses, a series of enzymes, preparations of cell nuclei. and a wheat germ homogenate on the Spaceship-Satellites and Vostok 1, no details on the methodology employed have been published.

Adverse effects on the human cell cultures were studied by: (1) macro- and microscopic examination of the cell culture (assessment of condition of cell layer on the glass); (2) transfer of cells to new vessels with normal culture methods or change in the culture medium without transfer; (3) determination of the number of living and dead cells in the initial nutrient, in the initial layer on the glass, and in cells in subsequent passages; (4) morphological changes in the cells; and (5) determination of antigenic properties of the cultures. The optical density, thermal stability, and serological properties of DNA were determined and compared to laboratory control samples for evaluation of space flight effects.

Genetic Studies

The Soviet space biology program has been concerned principally with the genetic effects of space flight, especially those attributable to cosmic radiation. Accordingly all of their biosatellites have carried Drosophila (fruit flies), the classic subject of genetics, flowering spiderwort (Tradescantia), air-dried seeds of various higher plants (principally edible or crop plants) and various microbiological materials. In addition, simple plants, i.e., Chlorella, yeasts, and actinomycete have been exposed on several flights, as were sprouts of *Allium fistulosum* (spring onion) and *Nigella damascena* (variously known as nutmeg flower, fennel flower, and Love-of-the-Mist) to determine genetic effects as well as viability.

Drosophila melanogaster

In the experiments with Drosophila, four tests were used to determine the rate of various types of mutations: (1) primary nondisjunction of chromosomes; (2) induced crossing over in males; (3) dominant lethal mutations, and (4) sex-linked recessive lethal mutations. In addition, studies of the reproductive processes of Drosophila under conditions of weightlessness were conducted on the Vostok 3 and Vostok 4 flights with the assistance of the Cosmonauts Nikolayev and Popovich. Similar experiments were conducted on the Vostoks 5 and 6 but the flies perished after landing before they could be returned to the laboratory.

For investigation of the nondisjunction of chromosomes, virgin females of the white-eyes strain were sent in flasks on the Vostoks 1, 2, and 3 and both Voskhod flights and crossed with red-eyed males after return from flight. The period of crossing in each of the experiments lasted 15 days. Every three days the males and females were transferred to new cultures. Thus five series of progeny were obtained in each experiment. The appearance of white-eyed females of the XXY genotype and red-eyed males of the XO genotype was an index of primary nondisjunction of the X-chromosomes.

Crossing over, i.e., the result of the breaking of the chromatids during divisions of premiotic cells, in the zygotes of *D. melanogaster* males was used as an index of genetic mutation resulting from flight factors other than cosmic radiation. Hybrid males, obtained from the breeding of females of laboratory line 50 with males of the "wild" Domododevo 18 strain, were carried aboard Spaceship-Satellite 5 and Vostoks 1 and 2. In the latter two flights, the tubes containing the flies were wrapped in a sheet of porolon (a white foamed plastic) which provided some protection against vibration. After 10 days, the males were crossed for three days with females homozygous as regards genes *b* (black), *cn* (cinnabar eyes), and *vg* (vestigial wings). Frequency of induced crossing over in the progeny was estimated from the number of crossover flies that developed. Double crossovers (black flies with vestigial wings but normal eye color, and flies with cinnabar eyes but normal wings and body color) were recorded as two cases of crossing over.

For the experiments on dominant lethal mutations two strains of Drosophila were used on all flights but Vostok 1. These strains were the D–18, characterized by a high degree of spontaneous mutability with respect to recessive lethals, and the D–32 strain, distinguished by low spontaneous mutability with respect to recessive lethals. The effects of space flight on both mature sperm and spermatids (at the time of flight) were determined by breeding the males which made the flight to females of the same strain over a period of several days. The frequency of induced dominant lethals was determined by comparing the number of undeveloped eggs between the experimental and control flies. The arbitrary assumption was made that the frequency of undeveloped eggs in the control group is equal to zero.

D–18 and D–32 strains of Drosophila were also used to determine the effects of space flight on recessive lethal mutations. Both strains were studied on the Spaceship-Satellites and Vostoks 1 and 2. For subsequent flights, only the low spontaneous mutability strain, D–32, was used. After flight, the males that had been exposed to space flight were crossbred with a yellow Muller–5 strain, which have yellowish body color and yellowish or apricot-colored eyes. The D–18 and D–32 strains have gray bodies and eyes. Females of the first generation of the crossbreeding were then crossbred again with males of the yellow Muller–5 strain. The percentage of these second generation (F_2) cultures containing no gray red-eyed males was used as an indicator of the frequency of occurrence of recessive lethal mutations in the X-chromosomes of the males which had been orbited in space. In the studies on Voskhod flights, the radioprotective drug, pentamethoxytryptamine was used to treat the flies, but no detailed results have been published to date.

The studies of reproductive processes of Drosophila on Vostoks 3 and 4 were carried out with 40 males and virgin females of the D–32 strain. The cosmonauts mixed the flies on the first orbit, permitting the flies to mate. Analysis and evaluation of the possible genetic disruptions were based on the rate of embryo development, on the number, sex, and weight of the progeny and on deviations from normal in body shape or size, eye size, color or shape, and veining of the wings. Mature sperm from hatched-out males was analyzed for the presence of lethal and visible recessive mutations in the sex (X) chromosome.

Tradescantia paludosa
(Flowering Spiderwort)

According to articles by A. A. Gyurdzhian[30] and V. V. Antipov, *et al.*[6], flowering spiderwort was exposed to flight conditions during the Spaceship-Satellite series. However, no details on the methodology employed or the results obtained were published by the Soviets until the experiments beginning with the Vostok 3 event. Presumably, the earlier studies were unsuccessful. In the studies on Vostoks 3 and 4, so-called "bio-cartridges", consisting of two cylinders, were used. One cylinder contained about 2 cc of water in which the freshly cut racemes of Tradescantia were placed. The other cylinder contained the fixing fluid which was a mixture of absolute alcohol and glacial acetic acid in a ratio of 3:1. Turning a screw connecting the two cylinders punctured a thin membrane and allowed the fixing fluid to pass into the cylinder with the inflorescences. (A photograph of this "bio-cartridge" is shown in Figure 14.1.) On Vostoks 5 and 6 some of the cylinders were separated by a diaphragm with five holes in it. Tradescantia stems were thrust through the holes into the water and sealed in place with collodion. The cosmonauts on the Vostok 4 and Vostok 5 flights fixed the anthers at various times throughout the flight. In the other experiments, the anthers were fixed after the flights. For analysis, crushed acetocarmine preparations of the microspores were subjected to microscopic analysis. All types of chromosome disturbances were counted, including those noted in metaphase, anaphase, and telophase.

Seeds

Cytogenic techniques used for study of seeds carried on all the Soviet biosatellites (see Table 14.2 for the identification of the seeds aboard each vehicle) were identical to those employed in the experi-

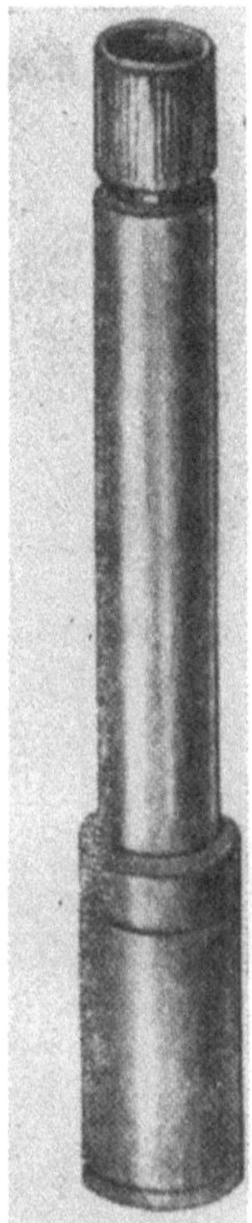

Fig. 14.1. Biocapsule Containing Cut Racemes of Tradescantia During Space Flight[8]

1. Knob of screw closing aperture of fixation fluid reservoir
2. Fixation fluid reservoir
3. Raceme container

ments with Tradescantia. The seeds were grown simultaneously with laboratory controls on filter paper in petri dishes. After the radicles had reached various lengths (determined by the plant being studied), they were fixed in the mixture of alcohol and glacial acetic acid for four hours, then transferred to 70 per cent alcohol. The cell root tips were crushed, stained with acetocarmine, and placed on temporary mounts.

Microscopic examinations were made to determine the mitotic index and chromosome aberrations in anaphase and telophase. The genetic effect was judged by the number of cells showing chromosomal re-arrangements.

For some crop plants, e.g., wheat, the sprouting and germination energy and the physiological (biochemical) properties of the plants produced from them were studied. Wheat seeds were soaked in water for 24 hours and then set out in metallic Wagner vessels. Vessels were

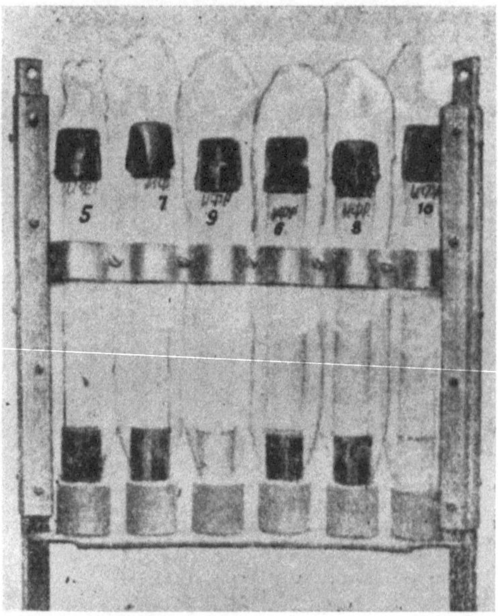

Fig. 14.2. Rack With Ampules Containing Chlorella Suspensions in a Liquid Medium[52]

filled with 5·5 kg of soil containing a mineral fertilizer. Cuttings were taken at various stages in the development of the plants for determination of water-soluble carbohydrates and nitrogen compounds, other nitrogenous substances, and starch.

Microbiology

The most outstanding research in the Soviet space biology program has been within the realm of microbiology. This effort was conducted by a team directed by N. N. Zhukov-Verezhnikov, who is Head of the Department of Immunology of the Institute of Experimental Biology of the Academy of Medical Sciences, USSR. All of the Soviet human and rabbit tissue culture studies were conducted by this group. The most significant efforts of this team have been the development of the AMN–1 biosensor (or, as termed by the Soviets, "bioelement") for the automatic detection of radiation danger to living organisms on long duration space flights, and their studies with the lysogenic bacterium *E. coli* K–12 (λ).

The mission of the Institute of Experimental Biology was to determine:

1. The degree to which space flight factors affect the viability of unicellular organisms (bacteria), primitive organisms (phages), and cells of higher organisms
2. The ability of uncellular organisms (bacteria) to multiply in space conditions
3. The effect of space flight on the genetics of microorganisms, i.e., auxotrophic mutations in bacteria, induced phage production in lysogenic bacteria, and colony-shape mutations in phages.

Two strains of *E. coli*, *Aerobacter aerogenes* 1321, *Staphylococcus aureus*, and *Cl. butyricum* were used to determine the effect of space flight on the viability of unicellular organisms. Similarly, filtrates of the T–2 coliphage, which lyses *E. coli*, and the phage 1321, which lyses *A. aerogenes* 1321, were exposed to space flight. Two ml of saline suspensions of the various bacteria, each containing 500 million cells per ml, were put into ampules. The number of viable cells per ml was determined by an agar layer method before and after flight. The viability of the butyric acid bacterium, *Cl. butyricum*, was determined by its gas-producing ability. Phage filtrates were also evaluated by the agar layer method. The experimental conditions used in the bacterial studies are shown in Table 14.3.

Study of bacterial multiplication in space was facilitated by the development of the AMN–1 biosensor. Criteria used in selecting the bacteria were that they be nonpathogenic, anaerobic, spore-forming, and gas-producing when cultivated in appropriate nutrient media. Among the bacteria investigated were *Cl. perfringens* 48 and 28, *Cl. bifermentes*, *Cl. sporogenes*, *Cl. putrefaciens*, *Cl. butyricum*, *Cl. acetobutylicum*, and *A. aerogenes* 59/9 and 1321. The gas-producing ability of these bacteria was determined by volumetric methods in cultures in 1% peptone water containing various sugars in 0·5–2% concentrations. The butyric acid bacterium (*Cl. butyricum*) rapidly formed the greatest amount of gas at 20° C and was selected for use in the AMN–1 bioelement. Experiments showed that this bacterium, in the temperature range 14 to 37° C, produced enough gas in a 6–24 hour incubation period to actuate the membrane which in turn signaled the operation of the bioelement.

TABLE 14.3[60]

Bacterial Samples on Board Spaceship-Satellite 2

Species	Nature of Sample	Initial Concentration of Viable Cells, 10^8 cells/ml	Number of Samples (Ampules, bottles)
Escherichia coli K–12 (λ)	Suspension	3·75	5
E. coli K–12 (λ)	Suspension + oxygen	3·75	2
E. coli B	Suspension	5·5	2
Aerobacter aerogenes 1321	Suspension	4·75	2
Staphylococcus aureus 0–15	Suspension	1·93	1
Clostridium butyricum	Suspension	Not determined	1

The AMN–1 bioelement, shown in Figure 14.3, consists of a cylindrical metal reservoir composed of two chambers, a 1-cc inoculum chamber which contains the culture and a 10-cc cultivation chamber that contains the nutrient medium. The chambers were separated by a transverse ground-glass partition. The top of the inoculum chamber and the bottom of the cultivation chamber consist of flexible membranes. A striking pin attached to the membrane is mounted vertically in the inoculum chamber. When a hammer strikes the membrane, the pin breaks the glass separating the chambers thus inoculating the medium. The gases produced in the cultivation chamber by the growing bacteria distort the membrane at the bottom of this chamber. The expanded membrane closes an electric circuit, which in turn transmits a signal indicating operation of the instrument. Four such instruments were flown aboard Spaceship-Satellite 2, two in an incubator maintained at 37° C and two in an unheated container.

Auxotrophic mutants, i.e., bacterial variants that have lost their ability to synthesize certain amino acids or vitamins, were determined

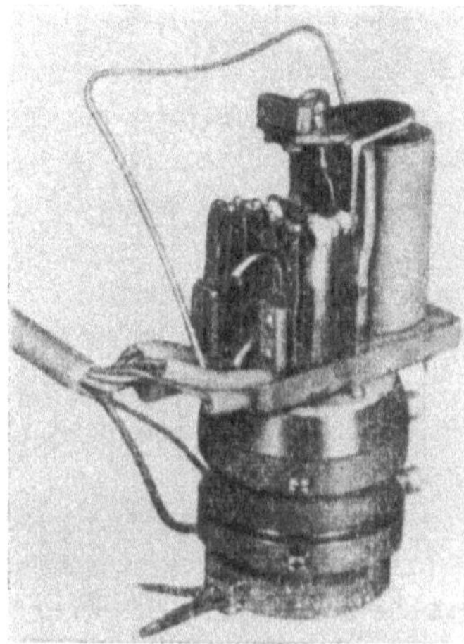

Fig. 14.3a. Bioelement AMN-1 With Striker Element (the latter with case removed)[60]

Fig. 14.3b. Bioelements in Packed Form[60]
Left—in incubator
Right—in unheated container

by the difference in the growth of exposed samples of bacteria in "minimal" and "complete" media. Auxotrophic mutants do not grow on minimal medium.

The experiments on bacterial lysogeny, i.e., the ability of bacteria to produce phages, a pathological heritable process, were carried out on the lysogenic strain *E. coli* K-12 (λ). On the second Spaceship-Satellite, the effect of space flight was determined by counting the phages produced after inoculation of fresh nutrient media with the samples exposed in space, and comparing results with control samples kept in the laboratory. In subsequent flights, the procedure was modified. Suspensions of *E. coli* K–12 (λ) cells exposed in space were diluted 1 : 10 in bouillon and sufficient antiphage serum was added to inactivate completely the free phage particles. After the end of the latent period (55–56 min) the serum was rendered inactive by 1 : 100 dilution of the bacterial suspensions. Streptomycin was then added and the samples were plated out by the agar layer method with streptomycin-resistant bacteria of the indicator strain of *E. coli*.

The bacteriophages carried on the Spaceship-Satellites were used to investigate the possibility of direct hits of corpuscular radiation on the phage particles. Evaluation of exposed samples was again conducted by the agar layer method with visual examination of the shape of the phage plaques.

Actinomycetes

Three strains of actinomycetes were flown on Spaceship-Satellite 2 and an additional strain on the fourth and fifth Spaceship-Satellites (see Table 14.2) to determine their viability and morphology after space flight. Viability was determined by inoculating petri dishes with spores and counting the number of colonies that developed. Morphology was determined by microscopic examination of cultures growing on a liquid maize medium at 27 to 28° C. Density of the spore suspensions was determined spectrophotometrically for the strain used on the fourth and fifth Spaceship-Satellites.

Chlorella

According to V. V. Antipov[4] and N. M. Sisakyan, *et al.*[56], various strains of Chlorella were flown on the Spaceship-Satellites, Vostoks 5 and 6, and Voskhod, but the only detailed results published to date concern the studies done in conjunction with the flight of Spaceship-Satellite 2. Two culture methods for *Chlorella pyrenoidosa* were used: (1) dormant in the form of a streak on an agar slant; and (2) actively growing cultures maintained in a liquid nutrient. The cultures in a liquid medium were poured into glass ampules, which were then sealed into bags made of a transparent film. The rack with the ampules (Figure 14.2) was positioned in the capsule in such a manner that the ampules were illuminated for photographic and television observation during the flight. The agar slant cultures were similarly exposed. The effect of space flight was determined by studies of the external characteristics of the cultures, microscopic appearance, growth, and photosynthetic activity as compared to laboratory control cultures.

Yeasts

Two types of yeast were used on the Vostok 2 flight. These were the natural haploid yeast *Zygosaccharomyces bailii* and the diploid *Saccharomyces vivi*, strain Megri 139–B. After cultivation the yeasts were suspended in sterile tap water at a density of $2\text{–}3 \times 10^6$ cells/ml. These suspensions were then transferred to test tubes at concentrations of 0·5 ml of aqueous cell suspension to 0·5 ml of distilled water or, in some tubes, 0·5 ml of oleic acid–alcohol solutions varying in concentration from 47×10^{-2} to 47×10^{-8} per cent. For evaluation of space flight effects, yeast cells from laboratory and experimental series of ampules were seeded on nutrient medium 20 days after preparation, and cells from the transported (to the launch site) control series 27 days after preparation. One ml of diluted cell suspension from each ampule was transferred to petri dishes containing wort agar medium. The dishes of seeded medium were incubated at 30° C for four days, and the effect of flight was determined by the number of visible colonies. The number of colonies growing after seeding of a suspension of cells taken from ampules without acid was taken as 100 per cent.

SUMMARY OF RESULTS

The biological experiments conducted by the Soviets on Spaceship-Satellites 2, 4, and 5, the entire Vostok series, and both Voskhod flights have established, according to Soviet space biologists, that five-day flights at orbital inclination of 65° and apogees up to 495 km do not have any seriously deleterious effects on the vital activity of various organisms. However, disruptions of the mitotic mechanisms were noted in certain biological specimens, e.g., bone marrow cells, HeLa cells subjected to repeated flights, Tradescantia, and seeds of wheat, pea, carrot, and tomato, which on prolonged flights could result in unacceptable alterations of closed or partially closed ecological systems. The specific results of experiments with various organisms are briefly summarized in Table 14.4.

TABLE 14.4

Results of Soviet Biological Studies in Space (1960–1965)

Biological Specimen (Space Flight)	Properties Studied	Results Reported	References
Dogs, rats, and mice (SS–2, 4, 5)	Biochemical changes of blood and urine	After a single period of acceleration or vibration under laboratory conditions simulating space flight, short-term changes were noted indicating activation of the liver and other systems, i.e., increase in nonspecific cholinesterase activity and decrease in concentration of serum mucoids in blood. After a 24-hour flight, the dogs Belka and Strelka showed transient biochemical changes characteristic of reversible and moderate stress reactions, i.e., short duration increase in α_2-globulin and serum mucoids (2–6 days after flight). No changes were detected in metabolism of the dogs Chernushka and Zvezdochka after 90-minute flight.	A. A. Gyurdzhian, *et al.* (1961)[32] (1962)[33]
Mice, C 57/B1 strain (SS–2)	Effect of space flight factors on hemopoietic organs	Cytological investigations of bone marrow cells showed a statistically significant increase of mitotic damage, lasting until 60 days after flight. Chromosomal aberrations were not similar to those caused by radiation damage. Up to 30 days after flight, bone marrow revealed an increase of myelopoiesis accompanied by the appearance of young white blood cell forms and an inhibition of erythropoiesis. Spleen showed a depression of hemopoiesis during first 10 days after flight, but return to normal activity by 30th day. Laboratory studies showed effects of flight similar to those caused by vibration and acceleration.	M. A. Arsen'yeva, *et al.* (1961)[9], (1962)[10,11]
Dogs and mice (Spaceship-Satellites 4 and 5)	Serotonin concentration of blood	Serotonin concentration dropped to 1/8 or 1/10 in mice and by a factor of 3·5 to 10 in dogs on the first and second days after landing as compared to control level (0·12–0·2 g/ml). After the 10th day returned to normal. Mice and guinea-pigs were subjected to vibrations of 35 and 70 cycles/min for 15 min at 10 g and an amplitude of 0·4 mm showed a similar drop in serotonin concentration for two days. Vibration was suggested as cause of lowered serotonin concentration resulting from space flight.	V. V. Parin, *et al.* (1965)[49] V. S. Shashkov, *et al.* (1962)[53]

Biological Specimen (Space Flight)	Properties Studied	Results Reported	References
Mice, C 57/B1 and white; guinea pigs (Spaceship-Satellites 2, 4, and 5)	Pathology of internal organs	Some degenerative changes in organs occurred which were largely reversible. Most affected were the brain, liver, myocardium, spleen, and suprarenal glands. The changes were related to tissue hypoxia.	V. G. Petrukhin (1962)[50]
Dogs (Spaceship-Satellites 2, 4 and 5)	Bactericidal activity of skin and blood	Dogs exposed to space flight factors exhibited a wave-like fluctuation of the phagocytic index. The ingestive ability of the neutrophils was most affected. Digestive ability of neutrophils and bactericidal function of plasma and skin were less affected. Appearance of *E. coli* in the oral cavity immediately after flight indicates a drop in the immunological activity. Similar results were found in laboratory tests after vibration (70 cps, 0·4 mm) for 13 minutes and accelerations (2, 6, 9, and 12 g) for one to two minutes. Moderate immunological changes persisted in dogs for months and sometimes longer.	O. G. Alekseyeva (1961)[2] O. G. Alekseyeva and A. P. Volkova (1962)[3]
Male albino rats (Spaceship-Satellite 2)	Higher nervous activity (conditioned reflex)	Seven animals were used, two experimental and five control. Six positive and one differentiated motor-alimentary conditioned reflexes from optical and acoustical analyzers were produced in all animals. For two and a half months after flight no changes could be detected in the higher nervous activity of the rats studied.	L. D. Luk'yanova (1961)[41] L. D. Luk'yanova, *et al.* (1962)[42, 43]
Human and rabbit skin (Spaceship-Satellite 2) Tissue cultures of cancerous human cells (Spaceship-Satellites 2, 4, and 5 Vostoks 1–6 Voskhods 1 and 2	Viability, morphological and antigenic properties of cells	Viable cells were found in the human and rabbit skin scraps after return from 24 hours in space. Reimplantation of the scraps into donors showed that degree of healing-in of the experimental samples was the same as that of controls. Repeated exposure of HeLa cancer cells on Vostoks 4 and 6 produced changes as compared with laboratory controls or HeLa cells exposed only on a single flight. A longer latent period of recovery of growth capacity and other characteristics were noted in twice-flown cultures. The coefficient of proliferation of HeLa cells on both Vostok 4 and Vostok 6 was one-half that for controls or cultures exposed to only one flight. The same cells exposed on Vostoks 4 and 6 were re-exposed on Voskhod 1, and a well-defined drop in the proliferation coefficient was observed in comparison with controls. Experimental colonies were more compact, and there were more dead cells in the thrice-exposed cultures. Normal fibroblast and amnion cells recovered their viability and growth potency more rapidly than the HeLa cells.	N. N. Zhukov-Verezhnikov, *et al.* (1961)[60] N. M. Sisakyan, *et al.* (1965)[56] N. N. Zhukov-Verezhnikov, *et al.* (1962)[61], (1964)[62], (1965)[65], (1966)[66]
Calf thymus DNA (Spaceship-Satellite 2)		No changes were found in the optical density, thermal stability, antigenic, or serological properties of calf thymus DNA.	N. N. Zhukov-Verezhnikov, *et al.* (1961)[60]

TABLE 14.4 (Continued)

Results of Soviet Biological Studies in Space (1960–1965)

Biological Specimen (Space Flight)	Properties Studied	Results Reported	References
Drosophila melanogaster D–18, D–32, white eyed, hybrid strains (all vehicles)	Recessive lethal mutations, dominant lethal mutations, and nondisjunction of chromosomes	Statistically significant differences in sex-linked recessive lethal mutations were found on Spaceship-Satellites 2 and 4 and on Vostok 1, but not on any of the other flights. This mutagenic effect accordingly seems to be independent of duration of flight.	Ya. L. Glembotskiy, *et al.* (1961)[21], (1962)[23], (1963)[25, 26] Ya. L. Glembotskiy and G. P. Parfenov (1962)[20]
		With regard to dominant lethal mutations, statistically significant effects were noted on the Spaceship-Satellites 2 and 4, and on Voskhods 1 and 2. A slight increase was also noted on the Vostok 3 flight, which was not statistically significant. In the remaining flights, no significant changes were noted. Again, this mutagenic effect is also independent of duration of space flight.	G. P. Parfenov (1961)[45], (1962)[46], (1964)[48] Yu. M. Volynkin, *et al.* (1966)[59]
		Primary nondisjunction of the X-chromosome in Drosophila females increased significantly during the Vostok 1 and 2 flights and also during both Voskhod flights, but not during the Vostok 3 and 4 flights. Laboratory investigations with X-rays indicate that the effects were due not to radiation but to vibration, weightlessness, or combined factors.	N. P. Dubinin and O. L. Kanovets 1962)[18] Ya. L. Glembotskiy, *et al.* (1963)[26] Yu. M. Volynkin, *et al.* (1966)[59]
	Crossing-over in males	A significant increase in the incidence of crossovers was found in the offspring of males flying in Spaceship-Satellite 5, but not in those on Vostoks 1 and 2. Vibration at 70 cps also induced crossovers, and there appears to be a linear relationship between duration of vibration and frequency of induced crossing-over.	E. A. Abeleva, *et al.* (1961)[1]
	Reproduction and growth in weightlessness	Copulation, oviposition, and normal development of Drosophila can occur normally during periods of weightlessness up to four days. Embryonic development of a normal strain of Drosophila under weightless conditions does not result in significant (statistically) increases in morphological changes or mutations in sex cells. In these experiments on Vostoks 3 and 4, however, the number of females in all cultures considerably exceeded the number of males, and the latter hatched somewhat later.	V. V. Antipov, *et al.* (1965)[8] G. P. Parfenov (1964)[47]

Biological Specimen (Space Flight)	Properties Studied	Results Reported	References
Tradescantia paludosa, microsopores (flowering spiderwort) (Vostoks 3–6, Voskhods 1 and 2)	Type and frequency of chromosomal rearrangements and mitotic disruptions	On Vostoks 3, 4, 5, and 6 and Voskhod 1, a significant increase in chromosomal rearrangements was noted. On Vostok 3 the total percentage of these rearrangements in metaphase and anaphase was $1{\cdot}89 \pm 0{\cdot}8$, and on the Vostoks 4, 5, and 6 they were $1{\cdot}49 \pm 0{\cdot}04$, $1{\cdot}24 \pm 0{\cdot}10$ and $1{\cdot}94 \pm 0{\cdot}09$, respectively. On the Voskhod 1, the percentage was $1{\cdot}99 \pm 0{\cdot}14$. Predominant among the chromosome disturbances was a new type of reorganization, i.e., spherical fragments were detected in prophase and interphase as well as in the active mitotic phases. Dynamic flight factors seem to be of primary significance in causing these chromosomal rearrangements for the maximum percentage was recorded shortly after the ship went into orbit and the minimum shortly before descent. After landing, the number of rearrangements increased. Disturbances of the mitotic mechanism, including tri- and quadripolar mitosis and change in the direction of the pivotal axis, were noted on all flights and the percentage of cells with mitosis disturbances was related to the flight duration.	V. V. Antipov (1964)[4] V. V. Antipov, *et al.* (1965)[8] N. L. Delone (1965)[13] N. L. Delone, *et al.* (1963)[15], (1964)[16], (1966)[14] N. M. Sisakyan, *et al.* (1965)[56] Yu. M. Volynkin, *et al.* (1966)[59]
Seed of higher plants (all vehicles)	Chromosomal aberrations in anaphase and telophase of primary rootlets (seedlings)	Air-dried seeds of various plants were carried on all the Soviet biosatellites. Space flight resulted in an increase in the percentage of chromosome aberrations in most of the seeds studied. This increase, however, proved statistically significant (P test) only in wheat on Spaceship-Satellite 2, Vostok 1, and Vostok 2; peas on Spaceship-Satellite 2 and Vostok 2; and carrots and tomatoes on the Vostok 3 through 6 flights. Experimental specimens showed more chromosome reorganization and recombinations while the control seeds showed more chromatid reorganizations and more fragments. No correlation was found between either the radiosensitivity of the seeds or flight duration and the amount of chromosomal disturbances. Furthermore, no difference was noted in the experiment with prickwood seeds on Vostok 5 between previously irradiated seeds (10 000 rad) and nonirradiated seeds.	V. V. Antipov (1964)[4] Ya. L. Glembotskiy, *et al.* (1963)[26] V. V. Khvostova, *et al.* (1962)[37], (1963)[38] N. L. Delone (1964)[13] N. L. Delone, *et al.* (1965)[17] Ya. L. Glembotskiy, *et al.* (1961)[22], (1962)[24] L. K. Gordon, *et al.* (1963)[27], (1965)[28] B. N. Sidorov and N. N. Sokolov (1961)[54], (1962)[55] N. N. Zhukov-Verezhnikov, *et al.* (1966)[67]

TABLE 14.4 (Continued)

Results of Soviet Biological Studies in Space (1960–1965)

Biological Specimen (Space Flight)	Properties Studied	Results Reported	References
Nigella damascena (nutmeg flower) and *Allium fistulosum* (spring onion), moistened seeds and sprouts (SS–5)	Chromosomal aberration in anaphase and telophase of primary rootlets	Moistened seeds and sprouts were sent up on Spaceship-Satellite 5. Although the radiosensitivity of moistened seeds and sprouts was 20 to 30 times greater than that of dry seeds, no differences were found in the number of cells with chromosomal rearrangements between the dry seeds on Spaceship-Satellite 2 and Vostok 1 and either the moistened seeds or sprouts on Spaceship-Satellite 5.	Ya. L. Glembotskiy, *et al.* (1963)[26] V. V. Khvostova, *et al.* (1962)[37]
Seed of higher plants (all vehicles)	Absolute germination of seeds, growth and development of seedlings	Space flight had a stimulating effect (acceleration of the rate of germination and an increase in ability to germinate) on germination of seed of *Allium fistulosum*, *Nigella damascena* and winter wheat (PPG–186), and an inhibitory effect on seed of Berlin lettuce, and Krasnozerna wheat seeds. Plants developed from Krasnozerna wheat showed no differences in nitrogen or starch content from controls.	L. K. Gordon, *et al.* (1963)[27] G. V. Il'ina, *et al.* (1966)[34] B. N. Sidorov and N. N. Sokolov (1961)[54] (1962)[55]
Chlorella (Spaceship-Satellite 2, Vostoks 5 and 6, Voskhod 1)	Viability, growth and photosynthesis	The *Chlorella pyrenoidosa* cultures exposed 24 hours in space on Spaceship-Satellite 2 retained their viability and did not show any irreversible changes. Immediately after landing the cultures contained a considerable number of dead cells and photosynthetic activity diminished for a period of six days. After six days of intensive cultivation, the flight specimens were identical to controls.	V. Ye. Semenenko and M. G. Vladimirova (1961)[51], (1962)[52]
		The various strains flown on Vostoks 5 and 6 and Voskhod 1 showed no noticeable mutagenic changes. The more radiosensitive strains, however, showed a tendency toward reduced viability.	V. V. Antipov (1964)[4] N. M. Sisakyan, *et al.* (1965)[56]
Actinomycetes (Spaceship-Satellites 2, 4, and 5)	Viability and morphology	The radioresistant strain *Actinomyces erythreus* 2577 showed an increased viability by a factor of six over controls after flight on Spaceship-Satellite 2. The radiosensitive strain, No. 8594, showed a 90 per cent reduction in viability. The strain of *A. aureofaciens*, LSB–2201, carried on Spaceship-Satellites 4 and 5, showed an approximate 75 per cent loss of viability on both flights. Cytological analyses of the strains on Spaceship-Satellite 2 showed a definite stimulation of mycelial growth.	Ya. L. Glembotskiy, *et al.* (1961)[22], (1962)[24] V. V. Khvostova, *et al.* (1962)[37] N. N. Zhukov-Verezhnikov, *et al.* (1964)[62]

Biological Specimen (Space Flight)	Properties Studied	Results Reported	References
Yeasts (Vostok 2)	Survival and effect of oleic acid on yeast	No effect of flight on Vostok 2 was noted in either unsensitized haploid or diploid yeasts. The haploid yeasts sensitized to flight factors by oleic acid perished while the sensitized diploid yeasts maintained their viability.	N. N. Kovyazin, *et al.* (1961)[39], (1962)[40]
Bacteria and bacteriophages (all vehicles)	Viability and genetic effects (auxtotrophic mutations in bacteria, colony-shape mutations in phages, and induced phage production in lysogenic bacteria)	No changes in viability or genetic effects were noted in the bacteria *E. coli* K–12 and *Aerobacter aerogenes* after flight on the second Spaceship-Satellite. Similarly no effects were noted on the bacteriophages T–2 and 1321. The AMN–1 bioelement on Spaceship-Satellite 2 functioned well, and validated the usefulness of this biosensor. Studies on the lysogenic bacterium *E. coli* K–12 (λ) revealed no increased phage production over controls resulting from flights on Spaceship-Satellites 4 and 5 or Vostok 1. On Vostok 2, the induced phage production was 1·2 times that of controls, but this difference was statistically insignificant. On Vostok 3 the index of induced phage production was 4·96, on Vostok 4 only 1·96, on Vostok 5, 3·67, and on Vostok 6, 1·70. On the Voskhods, phage production did not exceed that of controls, thereby preventing evaluation of the effectiveness of the radioprotective agent β-mercaptopropylamine. These results show that there is no linear relation between length of flight and induced phage production. Accordingly, laboratory experiments were conducted to test the effects of vibration, acceleration and irradiation on lysogenic bacteria both singly and in combination. Acceleration had no inducing effect. Vibration by itself or after irradiation did not induce phage production, but low frequency vibration (35 to 75 cps) increased the sensitivity of lysogenic bacteria to subsequent gamma irradiation. The induced phage production on Vostoks 3 through 6 corresponded to doses higher than measured during the flights. Accordingly, the combination of vibration and radiation on these flights was suggested as the cause of the increased phage production.	N. N. Zhukov-Verezhnikov, *et al.* (1961)[60], (1962)[61], (1964)[62], (1965)[63–65], (1966)[66, 67]

The major achievements of these studies have been the development and validation of the AMN–1 bioelement as an automatic sensor of the effects of prolonged space flight on living organisms, and the establishment of lysogenic bacteria as a radiation biodosimeter. Lysogenic bacteria appear to be particularly useful for rapid evaluation of potential radioprotective pharmaceuticals. In addition, Soviet scientists have engaged in intensive laboratory research to determine the relative biological effectiveness (RBE) of various types of ionizing radiation including high-energy protons. Discussion of this research is beyond the scope of this chapter, and interested readers are referred to the following publications:

1. N. N. Livshits, Editor-in-Chief, "Effects of Ionizing Radiation and of Dynamic Factors on the Functions of the Central Nervous System", Problems of Space Physiology, Izdatel'stvo Nauka, Moscow, 1964 (NASA TT F–354).
2. V. G. Bobkov, *et al.*, "Radiation Safety During Space Flights", Atomizdat, Moscow, 1964 (NASA TT F–356).
3. Yu. G. Nefedov, Editor-in-Chief, "Problems of Radiation Safety in Space Flights, Physical and Biological Studies With High-Energy Protons", Atomizdat, Moscow, 1964 (NASA TT F–353).

Numerous Soviet publications treat the general problems of radiobiology as related to space exploration. Those publications which refer directly to specific flight experiments have been included in the bibliography for this chapter[5, 7, 19, 29–31, 59]. Papers which best review the major results of Soviet biological space experiments are also included in the bibliography[4–6, 19, 29–31, 56–58, 67].

References

1. E. A. Abeleva, *et al.*, "Crossing Over in *Drosophila melanogaster* Males Caused by Cosmic-Flight Factors", Iskusstvennyye Sputniki Zemli, 1961, No. 13, pp. 119–122 (Artificial Earth Satellites, pp. 127–131).

2. O. G. Alekseyeva, "The State of Natural Immunity Factors in Dogs During Cosmic Flight", Iskusstvennyye Sputniki Zemli, 1961, No. 12, pp. 63–76 (Artificial Earth Satellites, pp. 69–82).

3. O. G. Alekseyeva and A. P. Volkova, "Influence of Space-Flight Factors on the Bacterial Activity of the Body", Problemy Kosmicheskoy Biologii, 1962, Vol. 1, pp. 181–189 (NASA TT F–174, pp. 201–209).

4. V. V. Antipov, "Biological Research Conducted on the Spaceships Vostok and Voskhod", Academy of Medical Sciences SSSR, 1964, pp. 1–33 (JPRS: 27 925, pp. 27–45).

5. V. V. Antipov, N. N. Dobrov, and P. P. Saksonov, "The Basic Trends in the Study of the Biological Effect of Cosmic Radiation and the Search for Means of Protection Against Radiation", Problemy Kosmicheskoy Biologii, 1964, Vol. 3, pp. 113–124 (JPRS: 25 287, pp. 115–128).

6. V. V. Antipov, *et al.*, "Some Results of Medical and Biological Investigations in the Second and Third Satellites", Problemy Kosmicheskoy Biologii, 1962, Vol. 1, pp. 267–284 (NASA TT F-174, pp. 295–314).

7. V. V. Antipov, *et al.*, "Ensuring Radiation Safety During Flights of Vostok 3 and Vostok 4", Kosmicheskiye Issledovaniya, 1963, Vol. 1, No. 2, pp. 303–308 (TT-64-42, pp. 186–195).

8. V. V. Antipov, *et al.*, "Results of Biological Investigations Conducted During Flights of Vostok-Type Vehicles With the Participation of Cosmonauts A. G. Nikolayev, P. R. Popovich, and V. F. Bykovskiy", Problemy Kosmicheskoy Biologii, 1965, Vol. 4, pp. 248–260 (NASA TT F-368, pp. 239–251).

9. M. A. Arsen'yeva, *et al.*, "Changes in Hemopoietic Organs of Mice Due to a Flight in a Spaceship-Satellite", Iskusstvennyye Sputniki Zemli, 1961, No. 10, pp. 82–92 (Artificial Earth Satellites, pp. 218–228).

10. M. A. Arsen'yeva, *et al.*, "Effects of Flight in the Second Soviet Spaceship-Satellite on the Hemopoietic Organs of Animals", Problemy Kosmicheskoy Biologii, 1962, Vol. 1, pp. 205–218 (NASA TT F-174, pp. 227–242).

11. M. A. Arsen'yeva, *et al.*, "Cytological and Histological Changes in the Hematopoietic Organs of Mice Under the Influence of Space Flight on Spaceships", Problemy Kosmicheskoy Biologii, 1962, Vol. 2, pp. 116–127 (JPRS: 18 395, pp. 123–135).

12. V. N. Chernov and V. I. Yakovlev, "Research on Animal Flight in an Artificial Earth Satellite", Iskusstvennyye Sputniki, Zemli, 1958, No. 1, pp. 80–94 (Artificial Earth Satellites, pp. 102–120).

13. N. L. Delone, "Utilization of Higher Plants as Indicators of the Effects of Orbital Space Flight Factors on the Living Cell", Problemy Kosmicheskoy Biologii, 1965, Vol. 4, pp. 304–307 (NASA TT F-368, pp. 290–293).

14. N. L. Delone, B. B. Yegorov, and V. V. Antipov, "The Sensitivity of the Mitotic Phases of *Tradescantia paludosa* Microspores to Voskhod 1 Space Flight Factors", Doklady Akademii Nauk SSSR, 1966, Vol. 166, No. 3, pp. 713–715 (Doklady Biological Sciences, Consultants Bureau, pp. 5–7).

15. N. L. Delone, *et al.*, "Effect of Space-Flight Factors on the Vostok 3 and Vostok 4 on the Microspores of *Tradescantia paludosa*", Kosmicheskiye Issledovaniya, 1963, Vol. 1, No. 2, pp. 312–325 (TT-64-42, pp. 200–224).

16. N. L. Delone, *et al.*, "The Effect of Space-Flight Factors on Microspores of *Tradescantia paludosa* on Vostok 5 and Vostok 6", Kosmicheskiye Issledovaniya, 1964, Vol. 2, No. 2, pp. 320–329 (TT-64-547, pp. 233–251).

17. N. L. Delone, *et al.*, "Influence of Space-Flight Conditions Aboard the Vostok 5 and Vostok 6 Spacecraft on the Primordial-Root Chromosomes of Embryos in Seeds of Certain Higher Plants", Kosmicheskiye Issledovaniya, 1965, Vol. 3, No. 3, pp. 480–487 (TT-65-828, pp. 225–238).

18. N. P. Dubinin and O. L. Kanovets, "Space-Flight Factors and Primary Nondisjunction of Chromosomes", Problemy Kosmicheskoy Biologii, 1962, Vol. 1, pp. 252–257 (NASA TT F-174, pp. 277–282).

19. G. M. Frank, *et al.*, "Radiobiological Problems of Space Flights", 1st International Symposium on Basic Environmental Problems of Man in Space (29 Oct–2 Nov 1962), Springer-Verlag, New York, 1965, pp. 240–266.

20. Ya. L. Glembotskiy and G. P. Parfenov, "The Effect of Space-Flight Factors on Some Biological Indices in Insects", Problemy Kosmicheskoy Biologii, 1962, Vol. 2, pp. 98–115 (JPRS: 18 395, pp. 104–122).

21. Ya. L. Glembotskiy, *et al.*, "The Effect of Cosmic-Flight Factors on the Frequency of Occurrence of Recessive Lethal Mutations in the X-Chromosome of *Drosophila melanogaster*", Iskusstvennyye Sputniki Zemli, 1961, No. 10, pp. 61–68 (Artificial Earth Satellites, pp. 197–204).

22. Ya. L. Glembotskiy, *et al.*, "Effect of Cosmic-Flight Factors on Heredity and Development of Actinomycetes and Higher Plants. I. Cosmic Genetics", Iskusstvennyye Sputniki Zemli, 1961, No. 10, pp. 72–81 (Artificial Earth Satellites, pp. 208–217).

23. Ya. L. Glembotskiy, *et al.*, "Effects of Cosmic-Flight Factors on the Incidence of Recessive Lethal Mutations in the X-Chromosome of *Drosophila melanogaster*", Problemy Kosmicheskoy Biologii, 1962, Vol. 1, pp. 219–231 (NASA TT F–174, pp. 243–254).

24. Ya. L. Glembotskiy, *et al.*, "Influence of Space-Flight Factors on Heredity and Development in Actinomycetes and Higher Plants", Problemy Kosmicheskoy Biologii, 1962, Vol. 1, pp. 236–247 (NASA TT F–174, pp. 259–272).

25. Ya. L. Glembotskiy, *et al.*, "The Effect of the Factors of Cosmic Flight on the Incidence of Sex-Linked Recessive Lethal Mutations in *Drosophila melanogaster*", Iskusstvennyye Sputniki Zemli, 1963, No. 15, pp. 113–119 (Artificial Earth Satellites, pp. 116–122).

26. Ya. L. Glembotskiy, *et al.*, "Effect of Space-Flight Factors on the Incidence of Sex-Linked Recessive Lethal Mutations in *Drosophila melanogaster*", Kosmicheskiye Issledovaniya, 1963, Vol. 1, No. 2, pp. 326–333 (TT–64–42, pp. 225–240).

27. L. K. Gordon, *et al.*, "Effect of Space-Flight Conditions in Vostok 3 on Seeds of Higher Plants", Kosmicheskiye Issledovaniya, 1963, Vol. 1, No. 1, pp. 182–185 (TT–63–968, pp. 326–331).

28. L. K. Gordon, *et al.*, "Influence of Space-Flight Factors on Physiological Processes in the Germination of Seeds of Certain Higher Plants", Kosmicheskiye Issledovaniya, 1965, Vol. 3, No. 3, pp. 473–479 (TT–65–828, pp. 215–224).

29. A. A. Gyurdzhian, "The Biological Action of Cosmic Radiation", Iskusstvennyye Sputniki Zemli, 1961, No. 12, pp. 77–104 (Artificial Earth Satellites, pp. 83–113).

30. A. A. Gyurdzhian, "Radiobiological Problems of Space Flight", Problemy Kosmicheskoy Biologii, 1962, Vol. 1, pp. 27–103 (NASA TT F–174, pp. 29–118).

31. A. A. Gyurdzhian, "Studies of the Biological Effect of Cosmic Radiation", Problemy Kosmicheskoy Biologii, 1962, Vol. 2, pp. 93–97 (JPRS: 18 395, pp. 100–103).

32. A. A. Gyurdzhian, *et al.*, "Some Aspects of Metabolism of Animals After a Space Flight", Iskusstvennyye Sputniki Zemli, 1961, No. 11, pp. 78–86 (Artificial Earth Satellites, pp. 82–90).

33. A. A. Gyurdzhian, *et al.*, "Biochemical Investigations of Blood and Urine of Animals That Flew in Satellites", Problemy Kosmicheskoy Biologii, 1962, Vol. 1, pp. 152–160 (NASA TT F–174, pp. 169–178).

34. G. V. Il'ina, *et al.*, "Effect of Space-Flight Factors on Wheat Seeds and the Plants Produced From Them", Kosmicheskiye Issledovaniya, 1966, Vol. 4, No. 2, pp. 320–323 (HT–66–244, pp. 241–247).

35. I. I. Kas'yan, *et al.*, "Changes in Certain Morphological and Biochemical Indices of the Peripheral Blood of Animals After Rocket Flights", Problemy Kosmicheskoy Biologii, 1962, Vol. 1, pp. 161–165 (NASA TT F–174, pp. 179–184).

36. I. B. Keirim-Markus, *et al.*, "Measurement of Radiation Doses on the Second, Fourth, and Fifth Spaceship-Satellites", Iskusstvennyye Sputniki Zemli, 1961, No. 12, pp. 47–50 (Artificial Earth Satellites, pp. 52–54).

37. V. V. Khvostova, *et al.*, "The Effect of Space-Flight Conditions on the Seeds of Higher Plants and Actinomycetes", Problemy Kosmicheskoy Biologii, 1962, Vol. 2, pp. 153–163 (JPRS: 18 395, pp. 161–172).

38. V. V. Khvostova, *et al.*, "A Further Study of the Effect of Space-Flight Conditions on the Chromosomes of the Primary Rootlets of Germs in Pea and Wheat

Seeds", Kosmicheskiye Issledovaniya, 1963, Vol. 1, No. 1, pp. 186–191 (TT-63-968 pp. 332–341).

39. N. V. Kovyazin, A. A. Lukin, and G. P. Parfenov, "The Effect of Cosmic-Flight Factors of the Satellite-Ship Vostok 2 on Microorganisms (Investigation of Haploid and Diploid Yeasts)", Iskusstvennyye Sputniki Zemli, 1961, No. 13, pp. 123–129 (Artificial Earth Satellites, pp. 130–142).

40. N. V. Kovyazin, A. A. Lukin, and G. P. Parfenov, "The Effect of Space-Flight Factors of the Satellite 'Vostok 2' on Haploid and Diploid Yeasts", Problemy Kosmicheskoy Biologii, 1962, Vol. 2, pp. 149–152 (JPRS: 18 395, pp. 156–160).

41. L. D. Luk'yanova, "Investigation of the Higher Nervous Activity of White Rats After Flight in the Second Spaceship-Satellite", Iskusstvennyye Sputniki Zemli, 1961, No. 12, pp. 51–55 (Artificial Earth Satellites, pp. 56–61).

42. L. D. Luk'yanova, "Observations on the Conditioned Reflex Activity of White Rats Some Considerable Time After Travelling on the Second Soviet Satellite", Problemy Kosmicheskoy Biologii, 1962, Vol. 1, pp. 171–180 (NASA TT F-174, pp. 191–200).

43. L. D. Luk'yanova, *et al.*, "Remote Effect of Space Flights on Higher Nervous Activity and Some Unconditioned Reflexes", Problemy Kosmicheskoy Biologii, 1962, Vol. 2, pp. 192–205 (JPRS: 18 395, pp. 203–218).

44. V. V. Matveyev and A. D. Sokolov, "Determination of Induced Radioactivity in the Second Soviet Artificial Satellite", Problemy Kosmicheskoy Biologii, 1962, Vol. 1, pp. 265–266 (NASA TT F-174, pp. 291–294).

45. G. P. Parfenov, "The Occurrence of Dominant Lethal Mutations in *Drosophila melanogaster*", Iskusstvennyye Sputniki Zemli, 1961, No. 10, pp. 69–71 (Artificial Earth Satellites, pp. 205–207).

46. G. P. Parfenov, "Incidence of Dominant Lethal Mutations in *Drosophila melanogaster* During a Satellite Flight", Problemy Kosmicheskoy Biologii, 1962, Vol. 1, pp. 232–235 (NASA TT F-174, pp. 255–258).

47. G. P. Parfenov, "Growth of Organisms Under Weightlessness", Kosmicheskiye Issledovaniya, 1964, Vol. 2, No. 2, pp. 330–334 (TT-64-547, pp. 252–260).

48. G. P. Parfenov, "The Causes of Germ Cell Mortality in Drosophila After Flights on Vostoks 3 and 4", Kosmicheskiye Issledovaniya, 1964, Vol. 2, No. 2, pp. 335–340 (TT-64-547, pp. 261–271).

49. V. V. Parin, *et al.*, "Effect of Ionizing Radiation and Space Flight on Serotonin Concentration in Animal Blood", Izvestiya Akademii Nauk SSSR, Seriya Biologicheskaya, 1965, No. 1, pp. 3–9 (JPRS: 29 156, pp. 1–10).

50. V. G. Petrukhin, "Pathological Changes in the Internal Organs of Animals Under the Influence of Flight on Spaceships", Problemy Kosmicheskoy Biologii, 1962, Vol. 2, pp. 128–139 (JPRS: 18 395, pp. 136–147).

51. V. Ye. Semenenko and M. G. Vladimirova, "First Results of Tests With a Chlorella Culture Exposed in Space on the Second Spaceship-Satellite", Iskusstvennyye Sputniki Zemli, 1961, No. 12, pp. 56–62 (Artificial Earth Satellites, pp. 62–68).

52. V. Ye. Semenenko and M. G. Vladimirova, "Effect of Space-Flight Conditions in the Satellite on the Preservation of the Viability of Chlorella Cultures", Problemy Kosmicheskoy Biologii, 1962, Vol. 1, pp. 190–204 (NASA TT F-174, pp. 211–226).

53. V. S. Shashkov, *et al.*, "Influence of Space-Flight Factors on the Level of Serotonin in the Blood of Animals", Problemy Kosmicheskoy Biologii, 1962, Vol. 1, pp. 258–264 (NASA TT F-174, pp. 283–290).

54. B. N. Sidorov and N. N. Sokolov, "Effect of Cosmic Flight Conditions on Seeds of *Allium fistulosum* and *Nigella damascena*", Iskusstvennyye Sputniki Zemli, 1961, No. 10, pp. 93–95 (Artificial Earth Satellites, pp. 229–231).

55. V. N. Sidorov and N. N. Sokolov, "Influence of Space-Flight Conditions on Seeds of *Allium fistulosum* (Spring Onion) and *Nigella damascena* (Nutmeg Flower)"

Problemy Kosmicheskoy Biologii, 1962, Vol. 1, pp. 248–251 (NASA TT F-174, pp. 273–276).

56. N. M. Sisakyan, O. G. Gazenko, and V. V. Antipov, "Satellite Biological Experiments—Major Results and Problems", Life Sciences and Space Research III, North-Holland Publishing Company, Amsterdam, 1965, pp. 185–206.

57. Yu. M. Volynkin and P. P. Saksonov, "Physical Conditions of Space Flight and a Biological Characterization of Them", Problemy Kosmicheskoy Biologii, 1964, Vol. 3, pp. 10–22 (JPRS: 25 287, pp. 7–20).

58. Yu. M. Volynkin and P. P. Saksonov, "Physical Conditions of Space Flight", Proceedings of the 1st International Symposium on Basic Environmental Problems of Man in Space (29 Oct–2 Nov 1962), Springer-Verlag, New York, 1965, pp. 77–102.

59. Yu. M. Volynkin, *et al.*, "Assurance of Radiation Safety During the Voskhod 2 Flights", Kosmicheskiye Issledovaniya, 1966, Vol. 4, No. 4, pp. 630–633.

60. N. N. Zhukov-Verezhnikov, *et al.*, "Results of First Microbiological and Cytological Experiments on Earth Satellites in Space", Iskusstvennyye Sputniki Zemli, 1961, No. 11, pp. 44–67 (Artificial Earth Satellites, pp. 47–71).

61. N. N. Zhukov-Verezhnikov, *et al.*, "Microbiological and Cytological Studies on Spaceships", Problemy Kosmicheskoy Biologii, 1962, Vol. 2, pp. 140–148 (JPRS: 18 395, pp. 148–155).

62. N. N. Zhukov-Verezhnikov, *et al.*, "Microbiological and Cytological Studies in the Conquest of Space", Problemy Kosmicheskoy Biologii, 1964, Vol. 3, pp. 184–192 (JPRS: 25 987, pp. 198–205).

63. N. N. Zhukov-Verezhnikov, *et al.*, "Study of Phage Production of *E. coli* K–12 (λ) Induced Under the Conditions of the Flight Made by the Vostok 3 and Vostok 4 Spacecraft", Kosmicheskiye Issledovaniya, 1965, Vol. 3, No. 3, pp. 488–491 (TT–65–828, pp. 239–245).

64. N. N. Zhukov-Verezhnikov, *et al.*, "Investigation of the Biological Effect of Spaceflight Factors Using Lysogenic Bacteria in the Experiments on Vostok 5 and Vostok 6 Spacecraft", Kosmicheskiye Issledovaniya, 1965, Vol. 3, No. 3, pp. 492–494 (TT–65–828, pp. 246–251).

65. N. N. Zhukov-Verezhnikov, *et al.*, "Results of Microbiological and Cytological Investigations Conducted During the Flights of Vostok-Type Vehicles", Problemy Kosmicheskoy Biologii, 1965, Vol. 4, pp. 261–269 (NASA TT F-368, pp. 252–259).

66. N. N. Zhukov-Verezhnikov, *et al.*, "Results of Study of Effects of Cosmic Radiation and Other Space-Flight Factors on Lysogenic Bacteria and Human Cell Cultures", Izvestiya Akademii Nauk SSSR, Seriya Biologicheskaya, 1966, No. 4, pp. 592–593.

67. N. N. Zhukov-Verezhnikov, *et al.*, "Biological Investigations on the Voskhod 1 and Voskhod 2 Spaceships", Kosmicheskiye Issledovaniya, 1966, Vol. 4, No. 4, pp. 634–640.

Part IV

Soviet Activities Associated with Artificial Earth Satellite Applications

The discussions in Part IV are concerned somewhat with systems concepts, as opposed to the study of the environment *per se*. In particular, this is true of Chapter 15, "Communications Satellites". Here the environment is of concern only with regard to its effects on radiowave propagation, a topic not considered. The discussion is appropriate, however, because communication via satellites is perhaps the most directly useful practical application of the fruits of space-science research relating to artificial earth satellites. It also is closely related to space communications in general, a topic of importance to nearly every aspect of space-science research. In addition, the Soviets have not only established a communications satellite activity, but have related it, by providing a photographic capability to their later Molniya 1 vehicles, to the study of the environment.

Chapter 16, "Satellite Meteorology", is less systems oriented, in part because the Soviets have only recently attained a meteorological satellite, Cosmos 122, and have yet to establish a meteorological satellite system.

The subject is considered separately with regard to its two basic aspects, terrestrial radiation research, and cloud photography. Included in the former is information with regard to their earlier radiation research involving rockets, since this is integral to their later satellite studies. The Soviet literature reflecting their studies based upon U.S. meteorological satellite data is also considered.

In Chapter 17, "Satellite Geodesy", it has been possible only to consider the literature, since the Soviets have not as yet launched a geodetic satellite. They have authored numerous theoretical papers, however, and have made some observations of U.S. satellites, a few of the results of which have been published. This material is considered under two headings, (1) the geometric method and (2) the dynamic method.

Chapter 15

Communications Satellites

R. C. Behn

GENERAL

A national need for long distance telecommunications has long been recognized in the USSR, and the Soviets have given consideration over the years to elevated platforms for extending their range. Consequently, after the launching in October 1957 of Sputnik 1 considerable interest in a communications satellite system developed. The first Soviet activity was a brief cooperative effort with the U.S. and Great Britain in conducting experimental passive communications satellite transmissions. This was part of a larger U.S.-USSR program for cooperation in space exploration, the groups making the technical arrangements for which were headed by the late Dr. Hugh L. Dryden for NASA and Academician A. A. Blagonravov for the Soviet Academy of Sciences. An agreement resulted to utilize the then projected U.S. Echo 2 satellite. Because the programmed altitude of the vehicle precluded simultaneous visibility in the U.S. and USSR, it was planned that transmissions be beamed via the satellite between the Radio Astronomical Observatory of the University of Manchester, at Jodrell Bank, England, and the radio astronomical facility at Zimenki of the Scientific Research Radio Physical Institute of Gor'kiy State University. Relay to the U.S. was to be via conventional cable. The vehicle was launched in January 1964 and the experiments were conducted during the succeeding February and into March. Only transmissions from Jodrell Bank to Zimenki were made, however.

On April 23 1965 the Soviets launched their first communications satellite, named Molniya 1 (pronounced "Mol′nee-ah", meaning lightning), into a highly elliptical orbit having an inclination of 65°. The vehicle's apogee was roughly 40 000 km in the Northern Hemisphere, and its period approximately 12 hours. According to the Soviets this permitted it to relay communications between points in the Soviet Union for the greater part of every other orbit traverse. During part of the intervening traverse, communications could be relayed between the European part of the USSR and North America. The satellite was equipped to relay television or multichannel telecommunications, and to measure radiation intensities in order to study their effects on the satellite system. The initial television relay was a three-hour program on the day of launch, from Vladivostok to Moscow, and the first experimental telephone transmissions between the two cities occurred May 7 1965. Subsequently the Soviets have launched three additional Molniya 1 vehicles at approximately six-month intervals. All have been launched into essentially the same orbit and have approximately the same orbital parameters. The vehicles apparently have been essentially identical except that each of the last two has been equipped with a television camera. According to the report of the Soviet Academy of Sciences made to COSPAR in May 1966 the main objectives of the Molniya 1 satellites are:

1. Perfection of experimental long-range two-way multichannel telegraph and telephone communications
2. Long distance relay of television programs
3. Perfection of vehicle design and onboard and ground instrumentation
4. Accumulation of experimental data for establishing a commercial communications satellite system.

The Soviets and the French have been engaged intermittently for somewhat more than a year in cooperative color television transmission experiments, utilizing the Molniya 1 satellites. On November 29 1965, after the launching of the second Molniya 1 vehicle, the first satellite-relayed color television program was transmitted from Moscow to Paris. The French SECAM 3 color television system was used. On May 28 1966, subsequent to the launching of the third Molniya 1 satellite, a color television transmission, this time from Paris to Moscow and also utilizing the SECAM 3 system, initiated a new series of experiments. Only the video was transmitted via the satellite, a surface link being used to transmit the audio. Sessions of this second series of transmissions apparently originated both in Paris and Moscow and the series continued for at least 13 consecutive days, the experiments being from one to two hours duration.

The third Molniya 1 vehicle was launched on April 25 1966, and the Soviets later reported that it was equipped with a television camera and described a 20-minute session on May 18 1966 during which pictures were televised to ground receiving points from an altitude of between 30 000 and 40 000 km. The Soviets have published sample photographs from this and a second televising session of May 30 1966. These are presented in Chapter 16. The fourth Molniya 1 satellite, launched October 20 1966, was also equipped with a television camera, presumably of the same design as that of the third vehicle, but no photographs from its transmissions have been published.

The USSR is engaged in a major program to extend and improve its communications system. According to the Soviets, during the seven-year plan which ended with 1965 major extensions of their long line telecommunication and television circuits were effected, and during the current Five Year Plan (1966–1970) extensive additional long distance circuit construction is predicted. Communications satellites are envisioned as contributing importantly to the provision of this integrated Union-wide network, and a number of new ground receiving terminals are currently under construction. These are widely distributed throughout the Soviet Union, extending from Syktyvkar in Europe to Yuzhno-Sakhalinsk on Sakhalin Island. Among other locations are Vorkuta, Yakutsk, Komsomol'sk, and Magadan.

Only a few papers have appeared in the Soviet literature discussing specific aspects of communications satellite technology. Among these is a 1960 review paper[10] in which the author considers the relation of orbital parameters to broadcast area coverage, and the approximate power requirements of both ground and vehicle transmitters. A synchronous system is emphasized. A second paper[11] presents an analysis of the relation of orbital motion to the duration of communications sessions. A third paper[12] considers the economic aspects of system design and the influence on the cost per information unit transmitted of factors such as antenna design, vehicle height, operating frequency, and others. Several Soviet books[2–4, 8] also consider communications satellite technology.

THE ECHO 2 EXPERIMENTS

Echo 2 was launched January 25 1964 from Vandenberg Air Force Base, California, into a near polar (inclination 81°.5), near circular (apogee/perigee about 1300 km/1000 km, respectively) orbit with a nodal period of approximately 108 minutes. According to the Soviets, periods of simultaneous visibility from Jodrell Bank, England, and Zimenki, 30 km from Gor'kiy in the Soviet Union, varied from about 12–22 minutes. The distance between terminals was about 3000 km. From February 21 through March 8 1964 a total of 34 experimental communications sessions via the satellite and 10 via the moon occurred. Three additional sessions via the moon were conducted April 27–29. Transmission modes included the unmodulated carrier, 400-cycle tone modulation, Morse telegraphy, teletype, telephone, and facsimile. Direct two-way telegraph communication via existing international circuits was established for control of the experiments.

The Communications Link[5–6]

The Jodrell Bank equipment consisted of a rated 1-kw transmitter working into a parabolic antenna having an aperture diameter of 76 m and an antenna gain of 40 db at the operating frequency of 162·4 mc. Beam width was approximately 1°.8, and the radiation was right circularly polarized. The satellite, with a diameter of 41 m had an effective scattering area of 1320 m^2 and a reflection factor of approximately unity. The Zimenki receiving antenna was a 15-m diameter parabola, having an effective area of 88 m^2 and a main lobe half-power width of about 9°. Figure 15.1 is a photograph of the antenna.

Figure 15.2 is a block diagram of the receiving apparatus. Crossed dipoles in combination with a simple switching arrangement permitted reception of either linearly polarized radiation or that circularly polarized in either sense. The polarization properties of the antenna, when tested with an auxiliary oscillator located 100 m from the reflecter, were such that reversing the sense of the receiving antenna's circular polarization was found to change the power level of the received signal by

Fig. 15.1. 15-Meter Radio Telescope at Zimenki[9]

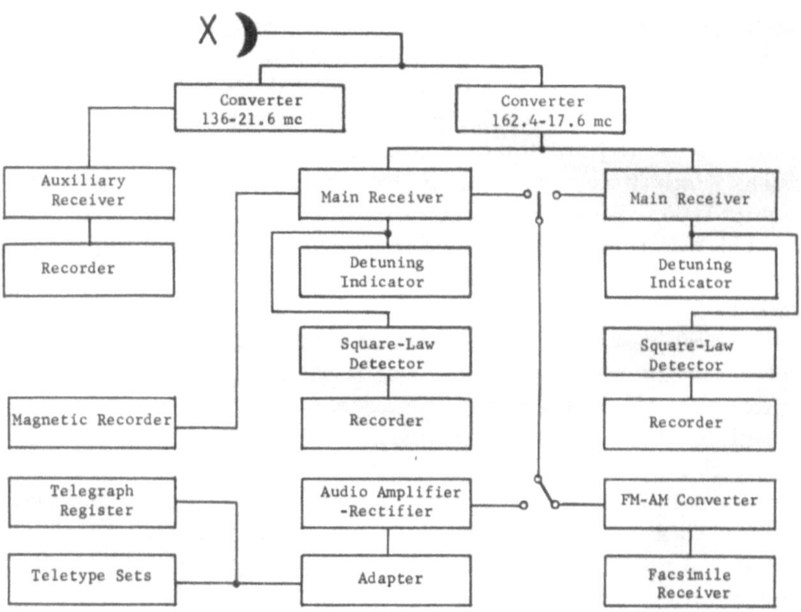

Fig. 15.2. Block Diagram of Zimenki Receiving Apparatus[6]

20 db. As shown, the equipment included an auxiliary converter and its associated receiver and recorder for receiving signals from transmitters mounted on Echo 2 and operating at 136·02 and 136·17 mc. The main transmissions at 162·4 mc were fed to a converter, physically located on the antenna structure and comprised of an RF amplifier, a crystal-stabilized local oscillator, a mixer, and a preamplifier of the 17·6 mc first IF signal. Bandwidth of the converter was 700 kc, its voltage gain 300, and its noise factor 2·6. The relative local oscillator frequency instability did not exceed 3×10^{-6}. The converted signal was then fed through cathode followers to two receivers located in the terminal building. One with its associated subsidiary components was a redundant standby.

The signal from the IF amplifier of the main receiver was applied through a square-law detector to an automatic recording potentiometer, where the signal was chart recorded. The same IF signal was simultaneously applied to the detuning indicator enabling tuning corrections to be made by adjustment of the local oscillator for Doppler or other effects. Tuning accuracy was stated as ± 50 cycles/sec. The outputs from the detector of the main receiver in the cases of 400-cycle tone modulated signals and the speech transmissions were fed to the magnetic tape recorder. In the case of Morse telegraphy, the output from the detector was received audibly and on the magnetic tape recorder, and was fed through the audio amplifier-rectifier (to produce two-polarity d–c) and adapter (to convert to single-polarity d–c) to the telegraph register (paper-tape recorder). Teletype transmissions were channeled similarly to two (tape and page) start-stop teletype receivers. Facsimile signals, which were generated by frequency-shift (400 cps) keying of a 1900 cps subcarrier, were changed to AM signals in the FM–AM converter, detected, and recorded on facsimile paper. All terminal equipment parameters were in accordance with CCITT standards.

The signal power at the receiver input for the optimum point on the orbit (minimum product of the ranges of the satellite from the transmitter and the receiver) was computed to be 6×10^{-16} w, neglecting any path attentuation. The equivalent noise temperature at the receiver input, including receiver and external noise, being roughly 800° K, the noise power for a bandwidth of 1 kc (employed for transmissions except the 400 cycle tone, and facsimile, which required 5 kc) was $1{\cdot}2 \times 10^{-17}$ w. Thus the optimum signal-to-noise ratio is computed to be 50, or 17 db, a figure much reduced in practice by increased distances to the satellite, inaccurate tracking, increased cosmic noise, varying receiver parameters, and other factors. Assuming a reflection factor of 0·07 for the moon, the signal power at the receiver input was computed to be $1{\cdot}3 \times 10^{-16}$ w, again neglecting path losses, giving an optimum signal-to-noise ratio of approximately 10 db. The sensitivity of the receiving system was calibrated regularly by homing the antenna first on the Pole Star and then on Cassiopeia A, and comparing the indicated with the known noise increase due to the latter. Signal-to-noise ratios for the various transmission modes were stated to be as follows:

Amplitude modulation by a 400 cps tone (50% modulation)	8 db
Amplitude modulation by speech signals, slowed by a factor of 8 to permit operation with a 1-kc bandwidth, with an assumed average modulation of 40%	6.3 db
Amplitude modulation by telegraph signals (100% modulation), the receiver utilizing a 100 cps bandpass filter	27 db
Amplitude modulation (50% modulation) by facsimile-signal frequency modulated subcarrier, demodulated by an 800 cps bandpass filter tuned to 1500 cps.	6 db

The receiving antenna was both program and manually controlled, the respective pointing accuracies being 3 and 15 minutes of arc. Ephemerides of Echo 2 were calculated with a BESM 2 computer from orbital elements provided to the Zimenki observatory daily by the Astronomical Council of the Soviet Academy of Sciences. These had been compiled from NASA-furnished data as well as from those of the Soviet optical tracking network. Computed ephemerides were checked against local optical observational data acquired during periods preceding communications sessions. They were also checked against radio signal strength data for specific orbital points, and by comparing the computed with the actual time of passage of the satellite through an orbital point to which both transmitter and receiver had been aimed and fixed in advance. It was determined that temporal rather than positional errors predominated. According to the Soviets, analysis of the ephemerides and the control data has shown that no pointing error was greater than 1°, making negligible any power loss in the received signal from that cause.

Results Utilizing Echo 2[5–6]

Table 15.1 contains data pertaining to the experimental communications sessions utilizing the Echo 2 satellite. The longest session (21) was 21 minutes, the shortest (19) was one minute. Tracking techniques were developed during the first seven sessions, after which signal-to-noise ratios at the receiver considerably improved. Sessions 7 and 19 were those in which the transmitting and receiving antennas were aimed and fixed in advance.

TABLE 15.1

Communications Sessions via Echo 2[5]

Session Number	Date, 1964	Time (GCT) Beginning	Time (GCT) Ending	Type of Transmission	Modulation, %	Transmitter Power, kw
1	2/21	2307	2310	Unmodulated carrier	—	1·00
2	2/22	0048	0107	Unmodulated carrier	—	1·00
3	2/22	1314	1317	Unmodulated carrier	—	1·00
4	2/22	2232	2246	Unmodulated carrier	—	1·00
5	2/23	1438	1455	Unmodulated carrier	—	1·00
6	2/23	2209	2224	Unmodulated carrier	—	1·00
7	2/24	0003 0012	0005 0014	Unmodulated carrier	—	1·00
8	2/24	1413	1430	Unmodulated carrier	—	1·00
9	2/24	2336	2351	400-cycle modulation	50	1·00
10	2/25	1350	1405	400-cycle modulation	50	0·98
11	2/25	2311	2325	Teletype	—	0·98
12	2/26	1324	1339	Teletype	—	0·98
13	2/26	2247	2301	Slowed speech	90	0·87
14	2/27	0037	0054	Slowed speech	100	0·85
15	2/27	1301	1316	400-cycle modulation	100	0·85
16	2/27	2221	2236	Slowed speech	90	0·83
17	2/28	0014	0030	Slowed speech	100	0·86
18	2/28	1238	1250	Slowed speech	100	0·85
19	2/28	1433	1434	Unmodulated carrier	—	0·85
20	2/28	2157	2210	Facsimile	50	0·84
21	2/28	2345	0006	Facsimile	70	0·85
22	2/29	1214	1225	Facsimile	100	0·87
23	2/29	1401	1417	Facsimile	100	0·87
24	2/29	2322	2341	Facsimile	90	0·87
25	3/1	1336	1352	Facsimile	90	0·85
26	3/2	1312	1326	Unmodulated carrier	—	0·90
27	3/2	2237	2249	Facsimile	100	0·93
28	3/3	1248	1302	Telegraph (Morse code)	—	1·00
29	3/3	2209	2223	Telegraph (Morse code)	—	1·00
30			No data published			
31	3/7	1258	1313	Telegraph (Morse code)	—	1·00
32	3/7	2030	2042	Telegraph (Morse code)	—	1·00
33	3/7	2220	2237	Facsimile	100	1·00
34	3/8	1233	1249	Facsimile	100	1·00

Figure 15.3 illustrates the signal level fluctuations during Session 13, February 26 1964. The true time t is determined by applying a chronometer correction (here zero) to the time T displayed on the horizontal axis. The vertical scale is graduated to display the signal-to-noise

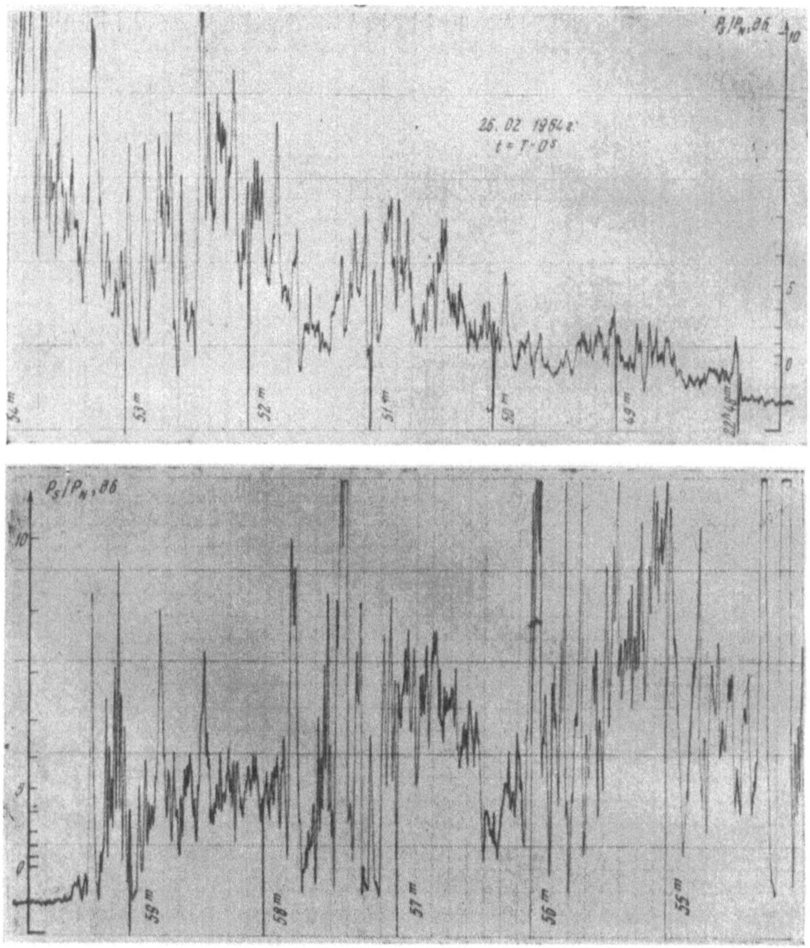

Fig. 15.3. Signal Level Fluctuations During Echo 2 Session 13[5]

ratio in db. Slow signal level variations of the order of a minute were attributed by the Soviets to the probable deviation of the satellite from a spherical shape, made evident by its rotation in orbit and to pointing errors. Rapid fluctuations of a few seconds duration were thought to be the result of local inhomogeneities of the satellite's surface, the result of shrivelling. Figures 15.4 and 15.5 are similar recordings on which the continuous line represents the theoretical signal level, plotted for convenience 8 db below its true value. Statistical analysis of the signal level records of individual sessions disclosed an approximate Rayleigh probability density distribution. Table 15.2 compares actual to theoretical signal-to-noise ratios for specific points in time during selected transmission sessions. The additional losses were thought to be caused

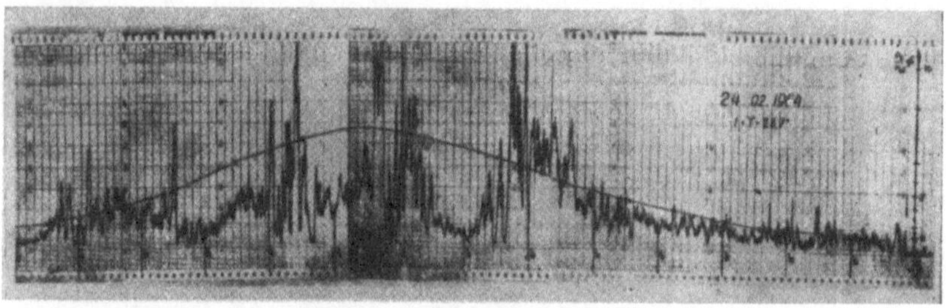

Fig. 15.4. Signal Level Fluctuations, Transmission via Echo 2, February 24 1964[6]

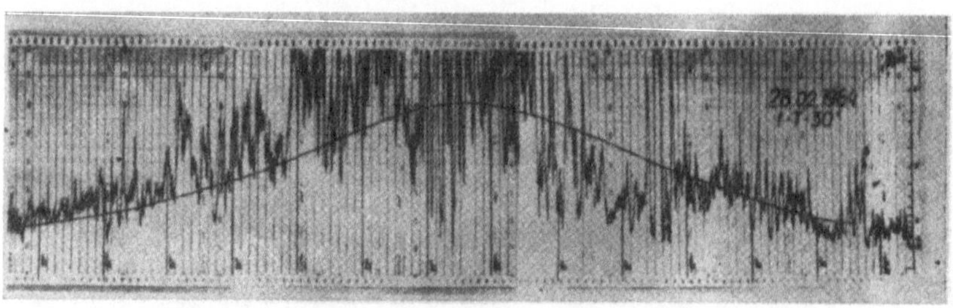

Fig. 15.5. Signal Level Fluctuations, Transmission via Echo 2, February 28 1964[6]

in part by noncorrespondence in the polarization of the transmitted signals and the receiver, and to a satellite scattering cross section smaller than calculated.

Speech transmissions, accelerated in the auditioning by a factor of 8, were just at the threshold of intelligibility. Teletype signals (Sessions 11 and 12) were transmitted in the international five-impulse code at a speed of 50 bauds to a letter-printing receiver. Signal levels were usually inadequate for reliable reception but when the signal-to-noise ratio was about 12 db good reception was attained. Morse code (Sessions 28, 29, 31, 32), transmitted at a rate of 60–100 characters per minute was clearly received. Facsimile transmissions (Sessions 20–25, 27, 33, 34) required a bandwidth of 5 kc, thus reducing the signal-to-noise ratio below those of other modes by 7 db. Figure 15.6 is a sample transmission displaying the Russian word for "peace" received during Session 24.

TABLE 15.2

Actual vs Theoretical Signal-to-Noise Ratios[5]

Session Number	Date, 1964	Time (GCT) hr	min	sec	Receiver Pass-band, kc	Signal-to-Noise Ratios, db Experimental	Theoretical	Additional Losses, db
5	2/23	14	52	40	1·0	0	5·5	5·5
11	2/25	23	11	45	1·0	2·0	6·1	4·1
13	2/26	22	52	10	1·0	7·5	12·5	5·0
14	2/27	00	45	45	1·0	11·5	14·1	2·6
17	2/28	00	24	15	1·0	6·5	10·3	3·8
20	2/28	22	07	15	5·0	1·0	6·0	5·0
21	2/28	23	59	00	5·0	0	4·5	4·5
22	2/29	12	21	00	3·5	3·5	7·4	3·9
23	2/29	14	02	00	3·5	1·0	3·8	2·8
24	2/29	23	30	10	3·5	4·5	9·4	4·9
25	3/1	13	45	00	5·0	3·0	7·5	4·5
33	3/7	22	26	30	5·0	3·5	7·5	4·0
34	3/8	12	35	30	5·0	−1·0	4·9	5·9

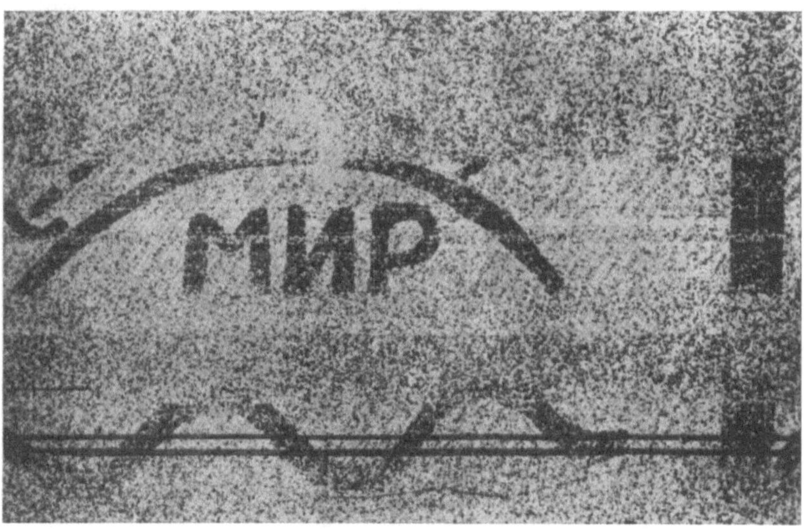

Fig. 15.6. Facsimile Transmission via Echo 2[6]

Results Utilizing The Moon[5]

Table 15.3 contains data pertaining to the experimental communications sessions utilizing the moon. Figure 15.7 illustrates the signal level fluctuations during Session 10, March 8 1964. Chart parameters are the same as those of Figure 15.3. The average value of the signal-to-noise ratio for a 1 kc bandwidth is stated to be about 5·5 db.

During Sessions 10 through 13 the receiving antenna polarization was varied to study the polarization of the moon-reflected signal. Linear, and circular polarization of both senses were utilized. The results indicated possible polarization losses of 3 db.

TABLE 15.3

Communications Sessions via the Moon[5]

Session Number	Date, 1964	Time (GCT) Beginning	Time (GCT) Ending	Type of Transmission	Modulation, %	Transmitter Power, kw
1	2/22	1330	1400	Unmodulated carrier	—	1·00
2	2/23	1500	1530	Unmodulated carrier	—	1·00
		1530	1600	400-cycle modulation	50	1·00
3	2/26	1610	1710	Teletype	—	0·95
4	2/27	2245	2252	Facsimile	50	0·84
5	2/28	2100	2140	Facsimile	100	0·875
6	2/29	2345	0010	Facsimile	100	0·875
7	3/1	2200	2230	Teletype	—	0·70
8	3/2	2252	2330	Teletype	—	0·93
9	3/3	2345	0030	Telegraph (Morse code)	—	1·00
10	3/8	0630	0700	Unmodulated carrier	—	1·00
11	4/27	2025	2221	Unmodulated carrier	—	0·80
12	4/28	2158	2341	Unmodulated carrier	—	0·80
13	4/29	2246	0050	Unmodulated carrier	—	0·80

With regard to other aspects of the complete series of experiments, reception of Morse telegraphy was satisfactory, speech and facsimile reception approximated that accomplished via Echo 2, but satisfactory reception of teletype signals was practically impossible.

The reflection coefficient of the moon was determined by comparing the received signal powers, using data from three sessions each via the moon and via the satellite. Only segments of recordings of the latter having the highest signal levels were used. The average value was computed to be 0·07 at 162·4 mc.

Transmissions of Session 10 were received on two antennas separated by 350 m in an east–west direction, and in addition to the usual recording, the signals were recorded by a loop oscillograph on a film moving 10

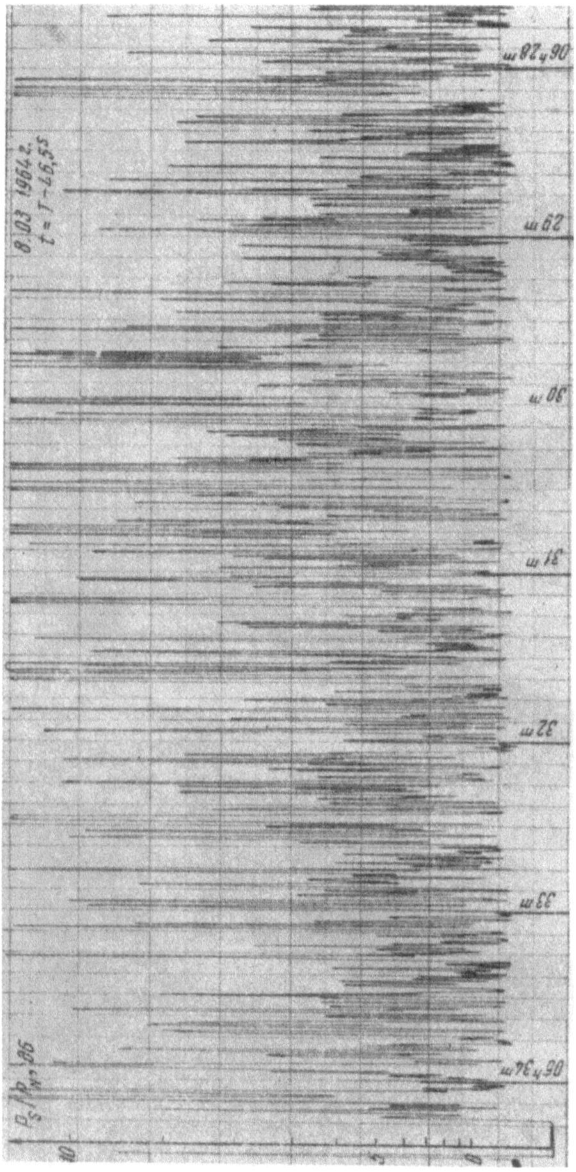

Fig. 15.7. Signal Level Fluctuations During Moon Session 10^5

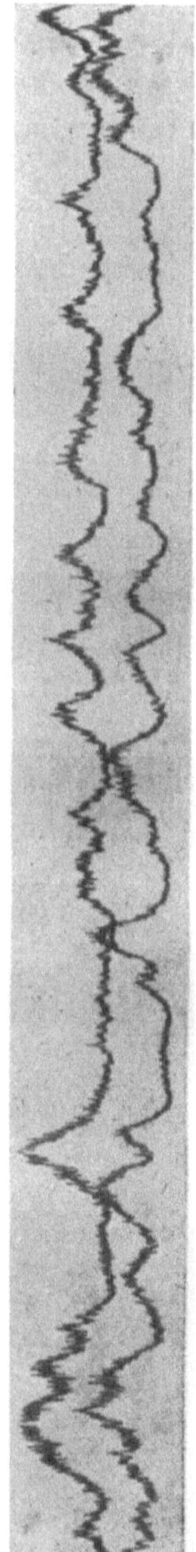

Fig. 15.8. Signal Level Fluctuations, Dual Reception, Moon Session 10^5

mm/sec. A sample trace is illustrated in Figure 15.8. According to the Soviets, analysis of the signal level trace for a three-minute interval indicated an approximate Rayleigh probability density distribution, an autocorrelation function near the theoretical only for time differentials greater than five seconds, and a maximum of the cross-correlation function of nearly 0·5, corresponding to a mean inclination of the lunar surface reflecting areas of 10°. The drift rate of the diffraction pattern in the east–west direction, as determined from the shift in the cross-correlation function maximum, was found to be 90 m/sec, corresponding to the libration rate of the moon.

THE MOLNIYA 1 SYSTEM

Orbital Information

Orbital data, as initially published, for the four Molniya 1 vehicles which have been launched by the Soviets are contained in Table 15.4.

TABLE 15.4

Molniya 1 Orbital Data

Molniya 1 Vehicles	Launch Date	Orbital Elements				Expected Lifetime, years
		Apogee, km	Perigee, km	Inclination, °	Period	
First						10
Initially	4/23/65	39 380	497	65	$11^h 48^m$	
After adjustment	5/2/65	39 957	548	65	12^h	
Second	10/14/65	40 000	500	65	$11^h 59^m$	14
Third	4/25/66	39 500	499	64·5	$11^h 50^m$	—
Fourth	10/20/66	39 700	485	64·9	$11^h 53^m$	6–1/2

According to the Soviets each of the first two vehicles, together with the final-stage booster, was first launched into a low intermediate orbit having an inclination of 65°. Apogees and perigees were 500 km and 200 km respectively, the former being in the southern hemisphere. From apogee the vehicles were injected into their high elliptical orbit. Presumably the last two vehicles were similarly launched. In addition to the advantage of the extended visibility time over the northern hemisphere provided by this orbital configuration, the orbital plane inclination is close to 63°.4, at which angle and in such an orbit it has been reported* to have been shown by two Britons, W. F.

* V. A. Altovsky, "Recent Evolution and Perspectives of Telecommunication Satellites", Thomson-Houston Company, Paris, France, July 1965 (in French).

Hilton and C. S. Dauncey, that the latitude of the apogee remains constant.

Other than the publishing of the early adjustment made in the orbit of the first Molniya 1, the Soviets have not discussed in their literature any changes in the orbital elements or the relative positions of the several Molniya 1 vehicles. A U.S. article* provides elemental data for the first three vehicles as of September 30 1966, which are here presented in Table 15.5. Orbital elements for the second Molniya 1 had not changed since the preceding April. The Soviets have also said very little about the increased available communication time in the Soviet Union which presumably results from the multiplicity of Molniya 1 vehicles in orbit. A figure of 14–16 hours per day has been mentioned.

TABLE 15.5

Molniya 1 Orbital Elements, September 30 1966

Molniya 1 Vehicles	Apogee, km	Perigee, km	Inclination, °	Period
First	39 268	1126	65·4	$12^{h}\ 0^{m}\cdot 1$
Second	39 590	560	64·9	$11^{h}\ 56^{m}\cdot 6$
Third	39 429	537	64·7	$11^{h}\ 49^{m}\cdot 8$

Ground Terminals

The principal existing Soviet ground terminals for satellite-relayed transmissions appear to be at Moscow and Vladivostok only. The Soviets have published little about them. Principally it is that they utilize automatic tracking parabolas and are equipped with highly sensitive, highly selective, and highly stable UHF receiving apparatus. A view of a terminal, presumably that near Moscow, appears in Figure 15.9.

The communications link for the Moscow–Paris color television transmissions is comprised of the Moscow Television Center, a relay link to the satellite ground terminal near that city, the satellite, the Pleumeur-Bodou ground terminal in France, and a relay link to the Buttes-Chamont transmitter in Paris. According to the French, the receiving antenna at Pleumeur-Bodou is that used in conjunction with NASA and Comsat communications vehicles, modified to function on the Soviet frequencies.

* Philip J. Klass, "Soviets Trim Comsat Orbit for Most USSR Coverage", Aviation Week & Space Technology, October 24 1966, p. 34.

Fig. 15.9. Molniya 1 System Ground Terminal[1]

Vehicle Description

The Molniya 1 vehicle is illustrated in Figures 2.17, 2.18, and 2.19 of Part I. The latter figure gives some indication of its size. Figure 15.10 illustrates a drawing of the vehicle with components in functioning position. The Soviets describe the satellite as having a body that is a hermetically sealed cylinder with conical ends. Six silicon-element solar battery panels and two parabolic antennas are mounted on the exterior. Prior to achieving orbit, these are folded and open automatically after separation of the last-stage booster. An orbit-correcting power plant and a system of attitude-correction jets are located on one end and solar and terrestrial orientation sensors on the other. Also, a cooling radiator of a cylindrical shape, and a heating panel shaped as a flat ring are mounted on the outside of the satellite. The structure of the heater is used for mounting some of the solar battery cells. The electronic apparatus and other instruments are located inside the satellite body. To aid performance, pressure and temperature inside the satellite are automatically maintained at proper values. Electrical power for the satellite is provided by the system of solar batteries, and chemical sources as well. An automatic device controls the electric power supply. During flight, the satellite's solar battery faces the sun. Besides vehicle orientation, one of the parabolic antennas, by means of a special drive

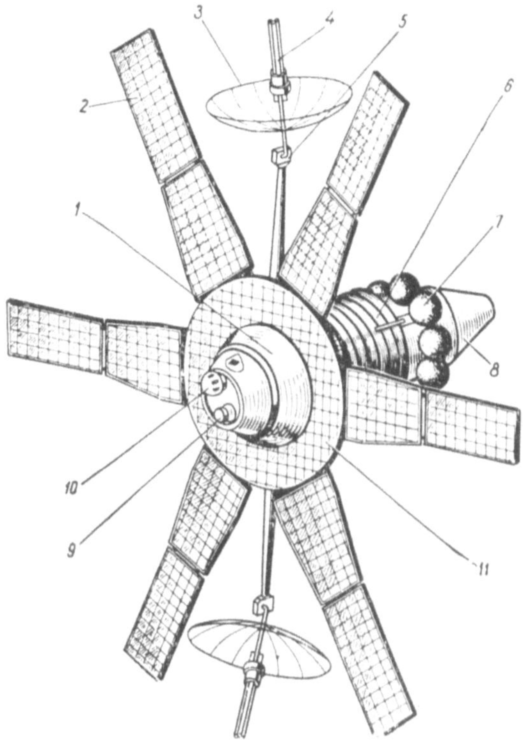

Fig. 15.10. Molniya 1 Satellite[7]

1. Gas-tight case
2. Solar battery panel
3. Directional antenna
4. Sensor which points the antenna to earth
5. Antenna drive
6. Cooling radiator
7. Attitude-correction apparatus
8. Orbit-correcting power plant
9. Terrestrial orientation sensor
10. Solar orientation sensor
11. Heating panel

controlled by a sensor fixed to the antenna, tracks the earth with a high degree of precision. The second antenna is a stand-by which to be used requires that the satellite be turned 180° on its longitudinal axis. The vehicle also carries instrumentation to study the effects of radiation on the apparatus. All systems are directed by an onboard electronic computer programmed periodically for several days in advance by telemetered instructions from the ground. Direct telemetric control is also possible.

The television camera of the third Molniya 1 is mounted externally on the surface of the vehicle, and is directed earthward by an autonomous system which functions independently of the satellite's orientation system. It is equipped with ground-controlled interchangeable lenses and filters to enable use throughout the entire altitude range traversed and the varying optical conditions encountered. A video pulse shaper associated with the camera is an additional equipment item. Presumably the fourth Molniya 1 is similarly equipped.

Electronic Characteristics

The Soviets have published very little about the detailed electronic characteristics of the Molniya 1 system. The satellites carry three relays, one operational, two said to be in reserve. This arrangement permits transmission of one television program, or its bandwidth equivalent in two-way multichannel telecommunications. Transmitter power is 40 watts, and amplification is linear. For telephony, special echo-extinguishing devices are used. For television, the scanning standard is 625 lines/frame, 25 frames/second. Audio transmission and the transmission of special monitoring impulses are integral to the video band. The relay frequencies utilized are in the 800–1000 mc range. Information on the components of the control and telemetry systems is not available. On those vehicles equipped with a television camera, camera signals are fed to a video pulse shaper and thence to the onboard relay equipment in which they are amplified and transmitted to ground reception points.

References

1. USSR National Reports to COSPAR, 1960–1966, presented by A. A. Blagonravov (in English).

2. F. Chestnov, "Radio Stations Above the Planet", Izdatel'stvo Ministerstva Oborony SSSR, Moscow, 1963, 96 pp.

3. M. P. Dolukhanov, "Long Range Propagation of Ultrashort Waves", Svyaz'izdat, Moscow, 1962, 177 pp.

4. I. A. Dombrovskiy, "Radio Communications Satellite Systems, 1963", Academy of Sciences, USSR, Moscow, 1964, 289 pp.

5. G. G. Getmantsev, N. I. Kalashnikov, V. L. Bykov, Ye. A. Benediktov, L. M. Yerukhimov, V. V. Belikovich, V. M. Bakhnin, L. Ya. Kantor, Yu. S. Korobkov, M. V. Kunilov, N. A. Mityakov, I. M. Puzyrev, V. O. Rapoport, A. G. Sigalov, V. A.

Cherepovitskiy, and E. A. Akim, "The Results of an Experiment on Radio Communications via 'Echo 2' and the Moon at a Frequency of 162·4 Megacycles Between the Observatories of Jodrell Bank and Zimenki", Kosmicheskiye Issledovaniya, 1965, Vol. 3, No. 4, pp. 618–629 (TT–65–1262, pp. 197–214).

6. N. I. Kalashnikov, L. Ya. Kantor, and V. L. Bykov, "International Experimental Radio Communication via a Satellite and the Moon", Elektrosvyaz', 1965, Vol. 19, No. 7, pp. 25–30 (Telecommunications and Radio Engineering, pp. 19–24).

7. V. D. Nikolayev, "Earth–Molniya 1–Earth", Zemlya i Vselennaya, 1965, No. 4, pp. 54–57.

8. N. T. Petrovich and Ye. F. Kamnev, "Problems of Cosmic Radio Communication", Sovetskoye Radio, 1965, 312 pp.

9. I. M. Puzyrev, "Experience in International Space Radio Communication", Zemlya i Vselennaya, 1965, No. 4, pp. 58–62 (JPRS 32 442).

10. P. V. Shmakov, "On the Employment of Earth Satellites for Television", Tekhnika Kino i Televideniya, 1960, No. 4, pp. 3–7.

11. R. M. Stetsevich, "Determination of the Duration of Communication in Systems Using Artificial Earth Satellites", Elektrosvyaz', 1964, Vol. 18, No. 12, pp. 1–8 (Telecommunications and Radio Engineering, pp. 1–6).

12. R. M. Stetsevich, "Consideration of the Economic Factors With the Selection of the Characteristics of a Television Communication System Using Artificial Earth Satellites", Trudy Uchebn. In-tov Svyazi. M-vo Svyazi SSSR, 1964, No. 22, pp. 79–89.

Chapter 16

Satellite Meteorology

R. C. Behn and R. A. Duffee

INTRODUCTION

Radiometric and photographic studies in the upper atmosphere were first done by the Soviets utilizing balloons and aircraft. In 1958 they began high altitude rocket investigations, which constituted a developmental phase of Soviet satellite meteorology. Radiometric studies were the first to be extended to their satellites, and have been conducted aboard Cosmos 45, 65, and 92, as well as on the only vehicle to date specifically designated by the Soviets as a meteorological satellite, the experimental Cosmos 122. Satellite meteorological photography other than on Cosmos 122 has been limited to very high altitude (20 000–40 000 km) work aboard the last two Molniya communications satellites. A listing of the vehicles associated with these radiometric and photographic activities appears in Table 16.1.

Launched on June 25 1966, and having been tested in orbit, Cosmos

TABLE 16.1

Soviet Rockets and Satellites Associated With Radiation Studies and Cloud Photography of the Earth (1958–1966)

						On-Board Observations			
				Approximate Orbital Elements		Radiation Measurement		Photographic	
Vehicle	Launching Date	Flight Terminated	Altitude, km	Apogee, km	Perigee, km	Terrestrial/ Atmospheric Infrared	Scattered Ultraviolet	Infrared	Visual
Geophysical Rockets	Aug. 27 1958	—	451	—	—	×	—	—	—
	July 10 1959	—	212	—	—	×	—	—	×
	June 15 1960	—	212	—	—	×	—	—	×
	June 24 1960	—	212	—	—	×	—	—	×
	Sep. 6 1960	—	100	—	—	×	—	—	—
	Sep. 19 1960	—	100	—	—	×	—	—	—
	Sep. 21 1960	—	100	—	—	×	—	—	—
	Feb. 15 1961	—	100	—	—	×	—	—	—
	Oct. 18 1962	—	500	—	—	×	—	—	—
	June 6 1963	—	515	—	—	×	—	—	—
	June 18 1963	—	510	—	—	×	—	—	—
	Sep. 20 1965	—	500	—	—	×	—	—	—
	Oct. 1 1965	—	500	—	—	×	—	—	—
Cosmos 45*	Sep. 13 1964	Sep. 18 1964	—	327	206	×	×	—	—
Cosmos 65*	Apr. 17 1965	Apr. 25 1965	—	342	210	×	×	—	—
Cosmos 92*	Oct. 16 1965	Oct. 24 1965	—	353	212	×	×	—	—
Third Molniya 1*	Apr. 25 1966	—	—	39 500	499	—	—	—	×
Cosmos 122*	June 25 1966	—	—	625	625	×	×	×	×
Fourth Molniya 1*	Oct. 20 1966	—	—	39 700	485	—	—	—	×

* All approximately 65° inclination.

122 in August 1966 was announced as an experimental Soviet meteorological satellite. The vehicle is illustrated in Figure 2.14 of Part I. Structurally, it is cylindrical and consists of two hermetically sealed compartments—an instrument module below, and a power supply and service systems module above. There are two solar battery panels which are folded during launch and are extended after separation from the booster rocket. The mechanism of the electrical drive for these panels is attached to the service module. The service systems include the power supply, systems control and monitoring equipment, vehicle orientation devices, and data processing apparatus. The vehicle is stabilized with its three axes oriented to the nadir, along the velocity vector, and perpendicular to the orbital plane, respectively. It is equipped with an electromagnetic damping system which interacts with the geomagnetic field. Operations are controllable by means of an on-board programming device, or from the ground. The satellite is equipped with visual range television cameras, apparatus for providing infrared imagery, and radiation sensing instrumentation.

According to the U.S. press* the Soviets began transmitting nephanalyses from Cosmos 122 over the Washington–Moscow meteorological communications line in August 1966. On September 11 1966 they began sending cloud photographs and continued into the following October, when transmissions ceased. The communications line mentioned became operational in late 1964, its establishment being part of the U.S.–USSR program for cooperation in space exploration, mentioned in Chapter 15.

Aside from their flight experimentation, the Soviets have published books and papers on the more theoretical aspects of satellite meteorology and numerous studies relating to the interpretation and use of data obtained from meteorological satellites, principally those of the U.S. Included among these, and not otherwise referred to in this discussion, are:

1. Books and papers relating to satellite meteorology in general—References 1†, 29–31, 37, 55, and 69.
2. Studies based upon meteorological satellite cloud photography—References 23, 25, 36, 52, 63–67, 70–73, 76, 84, 86–90, 93, and 97.

* "Soviet Weather Satellite Photos Sent to U.S.", William J. Normyle, Aviation Week and Space Technology, September 26 1966, pp. 26–27; "Washington Roundup", unsigned, Aviation Week and Space Technology, October 3 1966, p. 25.

† The Publishing House of Foreign Literature published this Russian translation of an English-language book.

3. Publications containing studies relating to both the cloud photographic and the radiation aspects of meteorological satellite observations—References 3 (a collection of 21 papers), 4 (a collection of nine papers), 5 (a collection of 16 papers), 16, 19, 38, 44–45, 56, 62, 68, 75, and 92.
4. Radiation or heat balance studies—References 2 (a collection of 14 papers dealing with the measurement of radiation, including applications to rockets and satellites), 13–14, 17–18, 20–22, 24*, 32–35, 39–42, 46–47, 53–54, 57–59, 77–79, 81–83, and 94–96.
5. Studies of miscellaneous topics—References 74, 80, and 85.

INSTRUMENTATION AND OBSERVATIONAL METHODS

Radiation Studies

The stated objectives of Soviet radiometric and photometric studies on geophysical rockets and on satellites are:

1. to obtain experimental data on the altitude variation of radiation parameters
2. to obtain statistical data on the global distribution of short-wave and long-wave radiation characterizing the radiation balance in the atmosphere
3. to obtain information relative to the distribution of ozone over the planet, the distribution of cloud and ice fields, and the distribution of air masses and the positions of fronts of synoptic formations
4. to obtain information on the nature of reflecting and scattering media, and quantitative means of determining atmospheric temperature profiles, the ozone and water vapor content of the atmosphere, and the height of the upper surfaces of clouds.

Initial measurements were carried out through the use of the so-called high-altitude optical stations (VOS), a modification of the high-altitude geophysical automatic station (VGAS) described in Part I and shown in Figure 1.11. The VOS was equipped with telephotometers, teleradiometers, telespectrometers, spectral analyzers, net radiometers,

* Reference 24 above reviews Soviet work but is by a non-Soviet author.

actinometers, and a panoramic circular-scan camera. Telephotometers are used to determine the angular distribution of radiative energy in the surrounding space in selected spectral regions between 0·2 and 1·2 μ. Teleradiometers are also used to determine the angular distribution of radiant energy in the surrounding space. They record the integral radiation in the range 0·3–40 μ, or only the infrared radiation in selected ranges such as 3–40 μ, or 9–40 μ. Like the telephotometers they can be used for panoramic scanning, or for narrow bands (2–10°) in 180° or 360° scans. Telespectrometers record the energy–wavelength relation within selected spectral regions between 0·3 and 20 μ. Spectral analyzers record the radiation flux in narrow spectral regions, or are used to determine the difference in radiation flux between adjacent narrow spectral regions within the intervals 0·2 to 0·35 μ or 0·3 to 1·2 μ. Net radiometers and actinometers both determine the integral flux of radiation. The radiometer measures separately the integral and scattered radiation along the Z-axis, and the long-wave radiation emitted by the atmosphere and the earth. Spectral sensitivity range is between 0·3 and 40 μ. Actinometers measure the integral flux of direct solar radiation between 0·2 and 40 μ.

Several types of telephotometers have been used by the Soviets on their high-altitude optical station. All of them are described in an article by A. M. Kasatkin[28]. The TP6, shown in Figure 16.1, scans in a 360° by 30° band with an angular resolution of 2°. The radiation receiver is a photomultiplier sensitive only to ultraviolet radiation between 0·2 and 0·3 μ. On an equatorial mount on the VOS, the TP6 provides data on the angular distribution of UV radiation over the planet from one horizon through the nadir to the other horizon in a 30° band. With a polar mount, a circular panorama of the angular distribution of UV radiation over the entire horizon of the earth is obtained. The TP9 telephotometer, shown in Figure 16.2, has an open aperture for measuring low intensities, an aperture covered by a neutral filter for high intensities, and a constant light source (phosphor) for periodic checking. The receiver is a highly sensitive photomultiplier with a broad spectral characteristic. Filters are used to measure specific spectral regions.

Kasatkin's article[28] also contains descriptions of a teleradiometer, a telespectrometer, a net radiometer and circular-scan photographic camera. The TR1 teleradiometer has an angular aperture of 3° and a 360° scanning range. The receiver is a semiconductor bolometer placed behind a chopper, which is a plate vibrating at the rate of 8–10 cps. The TS2 telespectrometer, a photograph of which is shown in Figure

Fig. 16.1. Photograph of the TP6 Telephotometer[28]

1. Rotating body
2. Oscillating mirror
3. Supporting tube which fits into the illuminator
4. Sealing ring
5. Holes for attaching the device to the flange
6. Multipin socket
7. Hermetically sealed body

16.3, is normally mounted on the equator of the VOS, with one of the six tubes facing the sun. This tube has a dense neutral filter designed to reduce direct solar radiation to equal reflected and scattered radiation. One cycle of measurement, i.e., six spectra, takes 12 seconds. The instrument measures transmitted, reflected, and scattered solar radiation in the range between 0·32 and 0·64 μ, with a wavelength resolution of about 40 Å. The ISTP1 net radiometer, shown schematically in Figure 16.4, uses a single black thermocouple which absorbs 97 per cent of all radiation. Another version, the ISTP2, has two additional thermocouples forming a single pile with a white coating which reflects 97 per cent of short-wave radiation. The VOS carries a number of such radiometers pointing upward, downward, toward the sun, toward the horizon and in the opposite direction to the sun. The circular-scan camera,

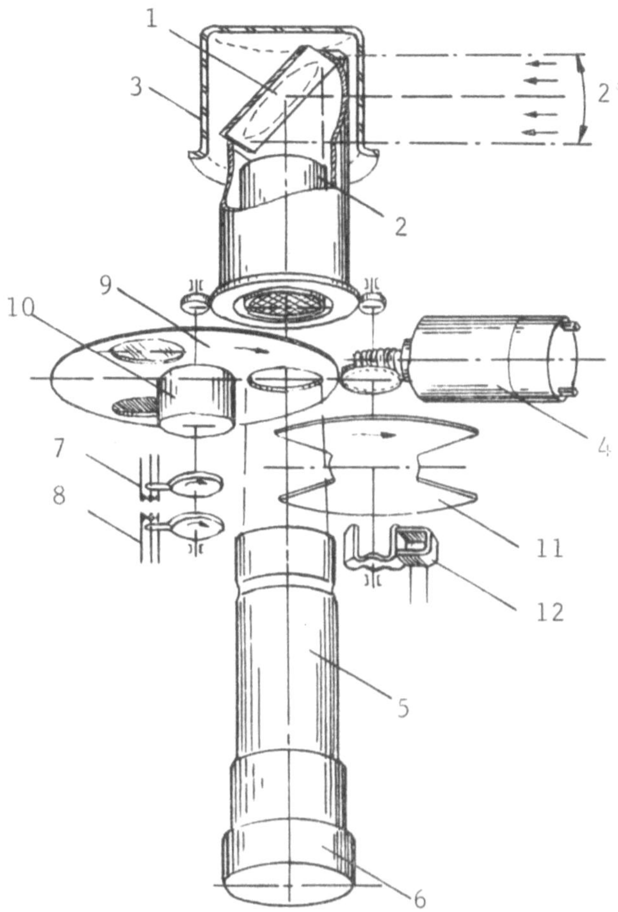

Fig. 16.2. Exploded View of the TP9 Telephotometer[28]

1. Rotating mirror
2. Tubular collimator
3. Transparent cap
4. Motor
5. Photomultiplier
6. Voltage divider
7. Switching contacts
8. Stopping contacts
9. Rotating disk
10. Disk reversing mechanism
11. Chopper
12. Standard pulse generator

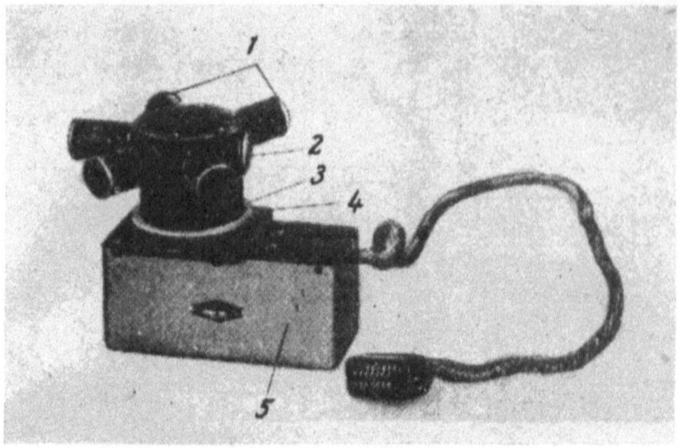

Fig. 16.3. Photograph of the TS2 Telespectrometer[28]

1. Tubes
2. Phosphor
3. Sealing ring
4. Plate for attachment to the flange
5. Container

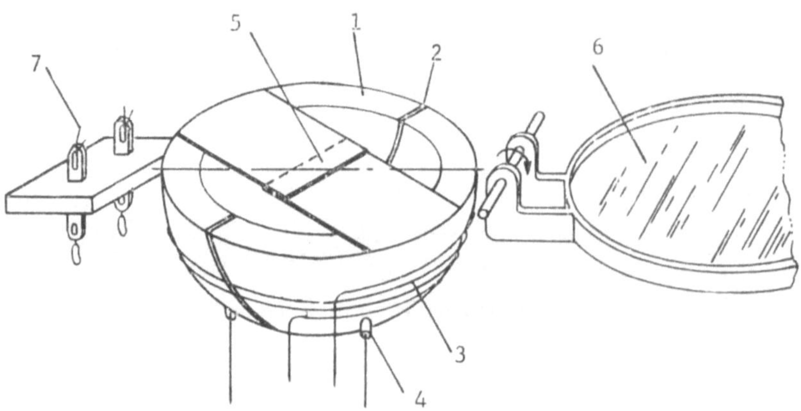

Fig. 16.4. Net radiometer[28]

1. Copper cup
2. Layer of adhesive
3. Resistance thermometer
4. Output leads (contacts) from the two parts of the cup
5. Thin film thermocouple
6. Shielding lid
7. Wire for fixing the lid (the wire burned off just before the measurements were begun)

shown in Figures 16.5a and 16.5b, may be used in continuous photography of a circular panorama at the rate of one per second. The camera is mounted on the VOS so that the photographed panorama coincides with the field of view of the scanning telephotometer or teleradiometer.

The first Soviet experimental measurement in space of the outgoing infrared radiation of the earth was conducted aboard a geophysical rocket launched to an altitude of 451 km on August 27 1958. On this flight an automatic infrared radiometer was used to measure outgoing radiation in the spectral range between 2·5 and 40 microns[51]. The instrumentation package reportedly consisted of: (1) an optical mirror system for focusing the intercepted thermal radiation onto a low inertia bolometer; (2) signal amplifiers; (3) magnetic and galvanometric recorders; and (4) a radiotelemetry system. Scanning angle was 180° and the scanning period was 30 seconds. The alleged spatial resolving power in the scanning plane was a fraction of a degree. Reflected solar short-wave radiation ($<2{\cdot}5\ \mu$) was filtered out by frosting the mirrors and covering them with a lead sulfide film.

A 1962 article by I. P. Aver'yanov, *et al.*[10], describes in detail the earth radiation recorder used on a geophysical rocket flight on February 15 1961, during a total solar eclipse, and presumably in other rocket flights. The instrument, shown in Figures 16.6 and 16.7, measures the integral emission in the 0·5 to 40 μ range, with a scan angle of 180° and a scanning rate of 6° per second. The visual angle of the optical system is reportedly 1 to 3°. Bolometer sensitivity was reported as 10^{-9} w/cps at a frequency of 80 cps, and the spectral sensitivity of the bolometers is reportedly uniform in the range of 1 to 40 μ.

The Soviets have conducted thermal and short-wave radiation studies aboard four satellites of the Cosmos series. Three of these had missions of comparatively short duration (Cosmos 45, 65, and 92), while that of the fourth, the experimental meteorological satellite Cosmos 122, was an extended flight[12, 15, 27, 48–49, 91]. The equipment used to study the energy distribution of the earth's thermal radiation on Cosmos 45, 65, and 92 consisted of a diffraction scanning spectrophotometer. It was designed to measure thermal radiation in two bands, 7–20 μ and 14–38 μ. The spectral resolution for the first band ranged from 1·4 μ for the 7 μ wavelength to 1·1 μ for the 18 μ wavelength. For the second band, the range was from 2·8 μ for the 14 μ wavelength to 2·1 for the 36 μ wavelength. The instantaneous field of view of the optical system was 1° 46′ × 2° 20′, encompassing a radiating-surface area of 7·5 × 10 km at the average altitude of 250 km. The instrument was capable of field of vision

Fig. 16.5a. Photograph of the Circular-Scan Camera[28]

1. Rotating optical head with objective
2. Fixed part
3. Contact markers

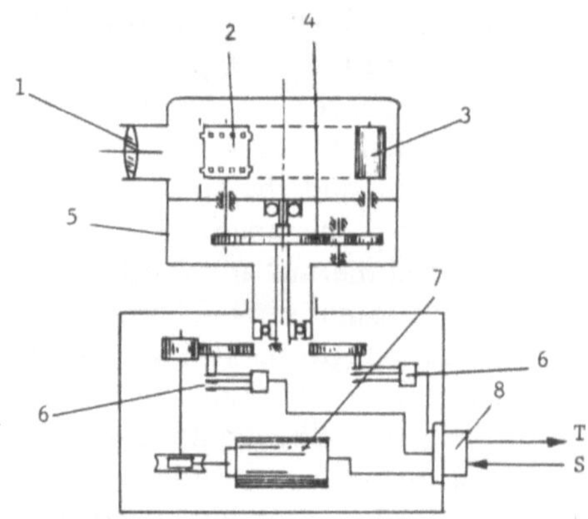

Fig. 16.5b. Schematic Illustration of the Circular-Scan Camera[28]

1. Objective
2. Fixed drum with sprockets
3. Receiving spool for the film
4. Reducing gear
5. Rotating optical head
6. Contact marker indicating the position of the objective
7. Motor
8. Multipin socket

S. Supplies
T. Telemetry

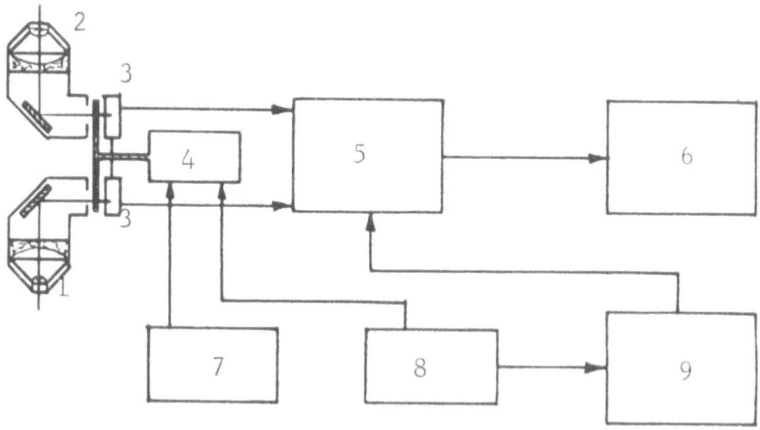

Fig. 16.6. Block Diagram of Instrument for Recording Earth's Heat Radiation[10]

1. Scanning optical head
2. Zero-signal optical head
3. Bolometers
4. Scanning mechanism
5. Amplifier
6. Telemetry
7. Motor supply
8. Command device on board
9. Power supply (independent)

Fig. 16.7. General View of Instrument for Recording Earth's Heat Radiation[10]

1. Optical system 2. Amplifier 3. Power pack

scanning within $\pm 8° 30'$. Spectral intensity measurements were carried out at $\lambda = 9{\cdot}5 \pm 0{\cdot}6\ \mu$ for the first band and $\lambda = 18{\cdot}5 \pm 1{\cdot}35\ \mu$ for the second. Semiconductor bolometers with a sensitive area of 1 mm^2 were employed as radiation sensors. Radiation detected by the bolometers was converted

Fig. 16.8. Spectrophotometric Apparatus Used on Cosmos Satellites[12]

1. Spectrophotometer
2. Oscillograph
3. Programming device

into 27 cps electrical signals which were amplified and converted into d–c voltages proportional to the radiation flux. Instead of the telemetry used in the rocket flights, the amplifier outputs were recorded on two magnetic oscillograph channels. The recording film transport speed was 25 mm/sec and the roll contained 100 m of film. Photographs of the spectrophotometer and oscillograph are shown in Figure 16.8. To measure cloud cover below the satellites, a photometer operating at 6000–8000 Å with a resolution of about 30 km was used.

On the Cosmos 122 satellite the actinometric devices consist of two narrow-sector scanning instruments, with an elementary field of view equal to a solid angle of 5°, and two wide-angle nonscanning instruments, whose field of view includes the entire area of the earth visible from the satellite. The scanning instruments measure the outgoing radiation in the spectral ranges of 0·3–3, 8–12, and 3–30 μ, the latter two being done by a single instrument. Scanning is accomplished on a plane perpendicular to the plane of the orbit, with a scanning angle of $\pm 60°$ from the direction to the nadir. The wide-angle instruments measure the outgoing radiation in the ranges of 0·3–3 μ and 3–30 μ[91].

Measurements of scattered solar UV radiation aboard the Cosmos 45, 65, and 92 satellites was accomplished by means of a UV spectrophotometer[15, 67, 73]. The instrument represented a double diffractional monochromator, which operates in a wavelength range 2250–3070 Å. The instrument's resolution was 15 Å, angle of view was $2{\cdot}5 \times 10^{-3}$ rad and field of vision 20 km in width. The instrument operated only on the day side of the earth. Switching and aiming of the UV spectrophotometer was accomplished via a photometer measuring radiation in the range 0·6 to 0·85 μ. Spectra were recorded on motion-picture film, and about 3500 spectra were recorded for 55 orbits during the Cosmos 65 flight.

Photography

The circular scan camera employed by the Soviets for photographing from rockets, used primarily for relating the orientation of the associated radiation sensing instruments to the acquired radiation data, is described above and information concerning the television equipment aboard the third Molniya 1 vehicle appears in Chapter 15. The available information[91] concerning the television apparatus aboard Cosmos 122 is presented below.

The visual range television instrumentation on this vehicle consists of two cameras, each of which views to one side of the flight path, there being a slight overlap. According to the U.S. press,* analysis at the U.S. National Environmental Satellite Center has indicated that these are 48° cameras, canted 16·5 degrees to each side of the orbital plane, and that their fields of coverage overlap by 1°, resulting in about a 10% overlap of adjacent frames. The width on the earth's surface of the

* "Soviet Weather Satellite Photos Sent to U.S.", William J. Normyle, Aviation Week and Space Technology, September 26 1966, pp. 26–27

strip photographed is about 1000 km, the resolution element being about 1–1/4 km square.

The infrared imagery is obtained in the 8–12 μ band by a detecting device which scans in a plane perpendicular to the plane of the orbit. The elementary view angle of the scanning sensor is 1°.5, and the scanning angle is $\pm 40°$ from the nadir. A continuous strip about 1100 km wide on the ground is observed, with a resolution element of about 15 km square at the nadir, from an altitude of 625 km. The infrared sensor is sensitive to a change of about 2–3° C for positive temperatures, and about 7–8° C for temperature changes near $-45°$ C.

Telemetered imagery is recorded both on magnetic tape and on photographic film. Nephanalyses are prepared by superimposing coordinate grid templates on the photographs and transferring the images to blank maps.

SUMMARY OF RESULTS

Radiation Studies

Detailed results of infrared terrestrial radiation studies for all rocket flights apparently have not been published by the Soviets. However, summaries of some of their rocket work (as well as balloon and aircraft

TABLE 16.2

Results of IR Measurements for Selected Flights

Rocket Flight and Description	Q, watts/cm^2	T_{eff}, °K	Cloud Conditions	Ground Temperature, °C
August 27 1958 $8^h\ 06^m$ Moscow Time 47 km	$1{\cdot}2 \times 10^{-2}$	216	Clear	20
July 10 1959 $4^h\ 12^m$ Moscow Time 200 km	$\cdot 9 \times 10^{-2}$	200	Medium overcast	16
June 15 1960 $5^h\ 42^m$ Moscow Time 200 km	$1{\cdot}4 \times 10^{-2}$	224	Medium overcast	15
February 15 1961 around noon 100 km	$1{\cdot}8 \times 10^{-2}$	238	Total overcast and solar eclipse	− 2
Data From Balloon Measurements				
Kondrat'yev and	$2{\cdot}2 \times 10^{-2}$	250	Clear	—
Filippovich	$2{\cdot}1 \times 10^{-2}$	248	Medium overcast	—
Simpson	$1{\cdot}9 \times 10^{-2}$	242	Medium overcast	—
F. Baur and H. Philipps	$1{\cdot}8 \times 10^{-2}$	238	Medium overcast	—

studies) have been included in their COSPAR reports and in several other articles[11, 15, 51, 60–61].

For several flights, using the recorded values of outward thermal radiation, the effective temperature of terrestrial radiation into space, T_{eff} (temperature of a black body emitting the same flux value), and the total long-wave radiation intensity, Q, were calculated through use of Lambert's law. The results are shown in Table 16.2 along with comparative data from balloon measurements.

The principal conclusions drawn by the Soviets from their geophysical rocket (and balloon) infrared radiation studies are[11, 15, 60–61]:

1. Intense infrared radiation is observed from atmospheric layers located at altitudes above 200 km. Maximum intensities were noted in the altitudinal regions of 250–300, 420–450, and about 500 km.
2. The radiation observed at these altitudes was primarily concentrated in the 2·5–8 μ spectral region and in the section of the atmosphere illuminated by the sun.
3. The upper atmospheric radiation has a maximum intensity of $(3–7) \times 10^2$ w·m^{-2} for sightings in a tangential direction, when the radiation is integrated along a line which is on the order of 100 km long. This corresponds to isotropic radiation of 1×10^{-3} erg·sec^{-1}·cm^{-3}.
4. Radiation intensity depends upon the influence of solar radiation upon the upper atmosphere and increases during periods of maximum solar activity.
5. The effective height of the radiating atmosphere is 150 km and the effective temperature is 270–280° K. The effective temperature is interpreted as the result of a major contribution to outgoing radiation by layers of the warm mesosphere (in particular H_2O or N_2O [sic]).
6. Changes in the intensity of the earth's thermal radiation at angles close to the nadir are relatively large scale (100–200 km) and generally correspond, especially in the atmospheric window between 8–12 μ, to meteorological inhomogeneities. Intensity variations reach 20–30 per cent for the integral radiation and 50 per cent in the atmospheric transparency window.

Over 10 000 spectrograms of outgoing radiation were obtained in the wavelength range 7–20 and 14–38 μ during the Cosmos 45 65, and 92 flights[15]. The measured characteristics of outgoing radiation were compared with calculated values by selecting points for which simultaneous aerological soundings were available. Measured vertical profiles of specific humidity and temperature were used in the calculations. The concentration of CO_2 was assumed constant and equal to 0·03 per cent by volume. The average vertical profiles of ozone concentration, as plotted for different latitudes and seasons by Ramanathan and Kulkarni were also used. Integration in the vertical was made in 0·5 km intervals to a height of 40 km, assumed to be the upper atmosphere boundary. The results of these comparisons were:

1. In the interval 10–14 μ the calculated values of radiation intensity matched the measured values within ± 10 per cent.
2. The ozone absorption band of 9·6 μ in the observed spectra is much stronger than in computed ones.
3. In the spectral interval 14–16 μ relating to the CO_2 absorption band, the deviations of the computed and measured intensities reach 25 per cent. These may be due to the choice of parameters of the calculation method, especially temperature in the 25–40 km layer, as well as adopted CO_2 transmission functions.

Other analyses, specifically of the Cosmos 45 data[49], showed that the intensity in the minimum absorption band near 15 μ is nearly constant. A close correlation between the intensities at other wavelengths was noted, indicating that the effective radiation levels differ but slightly for various regions of the spectrum within 8–35 μ. The lower layers of the troposphere are considered to be the basic source of the thermal radiation leaving the earth's atmosphere.

The atmospheric ultraviolet spectra obtained aboard the Cosmos satellites, specifically Cosmos 45, show a number of similarities with solar spectra, but differ in that intensities fall more steeply towards short wavelengths due to absorption in the ozone layer. The obtained spectra are determined primarily by Rayleigh scattering and ozone absorption. They agree well with the calculated spectra of some authors and with early rocket photometric studies but differ from Friedman's satellite results[15]. The measured radiation fluxes amounted to 7×10^{-2} erg/cm^2 for $\lambda = 3000$ Å and 5×10^{-3} erg/cm^2 for $\lambda = 2500$ Å.

Graphs of relative intensity for wavelengths 3160 Å, 2995 Å, 2977 Å, 2958 Å, 2906 Å, 2830 Å, and 2770 Å plotted according to the relationship of the sun's zenith angle and the latitude of the observation side show that for sun zenith angles exceeding 50° experimental (measured) curves are higher than values computed assuming one-time scattering. For the characteristic constant of the ozone concentration decrease a mean value of 4·5 km was deduced[15].

PHOTOGRAPHY

The Soviets have published no photographs taken aboard high altitude rockets, and with the exception of photographs taken from manned satellites, especially those of the album mentioned in Chapter 12, published Soviet meteorological satellite photographs are limited to examples taken by the third Molniya 1 communications satellite and by Cosmos 122.

Two from the first-mentioned vehicle are presented in Figures 16.9 and 16.10. The former was transmitted during a 20-minute television session while the satellite was in the 30 000–40 000 km altitude range, a long focal length lens with a light filter being used. The area in view is northern Eurasia, and according to the Soviets the extensive cloud cover visible is also evidenced by surface observational data. The latter photograph was televised from an altitude between 20 000 and 40 000 km. The cloud vortex in the lower center was associated with a deep cyclone over the Atlantic Ocean, some 1000 km west of the Iberian Peninsula. The horizontal cloud strip in the lower right separates a cool air mass over southern Europe from warmer air over northern Africa. The clear (dark) area above the cloud strip includes parts of the British Isles, France, and adjacent regions. Coastal outlines are said to be visible in the original. The cloud mass in the upper section is over the northern Atlantic Ocean and the Arctic.

Figure 16.11 is infrared imagery from Cosmos 122, reported as showing a cyclone over the southern Atlantic Ocean near the South Sandwich Islands. The boundary of Antarctic ice is said to be to the left of the cloud eddy. Figure 16.12 appears to be a mosaic showing weather systems over Siberia. However, it is described so as to imply that it is a composite television picture showing two views of the same cyclone. In the left-hand portion of the photograph as reported the cloud structure is seen when the cyclone lay somewhat west of the Yenisey River. The right-hand portion of the photograph is said to show the clouds associated with the system as it had developed when north of Lake Baykal. It is not clear which interpretation is correct.

Figure 16.13 as reported is a television mosaic of a multicenter depression southeast of the Fiji Islands. Figure 16.14, taken at 2350

Fig. 16.9. View of Earth Televised From the Third Molniya 1, May 18 1966[6]

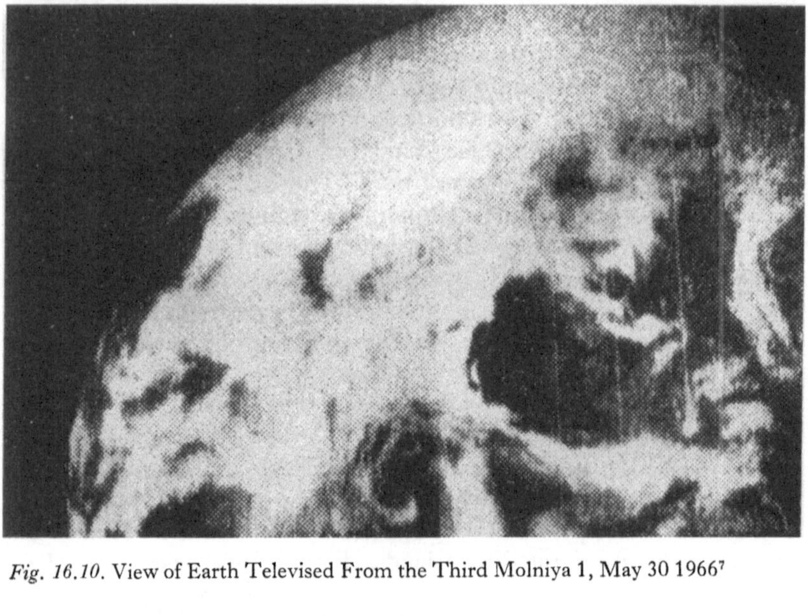

Fig. 16.10. View of Earth Televised From the Third Molniya 1, May 30 1966[7]

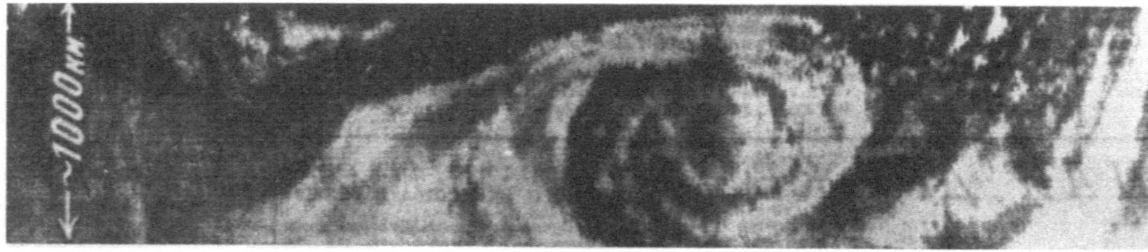

Fig. 16.11. Infrared Imagery From Cosmos 122, July 4 1966[26]

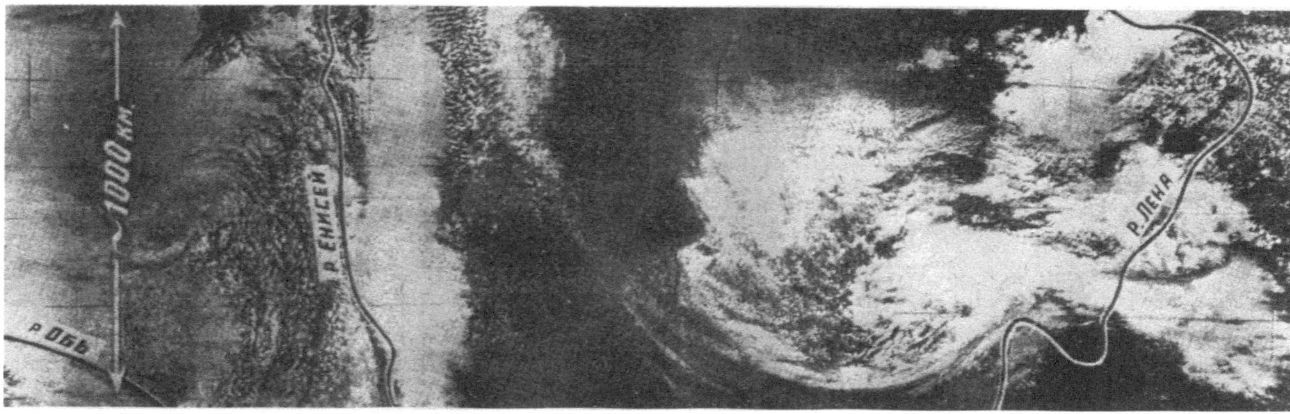

Fig. 16.12. Television Transmission From Cosmos 122, July 26 1966[8]
Area shown is western and central Siberia; in the lower left corner is the Ob' River, to the right the Yenisey River, and at the far right the Lena River.

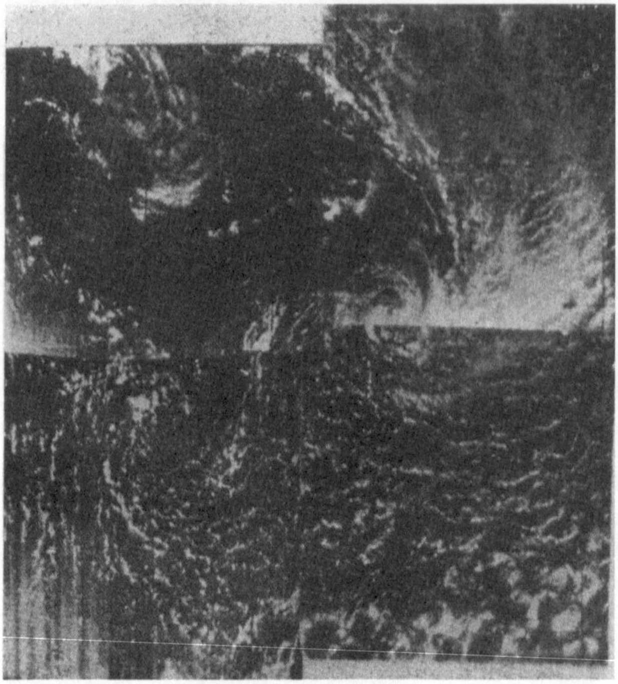

Fig. 16.13. Television Mosaic From Cosmos 122, August 22 1966[9]

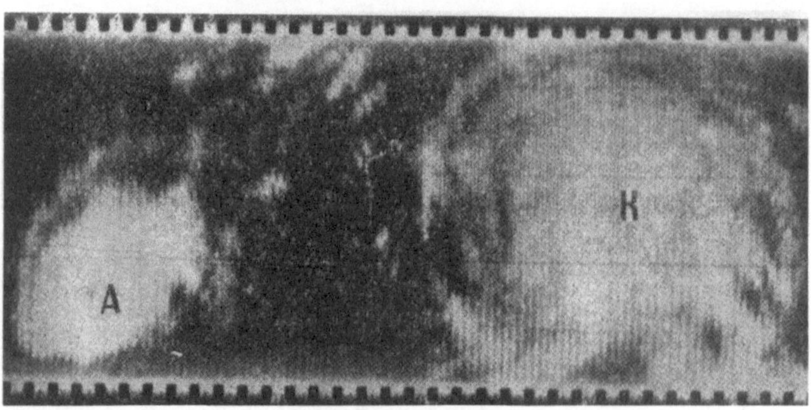

Fig. 16.14. Infrared Imagery From Cosmos 122, August 31 1966[9]
A. Typhoon Alice
K. Typhoon Cora

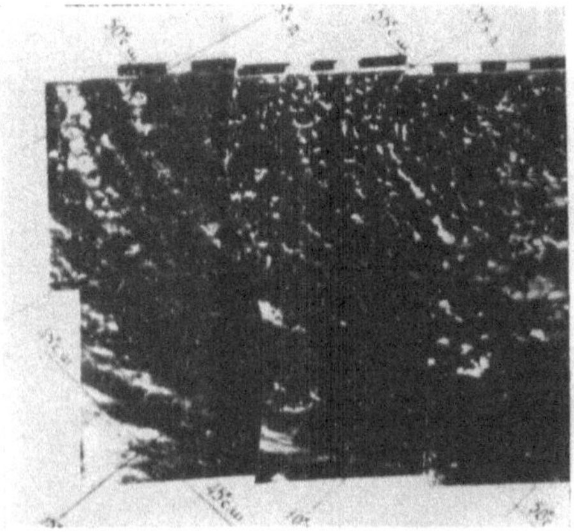

(a)

(b)

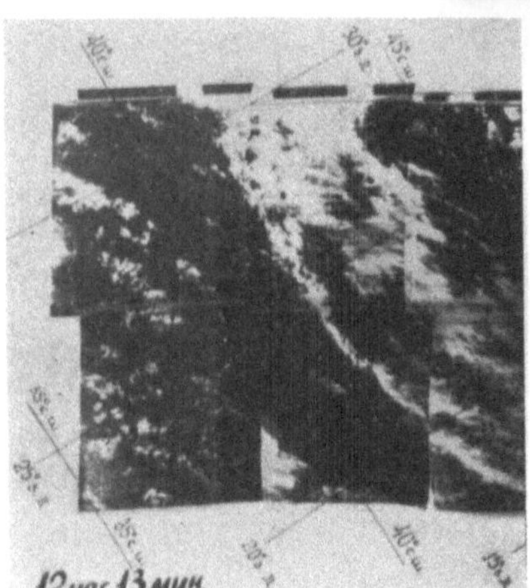

(c)

Fig. 16.15. Television Mosaics From Cosmos 122, September 1966

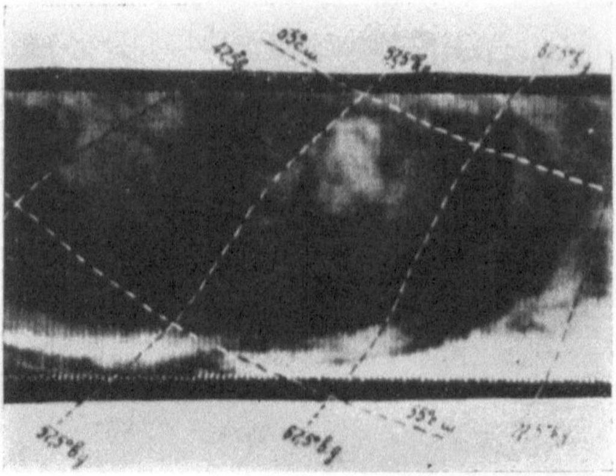

(a)

(b)

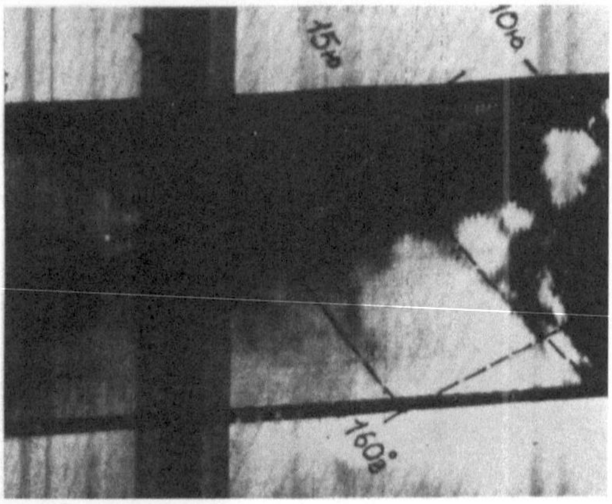

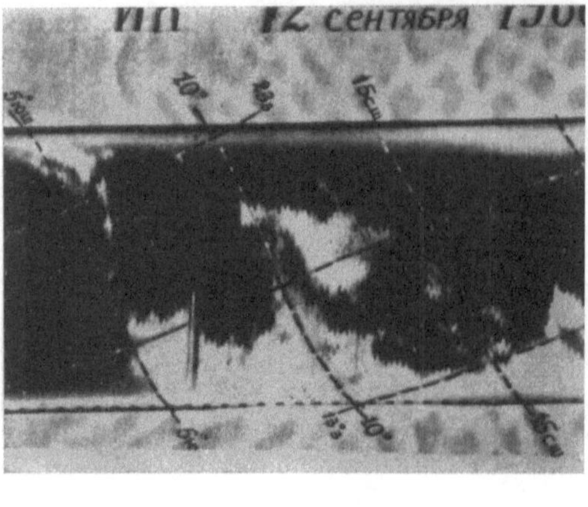

(c)

Fig. 16.16. Infrared Imagery From Cosmos 122, September 1966

Moscow time, August 31 1966 over the Pacific Ocean, is infrared imagery of typhoons Alice (25° N, 131° E) and Cora (15° N, 139° E).

Figures 16.15 and 16.16 are reduced copies of polaroid pictures of Cosmos 122 imagery as received in 8″ × 10″ facsimile format at the U.S. National Environmental Satellite Center, Suitland, Maryland, in September 1966. According to the U.S. press,* photos are received with 300 scan lines of resolution, and their quality is probably degraded by the extended communications link between Suitland and Moscow.

* "Soviet Weather Satellite Photos Sent to U.S.", William J. Normyle, Aviation Week and Space Technology, September 26, 1955, pp. 26–27.

References

1. "Rockets and Artificial Satellites in Meteorology", Izdatel'stvo Inostrannoy Literatury, Moscow, 1963, 286 pp.
2. Iskusstvennyye Sputniki Zemli, 1962, Vol. 14, 145 pp. (Artificial Earth Satellites).
3. "Problems in Interpretation of Data of Meteorological Satellites", Trudy Glavnoy Geofizicheskoy Observatorii imeni A. I. Voyeykova, Leningrad, 1964, No. 166, 300 pp.
4. Trudy Glavnoy Geofizicheskoy Observatorii imeni A. I. Voyeykova (Problemy Sputnikovoy Meteorologii), Gidrometeoizdat, Leningrad, 1966, No. 176, 140 pp.
5. Trudy Mirovogo Meteorologicheskogo Tsentra (Sputnikovaya Meteorologiya), Gidrometeoizdat, Leningrad, 1966, No. 11, 160 pp.
6. "The Earth From 40 000 km Away", Moscow News, May 28 1966, p. 16.
7. "A Picture From Space", Trud, June 1 1966, p. 4.
8. "Cosmos 122 Observes Clouds", Izvestiya, August 21 1966, p. 1.
9. "Observations From Space", Meteorologiya i Gidrologiya, 1966, No. 11, pp. 64–65 (JPRS: 39 987).
10. I. P. Aver'yanov, A. M. Kasatkin, A. V. Liventsov, M. N. Markov, Ya. I. Merson, M. R. Shamilev, and V. Ye. Shervinskiy, "Measurement of the Earth's Heat Radiation Into Space From a High-Altitude Geophysical Automatic Station During the Total Solar Eclipse of February 15 1961", Iskusstvennyye Sputniki Zemli, 1962, No. 14, pp. 49–56 (Artificial Earth Satellites, pp. 48–55).
11. D. A. Bazhulin, A. V. Kartashev, and M. N. Markov, "Angular and Spectral Radiation Distribution of the Earth in the Infrared Spectral Region", Trudy Vsesoyuznoy Konferentsii po Fizike Kosmicheskogo Prostranstva, Izdatel'stvo "Nauka", Moscow, June 10–16 1965, pp. 94–103 (NASA TT F-389, pp. 126–139).
12. P. A. Bazhulin, A. V. Kartashev, and M. N. Markov, "Study of the Angular and Spectral Distribution of Terrestrial Radiation in the Infrared Spectral Range From the Cosmos 45 Earth Satellite", Kosmicheskiye Issledovaniya, 1966, Vol. 4, No. 4, pp. 601–618.
13. P. N. Belov and A. F. Kivganov, "The Role of Radiation Processes in the Thermodynamics of the Atmosphere", Meteorologiya i Gidrologiya, 1965, No. 4, pp. 12–17 (JPRS: 30 277).
14. P. N. Belov and A. F. Kivganov, "Changes in Temperature and Geopotential Resulting From Heat Radiation Influx", Trudy Mirovogo Meteorologicheskogo Tsentra (Sputnikovaya Meteorologiya), Leningrad, 1965, No. 8, pp. 55–67 (JPRS: 39 599).
15. Soviet National Reports to COSPAR (1960 through 1966) presented by A. A. Blagonravov (in English).

16. Ye. N. Blinova, "Determining the Initial Pressure and Wind Fields From the Distribution of Temperature and Vertical Air Movements", Doklady Akademii Nauk SSSR, 1963, Vol. 149, No. 4, pp. 824–826 (Doklady of the Academy of Sciences USSR, Earth Science Sections, pp. 4–5).
17. V. G. Boldyrev, "Use of Radiation Observations From Satellites in Synoptic Analysis", Meteorologiya i Gidrologiya, 1962, No. 10, pp. 28–34.
18. V. G. Boldyrev, "The Calculation of Atmospheric Transfer Functions Within the 8–12 Micron Spectrum Interval for the Northern Hemisphere", Izvestiya Akademii Nauk SSSR, Fizika Atmosfery i Okeana, 1965, Vol. 1, No. 7, pp. 696–702 (Izvestiya Academy of Sciences USSR, Atmospheric and Oceanic Physics, pp. 401–405).
19. V. G. Boldyrev, "Measurements of Emergent Radiation for Computing Temperature of the Earth's Surface and Height of Upper Cloud Boundaries", Trudy Mirovogo Meteorologicheskogo Tsentra (Sputnikovaya Meteorologiya), 1965, No. 8, pp. 68–75.
20. V. G. Boldyrev, L. I. Koprova, and M. S. Malkevich, "The Role of Vertical Temperature and Humidity Profiles During the Determination of the Earth's Surface Temperature From Outgoing Radiation", Izvestiya Akademii Nauk SSSR, Fizika Atmosfery i Okeana, 1965, Vol. 1, No. 7, pp. 703–714 (Izvestiya Academy of Sciences USSR, Atmospheric and Oceanic Physics, pp. 406–411).
21. Ye. P. Borisenkov, Yu. P. Doronin, and K. Ya. Kondrat'yev, "The Structural Characteristics of the Radiation Field of the Earth as a Planet", Kosmicheskiye Issledovaniya, 1963, Vol. 1, No. 1, pp. 113–125 (TT 63–968, pp. 200–222).
22. M. I. Budyko and K. Ya. Kondrat'yev, "Thermal Balance of the Earth", Kosmicheskiye Issledovaniya, 1964, Vol. 2, No. 1, pp. 71–97 (TT–64–499, pp. 109–148).
23. V. A. Bugayev and L. S. Minina, "The Bank Structure of Cloud Cover", Meteorologiya i Gidrologiya, 1965, No. 5, pp. 29–36 (JPRS: 30 658).
24. B. Doncov, "Measurement of the Angular and Spectral Distribution of Atmospheric Infrared Radiation by the Cosmos 45 Satellite", Foreign Science Bulletin, Aerospace Technology Division, Library of Congress, October 1966, Vol. 2, No. 10, pp. 6–11.
25. V. A. Dzhordzhio, and O. A. Lyapina, "Cyclone Over the Caspian Sea Photographed From an Artificial Satellite", Meteorologiya i Gidrologiya, 1964, No. 8, pp. 23–24 (JPRS: 26 772).
26. Ye. K. Fedorov, "Meteorological Stations in Orbit", Izvestiya, August 21 1966, p. 5.
27. V. A. Iozenas, V. A. Krasnopol'skiy, A. P. Kuznetsov, and A. I. Lebedinskiy, "The Earth's Ultraviolet Spectrum Deduced From Cosmos 65", presented at the Ninth Plenary Meeting of COSPAR, Vienna, Austria, 1966.
28. A. M. Kasatkin, "High-Altitude Optical Station for the Atmosphere", Iskusstvennyye Sputniki Zemli, 1963, No. 15, pp. 3–21 (Artificial Earth Satellites, pp. 3–21).
29. K. Ya. Kondrat'yev, "Satellites and Meteorology", Uspekhi Fizicheskikh Nauk, 1961, Vol. 74, No. 2, pp. 193–222 (Soviet Physics Uspekhi, Vol. 4, No. 3, pp. 441–458).
30. K. Ya. Kondrat'yev, "Meteorological Investigations Using Rockets and Satellites", Gidrometeoizdat, Leningrad, 1962, 252 pp.
31. K. Ya. Kondrat'yev, "Meteorological Satellites", Gidrometeoizdat, Leningrad, 1963, 312 pp. (JPRS: 31 859).
32. K. Ya. Kondrat'yev, "Some Problems of Interpretation of the Results of Measuring Outgoing Radiation by Means of Meteorological Satellites", Kosmicheskiye Issledovaniya, 1963, Vol. 1, No. 1, pp. 98–112 (TT–63–968, pp. 172–199).
33. K. Ya. Kondrat'yev, "Practical Use of Radiation Data of Weather Satellites", Meteorologiya i Gidrologiya, 1965, No. 4, pp. 36–41 (JPRS: 30 277).
34. K. Ya. Kondrat'yev, "Outgoing Radiation and the Radiative Heat Influx", Meteorologiya i Gidrologiya, 1965, No. 11, pp. 57–61 (JPRS: 33 447).
35. K. Ya. Kondrat'yev, "Interpretation of Radiation Data From Meteorological Satellites", Trudy Vsesoyuznoy Konferentsii po Fizike Kosmicheskogo Prostranstva,

Izdatel'stvo "Nauka", Moscow, June 10–16 1965, pp. 555–568 (NASA TT F–389, pp. 741–757).
36. K. Ya. Kondrat'yev, "Synoptic Analysis of Meteorological Satellite Cloud Pictures", Idajoras, 1966, No. 4, pp. 198–210.
37. K. Ya. Kondrat'yev, "Meteorological Investigations Aboard Manned Spacecraft (A Review)", Izvestiya Akademii Nauk SSSR, Fizika Atmosfery i Okeana, 1966, Vol. 2, No. 7, pp. 740–757.
38. K. Ya. Kondrat'yev, Ye. P. Borisenkov, and A. A. Morozkin, "Utilization of Data of Meteorological Satellites", Gidrometeoizdat, Leningrad, 1966, 377 pp.
39. K. Ya. Kondrat'yev and M. P. Fedorova, "Outgoing Radiant Fluxes Onto Variously Oriented Surfaces at a Height of 300 km", Kosmicheskiye Issledovaniya, 1963, Vol. 1, No. 3, pp. 443–447 (TT–64–90, pp. 177–185).
40. K. Ya. Kondrat'yev and M. P. Fedorova, "The Outgoing Terrestrial Radiation Fluxes and the Problem of Satellite Heat Balance", Proceedings of the 15th International Astronautical Congress, Warsaw, Poland, 1964, Vol. 2—Satellite Systems, Gauthier-Villars, Paris/Polish Scientific Publishers, Warsaw, Poland, 1965, pp. 65–86.
41. K. Ya. Kondrat'yev and Yu. M. Timofeyev, "Fine Structure of the Thermal Radiation Spectrum of the Earth's Atmosphere", Kosmicheskiye Issledovaniya, 1964, Vol. 2, No. 4, pp. 603–609 (TT–64–892, pp. 154–164).
42. L. I. Koprova and M. S. Malkevich, "The Thermal Emission of a Spherical Atmosphere", Kosmicheskiye Issledovaniya, 1964, Vol. 2, No. 6, pp. 881–900 (TT–64–1316, pp. 102–134a).
43. V. A. Krasnopol'skiy, A. P. Kuznetsov, and A. I. Lebedinskiy, "Measurements of the Ultraviolet Spectrum of the Earth Made by the Satellite Cosmos 65", Geomagnetizm i Aeronomiya, 1966, Vol. 6, No. 2, pp. 185–189 (Geomagnetism and Aeronomy, pp. 145–148).
44. M. G. Kroshkin, "Investigation of High Atmospheric Layers and Meteorological Investigations Conducted by Means of Rockets and Satellites", Itogi Nauki (Geofizika, 1964), Moscow, 1965, 428 pp.
45. M. G. Kroshkin, "Investigations of Upper Atmosphere and Meteorological Investigations Made by Means of Rockets and Satellites", Itogi Nauki (Geofizika, 1965), Moscow, 1966, pp. 119–160.
46. Yu. V. Kurilova, "Possibilities of Meteorological Interpretation of Long-Wave Radiation", Trudy Mirovogo Meteorologicheskogo Tsentra (Sputnikovaya Meteorologiya), 1965, No. 8, pp. 76–86 (JPRS: 39 599).
47. V. I. Kushpil' and K. F. Khazak, "Measurements of the Brightness of the Earth as a Planet in the Water-Vapor Absorption ψ Band and in the 1·25 μ Transparency Window", Izvestiya Akademii Nauk SSSR, Fizika Atmosfery i Okeana, 1966, Vol. 2, No. 7, pp. 714–720.
48. A. I. Lebedinskiy, V. G. Boldyrev, V. I. Tulupov, G. N. Kudinova, A. D. Levchenko, T. E. Shvidkovskaya, "The Earth's Thermal Radiation Spectrum Deduced From Cosmos 45, Cosmos 65, Cosmos 92", presented at the Ninth Plenary Meeting of COSPAR, Vienna, Austria, 1966.
49. A. I. Lebedinskiy, D. N. Glovatskiy, V. I. Tulupov, B. V. Khlopov, A. A. Fomichev, and G. I. Shuster, "Infrared Spectrophotometry of the Earth's Thermal Radiation Balance", Trudy Vsesoyuznoy Konferentsii po Fizike Kosmicheskogo Prostranstva, Izdatel'stvo "Nauka", Moscow, June 10–16 1965, pp. 65–77 (NASA TT F–389, pp. 88–104).
50. A. I. Lebedinskiy, V. A. Krasnopol'skiy, A. P. Kuznetsov, and V. A. Iozenas, "Investigation of Terrestrial Atmospheric Radiation in the Visible and Ultraviolet Regions", Trudy Vsesoyuznoy Konferentsii po Fizike Kosmicheskogo Prostranstva, Izdatel'stvo "Nauka", Moscow, June 10–16 1965, pp. 77–88 (NASA TT F–389, pp. 105–119).
51. A. V. Liventsov, M. N. Markov, Ya. I. Merson, and M. R. Shamilev, "Experimental Determination of Outward Terrestrial Radiation", Doklady Akademi Nauk

SSSR, 1962, Vol. 146, No. 2, pp. 344–346 (Doklady of the Academy of Sciences, USSR, July 1964, pp. 6–7).

52. V. Ya. Lobanova, "Statistical Generalization of Cloud-Cover Information Obtained From Weather Satellites", Trudy Tsentral'nogo Instituta Prognozov (Planetarnaya Tsirkulyatsiya Atmosfery i Iskusstvennyye Sputniki Zemli), 1966, No. 152, pp. 19–24.

53. M. S. Malkevich, "Certain Problems in the Interpretation of Radiation Measurements From Artificial Satellites", Kosmicheskiye Issledovaniya, 1964, Vol. 2, No. 2, pp. 246–256 (TT–64–547, pp. 109–130).

54. M. S. Malkevich, "Some Problems of Interpretation of Radiation Measurements From Satellites", Proceedings of the 15th International Astronautical Congress, Warsaw, Poland, 1964, Vol. 2—Satellite Systems, Gauthier-Villars, Paris/Polish Scientific Publishers, Warsaw, Poland, 1965, pp. 123–137.

55. M. S. Malkevich, "Satellite Meteorology", Zemlya i Vselennaya, 1965, No. 6, pp. 31–35 (JPRS: 34 144).

56. M. S. Malkevich, I. P. Malkov, L. A. Pakhomova, G. V. Rozenberg, and G. P. Faraponova, "Determination of the Statistical Characteristics of Radiation Fields Over Clouds", Kosmicheskiye Issledovaniya, 1964, Vol. 2, No. 2, pp. 257–265 (TT–64–547, pp. 131–147).

57. M. S. Malkevich, A. S. Monin, and G. V. Rozenberg, "The Three Dimensional Structure of a Radiation Field as a Source of Meteorological Information", Izvestiya Akademii Nauk SSSR, Seriya Geofizicheskaya, 1964, No. 3, pp. 394–407 (Bulletin [Izvestiya] Academy of Sciences, USSR, Geophysics Series, pp. 237–244).

58. M. S. Malkevich and V. I. Tatarskiy, "Using Terrestrial Radiation Measurements by Artificial Satellites to Determine Atmospheric Temperature and Humidity", Trudy Vsesoyuznoy Konferentsii po Fizike Kosmicheskogo Prostranstva, Izdatel'stvo "Nauka", Moscow, June 10–16, 1965, pp. 104–111 (NASA TT F–389, pp. 140–152).

59. G. I. Marchuk, "Equation for the Evaluation of Information From Meteorological Satellites and Formulation of Inverse Problems", Kosmicheskiye Issledovaniya, 1964, Vol. 2, No. 3, pp. 462–477 (TT–64–770, pp. 169–192).

60. M. N. Markov, Ya. I. Merson, and M. R. Shamilev, "Study of the Angular Radiation Distribution of the Earth and the Earth's Atmosphere From Geophysical Rockets and Balloons", Trudy Vsesoyuznoy Konferentsii po Fizike Kosmicheskogo Prostranstva, Izdatel'stvo "Nauka", Moscow, June 10–16 1965, pp. 90–93 (NASA TT F–389, pp. 122–126).

61. M. N. Markov, Ya. I. Merson, and M. R. Shamilev, "Upper Atmosphere Layers Which Radiate in the Infrared Spectral Region", Trudy Vsesoyuznoy Konferentsii po Fizike Kosmicheskogo Prostranstva, Izdatel'stvo "Nauka", Moscow, June 10–16 1965, pp. 112–119 (NASA TT F–389, pp. 153–165).

62. L. T. Matveyev, "Requirements for Accuracy in the Determination of the Flux of Infrared Radiation by Artificial Earth Satellites and a Method for Computation of the Upper Cloud Boundary", Kosmicheskiye Issledovaniya, 1964, Vol. 2, No. 1, pp. 109–120 (TT–64–499, pp. 167–187).

63. L. S. Minina, "Vertical Structure of Cloud Cover Determined From Meteorological Satellite Data", Meteorologiya i Gidrologiya, 1964, No. 1, pp. 12–22 (JPRS: 24 288).

64. L. S. Minina, "The Use of Television Observations From Meteorological Satellites in the Synoptic Analysis Made in the Central Institute of Weather Forecasting", Meteorologiya i Gidrologiya, 1966, No. 4, pp. 45–58 (JPRS: 35 828).

65. L. S. Minina, "High Level Mesovortex and Some of Its Characteristics", Meteorologiya i Gidrologiya, 1966, No. 6, pp. 21–30 (JPRS: 37 281).

66. Sh. A. Musayelyan, "Problems in the Numerical Interpretation of Cloud Data Transmitted by Artificial Earth Satellites", Trudy Glavnoy Geofizicheskoy Observa-

torii (Voprosy Interpretatsii Danny'kh Meteorologicheskikh Sputnikov), 1964, No. 166, pp. 203–213.
67. Sh. A. Musayelyan, "Some Aspects of Interpretation and Use of Cloud Cover Data Obtained Using Meteorological Satellites", Doklady Akademii Nauk, SSSR, 1965, Vol. 163, No. 5, pp. 1134–1137 (JPRS: 33 405 and JPRS: 32 727; Doklady of the Academy of Sciences USSR, Earth Science Section, 1965, Vol 163, No. 4, pp. 10–13).
68. B. M. Novikov and V. A. Zyabrikov, "The Demand for Accuracy in Obtaining Meteorological Information From Artificial Earth Satellites", Kosmicheskiye Issledovaniya, 1963, Vol. 1, No. 2, pp. 249–255 (TT–64–42, pp. 97–107).
69. V. P. Petrov, "Space Weather-Stations", Moscow, Izdatel'stvo "Nauka", 1966, 119 pp.
70. N. Z. Pinus and S. M. Shmeter, "Macrostructure of Cloud Fields From Satellite Data", Aerologiya, Chast' II, Fizika Svobodnoy Atmosfery, Gidrometeoizdat, Leningrad, 1965, pp. 317–320 (JPRS: 35 749).
71. S. P. Pivovarov and I. A. Rosselevich, "TV Equipment on the Artificial Earth Satellites Intended for Observation of Clouds", Tekhnika Kino i Televideniya, 1966, No. 5, pp. 7–17.
72. T. P. Popova, "The Relationship Between Cloudiness and Vertical Movements", Trudy Mirovogo Meteorologicheskogo Tsentra (Sputnikovaya Meteorologiya), 1965, No. 8, pp. 15–28.
73. N. M. Potiyevskiy and N. I. Rumyantsev, "Experience in the Machine Compilation of Cloud-Cover Maps Using a Setun' Computer and Weather-Satellite Data", Trudy Tsentral'nogo Instituta Prognozov (Planetarnaya Tsirkulyatsiya Atmosfery i Iskusstvennyye Sputniki Zemli), 1966, No. 152, pp. 25–28 (JPRS: 41 294).
74. G. V. Rozenberg and Yu. A. R. Mullamaa, "The Possibility of Measuring Wind Velocity Above the Ocean Surface From Satellite Observations", Izvestiya Akademii Nauk SSSR, Fizika Atmosfery i Okeana, 1965, Vol. 1, No. 3, pp. 282–290 (Izvestiya Academy of Sciences, USSR, Atmospheric and Oceanic Physics, 1965, Vol. 1, No. 3, pp. 167–171).
75. G. V. Rozenberg and V. V. Tereshkova, "The Stratospheric Aerosol as Measured From a Spaceship", Trudy Vsesoyuznoy Konferentsii po Fizike Kosmicheskogo Prostranstva, Izdatel'stvo "Nauka", Moscow, June 10–16 1965, p. 61 (NASA TT F–389, p. 81).
76. V. P. Samrov, "The Relationship Between the Eddy Cloud Structures and Heavy Seas", Trudy Tsentral'nogo Instituta Prognozov (Planetarnaya Tsirkulyatsiya Atmosfery i Iskusstvennyye Sputniki Zemli), 1966, No. 152, pp. 11–18.
77. K. S. Shifrin and G. L. Shubova, "Statistical Characteristics of the Vertical Transparent Atmosphere", Izvestiya Akademii Nauk SSSR, Seriya Geofizicheskaya, 1964, No. 2, pp. 279–284 (Bulletin [Izvestiya] Academy of Sciences, USSR, Geophysics Series, pp. 161–164).
78. Sh. G. Shlionskiy, "Attenuation of Radiation From Artificial Earth Satellites in Above-Surface Trajectories", Geomagnetizm i Aeronomiya, 1965, Vol. 5, No. 6, pp. 1061–1067 (Geomagnetism and Aeronomy, pp. 832–837).
79. B. Ye. Shneyerov, "Computation of the Radiation Balance of the Earth's Surface-Atmosphere System and Its Components", Meteorologiya i Gidrologiya, 1963, No. 7, pp. 10–17 (JPRS: 20 716).
80. L. N. Shustova, "The Trajectory of the Trace of the Intersection Between the Optical Axis and the Surface of a Celestial Body", Spetsial'nyye Voprosy Fotogrammetrii, Izd-vo Nauka, Moscow, 1964, pp. 31–36 (JPRS: 32 203).
81. M. Ye. Shvets, "Computation of the Flux of Outgoing Long-Wave Radiation Using Data From an Artificial Earth Satellite", Kosmicheskiye Issledovaniya, 1964, Vol. 2, No. 2, pp. 272–275 (TT–64–547, pp. 157–162).
82. M. Ye. Shvets and B. Ye. Shneyerov, "Evaluation of Results of Radiation Measurements From Satellites in the Nonadiabatic Model of Atmospheric Move-

ments", Doklady Akademii Nauk SSSR, 1963, Vol. 152, No. 3, pp. 598–601 (Doklady of the Academy of Sciences, USSR, Earth Science Section, 1963, Vol. 152, No. 5, pp. 3–5).

83. M. Ye. Shvets and B. Ye. Shneyerov, "Calculation of Heat Flux Into the Ground", Izvestiya Akademii Nauk SSSR, Fizika Atmosfery i Okeana, 1965, Vol. 1, No. 2, pp. 167–174 (Izvestiya Academy of Sciences USSR, Atmospheric and Oceanic Physics, pp. 100–103).

84. D. M. Sonschkin, "Interpretation of Television Images of Cloudiness Obtained From Artificial Satellites", Meteorologiya i Gidrologiya, 1962, No. 9, pp. 30–33 (JPRS: 16 557).

85. B. N. Trubnikov, "Using Artificial Earth Satellite Orbital Data for Determining Wind Velocity in the Thermosphere", Trudy Vsesoyuznoy Konferentsii po Fizike Kosmicheskogo Prostranstva, Izdatel'stvo "Nauka", Moscow, June 10–16 1965, pp. 51–56 (NASA TT F–389, pp. 66–73).

86. L. A. Uranova, "Vortices of Cloud Systems According to the Data of Meteorological Satellites", Trudy Tsentral'nogo Instituta Prognozov, 1966, No. 152, pp. 3–10.

87. N. F. Vel'tishchev, "Processing TV Cloud Pictures Obtained With Meteorological Satellites", Meteorologiya i Gidrologiya, 1962, No. 11, pp. 39–42.

88. N. F. Vel'tishchev, "Interpretation of the Mesostructure of Cloudiness", Trudy Mirovogo Meteorologicheskogo Tsentra (Sputnikovaya Meteorologiya), Leningrad, 1965, No. 8, pp. 45–54 (JPRS: 39 599).

89. N. F. Vel'tishchev, "Structure of Cloud Cover in Atmospheric Vortices", Meteorologiya i Gidrologiya, 1965, No. 12, pp. 11–19 (JPRS: 39 599).

90. I. P Vetlov, "Some Results of Comparing Satellite and Terrestrial Observations of Cloudiness", Trudy Mirovogo Meteorologicheskogo Tsentra (Sputnikovaya Meteorologiya), 1965, No. 8, pp. 29–36.

91. I. P. Vetlov, "Experiment on the Investigation of Cloud Cover and Outgoing Radiation With the Satellite Cosmos 122", presented at the World Meteorological Organization Seminar on the Interpretation and Use of Meteorological Satellite Data, Moscow, October 1966 (JPRS: 39 188).

92. I. P. Vetlov, "Interpretation and Meteorological Use of Satellite Information", Meteorologiya i Gidrologiya, 1966, No. 4, pp. 41–44 (JPRS: 35 828).

93. I. P. Vetlov, "Experience in Analyzing the Infrared Image of Cloud Cover Obtained by the Meteorological Satellite Nimbus I", Meteorologiya i Gidrologiya, 1965, No. 9, pp. 20–26 (JPRS: 32 621).

94. A. I. Yefremov, A. L. Podmoshenskiy, I. M. Pribylovskiy, and V. S. Petrov, "Apparatus for Studying Short Wave Solar Radiation by the Filter Method With the Help of Earth Satellites", Optiko-Mekhanicheskaya Promyshlennost', 1964, No. 3, pp. 28–30.

95. A. P. Yurgenson, "Contribution to the Study of the Spectral Origin of Ascending Longwave Radiation in the Atmosphere", Trudy Arkticheskogo i Antarkticheskogo Nauchno-Issledovatel'skogo Instituta (Chislennyye Metody Issledovaniya Gidrometeorologicheskikh Usloviy v Arktike s Ispol'zovaniyem Elektronnykh Tsifrovykh Vychislitel'nykh Mashin; Sbornik Statey), 1964, Vol. 271, No. 1, pp. 19–33.

96. A. P. Yurgenson, "Spectral Absorption of Longwave Radiation in the Atmosphere", Trudy Arkticheskogo i Antarkticheskogo Nauchno-Issledovatel'skogo Instituta (Chislennyye Metody Issledovaniya Gidrometeorologicheskikh Usloviy v Arktike s Ispol'zovaniyem Elektronnykh Tsifrovykh Vychisletel'nykh Mashin), 1966, Vol. 277, pp. 11–19.

97. A. D. Zamorskiy and L. S. Minina, "Extinct Cyclone Photographed From a Satellite", Meteorologiya i Gidrologiya, 1965, No. 11, pp. 38–43 JPRS: 33 447).

Chapter 17

Satellite Geodesy

A. G. Mourad

INTRODUCTION

The available Soviet satellite geodetic literature is discussed under two major headings, the geometric method and the dynamic method. In considering the geometric method various techniques, such as photographic and electronic, as applied to the determination of relative or absolute positions on the earth's surface are discussed. With regard to the dynamic method, the discussion is limited to the determination of certain parameters of the earth and its gravitational field.

Since the beginning of the satellite era, published Soviet geodetic satellite discussions have dealt principally with theoretical concepts with practical considerations being limited to reviews of U.S. geodetic satellite activity. The Soviets do discuss advantages of geodetic satellites in the establishment of a world geodetic system, intercontinental ties, and other applications, but do not identify a satellite-established world geodetic system of their own.

The few published Soviet results on geodetic satellites deal mostly with western tracking data or with Soviet observations of U.S. satellites. The Soviets have identified an operational program in satellite geodesy, which uses simultaneous observations of at least the Echo 1 or Echo 2 satellites. It was initially limited to the USSR, but later was expanded to include Soviet Bloc and several European countries. Also, a plan for a world-wide space triangulation program composed of 13 stations and a

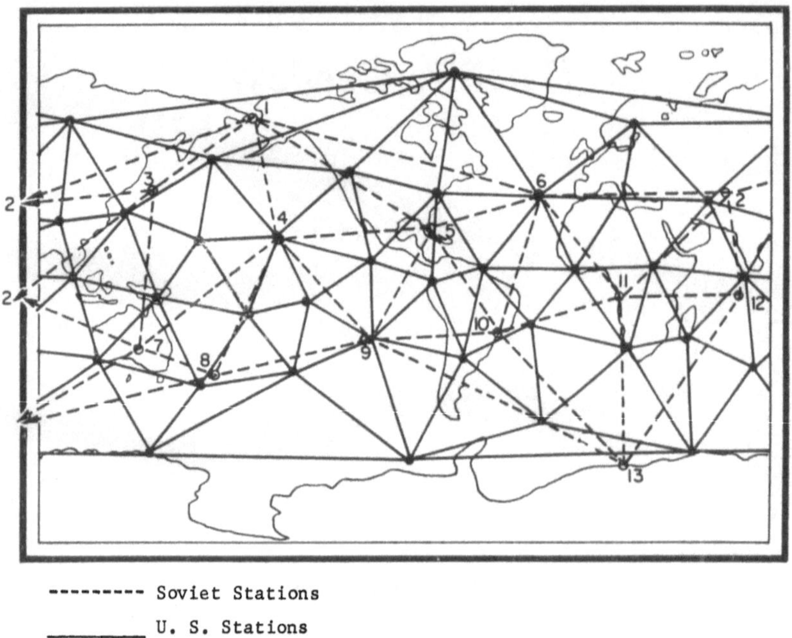

Fig. 17.1. Comparison of U.S. Geometric Satellite Network With a Planned Network Mentioned by the Soviets

geodetic satellite at an altitude of 12 000 km was proposed by Zhongolovich 1964[32, 38]. However, there is no evidence that this plan was put into effect. A diagram of this plan, compared with the U.S. geodetic satellite geometric network, is shown in Figure 17.1. The U.S. work predates corresponding Soviet efforts.

GEOMETRIC METHOD

The geometric method, particularly the space triangulation technique involving simultaneous photographic observations of satellites from several ground stations, is the one most discussed in Soviet literature.

The Soviets[26] favor this technique probably because of its simplicity and their familiarity with its operation, i.e., satellite photographic techniques have been routinely used by both the Soviets and East European countries as part of the INTEROBS (International Observations) satellite tracking effort[8, 29]. Also this technique became popular with the availability of large visible satellites such as U.S. Echo 1, Echo 2, and Pageos.

The first successful Soviet application of the photographic technique to satellites, according to Zhongolovich[35], was by Kiselev, Firago, and Shchegolev[15] in 1960. The photographic observation and reduction techniques were developed at the Main Astronomical Observatory of the Academy of Sciences, at Pulkovo.

To date, the results of geometric studies analyzed and reported upon by the Soviets, are those related to the U.S. passive Echo 1 and Echo 2 satellites[10, 20–21, 23]. It appears that extensive programs have been and are being conducted in the USSR and other East European countries to photograph these satellites against the stellar background for obtaining relative positions of the observation stations. The first test of the use of synchronous observations of Echo 1 satellite was carried out in 1961 in the Soviet Union[8, 17, 26, 29] at four stations, using the standard NAFA-3s/25, a modified aerial camera. These stations were located at Pulkovo, Khar'kov, Tashkent, and Nikolayev. The error in the position of the unknown station was reported to be about 130 meters[12]. However, the equipment used in 1964 improved the accuracy to $\pm$ 50 meters.[2]

As a result of this test, according to Cichowicz[8], a decision was made to incorporate several non-Soviet stations into this continuing experiment. In 1962 several East European countries—East Germany, Poland, Czechoslovakia, and Rumania—were included[8, 12, 26]. The progress in reducing these observations to determine station coordinates has been reported by Shchegolev, *et al*[12, 26, 29]. A 10 000 km separation distance is reported for another successful test made in May-June 1963, using Echo 1 and Echo 2. According to Yerpylev[32], the Soviets have expanded the satellite observation program to include, in addition to the above countries, Hungary, Bulgaria, Mongolia, Sweden, UAR, and Italy. The locations of most of the stations of this network are shown in Figure 17.2.

The use of satellite geodesy is often mentioned for the determination of absolute positions on the earth's surface, for studying gravity field parameters of the earth and celestial bodies, and for establishing intercontinental ties and the connection of widely separated points[1, 6, 19, 22, 32, 35, 37–38]. However, the only actual program carried out for intercontinental connection is that described by Pariyskiy[21]. He refers to the making of simultaneous observations, utilizing U.S. (Echo 1 and Echo 2), and Soviet unidentified satellites and NAFA–3s/25 wide-angle cameras for the interconnection of two widely separated stations. The program was conducted at Mali, Africa (December 1964–March 1965), as a cultural and scientific cooperative effort between Mali and the USSR. The main purpose of the program was reportedly to study the tidal deformation of the earth.

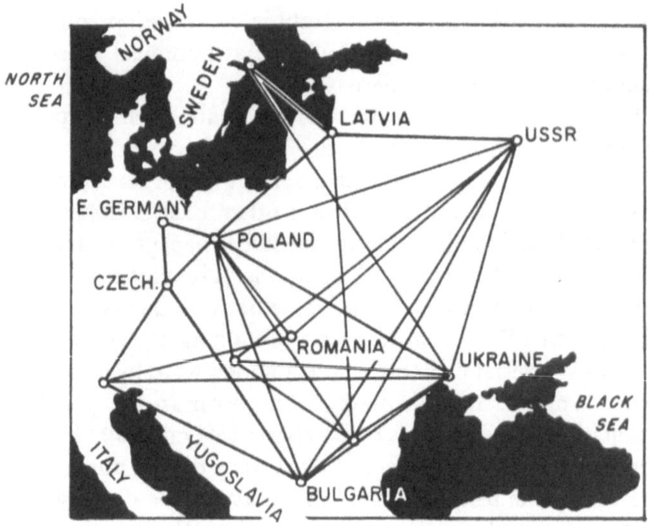

Fig. 17.2. USSR-Space Triangulation Program—1964

Other published Soviet data obtained from U.S. satellites include the synchronous observations of Echo 1 reported by Amelin[1], who also obtained the geocentric coordinates of the observation stations to ±40–60 meters. Later Kutuzov[17] reported that from May 20 to July 1 1963, 13 stations of the Soviet Union obtained more than 2,000 photographs of Echo 1. Some 970 of which were selected for geodetic processing. The results have shown that in practice, using NAFA 3s/25 cameras, the accuracy of determination of the satellite's position is ±4 seconds, and accuracy of time recording is approximately ±0·005 sec. The obtained accuracy in the determination of coordinates of space triangulation stations is of the order of ±100 meters, which is sufficient to provide horizontal control for mapping to a scale of 1:100 000 or less.

Zhongolovich's[33, 35, 37] work appears to provide the basis for most

Soviet published literature on both the geometric and dynamic satellite methods. Several other Soviet scientists have discussed space triangulation by simultaneous observations[1–4, 8, 10]. In most cases the work followed the same lines of development used in the West. An exception to this is, perhaps, a new approach, developed by Batrakov[4], to determine the relative position of the stations by using, in addition to the topocentric data (α, δ and ρ), their time derivatives ($\dot{\alpha}$, $\dot{\delta}$, and $\dot{\rho}$). A similar Western solution is known for $\dot{\rho}$, obtained from Doppler observations. Useful formulas relating to the use of synchronous satellites for geodetic azimuth control were developed and described by Razumov in 1963. For such satellites, however, to have practical geodetic applications, accurate ephemerides are required.

The accuracy of Soviet photographic observations was limited in 1962 to ± 100 meters in measurements of topocentric distances, and 5–10 seconds in angular measurements of satellites, according to Amelin[1]. Later [11, 13, 22] it was reported that errors in unknown station positions using NAFA–3s/25 cameras to be ± 30 to ± 50 meters including the remaining (after corrections) errors in reference station and synchronization. The synchronization problem can be avoided with the use of a flashing satellite. As early as 1958 Gindin, *et al.*[11], and Leykin[18] reported on such a satellite and calculated the minimum energy requirements, positioning accuracy, timing and field of view of both USSR NAFA–3s/25 and U.S. Baker-Nunn cameras. Although these paper predate similar U.S. discussions, the Soviets have not reported a launching of a satellite similar to those of the flashing light U.S. ANNA and GEOS series. Other authors reported the accuracy of observations and adjustment techniques[14, 22, 25, 30–31] to be of the order of about ± 3 μ in images of satellites and stars on the photographs[31], which corresponds to satellite angular coordinates of $\pm 1{\cdot}5$ to 2 seconds. The techniques and approaches used have been described previously by Western observers. It is believed that the Soviets have achieved about two to three seconds of arc in position and one millisecond in timing accuracy, which would limit tracking accuracy to about seven meters.

Several authors reported and reviewed various observations of other U.S. satellites, including Masevich[19] in 1961 and Fershtman[9] in 1964 on ANNA, Shpitsberg[27] in 1962 on Transit, and Polezhayev[23] in 1965 on SECOR. The Soviets continue to show interest in U.S. Transit navigation satellites, but they have not published any results from the use of these vehicles. Also, no Soviet publications are available indicating that the Soviets have launched a satellite for navigation purposes. However, in the New Five-Year Plan for Soviet Space Science, Siforov[28] states that

"Equipping satellites with superior electronic apparatus will lead to improved sea and air navigation. A vessel at any point in the ocean will be able to interrogate a satellite and obtain precise data on the position and velocity of motion of the satellite at this particular time. Using data on satellite motion and its orbital parameters the navigator will be able to precisely compute the positions of the vessel in the ocean".

In reference to the electronic method of satellite observation, the Soviets have published only reviews of U.S. work[9, 23, 27]. Zhongolovich[35] stressed the importance of radio measurements, and Polezhayev[23] derived the mathematics necessary to solve the simultaneous ranging equations, using the U.S. SECOR satellite. However, no observation techniques are discussed. Western work, however, predates such developments.

DYNAMIC METHOD

Soviet published results of the satellite dynamic method in terms of spherical harmonics describing the general gravity field of the earth are scarce. To date, the Soviets have not published any values for tesseral harmonics, and only a few for zonal harmonics, up to J^4. The accuracy of the results obtained from the dynamic method is dependent on the mathematical model for the various factors causing satellite perturbations. The U.S. has several groups working on the various types of zonal and tesseral harmonics from the analysis of earth satellites. Comparable groups in the USSR are not identifiable. It appears, however, that most of the Soviet geodetic efforts related to satellite harmonics are based on the earlier development of the gravity potential by Zhongolovich[33]. In 1960 he[34] published the results of observations of satellites 1957β2, 1958δ1, and 1958δ2. The approach to the problem and its solutions were well known to better accuracy at the time by Western scientists. Also in 1960, Batrakov and Sochilina[5] published some results for satellite harmonics. All the above results are reviewed in another paper by Zhongolovich[36] and constitute the total published Soviet satellite harmonic results. This information is summarized in Table 17.1, which contains also the results of Eastern European determinations of the zonal harmonics of the potential of the earth from satellite tracking data. These values appear to be quite large, as does the mean error, in comparison with U.S. results. For example, the best U.S. values at that time were those determined by Kozai in 1962, who gives the values for J_2, J_3, and J_4 as 1082·47 ± 0·06, −2·566 ± 0·012, and −1·84 ± 0·009, respectively. Kozai's standard errors, however, represent internal rather than absolute accuracy. The Eastern European nations have not published on higher harmonics. However, they emphasize possible errors caused by the customary neglect of the effect of harmonics above a certain degree[39].

Kochina[16] in 1962 used the dynamic method to determine the nature of the effect of different gravitational anomalies of the earth on the

TABLE 17.1

Zonal Harmonic (J_n) Determinations

Author	Buchar	Zhongolovich	Batrakov, Sochilina	Buchar	Buchar
Year	1958	1960	1960	1960	1962
Satellite	Sputnik 2	Sputniks 2, 3, Sputnik 3–r	Sputniks 3, 4	—	Sputniks 2, 3, Explorer 1, Vanguard 1, Vanguard 2
Disturbed Orbital Elements	Ω*	Ω, ω*	Ω	—	—
$J_2 \times 10^6$	1085·2 ±1·3	1083·3 ±0·7	1083·3 ±0·6	1083±0·4	1083·6±0·4
$J_3 \times 10^6$	—	−2±3	—	—	—
$J_4 \times 10^6$	[−2·4]*	−4·1±0·7	[−2·4]	−1·1±0·5	−1·1±0·5
Flattening	297·90	298·17	298·17	—	298·12
Reference	40	14,40·	40	14	40

* Brackets indicate assumed value
Ω = node length
ω = argument of perigee

motion of the satellite. A total of 23 harmonics were used in the expression of gravity potential, and numerical integration was carried out in all solutions. In the numerical solution of perturbations, she used an imaginary satellite moving in the earth's field at two different orbital inclinations, $i = 90°$ and $i = 65°$.

Batrakov[3] published a paper in 1963 on the use of resonant satellites to determine certain gravitational harmonics. The paper deals principally with the theory of the problem. The approach, the techniques, and the solutions are well known and have been considered previously by Kaula and others in the U.S.

References

1. V. M. Amelin, "The Possibility of the Establishment of Connection Between Different Triangulation Systems According to the Artificial Earth Satellite Observations", Byulleten' Stantsiy Opticheskogo Nablyudeniya Iskusstvennykh Sputnikov Zemli, 1962, No. 31, pp. 9–15.
2. V. M. Amelin, "Reduction of Simultaneous Satellite Observations", Byulleten' Stantsiy Opticheskogo Nablyudeniya Iskusstvennykh Sputnikov Zemli, 1964, No. 39, pp. 3–8.
3. Yu. V. Batrakov, "On the Use of Resonant Satellites for Determining the Constants of the Earth's Gravitational Field", Paper presented at the 14th International Astronomical Congress, UNESCO Palace, Paris, France, September, 1963.
4. Yu. V. Batrakov, "Determination of the Relative Position of Observation Stations Using Artificial Satellites", Astronomicheskiy Zhurnal, 1965, Vol. 42, No. 1, pp. 195–202 (Soviet Astronomy AJ, 1965, Vol 9, No. 1, pp. 149–154).
5. Yu. V. Batrakov and A. S. Sochilina, "Motion of the Rocket-Carrier of the Third Soviet Artificial Earth Satellite and the Degree of Oblateness of the Earth", Byulleten' Stantsiy Opticheskogo Nablyudeniya Iskusstvennykh Sputnikov Zemli, 1960, 7(17), pp. 6–12 (NASA TT F–74).
6. A. V. Butkevich, "Modern Geodetic Aides and Methods for Connecting Continents", Izvestiya Vysshikh Uchebnykh Zavedeniy Geodeziya i Aerofotos"yemka, 1960, No. 1, pp. 89–101 (JPRS: 3478).
7. A. V. Butkevich, "Studies in the Solution of Computing Problems in Spheroidal Geodesy", Izdatel'stvo "Nedra", Moscow, 1964, 260 pp. (JPRS: 32 750).
8. Ludoslaw Cichowicz, "Observations in Poland of Artificial Earth Satellites", Nauka Polska, 1964, No. 4, pp. 186–192 (JPRS: 27 667).
9. Yu. M. Fershtman, "Use of Artificial Earth Satellites for Geodetic Purposes", Geodeziya i Kartografiya, 1964, No. 11, pp. 67–77 (JPRS: 28 204).
10. B. A. Firago, "The Method of Astrometric Reduction of Synchronous Observation of Artificial Satellites", Byulleten' Stantsiy Opticheskogo Nablyudeniya Iskusstvennykh Sputnikov Zemli, 1963, No. 2, pp. 10–18.
11. Ye. Z. Gindin, G. A. Leykin, A. M. Lozinskiy, and A. G. Masevich, "Optical Observations of Artificial Earth Satellites", "Results of Scientific Research by Means of the First Soviet Artificial Earth Satellites and Rockets", Moscow, 1958, pp. 5–39.
12. M. I. Illenko, "Conference on Photographic Methods of Observing Artificial Earth Satellites", Astronomicheskiy Zhurnal, 1963, Vol. 40, No. 3, pp. 595–596 (Soviet Astronomy AJ, 1963, Vol. 7, No. 3, pp. 454–455).
13. T. V. Kasimenko and M. A. Lur'ye, "Conference of Observers of Artificial Earth Satellites", Vestnik Akademii Nauk SSSR, 1965, No. 6, pp. 94–95 (JPRS: 31 997).
14. A. A. Kiselev, "Application of a Smoothing Procedure of Satellite Tracking Data for the Estimation of Precise Positions", Byulleten' Stantsiy Opticheskogo Nablyudeniya Iskusstvennykh Sputnikov Zemli, 1962, No. 32, pp. 16–24.
15. A. A. Kiselev, B. A. Firago, and D. Ye. Shchegolev, "Instructions for Determining the Position of Artificial Earth Satellites From Photographs Obtained by Means of Camera NAFA–3s/25–C", Byulleten' Stantsiy Opticheskogo Nablyudeniya Iskusstvennykh Sputnikov Zemli, 1960, No. 3(13), 35 pp.
16. N. G. Kochina, "The Effect of Earth's Gravitation Anomalies on the Motion of Artificial Satellites", Trudy Instituta Teoreticheskoy Astronomii, 1962, No. 9, pp. 65–205.
17. I. A. Kutuzov, "Preliminary Reduction of Synchronous Photographic Observations of the Satellite Echo–1", Space Research V, Proceedings of the Fifth International Space Science Symposium, Florence, Italy, May 12–16 1964, North-Holland Publishing Company, Amsterdam, 1965, pp. 887–892.

18. G. A. Leykin, "Use of Light Flashes for Determining Locations of Artificial Earth Satellites", Izvestiya Akademii Nauk SSSR, Seriya Geofizicheskaya, 1958, No. 12, pp. 1520–1521 (Bulletin Academy of Sciences, USSR, Geophysics Series, pp. 889–890).

19. A. G. Masevich, "Optical Observation Techniques", "The Use of Artificial Satellites for Geodesy", North-Holland Publishing Company, Amsterdam, 1963, pp. 123–140 (In English).

20. L. I. Medvedeva and B. A. Firago, "Evaluation of the Topocentric Distances of Satellites", Byulleten' Stantsiy Opticheskogo Nablyudeniya Iskusstvennykh Sputnikov Zemli, 1962, No. 32, pp. 7–16.

21. N. N. Pariyskiy and Yu. S. Dobrokhotov, "Soviet Astronomical and Geophysical Research in Mali", Vestnik Akademii Nauk SSSR, 1965, No. 9, pp. 76–78 (JPRS: 32 786).

22. A. P. Polezhayev, "Use of Artificial Earth Satellites for Geodetic Purposes", Vestnik Akademii Nauk SSSR, 1965, No. 8, pp. 66–69 (JPRS: 32 389).

23. A. P. Polezhayev, "Methods of Calculating the Coordinates of Artificial Earth Satellites by the Measured Distances From the Initial Ground Tracking Stations", Second International Symposium on "The Use of Artificial Satellites for Geodesy", Athens, Greece, April 27–May 1 1965.

24. O. S. Razumov, "Possible Geodetic Use of a Stationary Artificial Earth Satellite", Izvestiya Vysshikh Uchebnykh Zavedeniy Geodeziya i Aerofotos"yemka, 1963, No. 6, pp. 15–34 (Geodesy and Aerophotography, pp. 324–331).

25. O. S. Razumov, "The Accuracy of Some Systems of Space Triangulation", Izvestiya Vysshikh Uchebnykh Zavedeniy Geodeziya i Aerofotos"yemka, 1964, No. 6, pp. 3–16 (Geodesy and Aerophotography, pp. 317–321).

26. D. Ye. Shchegolev, A. G. Masevich, and B. G. Afanas'yev, "Synchronous Observations of Man-Made Earth Satellite Echo I for Geodetic Purposes", Vestnik Akademii Nauk SSSR, 1964, No. 7, pp. 74–77.

27. I. P. Shpitsberg, "The Determination of a Ship's Position at Sea by Using Observations of Artificial Earth Satellites", Byulleten' Stantsiy Opticheskogo Nablyudeniya Iskusstvennykh Sputnikov Zemli, 1962, No. 31, pp. 16–34 (JPRS: 32 751).

28. V. I. Siforov, "Soviet Science in the New Five-Year Plan", Priroda, 1966, No. 6, pp. 2–3.

29. N. P. Slovokhotova, "Conference of Representatives of Socialistic Countries on Problems of Observing Man-Made Earth Satellites", Vestnik Akademii Nauk SSSR, 1964, No. 5, pp. 126–127 (TT–65–311).

30. G. A. Ustinoff, "Adjustment of a Space Triangulation", Byulleten' Stantsiy Opticheskogo Nablyudeniya Iskusstvennykh Sputnikov Zemli, 1963, No. 2, pp. 19–24.

31. L. L. Voskresenskiy, V. S. Plotnikov, and F. Ye. Sobolev, "Some Problems in Timing Photography in Determining the Coordinates of an Artificial Earth Satellite by the Method of Reference Stars", Izvestiya Vysshikh Uchebnykh Zavedeniy Geodeziya i Aerofotos"yemka, 1965, No. 1, pp. 81–90 (JPRS: 30 991).

32. N. P. Yerpylev, "The Geodetic Network and Artificial Earth Satellites", Zemlya i Vselennaya, 1965, No. 4, pp. 65–69 (JPRS: 32 431).

33. I. D. Zhongolovich, "Potential of the Terrestrial Attraction", Trudy Instituta Teoreticheskoy Astronomii, 1957, Vol. 6, No. 8(81), pp. 505–523 (ACIC TC-720).

34. I. D. Zhongolovich, "An Attempt to Determine Certain Parameters of the Earth's Gravitational Field From the Results of Observations of Satellites 1957 $\beta2$, 1958 $\delta1$, 1958 $\delta2$", Byulleten' Stantsiy Opticheskogo Nablyudeniya Iskusstvennykh Sputnikov Zemli, 1960, No. 2, pp. 1–24 (RAE 913).

35. I. D. Zhongolovich, "Earth Satellites and Geodesy", Astronomicheskiy Zhurnal, 1961, Vol. 38, No. 1, pp. 115–124 (Soviet Astronomy AJ, 1961, Vol. 5, No. 1, pp. 84–90).

36. I. D. Zhongolovich, "Review of Results of Determining Parameters of the Earth's Gravitational Field From Satellite Tracking Data", Byulleten' Stantsiy Opticheskogo Nablyudeniya Iskusstvennykh Sputnikov Zemli, 1962. No. 1, pp. 25–33 (Smithsonian Institution Astrophysical Observatory, 1964).

37. I. D. Zhongolovich, "Earth Satellites and Geodesy", Astronomicheskiy Zhurnal, 1964, Vol. 41, No. 1, pp. 156–169 (Soviet Astronomy AJ, 1964, Vol. 8, No. 1, pp. 117–126).

38. I. D. Zhongolovich, "Plan for a Common World-Wide Space Triangulation Network", Studia Geophisica et Geodetica, 1965, No. 9, pp. 185–200 (JPRS: 31 397).

39. I. D. Zhongolovich and L. P. Pellinen, "Certains Aspects de la Solution du Probleme Fondamental de la Geodesie Superieure", "The Use of Artificial Satellites for Geodesy", Proceedings of the First International Symposium on the Use of Artificial Satellites for Geodesy, Washington, D.C. April 26–28 1962, North-Holland Publishing Company, Amsterdam, 1963, pp. 341–345.

Part V

Related Topics

There are numerous related Soviet space activities which do not pertain directly to space-science research, but which are of current scientific and technical interest. Three such topics are treated in Part V. Chapter 18 reviews Soviet space tracking activities. Since information on Soviet radio and radar tracking is sparse, most of the chapter is concerned with optical tracking in the Soviet Union. Chapter 19 discusses Soviet attitudes concerning the controversial topic of the existence of extraterrestrial life. It is unique in that it is the only chapter of equal interest to both biological and physical scientists. The last chapter in the handbook, Chapter 20, treats a subject of increasing popularity, i.e., manned space stations. Although few details are available concerning planned Soviet manned space stations, the chapter reviews older configurations that have been pictured in Soviet publications.

Part V

Related Topics

Chapter 18

Space Tracking in the Soviet Union

A. G. Mourad and J. G. Stephan

INTRODUCTION

The Soviets recognized at the outset of the satellite era that the then existing equipment would not be adequate to meet space tracking requirements. Furthermore, a wide network of observing stations would be needed. The organization and fulfilment of the optical tracking portion of this requirement became the responsibility of the Astronomical Council of the USSR Academy of Sciences. Since Sputnik 1, the Soviets have used an impressive variety of optical tracking devices, from the largest astronomical reflectors equipped with modern electronic accessories to the simple binocular and naked eye. Specifically, they have established a network of nearly 100 visual observation stations and 23–30 photographic stations, located at universities, astronomical observatories, and pedagogical institutes, to carry out the optical observation program[32]. The photographic observation stations are located primarily at observatories and universities which have time service facilities.

Since the beginning of optical observation of satellites until the present (1966), an estimated total of 500 000 visual observations and 20 000 photographs (negatives) were made of Soviet and U.S. satellites in the USSR[3, 5, 6]. The extent of radio tracking activities is not known. Limited information available indicates that the Soviets do, however, employ various types of electronic tracking techniques. Figure 18.1 is a map showing the locations of the photographic stations (and a few radio tracking stations) as reported to COSPAR[2, 4].

In their optical tracking network the Soviets have used visual tracking techniques successfully to observe satellites up to the 9th magnitude, with reported position accuracies of the order of $0^\circ.1$ and timing accuracies of 0·1 second. Photographic observations have been accomplished primarily by using modified versions of the NAFA aerial camera. In general, these observations are limited to below the 5th magnitude, with the obtained position accuracies of 3–6 arc seconds and timing accuracies of 0·005 seconds. Higher accuracies reported have also been obtained with modified special cameras and medium range telescopes. The Soviets have tracked deep space probes up to the 19th magnitude with the 2·6 m telescope at the Crimean Observatory, and successful experiments have been conducted using television for tracking purposes. The reduction of photographic observations are made primarily according to the Kiselev and Deich methods. The Soviets are also quite active in making photometric observations of both U.S. and USSR satellites. The results of optical observations are published by the Astronomical Council in the Soviet bulletins "Results of Observations of Soviet Artificial Earth Satellites" and "Results of Observations of American Artificial Earth Satellites". A "Bulletin of the Stations of Optical Tracking of Artificial Earth Satellites" is also published by the same organization.

VISUAL OBSERVATIONS

Soviet visual observations of artificial satellites started in October 1957 and are still routinely performed by a network of some 80 observing stations[28, 36]. The participants are mainly students, who regularly observe both USSR and foreign satellites. These observations do not require extensive processing or great precision. Within an hour after the observations, the results are transmitted to the USSR computing center. Stations are informed of the next passage of the satellite by this computing center in a coded telegram, noting universal time, height of passage of satellite, and point of nearest approach[36]. Visual observations,

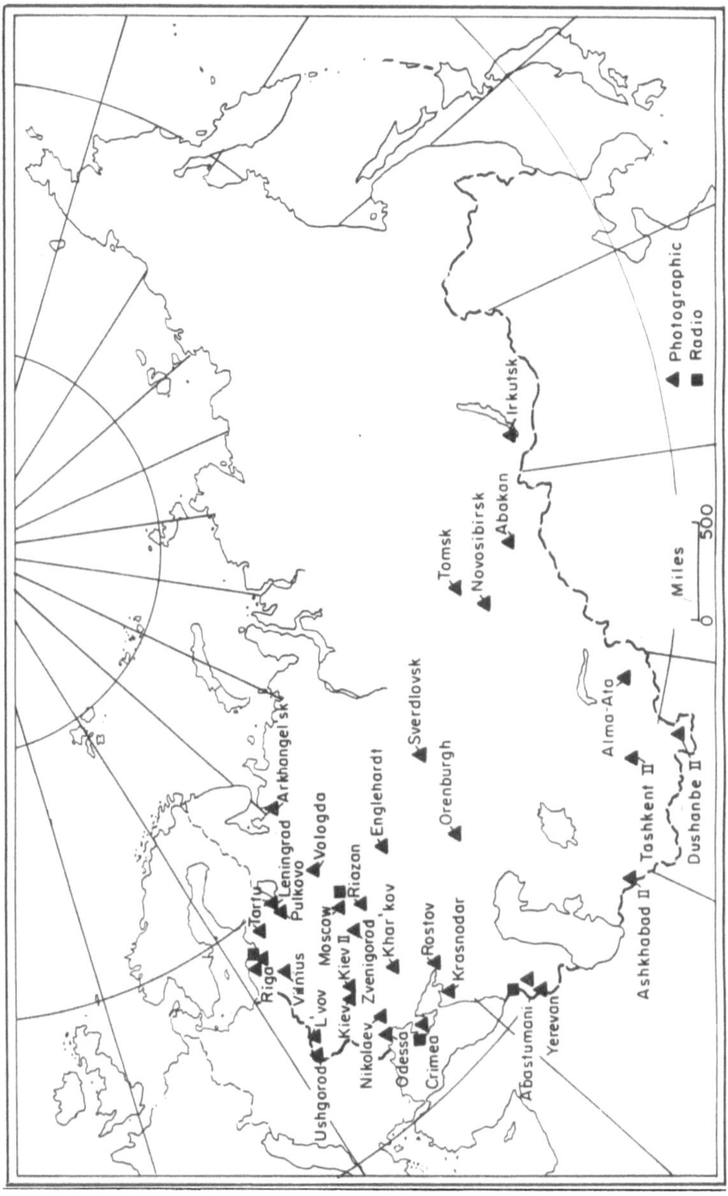

Fig. 18.1. Distribution of Tracking Stations in the USSR as Reported to COSPAR

which also serve as training and enthusiasm boosters to the Soviet youth, are primarily important in that they provide an ephemeris service. They play a substantial role, especially with respect to faint or passive satellites. However, good quality visual observations have been useful also in statistical determinations of atmospheric density, earth oblateness, etc[29]. Batrakov[11] calculated the orbital elements of Sputnik 2, using data from visual observations only, and found them to be in good agreement with King-Hele's (Britain) results based on cinetheodolite tracking data, and also with the results of the Smithsonian Astrophysical Observatory.

The Soviet requirements for visual observations are:

1. Position of satellite of second to eighth magnitude to be determined with a precision of $0\overset{\circ}{.}5$ to 1°
2. Time of passage to be determined to within 0·5 to 1 second
3. Results to be reported to the computing center with a minimum of delay[28, 36].

The practical accuracy achieved in visual observations is $0\overset{\circ}{.}2$ to $0\overset{\circ}{.}3$ in position and 0·2 to 0·3 seconds in timing[30], although $0\overset{\circ}{.}1$ in position and 0·1 second in timing are often achieved[29, 31].

Visual observations are generally made with special AT–1 telescopes and by means of binoculars "TZK"[30]. The AT–1 telescope is a small wide-angle instrument with an objective aperture of 50 mm, 6× magnification, and a field of view of about 11°. The telescopes are each equipped either with a tripod or a table base. In its field of view, there is a system of rings with a 1° interval and cross hairs with 20 minute graduations. The stations are equipped with stop watches, chronometers, two-speed magnetic tape recorders, PRV radio receivers, audio frequency oscillators, and telegraph keys, to help in recording and processing observations. Exact time signals are transmitted to the stations by radio, telephone, or wire[28, 36]. Time signals are recorded on high-speed magnetic tape. The same tape records simultaneously the signals given by the observer, who presses the telegraph key connected with the sound generator at the instant of the satellite's passing through a visible stellar configuration, or through the cross hairs of the telescope. When observations are completed, the tape recording is played at low speed and the instant of observation is determined with the aid of a stop watch. The coordinates of the satellite are determined by means of stellar maps, either of the Mikhaylov atlas, or of the Becvar (Bechvarzh) atlas.

In addition to visual observations, several stations are equipped with small tripod-mounted cameras[36]. At the instant of appearance of the satellite over the horizon the camera is quickly turned in that direction and the shutter is opened for a short period (2–5 sec) when the satellite passes in the field of view. Stop watches, chronographs, and magnetophones are used to record the time of beginning and ending of each exposure. The developed film is then projected on a stellar map of the Becvar atlas, where the scale of the map and projection are made to coincide. The equatorial coordinates of the satellite are thus determined.

The basic method of computing a satellite orbit consists of improving elliptical elements and their secular variations according to observations over 10 to 20 days[28, 36]. Each system of orbital elements is derived from tracking data, assuming a certain theory of the motion of artificial satellites, which may differ from one to another. An example of the expressions for secular motions of the node is given by Batrakov and Proskurin[12]. Orbit calculations are made at the Institute of Theoretical Astronomy in Leningrad. The method of deriving orbital elements developed at this Institute is given by Zhongolovich[45], Batrakov, *et al.*[12], and Masevich[30].

INTERNATIONAL OBSERVATIONS (INTEROBS)

The need for international participation in satellite tracking was recognized by the Soviets soon after the launching of the first satellite. As part of a limited international cooperative effort in making visual observations, the Soviets established a simultaneous visual tracking program, called INTEROBS. The purpose of this program is to observe low satellites of the "Cosmos" series for the determination of short-period variations in perigee[33], and thus corresponding atmospheric density variations. The program involves stations in the USSR, Poland, Bulgaria, Czechoslovakia, Hungary, and East Germany. According to Masevich[33], two independent programs, INTEROBS–1 and INTEROBS–2 have resulted. INTEROBS–1 is a cooperative effort of the following stations: Bautzen and Rodewisch (DDR), Brno and Skalnate-Pleso (Czechoslovakia), Krakow and Olsztyn (Poland), Cluj (Rumania), Stara Zagora (Bulgaria), Baja and Budapest (Hungary), Riga, Kiyev, Krasnodar, Dnepropetrovsk, and Chernovtsy (USSR). INTEROBS–2 consists of Riga, Arkhangelsk, Vologda, and Ryazan. Riga is a link to both programs.

The simultaneous tracking data are reduced according to a method proposed by Ill (Baja, Hungary)[33]. Kasimenko[19] compared Ill's method with the geometric method (used in reductions of synchronous photographic observations) using data from several stations which observed both methods. The comparison of the two reduction methods showed that the difference is of the order of ±1 km in the obtained radius-vector of the satellite. Stations from both INTEROBS–1 and INTEROBS–2 joined in 1963 in observing Cosmos 8, 11, 17, and 19, and the cabin of Spaceship-Satellite 1. These cooperative programs expanded in 1964 with respect to the number of stations and observers[33], including those of other countries. The first volume of the results of satellite tracking by the INTEROBS program was published in 1964 by the Hungarian Academy of Sciences[5]. The second volume was published in 1965[6].

PHOTOGRAPHIC OBSERVATIONS

Exact photographic observations of satellites are necessary to solve many problems associated with the investigation of the earth's atmosphere, the shape of the earth, and other geodetic problems. Because of the relatively high angular velocity (1°/sec) of a satellite, the development of sophisticated photographic equipment, accurate timing techniques, and specialized reduction procedures, has been required for space tracking in both the U.S. and the USSR. In the Soviet Union photographic observations of artificial satellites have been carried out systematically since at least 1958[36]. The first photograph of a space vehicle was that of the carrier rocket of Sputnik 1, obtained on October 10 1957[28] at Pulkovo Observatory. Initially, the accuracy of Soviet tracking cameras was reported to be about one minute of arc[29]. However, by 1961, according to Masevich, an accuracy of six seconds of arc in position and about $0\overset{s}{.}002$ in timing relative to USSR standard timing was achieved. To obtain this accuracy existing systems of shutters for astronomical telescopes and aerial cameras were modified and new systems were developed[29].

The backbone of the Soviet satellite tracking program are the NAFA-type aerial cameras, which range up to 25 cm in aperture. These are used in a variety of ways, from simple direct photography to the latest techniques, including the use of automatic shutters, timing devices, and a variety of mountings. In many instances these cameras are used in conjunction with the smaller aperture (up to 5 cm) refractors employed in the visual observation program. The detailed characteristics of the NAFA and other Soviet tracking cameras are listed in Table 18.1. Photographs of several of the instruments are shown in Figures 18.2 to 18.5.

TABLE 18.1

Soviet Optical Tracking Devices

Nomenclature/ Instrument Type	Optical Characteristics	Detection Capabilities, stellar magnitude	Positional Accuracy, seconds	Timing Accuracy, seconds	Station Location	Description	References
NAFA-3s/25/ modified aerial camera	Objective lens: "Uran 9" with f = 25 cm D = 10 cm Field of view: 32° × 52°	3–5, depending on the angular velocity of the satellite	4–6	$2–5 \times 10^{-3}$	Standard equipment at most photographic tracking stations	This camera photographs with high speed film (GOST rating of 1100) an object of $2^{m}.5$ with a tracking rate of 1°.5/sec. This camera uses a special modified shutter (T = 0·0002–0·0003 sec) to provide sharp satellite traces. To obtain several exposures on one frame automatic film cycling has been eliminated. The camera is equipped with a frame viewfinder and a special support, which permits its rotation around the horizontal axis. For marking the exact time, a PRV receiver and a printing chronograph are used. An azimuth mount for this camera was installed at the L'vov tracking station. As a result speed of observation and camera stability have been enhanced.	Abele (1962)[7] Logvinenko (1964)[25] Masevich (1961)[28–30] (1962)[31]
NAFA-6/50/ camera				NO DATA GIVEN			
NAFA-21/ camera	f = 21 cm Relative aperture: 1 : 4·5 Frame format 13 × 18 cm	Stars ≦9	—	—	Novosibirsk	This camera is mounted on the alidade of a Kern Universal instrument. The object finder is the search telescope from the TZK. Camera shutter operations proceed in one of two modes: Instantaneous exposure of ≦0·08 sec or time exposures are possible.	Gindin, *et al.* (1962)[18] Merkushev (1963)[34]
NAFA MK-75/ camera	D = 20 cm f = 75 cm	—	—	—	Zvenigorod	—	Abele (1962)[7]
TAFO-AL-75/ automatic camera	—	9·0–9·5	—	—	Designed in Riga and installed at the satellite observation station of the Latvian State University	This camera photographs the satellite by means of a cassette which oscillates along the direction of the satellite's motion, so that the satellite image appears as a dotted line. Data reduction of the photograph negatives is performed by electronic computer.	Lur'ye (1962)[26] Yerpylev (1962)[14]

TABLE 18.1 (Continued)
Soviet Optical Tracking Devices

Nomenclature/ Instrument Type	Optical Characteristics	Detection Capabilities, stellar magnitude	Positional Accuracy, seconds	Timing Accuracy, seconds	Station Location	Description	References
TAFO-AL-75/ three-axis automatic camera	Camera lenses D = 21 cm f = 75 cm Field of view: 4° × 5° Employs a Uran 16 lens with a resolving power up to 32 lines/ mm	8–10 10–11 (with Uran 12 lens and 1 sec exposure)	2–4	$1–2 \times 10^{-3}$	Developed at Riga University Observatory	This camera employs a hachuring film holder and an automatic guidance system. The speed of the satellite is compensated by the movement of the photographic plate which keeps the image on the same spot long enough to obtain sufficient exposure. The three-axis camera is utilized to eliminate the need for an accurate ephemeris. Frame motion, camera rotation and shutter operations are controlled by an electronic programing device. Up to 17 photographs are possible during each satellite passage with each 60-mm frame containing four to six images.	Abele (1962)[7] Abele, *et al.* (1962)[9] Gindin, *et al.* (1962)[18]
KPP/Camera	Utilizes Uran 12 objective	≦7·0	12	2×10^{-3}	Pulkovo Observatory	Moving film camera including: (1) use of "Uran 12" objective, (2) introduction of a fourth degree of freedom (rotation about the objective optical axis), (3) increase in the tracking rate from 0°.004/sec to 1°.5/sec, (4) electronic recording of time, (5) film motion occurs perpendicular to the satellite's motion.	Abele, *et al.* (1964)[9] Panaiotov (1962)[38] Panaiotov, *et al.* (1963)[39]
KPP/camera	D = 20 cm f = 25 cm	6–7	$\Delta\alpha \cos\delta = \pm 6$ $\Delta\delta \cos = \pm 4$	$\pm 2 \times 10^{-3}$	Uzhgorod, Pulkovo, Sverdlovsk	Moving film camera.	Chikarenko, *et al.* (1962)[14]
Zenith C/ camera	Uses Uran 9 lens f = 25 cm D = 10 cm Field of view: 30° × 50°	—	—	—	—	Mounted on equatorial equipment. Utilizes 35-mm photography.	Liygant (1963)[24] Mozhzherin (1963)[35]
Astrograph/ astronomical telescope	D = 40 cm f = 160 cm Field of view: 10° × 10°	—	—	—	Pulkovo and Crimea	Special film holder used. Satellite tracking hampered by inadequate accuracy of the ephemerides which are computed for use with wide-angle instruments.	

Nomenclature/ Instrument Type	Optical Characteristics	Detection Capabilities, stellar magnitude	Positional Accuracy, seconds	Timing Accuracy, seconds	Station Location	Description	References
Maksutov/50-cm meniscus telescope	f = 1200 mm D/f = 1 : 2·4	—	—	—	Alma Ata	—	
Television	Helios 53 lens D = 80 mm f = 200 mm	8–9	—	—	—	—	
Meniscus telescope coupled with highly sensitive television apparatus	f = 6·5 cm D = 50 cm Field of view: 15′	Observation of Cosmos 41: 14–15 stars: 13–14	—	—	Crimean Astrophysical Observatory	Employed image (brightness) amplifier, printing chronograph, and two kinescopes.	Agapov, *et al.* (1965)[10]
AT-1/standard visual (Moonwatch) wide-angle telescope	Aperture: 50 mm 6 × magnification Field of view: 11°	$\leqslant 9$	360	0·1	All visual tracking stations	In the field of view of the scope there is a system of rings with 1° intervals and cross hairs with 20 min graduation.	Masevich (1961)[28]
TZK/binocular on theodolite mounting	Field of view: 7°	$\leqslant 9$	360	0·1	—	Can be equipped for photographic recording of scale readings.	Blagonravov (1962)[13] Sentsova (1962)[11]
Leningrad Miniature Camera/ AT-1 and TZK telescopes	"Helios 40" objective f = 84·6 mm D/f = 1 : 1·5	Stars: $\leqslant 11{\cdot}0$ Satellites: $\leqslant 4$ with $v = 1°$/sec	$\alpha = 0{\cdot}2$ $\delta = 6$	—	Dnepropetrovsk	Time of the opening and closing of the shutter and the covering of the objective by a mechanical vane are recorded by a chronograph with a synchronous contact. Up to five breaks in a satellite trail on one photographic frame can be obtained, with 3–4 frames for a satellite passage.	Chikarenko, *et al.* (1963)[14] Chirtsov, *et al.* (1962)[15]

Fig. 18.2. NAFA MK–75 Camera[7]

Fig. 18.3. NAFA 3s/25 Camera[7]

Fig. 18.4. KPP Camera mounted on Twin Short-Focus Astrograph at Pulkovo[7]

Fig. 18.5. TAFO–AL–75 Camera[7]

To overcome the low light-gathering power of the NAFA camera, the Soviets have frequently used Maksutov-type cameras up to 70-cm aperture and focal ratio f: 2·5 to f: 3·0 to observe both earth satellites and intermediate-range space probes. This type of camera is fast, has great light-gathering power, and a fairly large field of view. It has the advantage of maneuverability, because its tube is relatively short. For example, a telescope with an effective focal length of 250 cm can have a tube length of 45 cm. When coupled with some of the electronic devices now available in the Soviet Union, this type of camera could be an outstanding tracking tool.

At the beginning of the satellite era the Soviets recognized that equipment at the astronomical observatories was inadequate for tracking purposes. Several innovations and devices had to be developed and tested. One of the most useful devices the Soviets have added to some equatorially mounted telescopes is a rotating or repeating rhythmic shutter. When the shutter opens and closes at a given rate, a series of dashes is superimposed on the star field of dots. If in addition the time of the beginning and ending of exposure is recorded, the time at which the satellite was at any particular point in its path can be determined. L. A. Panaiotov[37], at the Main Astronomical Observatory (Pulkovo), developed a camera with a moving film to record faint satellites. A photograph of the stellar field is taken on the fixed film before (and after) the passage of the satellite. Then the film is moved with the same speed and in the same direction as the image of the satellite in the focal plane, and photographs of the satellite with corresponding timing are taken. It is possible to obtain photographs of satellites up to sixth or seventh stellar magnitude with this camera[30]. Other refinements include the adaptation of such special devices as electron optical image converters and television cameras to photographic tracking.

Photographic observations for tracking deep space probes which are very faint can be made only by means of large telescopes. The problem here is no longer that of encompassing a wide field of view but rather that of the light gathering power, which is usually solved by the use of reflectors (one meter or larger). The Soviets have made use of some of their largest telescopes for photographing deep space probes. The recently installed 2·6-m telescope located at the Crimean Observatory has accurate stable mounts, electronic brightness amplifiers, and other automatic devices. This telescope, with an attached image converter, was successfully used to receive several good images of the Luna 4, and Mars 1 spacecraft and its rocket, with 1–3 seconds exposure time[6, 40]. The visual brightness of both objects was about $14–15^{m}$. The Soviets have reported[6] photographing space vehicles up to the 19th magnitude, moving with a speed on the order of one second of arc per second in the dark sky, or vehicles up to the 16th magnitude located at a distance of 20–30° from the moon in the first quarter. Coordinates of the Luna 7, Luna 8, and Zond 3 probes at distances of 100–150 000 km were determined using this telescope[6, 17]. Analysis of the data indicated a probable error of ± 5 seconds.

The Soviets are currently constructing a six-meter telescope which may find application to deep space tracking. In addition, the Soviets have long expressed an interest in applying such developments as image converters, television, and lasers to astronomical observations. The degree that such developments have been operationally incorporated into the Soviet tracking program is not known. Experiments using television for tracking satellites up to the 9^m using lens of only 80-mm diameter have, however, been reported in Soviet literature[5, 40], as has television detection of the Luna 7 payload and its rocket carrier[6]. Using this technique, the coordinates of Cosmos 41 were determined at the Crimean Astrophysical Observatory with a 500-mm meniscus telescope ($f = 6{\cdot}5$ m)[10]. The highly sensitive television apparatus has a field of vision of 15 minutes of arc and can photograph stars of the 13^m or 14^m. It also has two kinescopes, one for visual and the second for photographic tracking. After the detection of the satellite on the kinescope screen, the telescope is kept on the satellite, which is simultaneously photographed on the second kinescope. Using this technique, 13 positions of the satellite at an altitude of 40 000 km from the earth were obtained. The coordinates were determined by an interpolation method using three stars. The probable error of one determination is about 10″.

The Soviets employ several methods for determining the precise coordinates of satellites. The method of reduction used depends on the satellite and the star images. For example, with fixed cameras, the method used is that employed in meteor astronomy, in which the coordinates of crossing points of the satellite track are measured. The two methods apparently most commonly used are those developed by A. N. Deich[16] and by A. A. Kiselev[20–21]. Turner's and Schlesinger's methods of reduction are also used[18]. The computations are generally made on a Ural–1 computer. Deich's method involves the direct determination of the photographic position of an artificial satellite from formulas which give the equatorial coordinates (α and δ) of two reference stars, identified from a star catalog (Boss's 1950). In photographing a satellite, its track on the plate is obtained, with gaps of known time. A straight line is usually drawn connecting the two reference stars. The intersection of this line with the satellite track forms a nodal point, the coordinates of which α_0, δ_0 are determined by formulas given by Deich. Difficulties may arise in interpolating to obtain the time at the point of intersection. If this point is not centered in a gap, the systematic errors of these ends will distort the interpolated time. To avoid this, two other reference stars are chosen and the coordinates α_0 and δ_0 are determined for the second point of intersection of the track. Between two such points the coordinates of all gaps, formed by the opening and closing of the shutter at fixed intervals of time, can be determined. This method is convenient for calculation, and it can provide accuracies of one second in α and about 12 seconds in δ. For this accuracy, and with a declination of $\leqslant 60°$, the calculation of terms in the series up to the second order is sufficient. For higher declinations, however, it is necessary to include third order terms.

A new interpolation method of reducing photographic plates of wide-angle cameras in order to determine the positions of individual objects was developed by A. A. Kiselev[20], of the Pulkovo Observatory. This method corrects for the errors resulting in determining the position of the optical center, from the distortion of the objective lens, and from the refraction. In addition, the end and beginning of the tracks, and the corresponding time marks are determined by a smoothing technique involving the photographing of about 10 separate tracks near the center of the field of view. The advantage of this method is that it can be used to obtain the required precision, irrespective of the declination of the satellite.

Kiselev[20] developed equations with rigorous solutions to the problem of determining the position of a celestial object from photographs with two or three reference stars. His fundamental formulas of interpolation permit the determination of the position of the celestial object on the arc of a great circle connecting two reference stars. The coordinates of the reference stars are obtained from a star catalogue. The solution is then extended to the case of a central projection of the sphere on a plane, providing an ideal photographic projection on a plate. If approximate coordinates of the optical center are known, or distances from the optical center to the ends of the interpolation segment are measured, then the necessary formulas for the solution of the interpolation problem for an ideal photograph are given. The solution is further applied to an actual photographic projection. The interpolation method can be used for the determination of the position of a point object in a photograph by means of a double interpolation using three reference stars, as well as for the determination of the position of individual points in a photograph of a meteor trail or a satellite track. The latter is made with a minimum number of interpolations. Kiselev[20] has further developed equations describing the influence of errors in the adopted position of the optical center of the plate on the position of the object calculated by the interpolation method. These equations also allow the position of the optical center of the plate to be found from the measurement of several reference stars.

Kiselev[21] published another article in 1964 wherein he applied vector algebra to solve analytically the three main problems of photographic astrometry: (1) computing tangential stellar coordinates from spherical coordinates; (2) deriving the position of an isolated object in rectangular coordinates; and (3) determining rigorously the optical plate center through vectorial specification by four reference stars. The new solution of these problems is aimed at its immediate application to any working astronomical photograph, especially those of wide angle and short focus, and for use with electronic computers.

In 1965 Kiselev[22] published an article on the application of homography and photographic astrometry. In this article he considered the problem of expressing the measured coordinates on one plate in the coordinate system of another. He developed a solution, with an illustrative example, to this problem which does not require the knowledge of the optical center of the plate.

Several corrections are usually applied to the photographic measurements. These include:

> Diurnal correction for the right ascension of a satellite, for photographs taken with fixed cameras.
>
> Systematic error correction for reduced photographs. These errors are due to deviations of actual photo projection from its mathematical ideal, distortions due to refraction and aberration, and errors in determining the optical center.

For the NAFA-type cameras, corrections due to distortion of the objective lens and optical center are most important.

PHOTOMETRIC OBSERVATIONS

Most of the Soviet photometric observations of the apparent brightness of satellites and rockets are carried out in the USSR at the optical tracking stations. Visual, photographic, and photoelectric methods are used for such observations. Satellite brightness is determined photographically with fixed cameras by comparing satellite tracks with the surrounding stars[27]. In this method consideration must be given to the difference in velocities of the satellites and the stars and the relationship of photoemulsion sensitivity to the exposure time. The Soviets[28] proposed to determine satellite brightness visually by comparison of stars along a satellite orbit. For evaluation, he used the Pickering method of variable stars, where the brightness of the object is given in tenths of the brightness of the stars to which it is compared. For satellites this estimate is made over 2–3 seconds, which requires skillful observers.

Several useful studies resulted from photometric observations of bright satellites at the Odessa Observatory. Data on maximum brightness of the second Soviet satellite permitted the determination of the period of rotation about its axis[28]. This average period of rotation was found by scientists[27] at the above-mentioned observatory to be 208·5 seconds, as compared to 180 seconds at its start. Photometric observations also allow studies of the dependence of satellite brightness and color on atmospheric conditions, and also of the change in the conditions of the satellite surface due to solar and other effects. According to Masevich[28, 31], to obtain certain results from photometric observations, statistical reduction of large amounts of data collected over long orbits is needed. For example, M. Lur'ye at the Astronomical Council of the USSR Academy of Sciences statistically analyzed Sputnik 2 and 3 data from 70 Soviet tracking stations with respect to dependence of observing conditions on geographical latitude. The results showed seasonal correlations and clearly identified the effect of "white nights" on observing conditions, particularly in the northern latitudes.

The results of Echo 1 photometric observations permitted the detailed investigation of its surface conditions. For elongated cigar-like satellites there is evidence for a correlation between the period of its brightness variation and solar activity[29]. The Soviets conducted photometric observations of Echo 2 during its inflation in the portion of the orbit where it was visible only in the USSR. Several Soviet tracking stations which participated in the observation of the first three orbits sent their results to the U.S.[5]. The Soviets are continuing their photometric observations of Echo 1 and Echo 2 when they enter the earth's shadow to determine atmospheric ozone height distribution[6]. Also they observe rocket carriers of the Cosmos series to study the correlation of the rotation period and solar activity[3, 6]. Analysis of the observations obtained at various points on the earth is done at the Odessa Observatory[27–28].

RADIO TRACKING

Since the beginning of the satellite era, published Soviet information on radio tracking methods has been limited to the discussion of theoretical concepts, to reports of some basic experiments, and in connection with their IGY rocket and satellite program[1], to the discussion of several possible radio techniques. However, it is not clear to what extent these techniques are applied in the USSR. The Soviets do, however, freely discuss the results of U.S. radio tracking activity.

As of 1964 the Soviets had published the locations of only four radio tracking stations in the USSR[4]. The name of the station, the radio system used, and the timing accuracy, insofar as reported are: Riga (radio Doppler, 183 mc/s, ±0·1 sec), Moscow (radio Doppler, ±0·5 sec), Abastumani (20 mc/s, ±0·1 sec), Crimea (radio telescope, 200 mc/s, 3000 mc/s, and radio spectrograph, 100–150 mc/s, ±0·01 sec). Although requested by COSPAR, no associated positional accuracies for these stations have been reported.

A few published articles indicate the use of some of these techniques. For example, it is reported that Vitkevich, *et al.*[43], used the radio interferometer method (183·6 mc/s) to determine the coordinates of the first, second, and third Soviet space rockets. The intensity of the signal was also determined using radio astronomical methods. According to Kotel'nikov, *et al.*[23], theoretical as well as experimental work employing the Doppler effect to determine the orbital parameters of satellites was carried out by the Soviets, using a signal frequency of 40·002 mc. The choice of this frequency reportedly was made because it is least affected by the ionosphere and the Doppler effect is more pronounced. The accuracy of determining the time of closest approach was reported to be 0·2–1 second. The accuracy of velocity and inclined distances were determined to 3–5 per cent. The authors further indicate that errors in timing can be reduced to hundredths of a second if time markings are made using automatic frequency comparisons. Similar accuracy in results was also reported by Shmidt[42] on using the Doppler shift to determine the orbital elements of the third Soviet satellite.

The Soviets have established an amateur radio tracking network to receive satellite signals and to determine the orbit of the satellite by observing the records of the Doppler changes in the frequency of the received signals.

References

1. "The IGY Earth Satellite Program", Annals of the International Geophysical Year, Manual of Rockets and Satellites, Pergamon Press Ltd., London, New York, Paris, Los Angeles, 1958, pp. 115–452.

2. COSPAR Information Bulletin, August 1962, No. 10, 82 pp.

3. "Investigations of the Upper Atmosphere and Outer Space Carried Out in the USSR in 1963", Report to COSPAR presented by A. A. Blagonravov at the Fifth Meeting, Florence, Italy, May 1964.

4. "COSPAR World List of Satellite Tracking Stations. Part II—Radio Stations", COSPAR Information Bulletin, COSPAR Working Group on Tracking, Telemetry, and Dynamics, April 1964, No. 18, 55 pp.

5. "Investigations of the Upper Atmosphere and Outer Space Carried Out in the USSR in 1964", Report to COSPAR presented by A. A. Blagonravov at the Sixth Meeting, Mar del Plata, Argentina, May 1965.

6. "Investigations of the Upper Atmosphere and Outer Space Carried Out in the USSR in 1965", Report to COSPAR presented by A. A. Blagonravov at the Seventh Meeting, Vienna, Austria, 1966.

7. M. K. Abele, "A Three-Axis Automatic Photographic Camera for Satellite Tracking", Byulleten' Stantsiy Opticheskogo Nablyudeniya Iskusstvennykh Sputnikov Zemli, Spetsial'nyy Vypusk, 1957–1962, No. 1, pp. 55–61 (Smithsonian Institution Astrophysical Observatory, pp. 83–100).

8. M. K. Abele and K. K. Lapushka, "Observing Artificial Earth Satellites Using a Triaxial Tracking Camera", Byulleten' Stantsiy Opticheskogo Nablyudeniya Iskusstvennykh Sputnikov Zemli, 1962, No. 29, pp. 5–7 (JPRS: 25 462).

9. M. K. Abele and K. K. Lapushka, "A Device for Photographing Artificial Satellites, Class 57, No. 1626748", Byulleten' Izobreteniy i Tovarnykh Znakov, 1964, No. 10, pp. 84–85.

10. Ye. S. Agapov, V. F. Anisimov, S. G. Braunfeld, V. M. Mozhzherin, V. V. Prokof'yeva, and S. M. Sinenok, "Determining Coordinates of Far Off Satellites With the Aid of a Television System", Abstracts of the Scientific Papers to be presented at the 16th International Astronautical Congress, 1965.

11. Yu. V. Batrakov, Byulleten' Stantsiy Opticheskogo Nablyudeniya Iskusstvennykh Sputnikov Zemli, 1960, No. 7, p. 3.

12. Yu. V. Batrakov and V. F. Proskurin, Byulleten' Instituta Teoreticheskoy Astronomii, 1960, Vol. 7, No. 7, p. 537.

13. "Investigations of the Upper Atmosphere and Outer Space Carried Out in the USSR in 1962", Report to COSPAR presented by A. A. Blagonravov at the Fourth Meeting, Warsaw, Poland, June 1963.

14. A. L. Chikarenko and S. A. Filincheva, "Increasing the Accuracy of Processing of Artificial Earth Satellite Photographs Obtained With a Small Camera", Byulleten' Stantsiy Opticheskogo Nablyudeniya Iskusstvennykh Sputnikov Zemli, 1962 (1963), No. 35, pp. 13–20 (JPRS: 26 916).

15. A. D. Chirtsov and G. D. Koposov, "Photographing Interrupted Satellite Trails Using the Small 'Leningrad' Camera", Byulleten' Stantsiy Opticheskogo Nablyudeniya Iskusstvennykh Sputnikov Zemli, 1962, No. 28, pp. 7–9 (JPRS: 25 462).

16. A. N. Deich, "Determination of the Photographic Position of an Artificial Earth Satellite by Two Reference Stars", Astronomicheskiy Zhurnal, 1958, Vol. 35, No. 5, pp. 810–818 (Soviet Astronomy AJ, Vol 2, No. 5, pp. 758–766).

17. P. Dobronravin, "On Observations of Distant Artificial Space Objects", Paper presented at the Seventh International Space Science Symposium, Vienna, Austria, May 1966 (In English).

18. Ye. Z. Gindin, M. I. Illenko, and M. A. Lur'ye, "Organization of Optical Satellite Tracking in the Soviet Union", Byulleten' Stantsiy Opticheskogo Nablyu-

deniya Iskusstvennykh Sputnikov Zemli, 1957–1962, Spetsial'nyy Vyusk, 1962, No. 1, pp. 83–93 (Smithsonian Institution Astrophysical Observatory, pp. 149–170).

19. T. V. Kasimenko, "Observations of Artificial Earth Satellites", Warsaw, 1963.

20. A. A. Kiselev, "An Interpolation Method for the Photographic Determination of the Position of a Celestial Object", Astronomicheskiy Zhurnal, 1959, Vol. 36, No. 2, pp. 348–360 (Soviet Astronomy AJ, 1960, Vol. 3, No. 2, pp. 341–352).

21. A. A. Kiselev, "A Vector Formulation of the Basic Procedures of Photographic Astrometry", Astronomicheskiy Zhurnal, 1965, Vol. 42, No. 2, pp. 452–463 (Soviet Astronomy AJ, 1965, Vol. 9, No. 2, pp. 354–363).

22. A. A. Kiselev, "The Application of Homography to Photographic Astrometry", Astronomicheskiy Zhurnal, 1965, Vol. 42, No. 4, pp. 831–844 (Soviet Astronomy AJ, 1966, Vol. 9, No. 4, pp. 643–652).

23. V. A. Kotel'nikov, V. M. Dubrovin, V. A. Morozov, O. N. Rzhiga, and A. M. Shakhovskoy, "Using the Doppler Effect to Determine the Orbital Parameters of Earth Satellites", paper presented at the July 1958 CSAGI General Assembly, Moscow; Iskusstvennyye Sputniki Zemli, 1958, No. 1, pp. 50–61; Annals of the International Geophysical Year, 1960, Vol. 12, pp. 880–891.

24. M. K. Liygant, "Tartu Station for Observing Artificial Earth Satellites", Byulleten' Stantsiy Opticheskogo Nablyudeniya Iskusstvennykh Sputnikov Zemli, 1963, No. 33, pp. 26–29.

25. A. A. Logvinenko, "Azimuth Mount of a Photocamera for Observing Artificial Earth Satellites", Byulleten' Stantsiy Opticheskogo Nablyudeniya Iskusstvennykh Sputnikov Zemli, 1964, No. 40, pp. 3–6.

26. M. A. Lur'ye, "Photographic Methods of Observing Artificial Satellites", Priroda, 1962, No. 12, pp. 99–100.

27. A. G. Masevich, "Astronomical Observations on Artificial Earth Satellites", Vestnik Akademii Nauk SSSR, 1959, No. 5, pp. 85–94.

28. A. G. Masevich, "Results of Optical Observations of the Soviet Earth Satellites", July 1958 CSAGI General Assembly, Moscow; Annals of the International Geophysical Year, Pergamon Press, 1961, Vol. 12, pp. 833–847.

29. A. G. Masevich, "Optical Tracking of Satellites", Space Research II, Proceedings of the Second International Space Science Symposium, Florence, Italy, April 10–14 1961, North-Holland Publishing Company, Amsterdam, 1961, pp. 3–16.

30. A. G. Masevich, "Tracking of Artificial Satellites in the USSR", Space Research II, Proceedings of the Second International Space Science Symposium, Florence, Italy, April 10–14 1961, North-Holland Publishing Company, Amsterdam, 1961, pp. 102–114.

31. A. G. Masevich, "Optical and Radio Tracking of Satellites (and Interplanetary Probes)", Space Age Astronomy, Academic Press, New York, 1962, pp. 127–140.

32. A. G. Masevich, "Optical Observation Techniques", "The Use of Artificial Satellites for Geodesy", Proceedings of the First International Symposium on the Use of Artificial Satellites for Geodesy, Washington, D.C., April 26–28 1962, North-Holland Publishing Company, Amsterdam, 1963, pp. 123–140.

33. A. G. Masevich, "Some Results of International Cooperation on Visual and Photographic Simultaneous Tracking of Satellites at USSR and East European Tracking Stations in 1963", Space Research V, Proceedings of the Fifth International Space Science Symposium, Florence, Italy, May 12–16 1964, North-Holland Publishing Company, Amsterdam, 1965, pp. 839–848.

34. V. A. Merkushev, "Experience in Photographing Satellites With a NAFA–21 Camera", Byulleten' Stantsiy Opticheskogo Nablyudeniya Iskusstvennykh Sputnikov Zemli, 1962 (1963), No. 35, pp. 12–13 (JPRS: 26 916).

35. V. M. Mozhzherin, "Crimean Station for Observation of Artificial Earth Satellites", Byulleten' Stantsiy Opticheskogo Nablyudeniya Iskusstvennykh Sputnikov Zemli, 1963, No. 33, pp. 33–34.

36. K. F. Ogorodnikov, "Preliminary Summary of Optical Observations of Artificial Earth Satellites", Proceedings of the 9th International Astronautical Congress, Amsterdam 1958, Springer-Verlag, Vienna, 1959, pp. 434–444.

37. L. A. Panaiotov, "A Camera With a Moving Film for Photographing Faint Artificial Satellites—Design and Results of Preliminary Tests", Astronomicheskiy Zhurnal, 1961, Vol. 38, No. 1, pp. 145–156 (Soviet Astronomy AJ, Vol. 5, No. 1, pp. 106–113).

38. L. A. Panaiotov, "Design, Power, and Internal Accuracy of the Pulkovo Moving-Film Camera", Byulleten' Stantsiy Opticheskogo Nablyudeniya Iskusstvennykh Sputnikov Zemli, 1962, No. 29, pp. 12–17 (JPRS: 25 462).

39. L. A. Panaiotov and T. Ye. Syshchenko, "Results of Photographic Observations of Artificial Earth Satellites", Byulleten' Stantsiy Opticheskogo Nablyudeniya Iskusstvennykh Sputnikov Zemli, 1963, No. 35, pp. 20–25.

40. V. K. Prokof'yev, "Satellite Tracking", Transactions of the International Astronomical Union (1964), Academic Press, London and New York, 1966, Vol. XIIB, p. 383.

41. Yu. Ye. Sentsova, "Processing Satellite Negatives by the Kiselev and Turner Methods on a 'Ural' Electronic Computer", Byulleten' Stantsiy Opticheskogo Nablyudeniya Iskusstvennykh Sputnikov Zemli, 1962, No. 30, pp. 11–13 (JPRS: 25 462).

42. K. Kh. Shmidt, "A Method for Determining the Orbits of Artificial Earth Satellites by Measurements of the Doppler Effect", Problemy Dvizheniya Iskusstvennykh Nebesnykh Tel, AN SSSR, 1963, pp. 205–212.

43. V. V. Vitkevich, A. D. Kuz'min, R. L. Sorochenko, and V. A. Udal'tsov, "Results of Radio Astronomical Observations Obtained With Soviet Space Rockets", Iskusstvennyye Sputniki Zemli, 1961, No. 7, pp. 23–31 (Artificial Earth Satellites, Consultants Bureau, pp. 20–28).

44. N. P. Yerpylev, "Photographic Observations of Artificial Earth Satellites", Vestnik Akademii Nauk SSSR, 1962, No. 11, pp. 129–130.

45. I. D. Zhongolovich, Byulleten' Stantsiy Opticheskogo Nablyudeniya Iskusstvennykh Sputnikov Zemli, 1959, No. 5.

Chapter 19

Soviet Attitudes Concerning the Existence of Life in Space

*N. T. Bobrovnikoff**

INTRODUCTION

The idea that intelligent beings might exist outside of the earth was debated in antiquity (Anaxagoras, Pythagoras, Plutarch, Lucian, etc.). This speculation was frowned upon by the Catholic Church as contradictory to the Christian dogma of the uniqueness of man and his relation to the universe. During the Renaissance the idea of habitable worlds was again revived (Nicolaus Cusanus, Giordano Bruno, Kepler, etc.).

The telescope showed many details on the surface of the planets which generally favored the idea of habitability. It was assumed that man was the goal to which all creation moves, and consequently the celestial bodies did not have any reason to exist unless they served as homes for intelligent beings. In the 18th century, such scientists as

* Consultant, Columbus Laboratories, Battelle Memorial Institute; Professor Emeritus, The Ohio State University.

Huygens, Fontenelle, Swedenborg, and others wrote elaborate treatises on the supposed inhabitants of other planets, and even the great philosopher Kant thought that at least some of the planets besides the earth might be inhabited.

Further development of this idea occurred in the early 19th century. Sir William Herschel, perhaps the greatest observational astronomer of all times, deduced from his observations that the sun was really a dark body which very well might be inhabited. He theorized that the brilliant surface of the sun was actually its atmosphere and the so-called sunspots were simply the solid dark surface showing through the rifts of the atmosphere. The very influential French astronomer Arago, as late as 1850, could not find anything wrong with this theory. In 1832, von Littrow accepted the idea of J. Lambert (1750), that comets were undoubtedly inhabited and their extensive atmospheres had the purpose of mitigating and preserving the heat of the sun which must vary greatly along the eccentric orbits of those bodies. Both men were leaders in the mathematical theory of comets. The German astronomer Gruithuisen could see cities and railroads on the moon, and other astronomers speculated what function the rings of Saturn might have to make conditions more comfortable for the intelligent beings which were undoubtedly there.

In the second half of the 19th century, the science of astrophysics was born and quickly showed that the conditions on the sun, moon, comets, and the majority of the planets were such as to preclude the existence of any life there. The only possibly habitable planets were Venus and Mars, and even that was highly problematical. It became unfashionable to talk about inhabitants of other planets, and Lowell's ideas about the artificial origin of the canals on Mars were generally ridiculed. A few hardy souls here and there continued to maintain that Mars must be habitable, regardless of what scientific observations indicated. In the U.S., such were E. C. Slipher and W. H. Pickering, in the USSR, G. A. Tikhov and especially K. Ye. Tsiolkovskiy. Tikhov remained essentially a scientist and tried only to prove that terrestrial plants can adapt themselves to the conditions on Mars. Tsiolkovskiy was a dreamer who threw caution to the winds. One of his books, constantly quoted by Soviet astronomers, has the revealing title "Dreams About the Earth and the Heavens". With the development of rocket technology Tsiolkovskiy became in the USSR an almost infallible authority to be quoted alongside of Lenin and Marx.

The novelists, as usual, were years behind the scientists. H. G. Wells' "The War of the Worlds" did not appear until 1905. It was (and still is) extremely popular throughout the world, and many remember the panic in 1938 when this story was dramatized on the radio. Millions of people believed the Martians were landing in New Jersey and marching on New York City. However, scientists continued to be rather cool toward the possibility of life on Mars or elsewhere outside the earth. Perhaps the lowest point in the belief of extraterrestrial life was reached in the 1920's, when Sir James Jeans showed that the collision of two stars, according to him the only possible mode of the formation of a planetary system, is an extremely improbable event, and it may well be that the earth is a cosmic freak, with some kind of mold on it called life. Doubts were soon thrown on Jeans' theory of the origin of the solar system, and quiet investigations on the origin of life on the earth and other celestial bodies continued. In this respect, A. I. Oparin's work deserves to be mentioned. He is still Director of the Institute of Biochemistry, Academy of Sciences, USSR, and is the author of many articles and several books on this problem.

The situation changed radically with the postwar development of radio astronomy, when it became possible to think of a direct contact with extraterrestrial civilizations by means of radio. The beginning of the new approach was sharply marked by the appearance in the British periodical "Nature" of a letter by two U.S. scientists, G. Cocconi and P. Morrison, "Searching for Interstellar Communications" (1959). This letter fired the imagination of many people, including one of the remarkable Soviet scientists, I. S. Shklovskiy, the author of numerous articles and several books on the subject. Shklovskiy's first book, "The Universe, Life, and Intelligence", appeared in 1962, its second edition in 1965[21]. The first edition was revised by the author, translated by Paula Fern, annotated by the U.S. astronomer C. Sagan, and published in the U.S. in 1966 as "Intelligent Life in the Universe", by I. S. Shklovskiy and C. Sagan[22].

With the First Conference on "Extraterrestrial Civilizations" (Byurakan Observatory, May 20–23 1964)[3], which included all the leaders in Soviet radio astronomy and some optical astronomers, the problem can be said to have obtained the official recognition of the Soviet Union. As in the West, the Soviet scientists realize that the whole problem hinges on the answers to three general questions:

1. What is the origin of the solar system? Without knowing this answer it is not possible to decide whether planets are rare or common around the stars.
2. What is the nature of life?
3. What is the origin of life on the surface of the earth?

In spite of a very large amount of work, both in the East and West, no definite answers to these questions are available. We have to fall back on vague arguments such as "with so many stars some of them, at least, must have planets", etc. It is impossible at the present time to prove or disprove the existence of planets of the size of the earth even around the nearest stars, let alone life on these, and flourishing civilizations. Therefore, belief in the existence of intelligent life elsewhere in the universe must be based at the present time on faith rather than on scientific fact. In this respect, scientists are in exactly the same position as their predecessors were in the 18th Century, or even the ancient Greeks 2000 years ago. The only difference considered extremely significant by the proponents of life in the universe is modern man's possession of radio communication techniques capable of reaching out to 1000 light years and more. How to utilize this capability is the subject of animated discussion among the radio astronomers in the West and in the USSR.

EXISTENCE OF EXTRATERRESTRIAL LIFE

General Attitude

The Soviets are emphatic that their materialistic philosophy is in complete agreement with the idea of extraterrestrial civilizations. According to this philosophy life is a normal and inevitable consequence of the development of matter, and intelligence is a normal consequence of the existence of life. Even the best informed scientists in the USSR, like Oparin and Shklovskiy, must necessarily subscribe to this crude philosophy promulgated more than 100 years ago by Marx and Engels. However, once having stated their materialistic point of view they often introduce reservations. Thus Oparin thinks that the presence of oceans was the necessary factor in the appearance of life on the earth, and Shklovskiy is willing to accept the existence of life only on the earth, but this would be a "miracle". From the results of a questionnaire circulated by the editors of the Soviet periodical "Priroda"[2] among astronomers and biologists in the USSR, only two out of seven authorities (Shklovskiy and Tsitsin) believe in the existence of a large number of planetary systems. Astronomers, in general, are more optimistic in regard to the existence of extraterrestrial life. Biologists, knowing something of the complexity of life on earth, are more cautious.

What Kind of Life?

The Soviets seem to be committed to life based on the hydrocarbon compounds, that is essentially the same kind of life that exists on the earth, ranging in complexity from bacteria to man. Oparin considers any other basis of life sheer impossibility, and at any rate devoid of any physical meaning[17, 18]. Shklovskiy goes into considerable detail to show by energy considerations that life must necessarily be based on hydrocarbon reactions[21]. Speculations common in the West about the possibility of life based on ammonia, or even inorganic compounds (as in Hoyle's novel "The Black Cloud" which appears to be not only alive but even intelligent) seldom occur in Soviet scientific literature although not altogether absent. Of course in both the East and West, ignorant writers produce sensational novels in which astronomical or energetical considerations are neglected, but these writers are usually not scientists.

Persistence of Terrestrial Type of Life

As conditions on the moon, Venus, and Mars are known to be severe in terrestrial terms the problem arises whether even the simplest terrestrial organisms like bacteria can exist there. Experiments to test bacteria and other simple organisms under these conditions are conducted in both the East and West, on a comparable scale. In the USSR this is done at the Institute of Microbiology, Academy of Sciences, USSR, and probably other places. There is a recent report of that institute concerning the simulation of conditions on Mars for microbial growth (A. I. Zhukova and I. I. Kondrat'yev, 1965)[32]. Experiments are also conducted on board space probes and artificial satellites. The problem has assumed considerable importance as terrestrial bacteria have been shown to possess remarkable endurance and adaptability in planetary conditions. The danger of contamination of planets by terrestrial micro-organisms therefore exists and has required international cooperation since the introduction of space exploration.

Search for Life on Mars

Mars is the only planet where conditions remotely approach those on earth. It was therefore natural that Mars became the focus of attention of astronomers and biologists looking for evidence of life elsewhere in the solar system. In the U.S. the center of the study of Mars for a long time was the Lowell Observatory, Arizona, where Percival Lowell's work was continued by E. C. Slipher. In the USSR an indefatigable searcher for evidence of life on Mars was Tikhov (1875–1960), a Pulkovo astronomer who attained a considerable international reputation for his excellent observational work. In 1909, during one of the great oppositions of Mars, he studied that planet through filters and proved the existence of snow near its poles and clouds in its atmosphere, in spite of the low position of the planet during observations. This work remained little known in the West and was repeated at the next great opposition in 1924, with substantially the same results, by W. H. Wright at Lick Observatory, California.

After his retirement from Pulkovo, Tikhov settled down in Alma-Ata, Kazakh SSR, and in 1947 formed there a "Sector of Astrobotany" at the Institute of Physics and Astronomy of the Academy of Sciences of the Kazakh SSR. The idea of this sector (or section) was to study the behavior of plants under conditions approaching those of the planet Mars, that is in the Arctic tundra and high mountains. Many astronomers and botanists worked at this section, which published five volumes of its proceedings (1947–1960)[27]. Although this work did not resolve the question of life on Mars, it nevertheless uncovered many remarkable instances of adaptation of plants to extreme climatic conditions. Tikhov's method of obtaining spectra of plants in reflected light, to compare them with the spectrum of Mars, was later employed in the West, especially with the development of the infrared techniques. With Tikhov's death his section was absorbed by the Institute of Astronomy. Tikhov's works were published in five volumes by the Academy of Sciences, Kazakh SSR[27]. They contain 33 of his own papers on the problems of terrestrial plants and existence of life on Mars.

The results of Tikhov and his collaborators were indecisive so far as the existence of plants on Mars was concerned, paralleling similar results in the West. They simply increased the probability in favor of the existence of such life. The occurrence of intelligent life on Mars is even more difficult to prove. Shklovskiy's point of view is that Mars once had a civilization which launched artificial satellites, but is now a dead body. Tikhov's work is criticized by the Soviet astronomer K. A. Lyubarskiy[15], as being based on erroneous assumptions. Lyubarskiy's book is one of the few serious attempts to come to a conclusion regarding the possibility of life on Mars. He finds that Tikhov simplified the problem by assuming that life of Mars would be of a terrestrial type. This is hardly possible, as the physical environment and consequently cosmic evolution of the earth and Mars are entirely different. In particular, Lyubarskiy insists that Martian vegetation, if present at all, must be of red color. For this he is in turn criticized by other Soviet investigators[16].

The question of life on Mars will be definitely resolved only with an actual visit there, either by instrumented or manned vehicles. For this reason, current emphasis is being given to the development of techniques for detecting the existence of life on Mars in both the U.S. and USSR planetary exploration programs. The fine discovery by Mariner 4 of craters on the surface of Mars, however, has little direct bearing on the problem of life there. The same can be said of the presumed absence of the Martian Canals.

Few astronomers believe that there can be any life on Venus or the moon. An exception is N. A. Kozyrev, a Soviet astronomer famous for his observations of the moon, who thinks that the high temperature of Venus refers to its ionosphere, and the condition of the surface may allow the development of life. But even the moon cannot be assumed to be entirely devoid of life. Such is the opinion of A. I. Oparin, the greatest authority on such matters in the USSR. Oparin thinks that organic substances either alive or dead are possible on the moon. Such an idea would probably be unacceptable in the West, but it was only 30 or 40 years ago that W. H. Pickering, an American astronomer, tried to explain various changes of tint in the moon by colonies of insects appearing and disappearing during the progress of the lunar day.

Meteorites and Life

Meteorites are the only bodies of extraterrestrial origin that are available for study in our laboratories. In connection with the problem of extraterrestrial life, a large number of mineralogists, chemists, physicists, biologists, etc., everywhere are studying meteorites. The proof of the existence of organic substances in meteorites would support the existence of life outside the earth, no matter what the ultimate origin of meteorites might be. But in this problem, as in all other problems concerning extraterrestrial life, there is no simple answer and no convincing proof of the existence of life. The problem has recently been reviewed by A. A. Imshenetskiy (1966)[9], Director of the Institute of Microbiology, Academy of Sciences, USSR, where many investigations of this nature are being carried out.

There are three facets in the study of meteorites which must be considered in this connection:

1. Carbonaceous chondrites are stony meteorites which have some carbon matter (up to five per cent of weight) of possible organic origin. At the present time there are 30 meteorites of this class which can be divided into three sub-classes quite different from each other. At first it seemed that this is indisputable proof of the cosmic origin of organic matter, but later researches proved this improbable. The carbonaceous matter is now considered to be of inorganic origin and similar to matter found in terrestrial rocks.

2. "Organized elements" in the same meteorites are small round grains which have been considered as possibly produced by plant spores. The best authority in the USSR on these problems, G. P. Vdovykin[30], does not think they are of organic origin at all.
3. Bacteria in meteorites have been reported time and again both in the East and West. In every case they were proved to be introduced into the meteorite after its fall on the surface of the earth.

SOVIET ATTITUDE TOWARD SCIENCE FICTION

The idea of inhabited worlds naturally evokes in people all sorts of emotions which are not always amenable to scientific treatment. In the Soviet philosophy, scientific fiction occupies an honorable place provided it is not misrepresented as solid achievements of science. Much of what Tsiolkovskiy wrote, for instance, can be characterized as science fiction, and one of the outstanding Soviet writers, Alexis Tolstoy, was famous for his fantastic stories. Academician Obruchev, the explorer of Siberia, was also a science fiction writer. However, some Soviet scientists, principally astronomers, are busy refuting and criticizing sensational writers who exhibit more exuberance than knowledge. One such writer is Kazantsev, the author of a fantastic tale, "Guest of out the Cosmos" (1958)[11], which has had its repercussions abroad, also. The main idea is that the Tunguska meteor, which landed in Russia in 1908, was in reality a spaceship from Mars supplied with a hydrogen bomb. This ship blew up over Siberia thus saving the earth from conquest by the Martians. Astronomer Yu. G. Perel' (1959)[20] concedes that a fiction writer may invent anything he pleases, but Kazantsev represents his wild surmises and ignorant theories as scientifically established facts. Kazantsev, however, proceeded to attack official science as concealing from the public the true situation, etc., thus closely paralleling the UFO enthusiasts in the U.S. who accuse the Air Force of suppressing evidence supporting flying-saucer visitations.

Another variety of pseudo-scientific effort is directed toward the discovery of traces of contact of higher civilization with the earth. In the USSR, M. M. Agrest in 1959 put forward an idea that classical myths and biblical stories contain vague reminiscences of visits by extraterrestrial, highly civilized beings. These are gods coming down to earth, angels flying through the air, destruction of Sodom and Gomorrah (evidently by an atomic bomb), kidnapping of people (the biblical Enoch) by the intruders, etc. Little of value is, however, expected to come from such considerations. The search for information, however, is not restricted to the Bible. Anything is good if it points toward the existence of extraterrestrial civilizations; crude images on rocks in the Sahara, mythical small men in China, Peruvian fairy tales, are examples. More recently, in a Soviet popular magazine there is an article by Vyacheslav Zaytsev[31], which is full of such stories. It is stated that the writer spent 30 years of his life collecting this information. He is also the author of the fantastic book "Myths, Legends, and the Cosmos" (1965).

To the credit of Shklovskiy (second edition of his book, Chapter 23)[21] he refutes many of the ridiculous stories which have been propagated very assiduously in the West, particularly in the U.S., and which have been adopted by the adherents of the UFO cult. Other serious Soviet writer-scientists like V. N. Komarov (1966)[13] also exhibit an exemplary caution. In general, it appears that the problem of sensationalism in science is exactly the same both in the USSR and the U.S. There are scientists interested in the problem of extraterrestrial civilizations, and there are writers who want to publish a breath-taking book. There are even combinations of the two. Also, modern science is so fantastic that the boundary between possible and impossible is fairly indistinct. Some people, sometimes even bona fide scientists, simply cannot discern this boundary and mix up solid science, their unconscious desires, and fairy tales into the nightmarish whole. The Soviets cannot escape this situation any more than the Americans and Europeans.

POSSIBILITY AND MEANS OF ESTABLISHING CONTACT

In view of the complete absence of concrete data on extraterrestrial civilizations, the only possible formulation of the problem is this: Assuming that there are extraterrestrial civilizations, what would be the best way of getting in touch with them? This problem is twofold: (1) how can understandable signals be transmitted and (2) how can signals from outer space be detected and interpreted? Radio signals from other civilizations, no matter how clear and strong, would have had no significance for us 50 years ago, as nobody on earth could intercept, let alone interpret them. According to modern astrophysics the development of stars is a continuous process and they certainly were not created all at the same time. If there are planets around them, if there is life on these planets, and if there are civilizations, they must be in all stages of development. The extraterrestrial civilizations obviously should be in the same or higher state of development than our own in order to make a contact possible. On the earth, life has existed for something like two or three billion years, written documents can be traced for some 6000 years, and in contrast the capability to use radio for interstellar communications is less than 20 years old. In other words, the time during which a civilization like ours is in a position to communicate with other civilizations is infinitesimally short in comparison with the duration of life on the planet, and the age of the stars.

The next question is how long shall we have this ability to communicate with other civilizations, or in other words, how long is our civilization likely to endure? The answer to this can be based only on faith and temperament. Shklovskiy thinks that a civilization cannot last longer than 10 000 years, for which he is taken to task by his Soviet colleagues. According to the Communist philosophy, civilization, once reorganized by the adherents of Marx and Lenin, will go on forever, as all sources of internal friction will be removed. Therefore, the duration of a civilization should be put down as 10^9 rather than 10^4 years. Western writers would tend to the longer time scale. It is, however, clear that the duration of a civilization is something that cannot be decided a priori. Our own civilization may be said to be 6000 years old, and whether it will survive for another 4000 years, or 400 years, or even 40 years is anybody's guess. Some thinkers, notably H. G. Wells and O. Spengler, were very pessimistic in this respect. It is well known that our civilization has had its ups and downs. The ancient Romans, for instance, were much more highly civilized than their descendants a thousand years later. Therefore, there is no need to postulate a complete destruction of our civilization in order to lose our ability for interstellar communication. The duration of any civilization is accordingly a guess, and this factor makes all discussions about interstellar contacts very nearly a pure exchange of verbiage. Shklovskiy[23], for instance, develops a formula for the average distance between civilizations d depending on the time T of the duration of the existence of stars, and t the duration of civilization:

$$d = 5{\cdot}2\left(\frac{T}{t}\right)^{1/3} \text{ parsecs}$$

If we put $T = 10^{10}$ years as commonly accepted, and $t = 10^4$ years, we have the average distance between two civilizations in our galaxy 520 parsecs or about 1700 light years. He is evidently afraid of his own result and is willing to take $t = 10^5$ to 10^6 years. Even in this case the distances come out on the order of 100 parsecs or 326 light years.

The tremendous distances between the stars is another serious difficulty. They average out about three parsecs or 10 light years, not to speak of the millions of light years separating us from other galaxies. The situation is thus not very encouraging, even with the most favorable assumptions about the frequency of the planets and a simultaneous existence of highly developed civilizations on these planets. Soviet radio astronomers such as Troitskiy[29] and Kotel'nikov[14] think that 1000 light years is the maximum distance at which interstellar communications have any meaning at all, and at this distance the existence of only one civilization similar to ours at most can be expected. As is well known, Project Ozma in the U.S. was based on a much greater restriction of the problem. Only the nearest stars were considered, and among them only those that were more or less in the same physical class as our sun, so that their civilization would be in a similar state of development as on the earth. Only two stars τCeti- and ϵEridani, about 11 light-years distant, were tried. Signals to these stars in the hydrogen line 1420 Mc were sent from the National Radio Observatory in May–July 1960, and characteristics of the radio emission from these stars analyzed. No evidence of any artificial signals were discovered, and the answer to our own signals, if any, cannot be expected until 1982. It is not known whether the Soviets ever attempted a similar experiment. They all quote the Ozma project, and the book "Interstellar Communications" published by NASA in 1963 (in which the Ozma project is described) appears to be one of their fundamental information sources, although the Soviet expert Khaykin considers Ozma a waste of time and resources[12]. The inference in most of the Soviet papers, however, seems to be that the Soviets have nothing to offer in the experimental line comparable even to the modest Project Ozma[8, 12, 21, 28].

How can the existence of civilizations like ours be discovered? Shklovskiy[23] points out that at least one indication of intelligent activity is available, i.e., the generation of electromagnetic energy by planets which, of course, at stellar distances would merge with their stars. He notes that there are several thousand radio and television stations on the earth, and taking their power into consideration, concludes that the brightness temperature of the earth in television wavelengths is some millions of degrees. Moreover, this temperature started rapidly increasing since about 1940. He speculates, therefore, that if a similar situation can be associated with one of the nearest stars, it would be prima facie evidence of the existence of intelligent life there. He cautions, however, that this possibility requires a long and careful survey of all sources of cosmic origin, something that is not very easy to organize.

Developing the idea of energy criterion, Kardashev[10] points out that the earth's civilization is currently utilizing 4×10^{19} ergs/sec, and this quantity is rapidly increasing in an exponential way. By extrapolation he concludes that by the year 5000 A. D. humanity will consume 4×10^{33} ergs/sec, which is equal to the output of the sun and by the year 8000 A. D. to the energy output of the whole galaxy, that is 4×10^{44} ergs/sec. Obviously such possibilities require the harnessing of the whole energy of the sun, of which the earth intercepts now only one part in two billion. Projects of this sort are in existence, one of them being Dyson's sphere to capture and retain the energy of the sun. The utilization of the galaxy will then be the next problem.

Kardashev sets up a classification of civilizations according to the energy criteria as follows:

1. Technological level approaches that of terrestrial civilization; consumption of energy 4×10^{19} ergs/sec.
2. Civilization utilizing the whole energy of the star, that is, of the order 4×10^{33} ergs/sec.
3. Civilization, having at its disposal the energy of its galaxy, is about 4×10^{44} ergs/sec.

Further, Kardashev, basing his argument on our own experience, thinks that Stage 1 is reached in a few billion years. Stage 2, according to him, should develop within several thousand years after Stage 1 had been reached. Stage 3 should be developed in not more than 10 million years after Stage 2, thus indicating that the 10 000 years postulated by Shklovskiy for the existence of a civilization is not satisfactory to at least some Soviet astronomers. The evidence of the existence of civilization of Type 3 would consist of radio phenomena which could not be explained in any rational way. All this setting up of criteria is highly arbitrary as it presupposes complete understanding of radio astronomical processes which is hardly the case.

An illustration of this humble truth is the controversy produced by Soviet astronomers over STA–21 and STA–102, that is, Nos. 21 and 102 in the California Institute of Technology Catalogue of Cosmic Radio Sources. They were hastily declared satisfying the requirements of civilization of Type 3, and some more of such, LHE–210, LHE–459, and LHE–523 were found at the Shternberg Astronomical Institute (GAISh) of Moscow State University. So far as the situation with STA–102 is concerned, much doubt has been thrown on Kardashev's claim that its period variation in radio frequency should be considered as an artificial signal with a period of 100 days, drawing our attention to this galaxy. Astronomers in the West failed to confirm its periodic variation and it is generally considered now of the quasar type, that is, a perfectly natural, although not yet perfectly understood, object. Yu. N. Pariyskiy[19] investigated, on Kardashev's request, sources STA–21 and STA–102 with the Pulkovo radio telescope, but his conclusions are hardly in favor of the artificial origin of the radio emission from these two sources. He finds that their radio properties are similar to those of some other cosmic sources and the strength of the signals under the most favorable assumption exceeds by several orders of magnitude the strength that we can reasonably expect from civilizations of Class 2 or 3.

The criteria which an artificial signal from another civilization should satisfy, according to Kardashev, are:

1. The small angular size of the source (This he considers an extremely important if not a decisive indication of the artificiality of the source.)
2. Maximum intensity of signal in the range 3–10 cm.
3. Variability of the signal in time.

Much of the discussion at the Byurakan Conference was centered on these criteria, some participants declaring that many natural objects could satisfy them. V. I. Slysh[25] thinks that a simultaneous fulfillment of these criteria by a cosmic source would constitute a presumption (but not a proof) of its cosmic origin. The question whether a cosmic radio source is artificial or not can be settled, according to Slysh, only by a systematic survey of the whole sky by means of a radio interferometer with a resolving power $0''.1$. This at least would eliminate all sources that are clearly natural, so that our attention could be concentrated on a few suspicious objects. He does not indicate whether the Soviet technical capacity is adequate to meet this challenge.

Assuming that there are extraterrestrial civilizations willing to communicate, consideration must be given to how this may be accomplished. There are three possible ways of doing this:

1. Direct contact, that is, interstellar travel, seems to be excluded from serious consideration despite the fact that this mode of communication is the most appealing to human imagination. Even assuming that physiological requirements of inhabitants of various planets are identical, the problem of travel, aggravated by tremendous distances, still remains. The various proposals of photon rockets, etc., (for which Dr. Stanyukovich is famous in the USSR), taking advantage of

the relativistic dilatation of time, will not be of much use even when they are technically possible. According to Sagan the flight with acceleration of 10 m/sec would allow a trip to the Andromeda galaxy in 28 years so far as the passengers in the rocket are concerned. However, for the home civilization that sent them this would be equivalent to 1.5 million years. A round trip taking three million years is of doubtful value. Besides nothing is known of human physiological reaction to such a flight.

2. Radio contact is a method of exchange of signals which is now technically possible but the distances at which it is effective are very small in comparison with the size of the universe. One-way radio contact, of course, is not limited by distance. We may imagine a civilization in the Andromeda galaxy that sent out signals "to whom it may concern" a million and a half years ago. We would be now just receiving them.
3. Possible contact by means of masers, lasers, and other modern electronic means.

L. M. Gindilis (1965)[6] in his survey of the problem gives a tabulation summarizing the present situation.

TYPES OF CONTACT BETWEEN CIVILIZATIONS

Distance Between Civilizations, light years	Possible Types of Contact
$d<100$	All types are possible.
$100<d<1000$	1. One-way radio communication 2. Two-way radio communication possible 3. Direct contact by bodily visits possible but unlikely
$1000<d<t_c$	1. One-way radio communication 2. Direct contacts, if possible, will be only one way
$d>t_c$	Only one-way radio communication possible.

Where d denotes the distance between civilizations in light years, and t_c

the lifetime of a civilization. This t_c, as has already been remarked, is of a highly speculative nature. Shklovskiy takes it to be of the order of 10 000 years, Gindilis thinks it should be billions of years, comparable to the lifetime of the planets themselves.

The bulk of discussion in the USSR (as well as in the West) is in the selection of suitable radio frequencies and other characteristics of radio waves for interstellar communications. Use of the 21–cm hydrogen wavelengths, originally proposed as having a universal meaning and actually used in the Ozma project, is objected to by many scientists both East and West. The reason for this is the abundance of interstellar hydrogen and, therefore, the high threshold of radio noise which lies exactly in this line. The choice of the wavelength for communication is, of course, badly restricted by the known properties of the earth's atmosphere. Moreover, it may be equally restricted by the unknown properties of the other bodies' atmospheres. It is easy to imagine a planetary atmosphere suitable for life having argon instead of nitrogen which would radically change its transmission properties.

Perhaps the most thorough discussion of this problem was given by Kotel'nikov[14] in the Byurakan Symposium. The 21-cm hydrogen wavelength is assumed to be impractical for the above-mentioned reasons. He proposes a multi-channel receiver containing a large number of narrow-band filters. If a monochromatic signal of a certain frequency reaches the antenna it will be automatically recorded and an appropriate channel tuned to that frequency. Even with this device the coverage of the whole sky is not an easy undertaking. Assuming a limiting distance of 1000 light years the number of stars in this space will be of the order of 10 million. To cover the whole sky, including all these stars, would take exactly one year utilizing the antennas and recorders recommended by Kotel'nikov. Further, what guarantee is there that the signal would be available for detection on exactly the date programmed for observation? Kotel'nikov's final conclusion is that it may be possible to discover a civilization of our type by our present radio means if it exists on one star out of 10^6. If this figure is one star out of 10^7 the discovery will be almost impossible, and if a civilization exists on only one star out of 10^8 its discovery will be impossible unless the radio apparatus becomes much more efficient. The criteria of one civilization per 10^6 stars corresponds statistically to the limiting distance of 500 light years. Thus a distance of only 500 to 1000 light years must be considered as the limiting distance for interstellar communications. V. S. Troitskiy[29] by an entirely different line of reasoning comes to the same conclusion, that even with a narrow direction signal the limiting distance of a civilization detectable by radio is about 1000 light years. He estimates a power requirement for this distance on the order of 1.6×10^{16} watts.

The problem of what to transmit to stellar civilizations and how to interpret signals received from them was only briefly treated at the Byurakan Conference[7, 26]. A. V. Gladkiy[7] expressed only general ideas as to the form a language can take under different conditions. He is a member of the Institute of Mathematics, Siberian Section of the Academy of Sciences, USSR, and being a mathematician he declares that it should not be assumed that the mathematics of our stellar correspondents will be the same as ours. A short discussion of the artificial language, Lincos, developed by the Dutch mathematician Hans Freudenthal, does not indicate any Soviet originality in this direction. The attempt to unravel the meaning of the Mayan inscriptions of Yucatan by a mathematical analysis, carried out by the same Mathematical Institute of Siberia, was not well received in the West, and the Mayan language is probably much simpler than the language of a planet *X* attached to star *Y* in galaxy *Z*. The understanding of stellar language may possibly prove to be a more difficult problem than sending or receiving stellar communications. Resolutions of the Byurakan Conference emphasize the importance of linguistic studies in this connection.

As to other than radio communications with stellar civilizations, the only promising means is an apparatus of laser type. Shklovskiy discusses it in considerable detail (second edition, Chapter 20)[21], but he cautions that it requires space platforms for its use, which are not yet available. As Shklovskiy notes in the introduction to his book, the present rapid development of radio astronomy, gamma-ray astronomy, X-ray astronomy, etc., indicates possibilities never dreamed of just a few years ago. What is said about stellar civilizations today may become obsolete tomorrow. The fundamental question, whether extraterrestrial civilizations (or even life in general) exist at all, has not been answered in these papers, nor in similar papers in the West. The next question, whether mankind is willing to put so much effort into a search which may well prove futile, likewise has not been settled.

RESOLUTIONS OF THE FIRST ALL-UNION CONFERENCE DEVOTED TO THE PROBLEM OF EXTRATERRESTRIAL CIVILIZATIONS (MAY 20–23 1964)

A further indication of Soviet interest in and research plans for this topic is contained in the resolutions of the Byurakan Conference, a summary of which is presented below.

1. Although materialistic philosophy favors the existence of intelligent extraterrestrial life, at the present time there is no valid proof of such life. However, there are strong indications that such life might exist and might develop civilizations. A contact with these would be of the highest importance and interest, but until very recently such a contact was clearly impossible. At the present time, however, there is a possibility of establishing interstellar communications by means of electromagnetic waves. The best range of frequencies for this purpose is from 10^9 to 10^{11} cycles, that is the region of centimeter and decimeter waves. Present-day technology allows the registration of radio signals across stellar distances. A rapid development of cybernetics makes it possible to formulate the problem of cosmic linguistics. The rapid growth of scientific literature on these subjects, and the first practical steps made in the U.S. to contact extraterrestrial civilizations clearly show that interstellar communication is an actual scientific problem.

2. It is therefore necessary to undertake the development of an experimental as well as a theoretical approach to this problem.

A. Experimental work should be conducted along the following two lines of effort:

(1) A systematic survey of the sky in order to detect signals from objects within 1000 light years, and the sending of signals within that distance to possible cosmic correspondents.

(2) A search for signals from civilizations, substantially more developed than our own, by applying a careful analysis to discrete cosmic radio sources suspected to be of artificial origin. To carry out these projects it is necessary to utilize the already existing apparatus and set up radio interferometers with long base lines of the order of 10^6 to 10^7 λ, in the centimeter wavelengths.

B. It is necessary to continue and intensify optical investigations having a bearing on the above-mentioned programs. These would include work on planetary and stellar cosmogony, a search for planetary systems, identification of radio sources, and the organization of special investigations outside the atmosphere of the earth.

C. Along with these programs there should be organized studies in adjacent fields:

(1) A theoretical study of statistical properties of artificial radio sources, that is, the establishment of criteria for the artificiality of signals and the development of methods for the discovery of artificial signals. Further, it is necessary to develop methods of analysis of the statistical properties of radio signals and to apply these methods to cosmic sources of suspected artificial origin.

(2) Development of methods of establishing contact and of a cosmic language on the basis of the general theory of linguistics. Also, the development of the theory of decipherment and of the basic principles of the theory of learning.

3. To carry out these programs it is desirable to establish special working groups in a number of scientific organizations. The institutions recommended for this purpose are:

Shternberg Astronomical Institute, Moscow State University (GAISh)
Pulkovo Astronomical Observatory (GAO AN SSSR)
Byurakan Astronomical Observatory (BAO AN ArmSSR)
Radio-Physical Institute at Gor'kiy University (NIRFI)
Institute of Radio Technology and Electronics, AN SSSR (IRE)
Siberian Section of the Academy of Sciences, USSR
Mechanical-Mathematical Faculty of Moscow State University.

4. For coordination of research work in various organizations, the Astronomical Council and the Council for Radio Astronomy of the Academy of Sciences, USSR, are asked by the Byurakan Conference to organize a special Commission for Interstellar Communications. This Commission should be empowered:

A. To work out for the next conference a program of search for artificial cosmic sources, using available optical and radio astronomy information. A possibility of international cooperation in this task should be considered.

B. To work out during 1964–1965 a plan for technical and financial assistance in the problem of interstellar communications, paying attention to the recommendations of the present conference. This plan should include the construction of appropriate radio telescopes and of receiving and analyzing apparatus.

The personnel of the proposed commission is recommended as follows:

I. S. Shklovskiy, GAISh, MGU
V. S. Troitskiy, NIRFI, Gor'kiy University
G. M. Tovmasyan, Byurakan Observatory, Armenian AN
Yu. P. Pariyskiy, GAO AN SSSR (Pulkovo)
N. S. Kardashev, GAISh, MGU

L. M. Gindilis, GAISh, MGU
B. N. Panovkin, Council for Radio Astronomy, AN SSSR.

5. It is considered desirable to call the next conference on the problems of extraterrestrial civilizations and interstellar communications in 1965. (There is no further reference to this second conference in available Soviet literature.)

References

1. Symposium on the Origin of Life in the Universe, Academy of Sciences, USSR, 1963.
2. "Are There Many Inhabited Worlds?", Priroda, 1963, Vol. 52, No. 12, pp. 80–88.
3. "The First All-Union Conference Dealing With the Problem of Extraterrestrial Civilizations" (Trudy Vsesoyuznoye Soveschaniye Posvyashchenoye Probleme Vnezemnykh Tsivilizatsiy, Izd-vo AN Arm. SSR, Yerevan), 1965, 151 pp.
4. "Interstellar Communications", W. A. Benjamin, Inc., New York, 1963.
5. A. D. Fortushenko, "70 Years of Radio", Svyaz', Moscow, 1965.
6. L. M. Gindilis, "The Possibility of Communication With Extraterrestrial Civilizations", Zemlya i Vselennaya, 1965, No. 1, pp. 18–27 (HT–66–517).
7. A. V. Gladkiy, "On Possible Languages for Contact Between Different Civilizations", Trudy Vsesoyuznoye Soveschaniye Posvyashchenoye Probleme Vnezemnykh Tsivilizatsiy, Izd-vo AN Arm. SSR, Yerevan, 1965, pp. 145–146.
8. L. I. Gudzenko and B. N. Panovkin, "On the Problem of Reception of Signals From Extraterrestrial Civilizations", Trudy Vsesoyuznoye Soveschaniye Posvyashchenoye Probleme Vnezemnykh Tsivilizatsiy, Izd-vo AN Arm. SSR, Yerevan, 1965, pp. 68–71.
9. A. A. Imshenetskiy, "Meteorites and the Problem of Existence of Extraterrestrial Life", Vestnik Akademii Nauk SSSR, 1966, No. 1, pp. 36–45.
10. N. S. Kardashev, "Transmittal of Information by the Extraterrestrial Civilizaations", Trudy Vsesoyuznoye Soveschaniye Posvyashchenoye Probleme Vnezemnykh Tsivilizatsiy, Izd-vo AN Arm. SSR, Yerevan, 1965, pp. 37–53.
11. A. P. Kazantsev, "Guest Out of the Cosmos", Geographgiz, Moscow, 1958, 129 pp.
12. S. E. Khaykin, "On the Problem of Contact With Extraterrestrial Civilizations", Trudy Vsesoyuznoye Soveschaniye Posvyashchenoye Probleme Vnezemnykh Tsivilizatsiy, Izd-vo AN Arm, SSR, Yerevan, 1965, pp. 83–94.
13. V. N. Komarov, "Man and Mysteries of the Universe", Mysl', Moscow, 1966.
14. V. A. Kotel'nikov, "Contact With Extraterrestrial Civilizations in the Radio Range", Trudy Vsesoyuznoye Soveschaniye Posvyashchenoye Probleme Vnezemnykh Tsivilizatsiy, Izd-vo AN Arm. SSR, Yerevan, 1965, pp. 113–120.
15. K. A. Lyubarskiy, "Introduction Into Astrobiology", Akademiya Nauk SSR, 1962.
16. D. A. Maksimik, "On the Probable Vegetation on Mars", Astronomicheskiy Vestnik, 1967, Vol 1, No. 1, pp. 34–43.
17. A. I. Oparin, "The Origin of Life on Earth", Oliver and Boyd, London, 1957.
18. A. I. Oparin, "The Origin of Life in Space", Space Science Reviews, 1964, Vol 3, No. 1, pp. 5–26 (In English).
19. Yu. N. Pariyskiy, "Observations of Peculiar Radio Sources STA–21 and STA–102 in Pulkovo", Trudy Vsesoyuznoye Soveschaniye Posvyashchenoye Probleme Vnezemnykh Tsivilizatsiy, Izd-vo AN Arm. SSR, Yerevan, 1965, pp. 54–60.

20. Yu. G. Perel', Review of Kazantsev's "Guest out of the Cosmos", Astronomicheskiy Zhurnal, 1959, Vol. 36, No. 2, pp. 381–384 (Soviet Astronomy AJ, 1959, Vol. 3, No. 2, pp. 375–379.)

21. I. S. Shklovskiy, "Universe, Life, and Intelligence", Second Edition, Izd-vo "Nauka", Moscow, 1965.

22. I. S. Shklovskiy and Carl Sagan, "Intelligent Life in the Universe", Holden-Day, San Francisco, 1966 (in English).

23. I. S. Shklovskiy, "Multiplicity of Inhabited Worlds and the Problem of Establishing Contact Between Them", Trudy Vsesoyuznoye Soveschaniye Posvyashchenoye Probleme Vnezemnykh Tsivilizatsiy, Izd-vo AN Arm. SSR, Yerevan, 1965, pp. 15–34.

24. V. I. Siforov, "Some Problems of Search and Analysis of Radio Emission From Other Civilizations", Trudy Vsesoyuznoye Soveschaniye Posvyaschenoye Probleme Vnezemnykh Tsivilizatsiy, Izd-vo AN Arm. SSR, Yerevan, 1965, pp. 121–128.

25. V. I. Slysh, "Radio Astronomy Criteria of Artificiality of Radio Sources", Trudy Vsesoyuznoye Soveschaniye Posvyashchenoye Probleme Vnezemnykh Tsivilizatsiy, Izd-vo AN Arm. SSR, Yerevan, 1965, pp. 61–67.

26. N. A. Smirnova and N. L. Kaydanovskiy, "Influence of Conditions of Radio Wave Propagation in Cosmic Medium and Atmosphere of the Earth on the Angular Size of the Source", Trudy Vsesoyuznoye Soveschaniye Posvyashchenoye Probleme Vnezemnykh Tsivilizatsiy, Izd-vo AN Arm. SSR, Yerevan, 1965, pp. 129–135.

27. G. A. Tikhov, Collected Works, Trudy, Akademiya Nauk Kazakh SSR, 1959–1960.

28. G. M. Tovmasyan, "Ring Radio Telescope for the Establishment of a Contact With Extraterrestrial Civilizations", Trudy Vsesoyuznoye Soveschaniye Posvyashchenoye Probleme Vnezemnykh Tsivilizatsiy, Izd-vo AN Arm. SSR, Yerevan, 1965, pp. 95–96.

29. V. S. Troitskiy, "Some Considerations on the Search of Intelligent Signals From the Universe", Trudy Vsesoyuznoye Soveschaniye Posvyashchenoye Probleme Vnezemnykh Tsivilizatsiy, Izd-vo AN Arm. SSR, Yerevan, 1965, pp. 97–112.

30. G. Vdovykin, "On the Origin of Carbon Chrondrites", Meteorika, 1965, Vol. 26, pp. 152–168.

31. V. Zaytsev, "Visitors From Outer Space", Sputnik, 1967, No. 1 (In English).

32. A. I. Zhukova and I. I. Kondrat'yev, "A Camera Imitating Conditions on Mars for Microbiological Investigations", Kosmicheskiye Issledovaniya, 1965, Vol. 3, No. 2, pp. 330–333 (TT-65-414, pp. 251–258).

Chapter 20

Manned Space-Station Concepts in the USSR

G. E. Wukelic and N. T. Bobrovnikoff†*

INTRODUCTION

Since the beginning of the space age, numerous items have appeared in both the U.S. and the USSR relating to the establishment and scientific use of manned space stations, or "interplanetary space stations" as they are referred to in early Soviet writings. A perusal of Soviet literature (1955–1966) reveals a sincere and increasing interest in developing such stations but little about Soviet plans or progress in this direction. Even complete books on the topic, such as the 200-page item by I. N. Bubnov and L. N. Kamanin (1964)[5], appear to be based primarily on foreign (mostly U.S.) material. Also, since most of the some 35 Soviet items available on this topic are semipopular, technical information concerning such important space-station parameters as crew size, mission

* Columbus Laboratories, Battelle Memorial Institute.
† Consultant, Columbus Laboratories, Battelle Memorial Institute; Professor Emeritus, The Ohio State University.

duration, functions, and orbital and physical characteristics is meager. This chapter, therefore, is limited for the most part to generalized descriptions of the space-station concepts discussed or illustrated in available Soviet literature.

REVIEW OF SOVIET SPACE-STATION CONCEPTS

Most Soviet items on this topic boast that the idea of manned cosmic stations dates back to Tsiolkovskiy, who first announced the basic principles in 1903. His initial proposal called for the creation of an "aerial settlement", or manned space station, to orbit 1000 or 2000 km above the earth. Tsiolkovskiy is credited with having posed and solved a number of technical tasks involved in the creation of such a station. Most frequently quoted tasks are the following:

1. Use of special slots with a shiny reflecting surface on one side and a dark surface on the other to regulate the temperature of space apparatus.
2. Design of a closed cycle for metabolism made up of plants which absorb CO_2 and produce O_2.
3. Achievement of fractional, but not complete terrestrial gravity, by revolving the station at a definite angular velocity.
4. Use of space suits with compressed oxygen for working outside the station.
5. Use of solar energy as the primary power source.
6. Protection of the station from meteorite danger.
7. Assembly of a station in a vacuum free of gravity.

However, since Tsiolkovskiy's writings, a multitude of projects involving stations for only a few people, up to towns with several thousand inhabitants, have been suggested. The Soviet scientist most prolific at suggesting concepts for large inhabited earth satellites or manned space stations has been A. Shternfel'd.* Many of his concepts published in 1955–1958 are shown in Figures 20.1–20.8. The most popular of his proposals is the one introduced in 1956 and shown in Figures 20.1 and 20.2. This concept consists of joining together a number of spent rockets or specially designed satellites. Weightlessness is achieved in the main portion of the station and artificial gravity is created in the rotating sec-

* Recipient of "Honored Scientist and Engineer of the Russian Soviet Federated Socialist Republic" award, 1965.

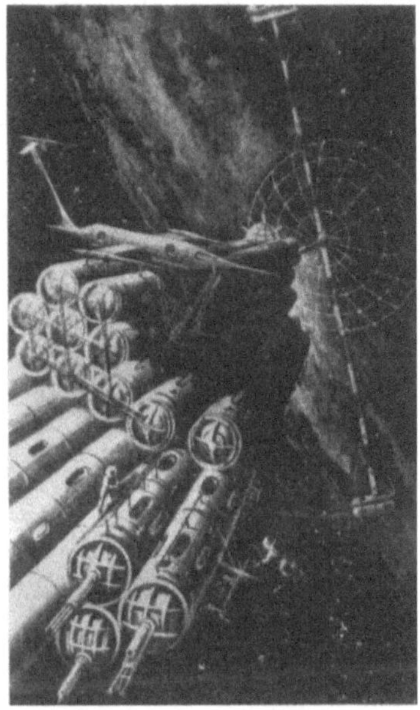

Fig. 20.1. Possible Design Variation of Large Inhabited Satellite (A. Shternfel'd, 1956)[5, 28]

Fig. 20.2. Approximate Design of a Large Inhabited Satellite (A. Shternfel'd, 1957)[24, 29, 31]

tion. What is believed to be a model of this station was exhibited in 1963–1964 at the Moscow Polytechnical Museum (Figure 20.3). Its description in "Pravda" mentions a number of silvery cylinders attached to the astronauts' cabin. Telemetry and communications antennas are placed at the sides of the cylinder cluster.

Fig. 20.3. A Model of an Assembled Interplanetary Station Exhibited at the "Cosmos" Hall of the Polytechnical Museum in Moscow (Shternfel'd, 1957)[3]

Although pictured in several Soviet publications[18, 30], little is available regarding the Shternfel'd concepts shown in Figures 20.4–20.8. Actually, Figures 20.4 and 20.5 appear to be preliminary steps in assembling the completed concept as shown in Figure 20.6. Figures 20.7 and 20.8 show Shternfel'd's concept relative to a circular or toroidal-shaped space station. All that has been noted regarding this concept is that interplanetary spaceships will be launched from such a space station, and that rotation of the satellite about its axis will create the required artificial gravity.

Fig. 20.4. First Components of an Artificial Station in Orbit (A. Shternfel'd, 1957-1958)[30]

Fig. 20.5. Assembling an Interplanetary Station (A. Shternfel'd, 1957–1958)[30]

Fig. 20.6. Interplanetary Station (A. Shternfel'd, 1957–1958)[30]

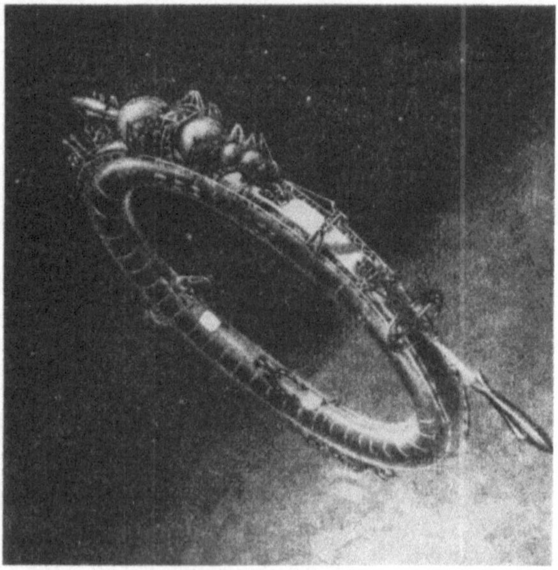

Fig. 20.7. Approximate Design of a Space Station (A. Shternfel'd, 1958)[18, 30]

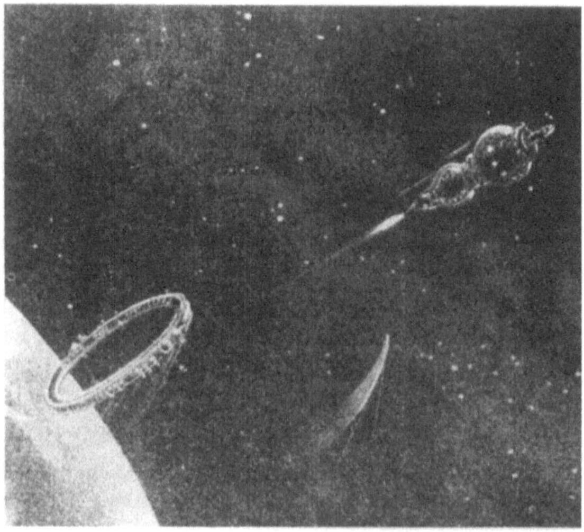

Fig. 20.8. A Spaceship is Leaving an Artificial Earth Satellite for Deep Space (A. Shternfel'd, 1958)[30]

Fig. 20.9. This is How the First Interplanetary Station Will be Constructed by Man Beyond the Earth's Atmosphere (1957)[33]

The circular space station concept appears to be very popular, and additional illustrations, as noted in Soviet literature, are shown in Figures 20.9–20.11. Figure 20.9 is an illustration contained in an item by N. A. Varvarov titled "Artificial Earth Satellite" (1957)[33]. It is a large wheel with a diameter of at least 100 feet, and a capacity of about 30 people. One of the earliest illustrations of a circular space station is that noted in Figure 20.10, first published in "Znaniye-Sila" (1954)[14]. This concept contains a mirror at the top, a wheel, and a round part at the bottom. Artificial gravity is created by rotation. The saucer-like mirror concentrates the sun's rays, the heat of which powers the steam generator of the electric plant. This design appears to be the predecessor to the model of a large circular manned cosmic station exhibited at the Exposition of the Achievements of National Economy of the USSR in Moscow (Figure 20.11). The date of the exposition has not been specifically mentioned, but the concept has been pictured in items by B. V. Lyapunov published in 1962[16] and 1963[17, 18]. No detailed description of this station is given beyond a statement that its purpose is:

> "Research in cosmic biology and medicine, study of cosmic and solar radiation, study of properties of interplanetary space, and observations of the sun, moon, planets, and the earth."

The station does not appear to be composed of rockets, as in the Tsiolkovskiy or Shternfel'd design. There is a long axis of rotation with a convex mirror on top (for collecting solar energy?) and a sphere at the bottom. Perpendicular to this axis is a large wheel for the inhabitants. Judging by the number of windows in this wheel, the capacity of the station should also be about 30 people.

The latest Soviet illustration of a model of a circular manned space station was released by TASS in 1965.* In this model the space station has six hermetically sealed compartments mounted around a central core. These compartments include a control station, laboratory, garden, navigation system, radar installation, and a solar observation station. In addition, the station has an installation for communication with other spaceships.

Several other concepts have been pictured in Soviet literature, but information is insufficient to determine whether they represent Soviet or foreign ideas. Examples are shown in Figures 20.12–20.14. Pictured in Figure 20.12 is an illustration of a cosmic station published as early as 1953 in "Tekhnika-Molodezhi"[13], and later in Lyapunov's 1963

* Wissenschaft und Technik, July 25 1965, p. 4.

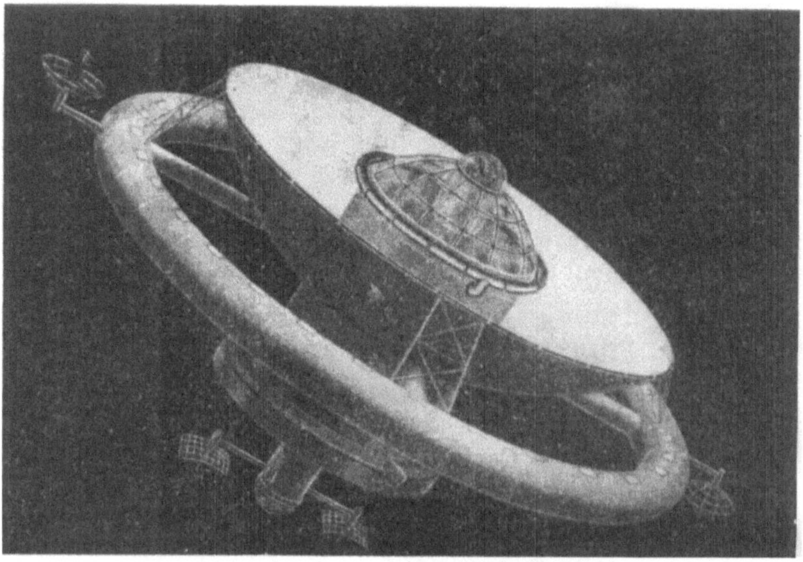

Fig. 20.10. Space Station Containing Living Quarters and Laboratories (1954)[16, 18]

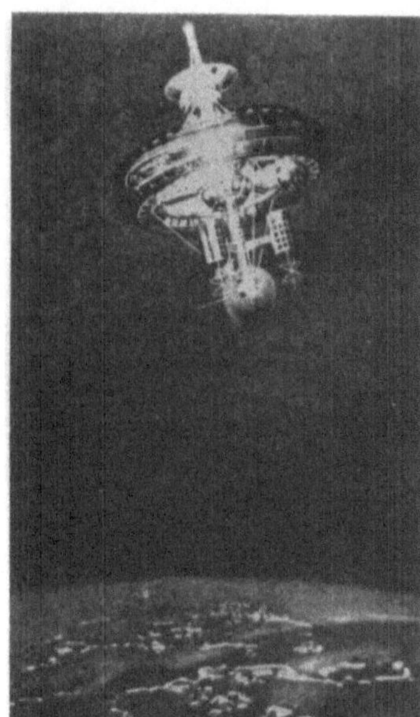

Fig. 20.11. Mock-Up of a Large Inhabited Satellite-Station of the Earth Exhibited at the Fair of the Achievements of National Economy, USSR, in Moscow (1962)[5, 16, 18]

book[18]. Like the early Soviet concepts by Tsiolkovskiy and Shternfel'd, this station consists of rocket parts and contains provisions for artificial gravity. In the 1963 item Lyapunov states that the space station is assembled from parts of rockets which were launched to orbit the earth and which have become satellites. The rockets serve as the principal building material for the station. Assembled rocket cases form the first "belt" of the cylinder in which living and auxiliary quarters and laboratories are located. The rotation of the station produces a centrifugal force which substitutes for gravity. The second "belt" is a warehouse for fuel

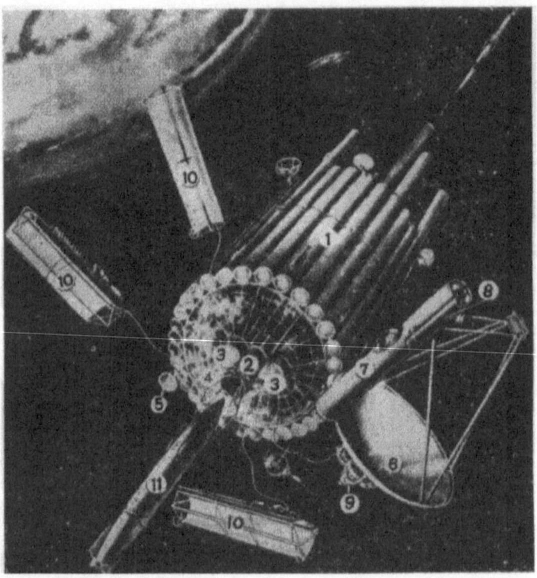

Fig. 20.12. Space Station According to Design Published in "Tekhnika Molodezhi" 1953, No. 8

needed for rockets setting out from the station on space trips. Rockets are used to construct additional "belts" as required. A hothouse is located inside the cylinder. The base of the cylinder facing the sun is covered with glass, and sufficient pressure is maintained to sustain life and permit the growth of plants in the hothouse. The plants are arranged on the cone surface which assures the most advantageous lighting. The hothouse has passages leading into the living quarters. Parallel to the axis of the cylinder is a tube for landing and launching rockets. Next to it are shops and hangars where rockets are repaired and fueled. Radars help to navigate the rockets making trips between the station and the earth. Radio and telegraph are used for communication with the earth. Power installations, consisting of mirrors, steam boilers, and turbo-generators are set up outside of the station. The astronomical observatory is also outside.

Figure 20.13 shows a space-station concept (also reportedly based upon the ideas of Tsiolkovskiy) contained in another article by Lyapunov, published in 1954[14]. The space station contains a hothouse,

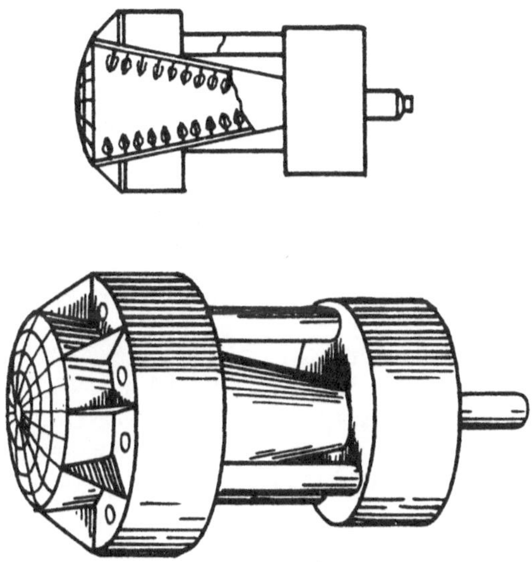

Fig. 20.13. Space Station Based Upon the Ideas of Tsiolkovskiy[14, 16]

living quarters and laboratory, passage ways, auxiliary quarters, and an air lock.

Finally, in a 1965 article by Ye. Sergeyev and V. Alekseyev[25] entitled "The Cosmonaut Leaves the Spaceship", there is an illustration (Figure 20.14) described as one of the designs of an "Orbital Cosmic Laboratory", without any further explanation. The main section of the station is a cylinder with six floors, and there is an extension on the rotational axis that looks like a concave mirror for collecting solar radiaticn. The same indefiniteness of origin applies to another illustration in the article titled "Ship Assemblage in Orbit"[25].

A 1964 Polish article[20] contains a similar drawing of a six-floor space station concept resembling very closely the 1965 Soviet illustration shown in Figure 20.14. This drawing indicates the station to be about 10 meters wide and 30 meters long. In this item the author notes that the construction of orbital stations for servicing spaceships and space laboratories as now envisioned will take place in two stages: first, a large ship, which will serve as a scientific laboratory for materials testing and biological and astronomical research, will be built and orbited; and second, the orbital laboratory will be enlarged and equipped to control and provide the maintenance for spaceships, to serve as a station for direct rendezvous in space, and as a center for the construction of huge spaceships which cannot be launched directly from the earth. It will be equipped with a collapsible hangar and a unit to monitor and stabilize the temperature and atmospheric conditions within the station. Each segment of the space station will be completely equipped before orbiting

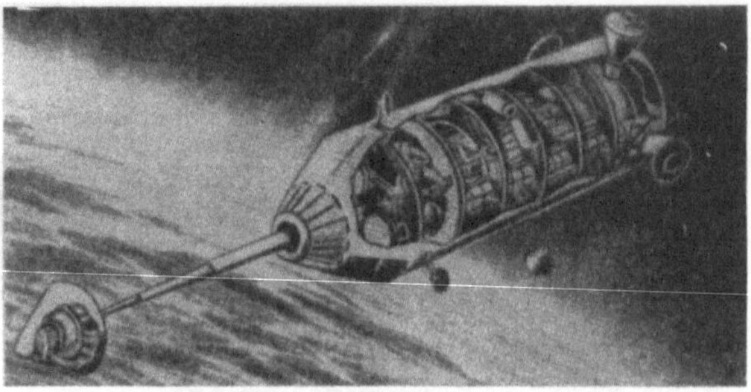

Fig. 20.14. One of the Designs of a Space Laboratory[25]

and, after all functions have been thoroughly checked, the crew (21–25 persons) will be sent up. It is planned that 10 of the crew members will operate the service unit, and the rest will do scientific research. For maximum protection against cosmic rays, the ship will be launched at a 40° angle of inclination with an apogee of 550 km. The source of power will be a nuclear reactor.

In discussing space-station concepts the Soviets feel free to use foreign material, sometimes even without proper acknowledgment. For example, B. V. Lyapunov in his "Rockets"[15] has an illustration of a manned station being assembled in space without any further comment, so that the reader assumes it to be a Soviet concept. However, the identical picture is contained in an article by V. Petrov[23] in which the idea is attributed to the U.S. author Tinsley. Other Soviet items on manned space stations contain references to numerous U.S. proposals. Those referred to most frequently include the earlier proposals by

Wernher von Braun (1953), Darell Romick (1956), and more recent concepts (1960) proposed by major U.S. aircraft companies—Martin, Lockheed, Douglas, and North American. Of the British proposals, that of Smith and Ross (1949) is most frequently quoted. References to the MOL are infrequent and always critical because of its association with the Department of Defense.

References

1. "Problems of Motion of Artificial Celestial Bodies", Doklady of the Conference on General and Applied Problems of Theoretical Astronomy, Moscow, November 20–25 1961, Izdatel'stvo Akademii Nauk SSSR, Moscow, 1963, 295 pp.

2. Pravda, February 19 1964.

3. I. Artem'yev, "Artificial Earth Satellite", Gosudarstvennoye Izdatel'stvo Detskoy Literatury Ministerstva Prosveshcheniya RSFSR, Moscow, 1958, 136 pp.

4. I. Artem'yev, "The First Artificial Satellite of the Sun", Gosudarstvennoye Izdatel'stvo Detskoy Literatury Ministerstva Prosveshcheniya RSFSR, Moscow, 1959, 64 pp.

5. I. N. Bubnov and L. N. Kamanin, "Manned Space Stations", Voyennoye Izdatel'-stvo Ministerstva Oborony SSSR, Moscow, 1964, 189 pp.

6. K. A. Gil'zin, "Electric Interplanetary Ships", Izdatel'stvo "Nauka", Moscow, 1964, 320 pp.

7. M. G. Kroshkin, "A Satellite of the Sun", Izdatel'stvo "Znaniye", Moscow, 1959, 32 pp.

8. M. G. Kroshkin, "A Rocket Leaves the Earth", Izdatel'stvo VTsSPS Profizdat, 1959, 88 pp.

9. Yu. Kryuchkov, "Interplanetary Ships", Izdatel'stvo DOSAAF, Moscow, 1958, 72 pp.

10. V. Levantovskiy, "The Story of Artificial Earth Satellites", Gosudarstvennoye Izdatel'stvo Tekhniko-Teoreticheskoy Literatury, Moscow, 1957, 96 pp.

11. V. I. Levantovskiy, "Roads to the Moon and to Planets of the Solar System", Voyennoye Izdatel'stvo Ministerstva Oborony SSSR, Moscow, 1965, 206 pp.

12. V. Levantovskiy, V. Leshkovtsev, and I Rakhlin, "Soviet Rocket Investigates the Cosmos", Gosudarstvennoye Izdatel'stvo Fiziko-Matematicheskoy Literatury, Moscow, 1959, 128 pp.

13. B. V. Lyapunov, "Laboratories in Space", Tekhnika-Molodezhi, 1953, No. 8, pp. 33–37.

14. B. V. Lyapunov, "Stations Outside the Earth", Znaniye-Sila, 1954, No. 9, pp. 11–15.

15. B. V. Lyapunov, "Rockets", Voyennoye Izdatel'stvo Ministerstva Oborony SSSR, Moscow, Second Edition, 1960, 236 pp.

16. B. V. Lyapunov, "Rockets and Interplanetary Flight", Voyennoye Izdatel'stvo Ministerstva Oborony SSSR, Moscow, 1962, 125 pp.

17. B. V. Lyapunov, "The Wings of Dreams", Kaluzhskoye Knizhnoye Izdatel'stvo, 1963, 122 pp.

18. B. V. Lyapunov, "A Station Outside the Earth", Voyennoye Izdatel'stvo Ministerstva Oborony SSSR, Moscow, 1963, 147 pp.

19. B. V. Lyapunov, "Settlements in Space", Kryl'ya Rodiny, 1964, No. 8, pp. 23–24.

20. Capt. M. Makulski, "Orbital Station for Servicing Spaceships", Wojskowy Przeglad Lotniczy, 1964, No. 7, pp. 69–71.

21. I. A. Merkulov, "Flight of Rockets Into Space", Izdatel'stvo DOSAAF, Moscow, 1958, 89 pp.

22. P. I. Nikitin, "Artificial Earth Satellite", Izdatel'stvo DOSAAF, Moscow, 1958, 88 pp.

23. V. Petrov, "Iskusstvennyye Sputniki Zemli", Voyennoye Izdatel'stvo Ministerstva Oborony SSSR, Moscow, 1958, 306 pp.

24. Yu. A. Pobedonostsev, "Artificial Earth Satellite", Izdatel'stvo "Znaniye", Moscow, 1957, 72 pp.

25. Ye. Sergeyev and V. Alekseyev, "Cosmonaut Leaves the Ship", Nauka i Zhizn', 1965, No. 4, pp. 14–18.

26. N. F. Shibayev, "Antirocket Defense", Voyennoye Izdatel'stvo Ministerstva Oborony SSSR, Moscow, 1965, 131 pp.

27. A. Shternfel'd, "Interplanetary Flight", Gosudarstvennoye Izdatel'stvo Tekhniko-Teoreticheskoy Literatury, Moscow, 1955, 56 pp.

28. A. Shternfel'd, "Artificial Satellite of the Earth", Gosudarstvennoye Izdatel'stvo Tekhniko-Teoreticheskoy Literatury, Moscow, 1956, 181 pp.

29. A. Shternfel'd, "Interplanetary Travel", Foreign Languages Publishing House (English), Moscow, 1957, 59 pp.

30. A. Shternfel'd, "Artificial Satellites", Gosudarstvennoye Izdatel'stvo Tekhniko-Teoreticheskoy Literatury, Moscow, 1958, 297 pp.

31. A. Shternfel'd, "From Artificial Satellites to Interplanetary Flights", Gosudarstvennoye Izdatel'stvo Fiziko-Matematicheskoy Literatury, Moscow, 1959, 204 pp.

32. B. I. Smagin, "Outer Space and Science", "Naukovo Dumka", Kiyev, 1964, 99 pp.

33. N. Varvarov, "Artificial Earth Satellite", Izdatel'stvo "Sovetskaya Rossiya", Moscow, 1957, 16 pp.

34. Ye. Yu. Yur'yev, "Radio Communication With Cosmic Rockets", Voyennoye Izdatel'stvo Ministerstva Oborony SSSR, Moscow, 1963, 78 pp.

35. G. Zhdanov and I. Tindo, "Laboratories in the Cosmos", Izdatel'stvo Tsk–VLKSM "Molodaya Gvardiya", 1959, 191 pp.

Appendix A

List of Major Soviet Publications and Translation Availability

The translation availability of individual items referenced in the handbook is noted in parentheses after each bibliographic entry. This Appendix contains a list of the availability of the translations of major Soviet publications available in translated editions. In the listing, the transliterated Russian title of the item is given first followed by the English title, the translating organization, and the year translation began (in the case of journals). Translations issued by the USAF Systems Command are identified in the bibliographic references by TT-, MT-, HT, and MCL-numbers. Also, to assist interested readers in obtaining copies of referenced translations, the addresses of the publishers and/or translators are provided.

Journal/Monograph	Title of Translation	Publisher/Translator	Year Began
Astronomicheskiy Zhurnal	Soviet Astronomy—AJ	American Institute of Physics	1957
Byulleten' Eksperimental'noy Biologiy i Meditsiny	Bulletin of Experimental Biology and Medicine	Consultants Bureau	1956
Doklady Akademii Nauk SSSR	Earth Sciences:		
	Doklady Earth Sciences Sections	American Geological Institute	1959
	Physics:		
	Soviet Physics—Doklady	American Institute of Physics	1956
Elektrosvyiaz'	Telecommunications	Pergamon Press	1957
		American Institute of Electrical Engineers	1961
Geodeziya i Kartografiya	Geodesy and Cartography	American Geophysical Union	1959
(Superseded by: Izvestiya Vysshikh Uchevnykh Zavedeniya, Geodeziya i Aerofotos"emka)	Geodesy and Aerophotography	American Geophysical Union	1962
Geomagnetizm i Aeronomiya	Geomagnetism and Aeronomy	American Geophysical Union	1961
Iskusstvennyye Sputniki Zemli	Artificial Earth Satellites	Consultants Bureau and USAF Systems Command	1958
Izvestiya Akademii Nauk SSSR, Fizika, Atmosfery i Okeana	Izvestiya Academy of Sciences, USSR, Atmospheric and Oceanic Physics	American Geophysical Union	1965
Izvestiya Akademii Nauk SSSR, Seriya Fizicheskaya	Bulletin of the Academy of Sciences of the USSR: Physical Series	Columbia Technical Translations	1954
Izvestiya Akademii Nauk SSSR, Seriya Geofizicheskaya	Bulletin of the Academy of Sciences of the USSR: Geophysics Series	American Geophysical Union	1957
Izvestiya Vysshikh Uchebnykh Zavedeniy, Radiofizika	Izvestiya VUZov: Radiophysics	Clearing House, Department of Commerce (JPRS)	1958
	Soviet Radiophysics	The Faraday Press, Inc.	1965
Kosmicheskiye Issledovaniya	Cosmic Research	Consultants Bureau and USAF Systems Command	1963

Journal/Monograph	Title of Translation	Publisher/Translator	Year Began
Meteorologiya i Gidrologiya	Meteorology and Hydrology (Selected translations)	Clearing House, Department of Commerce (JPRS)	—
Pribory i Tekhnika Eksperimenta	Instruments and Experimental Techniques	Instrument Society of America	1958
Problemy Kosmicheskoy Biologii	Problems of Space Biology	Clearing House, Department of Commerce (JPRS)	—
Radiotekhnika i Elektronika	Radio Enginering and Electronic Physics	American Institute of Electrical Engineers	1961
Uspekhi Fizicheskikh Nauk	Soviet Physics—Uspekhi (Partial translation)	American Institute of Physics	1958
Vestnik Akademii Meditsynskikh Nauk SSSR	Vestnik of the USSR Academy of Medical Sciences	Clearing House, Department of Commerce (JPRS)	1962
Vestnik Akademii Nauk SSSR	(Selected articles)	Clearing House, Department of Commerce (JPRS)	—
Zhurnal Eksperimental'noy i Teoreticheskoy Fiziki	Soviet Physics—JETP	American Institute of Physics	1955
Zhurnal Vysshey Nervnoy Deyatel'nosti imeni I. P. Pavlova	Pavlov Journal of Higher Nervous Activity	Pergamon Press	1958
Vtoroy Gruppovoy Kosmicheskiy Polet i Nekotoryye Itogi Poletov Sovetskikh Kosmonavtov na Korabyakh "Vostok", Izdatel'stvo "Nauka", Moscow, 1965	Second Group Space Flight and Certain Results of Flights of Soviet Cosmonauts on the Vostok Ships	USAF Systems Command	—
Trudy Vsesoyuznoy Konferentsii po Fizike Kosmicheskogo Prostranstva, Izdatel'stvo "Nauka", Moscow, June 10–15 1965	Space Research—Transactions of the All-Union Conference on Space Physics	National Aeronautics and Space Administration	—
Pervyye Panoramy Lunnoy Poverkhnosti, Izdatel'stvo "Nauka", Moscow, 1966, 129 pages	The First Panoramic Views of the Lunar Surface	National Aeronautics and Space Adminstration	—

Journal/Monograph	Title of Translation	Publisher/Translator	Year Began
Sovetskiye Sputniki i Kosmicheskiye Korabli, S. G. Aleksandrov and R. Ye. Fedorov, Izdatel'stvo Akademii Nauk SSSR, Moscow, 1961, Second Revised Edition, 440 pages	Soviet Satellites and Spaceships	Clearing House, Department of Commerce (JPRS)	—
Sovetskiy Sputniki i Kosmicheskaya Raketa, S. G. Aleksandrov and R. Ye. Fedorov, Izdatel'stvo Akademii Nauk SSSR, Moscow, First Edition, 1959, 233 pages	Soviet Satellites and Cosmic Rockets	USAF Systems Command	—
Vysokiye Sloyi Atmosfery, I. A. Khvostikov, Gidrometeorologicheskoye Izdatel'stvo, Leningrad, 1964	The Upper Layers of the Atmosphere	National Aeronautics and Space Administration	—
Metody Issledovaniya Atmosfery s Ispol'zovaniyem Raket i Sputnikov, A. A. Kmito, Gidrometeorologicheskoye Izdatel'stvo, Leningrad, 1966, 366 pages	Methods of Investigating the Atmosphere With Rockets and Satellites	National Aeronautics and Space Administration	—
Gazovyy Sostav Atmosfery Zemli, Metody Ego Analiz, B. A. Mirtov, Izdatel'stvo Akademii Nauk SSSR, Moscow, 1961	Gaseous Composition of Earth's Atmosphere and Methods of its Analysis	National Aeronautics and Space Administration	—
Atlas Obratnoy Storony Luny, Izdatel'stvo Akademii Nauk SSSR, Moscow, 1960, 149 pages	An Atlas of the Moon's Far Side; The Lunik 3 Reconnaissance	Interscience Publishers, Inc., and Sky Publishing Company	—

Publisher/Translator	*Address*
American Geophysical Union	2100 Pennsylvania Avenue, N.W. Washington, D.C. 20037
American Geological Institute	Subscription Manager 1444 N. Street, N.W. Washington, D.C. 20005
American Institute of Electrical Engineers, now the Institute of Electrical and Electronics Engineers	Box A, Lennox Hill Station New York, New York 10021
American Institute of Physics	335 East 45th Street New York, New York 10017
Clearinghouse for Federal Scientific and Technical Information	Joint Publications Research Service U.S. Department of Commerce Building Tempo E Adams Drive, 4th and 6th Streets, S.W. Washington, D.C. 20443
Columbia Technical Translations	5 Vermont Avenue White Plains, New York 10606
Consultants Bureau	Consultants Bureau Enterprises, Inc. 277 West 17th Street New York, New York 10011
The Faraday Press, Inc.	84 Fifth Avenue New York, New York 10011
Instrument Society of America	530 William Penn Place Pittsburgh, Pennsylvania 15219
Interscience Publishers, Inc.	605 Third Avenue New York, New York 10016
National Aeronautics and Space Administration	Clearinghouse for Federal Scientific and Technical Information Springfield, Virginia 22151
Pergamon Press, Inc.	44–01 21st Street Long Island City, New York 11101
Sky Publishing Company	49–50–51 Bay State Road Cambridge, Massachusetts 02138
USAF Systems Command	Wright-Patterson Air Force Base, Ohio 45433

Author Index

Italic figures indicate Bibliographic References

R

Subject Index

Because of the handbook's detailed table of contents, the index has been made fairly short. Note that space payloads have been indexed by series only